Handbooks for the Identification of British Insects
Vol. 7. Part 12

Ichneumonid Wasps (Hymenoptera: Ichneumonidae): their Classification and Biology

Gavin R. Broad
Department of Life Sciences, Natural History Museum

Mark R. Shaw
National Museums of Scotland

Michael G. Fitton
Department of Life Sciences, Natural History Museum

Line drawings by
Dawn Painter

Habitus photos by
Olga Retka
Natural History Museum

Gavin R. Broad, Mark R. Shaw and Michael G. Fitton

© Royal Entomological Society, 2018.

Text: © The Trustees of the Natural History Museum, London; and Mark Shaw.

Illustrations: © The Trustees of the Natural History Museum, London; Dawn Painter; and photographers as credited on page v.

Published for the   Royal Entomological Society
                    The Mansion House
                    Bonehill
                    Chiswell Green Lane
                    Chiswell Green
                    St Albans
                    AL2 3NS
                    www.royensoc.co.uk

By the   Field Studies Council
                    Unit C1
                    Stafford Park 15
                    Telford
                    TF3 3BB
                    www.field-studies-council.org

ISBN: 978-1-800628-28-1

All rights reserved. No part of this book may be reproduced or translated in any form or by any means, electronically, mechanically, by photocopying or otherwise, without written permission from the copyright holders.

Contents

Abstract	iv	Cremastinae	161
Acknowledgements	v	Cryptinae	165
Dedication	v	Ctenopelmatinae	175
		Cylloceriinae	186
Introduction	1	Diacritinae	189
General biology	3	Diplazontinae	191
Differences between Braconidae and Ichneumonidae in the British fauna	5	Eucerotinae	197
		Hybrizontinae	201
Conservation	6	Ichneumoninae	204
Classification and phylogeny	6	Lycorininae	217
Trends in host associations	7	Mesochorinae	220
Collecting, rearing and preservation	13	Metopiinae	228
Checklist of the British genera of Ichneumonidae	17	Microleptinae	236
		Neorhacodinae	239
Morphological terminology	23	Ophioninae	242
		Orthocentrinae	248
Identification of the subfamilies of British Ichneumonidae	38	Orthopelmatinae	254
		Oxytorinae	257
Identifying ichneumonids	38	Phygadeuontinae	260
Notes on use of the keys	38	Pimplinae	268
Recognition of Ichneumonoidea	38	Poemeniinae	284
Key for the separation of British and Irish Braconidae and Ichneumonidae	39	Rhyssinae	288
		Stilbopinae	293
Key for the identification of British and Irish subfamilies of Ichneumonidae	41	Tersilochinae	298
		Tryphoninae	305
List of the species illustrated in the keys	106	Xoridinae	321
		Extralimital and fossil subfamilies	326
Accounts of British subfamilies		**Glossary**	332
Acaenitinae	111	**References**	345
Adelognathinae	115	**Index to Icheumonoidea**	395
Agriotypinae	119	**Index to hosts of Ichneumonidae, food plants and associated organisms**	411
Alomyinae	122		
Anomaloninae	125		
Ateleutinae	131		
Banchinae	134		
Brachycyrtinae	143		
Campopleginae	145		
Collyriinae	158		

Gavin R. Broad, Mark R. Shaw and Michael G. Fitton

Abstract

An extensively illustrated key is provided for the identification of the 35 subfamilies of Ichneumonidae known to occur in Britain and Ireland (plus Brachycyrtinae, not yet found here). This is supported by chapters for each subfamily (completed by briefer accounts of extralimital ones, in order to cover the entire world ichneumonid fauna), giving a comprehensive review of current knowledge on systematics, biology and host relations together with notes on useful recognition features and references to the most relevant species-level identification literature concerning the British Isles. Included in each British and Irish subfamily account is a folio of photographs covering all recognised tribes. Also given are an illustrated scheme for preferred morphological terminology, with cross referencing to more archaic schemes that have been used in some of the recommended identification literature on Ichneumonidae, a list of ichneumonid genera found in Britain and Ireland, and an extensive glossary of both morphological and biological terms.

Acknowledgements

Line drawings for the morphological terminology section are the work of Dawn Painter and the whole insect habitus photos were taken by Olga Retka. We are very grateful to Dr Andrew Bennett, Sir Anthony Galsworthy and Dr Richard Dickson for proof-reading and commenting on most of the chapters. Thank you to Andrew Bennett for also sharing his unpublished findings on the phylogeny of Ichneumonidae. Laurence Counter, Phil Adams, Mark Boddington, Tony Broome, Bob Brown, Larry Doherty, Tony Edwards, Geoff Nobes, Richard Revels, Stewart Sexton, Sue Taylor and Jeroen Voogd donated lovely images to enliven the text. Stefan Schmidt provided a photograph of the late Klaus Horstmann. Other photographs are the work of the authors. Photographed specimens are in the Natural History Museum, London, except for *Acaenitus dubitator* and *Coleocentrus excitator*, which are in the National Museums of Scotland, Edinburgh. Lucy Broad kindly helped with editing images and Jonathan Jackson scanned and digitally cleaned MRS's photographic slides. The participants of many workshops on ichneumonid identification have greatly improved the key to subfamilies; we also thank the British Entomological and Natural History Society for hosting and advertising these workshops. And we are very grateful to Dr Rebecca Farley-Brown for her assiduous editing of this unwieldy manuscript.

Dedication

We dedicate this book to the memory of Klaus Horstmann (1938–2013), who vastly improved our knowledge of European Ichneumonidae and who was a wonderful colleague and friend.

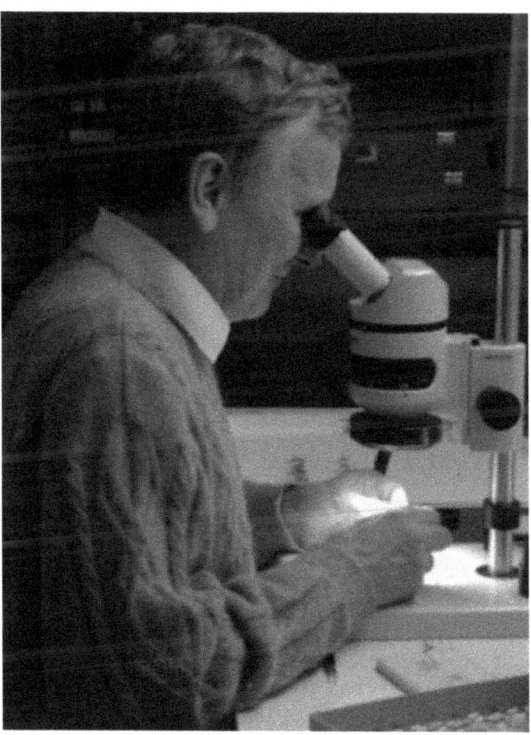

Introduction

Ichneumonids abound. The hymenopteran family Ichneumonidae comprises one of the largest single animal families, a huge radiation of parasitoids of a wide variety of holometabolous insects and to a much lesser extent spiders. With around 2,500 British species, almost 10% of British insect species are ichneumonids. They can be found everywhere and studying them is very rewarding, but it is not without its difficulties. The family is poorly served by accessible literature; there has been no summary of the identification of British ichneumonids since Morley's (1903-1915) volumes, which are not only difficult to use, but also very out of date and riddled with errors, and the only remotely comprehensive summary of the biology of Ichneumonidae (Quicke, 2015) deals with the Ichneumonoidea as a whole and at a global level. In aiming to provide a concise introduction to the identification, taxonomy and biology of this family of parasitoid wasps we hope to begin to plug that gap. Because the internal classification of the family is such an important framework we pay it good attention, but then the emphasis is on biology as it is difficult for the non-specialist to build up a general impression of the host spectra and other attributes of the various ichneumonid subfamilies from the vast, scattered and error-ridden literature. In various aspects of their biology ichneumonids are fascinating insects, and it is from their host relations that they are likely to present themselves to entomologists with other principal interests.

What are ichneumonids?

The Ichneumonoidea may comprise only two extant families, Braconidae and Ichneumonidae, but this is one of the more successful superfamilies of Hymenoptera (Quicke, 2015). There is ample evidence (Grissell, 1999; Noyes, 2012, Forbes *et al.*, 2018) that the Hymenoptera is the most species-rich order of insects, mainly because of the vast numbers of parasitoids in the superfamilies Ichneumonoidea and Chalcidoidea, although the number of described species lags behind Coleoptera. Within the Hymenoptera, the major radiations can be characterised as sawflies and woodwasps, aculeates, and the parasitoid wasps. These are not reciprocally monophyletic groups so we refrain from using the formal subordinal names that used to be deployed (respectively, Symphyta, Aculeata and Parasitica) although the wasp-waisted Apocrita (parasitoids including aculeates) and the Aculeata are each monophyletic. Parasitoid wasps arose from within the phytophagous sawflies and woodwasps and the aculeates are a specialised branch of the parasitoid wasps (e.g. Rasnitsyn, 1980; Vilhelmsen, 1997a; Klopfstein *et al.*, 2013). Basal hymenopteran relationships, with fossils dating back to the Triassic and an origin probably dating to the Permian (Zhang *et al.*, 2016), are fairly well understood (see Vilhelmsen, 2001; Klopfstein *et al.*, 2013) but those within the Apocrita have proved much more difficult to reconstruct, presumably because there was a rapid radiation of parasitoid lineages, reflecting the rapid diversification of other holometabolous insect orders alongside that of the angiosperm plants. The largest analyses to date (Klopfstein *et al.*, 2013) agree on the outline topology shown in Fig. 1. The aculeates were for some time thought to be the sister-group to the Ichneumonoidea, based largely on the shared presence of small flaps called valvilli in the ovipositor canal (Quicke *et al.*, 1992a), but the majority of molecular data now suggest that the Ichneumonoidea is the sister group to the huge lineage known as the Proctotrupomorpha, which includes the species-rich superfamily Chalcidoidea as well as the Cynipoidea, Diaprioidea, Mymarommatoidea, Platygastroidea and Proctotrupoidea (Sharkey, 2007). The Ichneumonoidea together contain over 44,000 described species, more than 24,000 of them in the Ichneumonidae (Yu *et al.*, 2012), and many more await description.

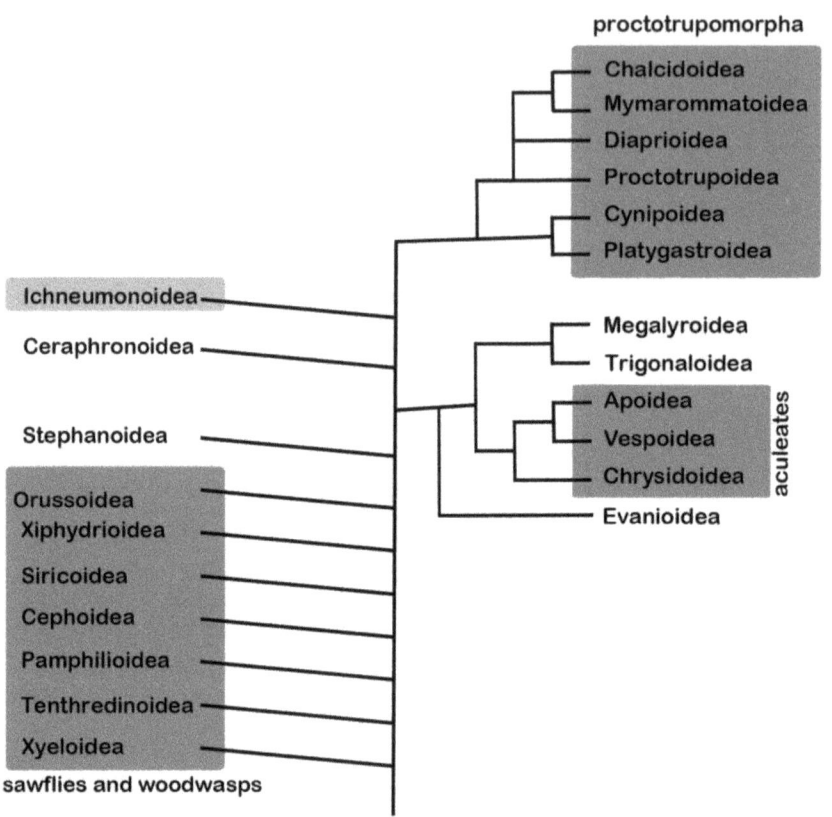

Figure 1. Overview of the phylogeny of the superfamilies (and higher assemblages) of Hymenoptera, derived from Klopfstein et al. (2013). Peters et al. (2017) presented a broadly similar picture of relationships, but with Ceraphronoidea as sister-group to Ichneumonoidea, and with Vespoidea split into several superfamilies to reflect Vespoidea's paraphyly with respect to Apoidea.

Identification of ichneumonids

There is no simple or comprehensive way of identifying British ichneumonids through to species level. Royal Entomological Society (RES) Handbooks cover a few subfamilies (see subfamily treatments) and some small papers deal with the British fauna of a few other small subfamilies, such as Cremastinae and Xoridinae, but the majority of species are not included in comprehensive identification guides or taxonomic revisions. There are, though, a great many papers dealing with the identification and taxonomy of small (occasionally large) parts of subfamilies, such as keys to the species of a particular genus. All relevant useful taxonomic papers (to the best of our knowledge) are given in the subfamily accounts, in the identification sections. Older compilatory works, particularly those of Schmiedeknecht (1902-1914), may be found useful in at least pointing towards a possible identification, although the taxonomy and classification of European Ichneumonidae has changed considerably since then. Recognising a specimen as an ichneumonid and identifying the subfamily are the first steps and these are covered in this handbook. Other keys to ichneumonid subfamilies may be found to be useful, particularly Wahl's (1993c), although Perkins's (1959) and Townes's (1969) subfamily keys are off-putting at best (impossible at worst) as well as being outdated. Identifying genera is another hurdle but the four volumes by Townes (1969-1971) on the world fauna are very useful. Although Townes did not treat the genera of Ichneumoninae, Perkins's work on British Ichneumoninae (1959, 1960) fills that gap for the British fauna, with necessary allowances made for changes in classification and some additions to the fauna. Some subfamilies, particularly Campopleginae and Cryptinae, have had their genera substantially reorganised since Townes's generic reclassifications: we have referenced papers introducing and employing these changes. Some other subfamilies have changed remarkably little, which is probably a sign of the amount of work that still remains to be done on their classification.

Most ichneumonid subfamilies (or tribes) have characteristic appearances, and are simple to recognise with practice. However, at first, when struggling through the keys, it will appear that subfamilies generally lack definitive identification features. Instead, determination to subfamily is usually clinched through a combination of characters. Identification of ichneumonids generally requires various parts of the body to be visible so mounting the specimen appropriately (see below) is very important. Happily, it is very rarely necessary to examine genitalia, or open mandibles or count numbers of bristles. We recommend that when starting on ichneumonids you build up a good number of specimens before launching into keys, which can help by providing specimens that key out on both sides of a couplet. Seeing the differences between different groups of ichneumonids is very helpful from the outset so having even a few reliably determined specimens to hand is a boon. Building up a collection is also very important given the paucity of collections of reliably named ichneumonids in local museums. Be aware that reliable identification of species from photographs is rarely possible. Taxapad (Yu *et al.*, 2012; recently updated by Yu *et al.*, 2016) is a comprehensive catalogue of the ichneumonoid literature and we have relied a lot on this resource. As discussed below, all catalogues involving host relations need to be treated with appropriate (in fact massive!) caution.

A common name for the family

There is no widely employed English common name for the Ichneumonidae. We simply use 'ichneumonids' in this handbook in preference to 'ichneumons' or 'ichneumon-flies'. The latter was employed frequently in the North American literature but not in recent decades, although, in an effort not to be off-putting, Fitton *et al.* (1988) referred to 'pimpline ichneumon-flies' in a previous handbook. Sadly, there is no equivalent of the Japanese *Himebati*, which can be translated as 'princess wasps', or the German *Schlupfwespen*, 'slippery' (in the sense of 'sly' or 'devious') wasps, for the Ichneumonoidea as a whole.

General Biology

Various introductions to parasitoid biology are given by Askew (1971), Gauld & Bolton (1988), Godfray (1994, emphasising behavioural ecology), Hanson & Gauld (1995, 2006), Shaw (1997), Quicke (1997, 2015) and Shaw *et al.* (2009). The basics of ichneumonoid biology are well covered by Shaw & Huddleston (1991) from a braconid perspective but there are differences between braconids and ichneumonids in their host foci and frequency of particular modes of parasitism, which are explored here and by Quicke (2015). Rather than repeating a lot of information we give a brief summary of parasitoid biology mainly to set the subfamily accounts in context.

There are two major dichotomies in life history that together are very important in circumscribing potential host ranges of parasitoids (for original definitions, see Haeselbarth, 1979; Askew & Shaw, 1986):

    ectoparasitoid / endoparasitoid
    idiobiosis / koinobiosis.

As the word suggests, *ectoparasitoids* develop externally on the host, puncturing the cuticle with their mandibles and extracting haemolymph before usually consuming much of the host tissue. *Endoparasitoids* develop within the body of the host with the egg placed internally by the ovipositing female. In contrast to several groups of braconids, endoparasitoids that have a final ectoparasitoid feeding phase are almost unknown in ichneumonids.

Idiobionts arrest the host's development at or very soon after oviposition, usually with permanently paralysing venoms. The host is consumed by the parasitoid larva as essentially immobilised meat. Because the host is immobilised, and therefore vulnerable to predation, idiobionts almost invariably parasitize concealed hosts, in a substrate or protected within a case or cocoon. Idiobionts are most often ectoparasitoids, but there are significant numbers of idiobiont endoparasitoids. *Koinobionts* allow the host to continue their movements and development after oviposition. Frequently the host is temporarily paralysed to facilitate egg placement but this is not invariably the case. Koinobionts are able to exploit host stages that may be much more accessible than later stages, and some delay their development until the host has concealed itself for pupation. Most koinobionts are endoparasitoids but koinobiont ectoparasitoids are well represented.

There is a tendency to think of koinobionts as being more 'specialised', with narrower host ranges, and of idiobionts as being more 'generalised', with wider host ranges. There is some empirical supporting evidence for this, though the concept is best seen in terms of potential (Askew & Shaw, 1986). Idiobionts generally have little to contend with in the way of host immune responses so can frequently develop on a variety of potential hosts, including primary parasitoids of the host that was originally detected, and can frequently develop on several stages of the host, from larva through to unemerged adult. Koinobionts have a more demanding relationship with their hosts: endoparasitoids will have to contend with (or evade) the host immune response, and ectoparasitoids may need to keep their place on a host through its ecdyses, and in both cases the parasitoid has to adapt to feeding on living tissues. As Haeselbarth (1979) stressed, they must also be more in tune with their hosts' life histories, to be feeding on hosts of the right developmental stage at the right time. Despite these constraints, we actually have very little comparative data (free of phylogenetic bias) that would allow us to compare relative host specificity of idiobionts and koinobionts (e.g. Shaw, 1994, although Askew & Shaw, 1986, present some compelling data). Whilst there are essentially host-specific koinobionts, most will attack a range of hosts, usually circumscribed by phylogeny, but also by ecology. Some of the more interesting examples of disparate host ranges arise when plurivoltine koinobiont endoparasitoids adopt very different univoltine host taxa in each generation (Shaw, 2003; Shaw *et al.*, 2016). On the other hand, idiobionts can be very specialised by being essentially ecological niche specialists, concentrating their searches in very narrow habitats in which they may parasitise a phylogenetically wide span of hosts, or become closely adapted as taxon specialists. Specific, referenced examples of these various life history patterns in ichneumonids are presented in the subfamily sections of the handbook.

Differences between Braconidae and Ichneumonidae in the British fauna

Both families include idiobionts and koinobionts, which each can be ectoparasitoids or endoparasitoids (though in Britain idiobiont endoparasitism may occur only in Ichneumonidae). Overall, Ichneumonidae and Braconidae use a partly overlapping span of hosts but each has a part of its host spectrum and/or behaviour not covered by the other:

1. Braconidae do not oviposit into cocoons or the pupal stages of hosts. These habits have, however, been important traits in the evolution and diversification of Ichneumonidae, and in this family oviposition through silk (many Phygadeuontinae and Cryptinae, some Pimplinae) has even led to the exploitation of spiders' egg sacs (all three subfamilies) and thence mobile spiders as such (evolved once, in Pimplinae). Oviposition into host Lepidoptera pupae (or prepupae) occurs widely in Ichneumoninae and one group of Pimplinae, whereas oviposition into Diptera puparia is known in some Phygadeuontinae.

2. Apart from the subfamilies Opiinae and Alysiinae, which do emerge as adults from cyclorrhaphous Diptera puparia, very few (no British) Braconidae emerge as adults from the host pupa, even if it was attacked as a larva. This is in strong contrast with Ichneumonidae, in which several subfamilies (Metopiinae, Anomaloninae and much of Ichneumoninae) are larva-pupal parasitoids (i.e. oviposit into the host larva and emerge as an adult from its pupa).

3. Very few Braconidae (the very small subfamily Ichneutinae; also a few Exothecinae, Braconinae and Doryctinae) are associated with sawflies, though they are the main hosts of two large subfamilies of Ichneumonidae (Tryphoninae and Ctenopelmatinae) and several genera of Campopleginae, as well as some smaller and even more specialised ichneumonid subfamilies (Collyriinae, Adelognathinae; and Rhyssinae on Siricidae and Xiphydriidae). In addition quite a few odd species in other subfamilies (Pimplinae, Cryptinae [in fact perhaps much of the moderately large tribe Aptesini]) also use concealed or cocooned sawflies.

4. Braconidae are not at all associated with apocritan Hymenoptera (apart from the special case of certain Euphorinae, see below). Several groups of Ichneumonidae, on the other hand, are parasitoids of aculeate Hymenoptera (the small specialised subfamilies Hybrizontinae and Neorhacodinae; also seen in Pimplinae, Poemeniinae and Cryptinae); one quite large subfamily (Mesochorinae) and one very small subfamily (Eucerotinae) are true hyperparasitoids of (especially) ichneumonoids as they develop in their own hosts; and pseudohyperparasitism (see glossary), by oviposition into ichneumonoid cocoons, is practiced by a number of Phygadeuontinae and Pimplinae, sometimes as specialists.

5. Because of (4), Braconidae are never hyperparasitoids in either of the above senses.

6. One group of Braconidae (Aphidiinae) parasitizes aphids, a group never exploited by Ichneumonidae (which only parasitize holometabolous insects – and in a few cases arachnids).

7. No Ichneumonidae parasitize adult insects, but one subfamily of Braconidae (Euphorinae) parasitizes adult insects (and additionally the nymphal stages as well as adults of some Hemiptera and Psocoptera). These include a few genera that parasitize adult Hymenoptera (including Ichneumonoidea). The essential requirement is that the host adult is quite long-lived, and feeds extensively; this may explain why, although several orders are attacked, no euphorines have specialised on Diptera or Lepidoptera.

8. All Ichneumonidae kill the host larva as they complete their feeding (if it is not dead before then), while several koinobiont endoparasitoid subfamilies of non-cyclostome Braconidae include taxa that are haemolymph-feeders throughout their larval life and leave their host still alive (though doomed) as they exit (most Euphorinae, many Meteorinae and Microgastrinae).

9. Gregariousness is rather uncommon in Ichneumonidae (scattered over several subfamilies) and usually involves only small brood sizes, but in some subfamilies of Braconidae it is spectacularly practised by a fairly high proportion of the species in some genera (particularly of Microgastrinae and Macrocentrinae) with typical brood sizes of several tens.

Conservation

Shaw & Hochberg (2001) drew attention to the disparity in conservation effort towards parasitoid wasps in comparison with other insects in Britain, a lack of all kinds of knowledge being the principal reason for a failure to engage with what must be a particularly endangered guild considering their generally high levels of specialisation and high position in the invertebrate food chain (see also LaSalle & Gauld, 1993). We still know far too little about habitat preferences of parasitoid wasps (Shaw, 2006a) to be able to make many recommendations for their conservation, other than that they need healthy populations of hosts in the correct (but often unknown) habitat. There have been some attempts to charaterise ichneumonid faunas of particular habitats and habitat patch sizes (e.g. Fraser et al., 2007; Mayhew et al., 2009) and useful data on ecological associations and distribution are contained in the series of papers cataloguing the collections of the National Museums of Scotland (e.g. Diller & Shaw, 2014; Kasparyan & Shaw, 2008, 2009; Schwarz & Shaw, 1998, 1999, 2000, 2010, 2011; Shaw & Kasparyan, 2003, 2005; Shaw et al., 2016). Basic data on population trends and georeferenced distribution data for ichneumonids are almost entirely lacking although the first author's Nocturnal Ichneumonoidea Recording Scheme (http://nocturnalichs.myspecies.info/) is gathering data on certain koinobiont, nocturnal groups (see Broad & Shaw, 2016, for preliminary distribution maps and phenology data for British *Enicospilus*). Long-term monitoring of parasitoid numbers would give us an idea of how vulnerable these species are, at their high trophic levels, and how the fauna has changed and is changing in the face of massive land use, vegetation and climate change and air-borne pollution. With some fairly well collected and conspicuous taxa such as Pimplinae we can point to a few potential extinctions and recent colonisations (e.g. Shaw, 2006b) but for most ichneumonids, as for most parasitoids, we simply lack the baseline knowledge (taxonomic and distributional) or community of amateur enthusiasts to gather the necessary data.

Classification and phylogeny

There is no definitive subfamily classification of the Ichneumonidae that can be regarded as settled. We accept 42 subfamilies (Table 1) with only seven of these extralimital to the British Isles. This classification largely follows the only large-scale phylogeny of the Ichneumonidae, by Quicke et al. (2009), and we frequently refer to this study in the individual subfamily accounts. Forthcoming publications by A.M.R. Bennett et al., S. Klopfstein et al., B. Santos et al. and B. Sharanowski et al. will alter some aspects of the classification but will not change the composition of the majority of subfamilies. The most recent changes to the subfamily-level classification affect Cryptinae *sensu lato*, with Santos (2017) recognising three subfamilies – Ateleutinae, Cryptinae and Phygadeuontinae – in recognition of the fact that Cryptinae in the traditional sense was paraphyletic with respect to Adelognathinae and Ichneumoninae. Santos's (2017) study combined morphological and molecular data with wide taxon sampling and was congruent with many of the conclusions of Gokhman (1992) and Laurenne *et al.* (2006). However, the restricted subfamily Phygadeuontinae (formerly the tribe Phygadeuontini of Cryptinae) still appears to be paraphyletic with respect to Cryptinae and Ichneumoninae. Note that we recognise the Neorhacodinae (again) as a separate subfamily, following Bennett *et al.* (in prep.), contrary to Quicke *et al.* (2009), who synonymised Neorhacodinae with Tersilochinae. We have also incorporated the changes made by Klopfstein *et al.* (in press) to the classification of some pimpliformes, recognising the tribe Theroniini of Pimplinae, transferring *Pseudorhyssa* from Poemeniinae to Pimplinae (tribe Delomeristini) and transferring *Hemiphanes* from Orthocentrinae to Cryptinae (tribe Aptesini). The broad pattern of relationships between the ichneumonid subfamilies is shown in Fig. 2, simplified from Quicke *et al.* (2009) (note that several different analyses were employed by Quicke *et al.* and that the relationships depicted here are those robust to most analysis parameters). Monophyletic assemblages of subfamilies have been arranged in informal groups, the largest of which are the ichneumoniformes, ophioniformes and pimpliformes, proposed in Gauld's (1985) and Wahl's (1986, 1990, 1993a) works on inter-subfamily relationships. The modern classification of Ichneumonidae originates in the pioneering work of Henry Townes, which was expressed particularly in his four volume 'The Genera of Icheumonidae' (Townes, 1969-1971), works which omitted only the Alomyinae, Ichneumoninae and Hybrizontinae (the latter he regarded as belonging to the Braconidae). Townes's reclassification of the Ichneumonidae placed most genera within a workable classification that recognised relationships based on shared biology as well as morphology; although Townes did not employ strictly cladistic or, indeed, phylogenetic hypotheses, his work was a huge advance on the attempts of earlier authors to define subfamilies based mainly on overall similarity in appearance, which is a very poor indicator of evolutionary history within the family.

In the literature there are various different names for some subfamilies, including variant spellings and different interpretations. Some of the 'traditional' large subfamilies, particularly Ophioninae and Pimplinae, have been progressively dismantled. Townes (1969-1971) employed a rather idiosyncratic approach to family-level nomenclature (see Fitton & Gauld, 1976) but his volumes on the genera of Ichneumonidae are in many ways the most definitive to date. Perkins (1959) included a key to the British subfamilies which was progressive at the time but is out of date now. As these are the 'standard works' in a British context, their names and differing concepts are listed in Table 1, along with our current interpretations of subfamily names and status, together with some of the other names encountered in the wider literature.

Trends in host associations

A robust knowledge of sister-group relationships should shed light on the ancestral biologies of major clades and thus the constraints that might explain why endoparasitism and ectoparasitism are more or less prevalent in some groups, why some families are much more closely linked to certain host taxa, and other questions that can throw some light on the evolution of the intricate relationships of parasitoids to their hosts. Although there have been few studies of ichneumonid phylogeny that would allow us to confidently reconstruct the evolution of host associations, some groupings of subfamilies (as discussed above in 'classification and phylogeny') seem to be robust enough to give us an idea of the wider trends.

General trends in host utilization patterns were explored by Gauld (1988b), which is still a very useful reference although there have been changes in our understanding of phylogeny. The basal Ichneumonidae are thought to be represented by Xoridinae and the extralimital Labeninae. These are idiobiont ectoparasitoids of concealed coleopteran and hymenopteran larval/pupal hosts (with some interesting digressions in Labeninae; see 'extralimital and fossil subfamilies'). This is generally assumed to be a plesiomorphic strategy across the parasitoid Hymenoptera although it

Table 1. Nomenclature of ichneumonid subfamilies, arranged in informal groupings: as employed here; as employed by Townes (particularly 1969-1971); and by Perkins (1959). Other frequently encountered spellings and alternative classifications are listed. Blank entries mean that the author used the same name as we do.

current name	Townes (1969-1971) if different	Perkins (1959) if different	other names
xoridiformes			
Xoridinae		tribe of Pimplinae	
labeniformes			
Labeninae	Labiinae	n/a	
pimpliformes			
Acaenitinae		tribe of Pimplinae	
Collyriinae			
Cylloceriinae	part of Orthocentrinae (as Microleptinae)	part of Orthocentrinae (as Plectiscinae)	tribe of Orthocentrinae[4]
Diacritinae	tribe of Pimplinae (as Ephialtinae)		part of Orthocentrinae[5]
Diplazontinae		Diplazoninae	
Orthocentrinae	Orthocentrinae and Microleptinae	Orthocentrinae and Plectiscinae	Oxytorinae, Helictinae
Pimplinae	Ephialtinae		Pimplinae in part
Poemeniinae	tribe of Pimplinae (as Ephialtinae)	tribe Neoxoridini of Pimplinae	
Rhyssinae	tribe of Pimplinae (as Ephialtinae)	tribe of Pimplinae	
brachycyrtiformes			
Brachycyrtinae	tribe of Labeninae (as Labiinae)	n/a	
Claseinae	tribe of Labeninae (as Labiinae)	n/a	tribe of Cryptinae[3]
Pedunculinae	part of Brachycyrtini of Labeninae (as Labiinae)	n/a	
ichneumoniformes			
Adelognathinae			
Agriotypinae			Agriotypidae[1]
Alomyinae	[not treated]		tribe of Ichneumoninae[2]
Ateleutinae	subtribe of Cryptinae (as Gelinae)	n/a	

[1]Mason (1971) argued, unconvincingly, that *Agriotypus* belongs in its own family in the Proctotrupoidea. [2]Variously regarded as a separate subfamily, tribe or as part of what we refer to as the Phaeogenini (which then took the name Alomyini). [3]Gauld (1983). [4]Humala (2003, 2007a). [5]Not mentioned by Perkins (1959) but earlier (Perkins, 1940) he had treated *Diacritus* as part of the Plectiscinae (equivalent to part of Orthocentrinae). [6]Quicke *et al*. (2009). [7]Placed within the ichneumoniformes by Santos (2017).

Table 1 continued.

current name	Townes (1969-1971) if different	Perkins (1959) if different	other names
Cryptinae	tribes Echthrini and Mesostenini of Cryptinae (as Gelinae)		Mesosteninae
Ichneumoninae			
Phygadeuontinae	tribe of Cryptinae (as Gelinae)	Cryptinae in part	Hemitelinae, Gelinae
orthopelmatiformes			
Orthopelmatinae			
ophioniformes			
Anomaloninae	Anomalinae	part of Ophioninae	
Banchinae		Lissonotinae	
Campopleginae	Porizontinae	part of Ophioninae	
Cremastinae		part of Ophioninae	
Ctenopelmatinae	Scolobatinae		
Hybrizontinae	[Braconidae]	[Braconidae]	Paxyllomatinae Paxyllomatidae Hybrizoninae
Lycorininae			
Mesochorinae			
Metopiinae			
Neorhacodinae	tribe of Banchinae	tribe of Banchinae (as Lissonotinae)	part of Tersilochinae[6]
Nesomesochorinae	tribe Nonnini of Campopleginae (as Porizontinae)	n/a	
Ophioninae		Ophioninae in part	
Oxytorinae	part of Orthocentrinae (as Microleptinae)	part of Orthocentrinae (as Plectiscinae)	part of Orthocentrinae (as Oxytorinae)
Sisyrostolinae	part of Phrudinae	n/a	Brachyscleromatinae
Stilbopinae	tribe of Banchinae		
Tatogastrinae	part of Orthocentrinae (as Microleptinae)	n/a	
Tersilochinae	Phrudinae and Tersilochinae	Phrudinae and Ophioninae (in part)	
Tryphoninae			
unplaced			
Eucerotinae	tribe of Tryphoninae	Euceratinae	Eucerinae
Microleptinae[7]	part of Orthocentrinae (as Microleptinae)	part of Orthocentrinae (as Plectiscinae)	

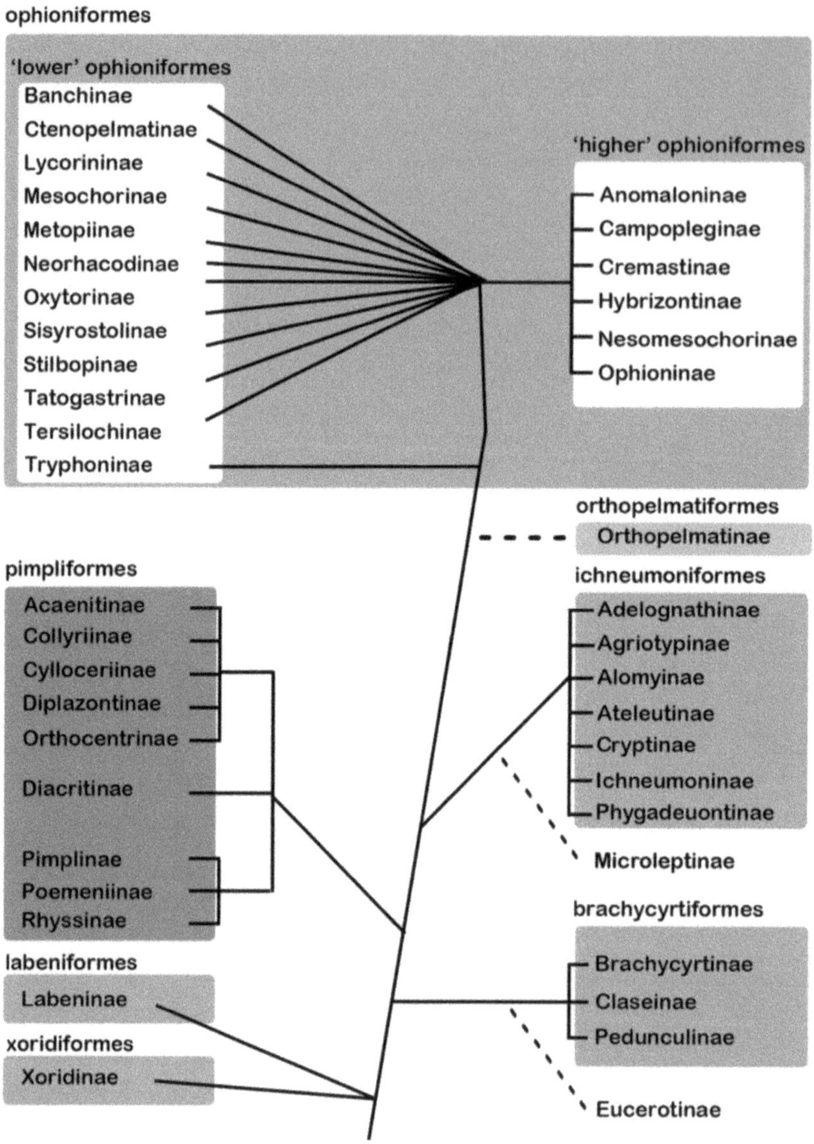

ophioniformes

'lower' ophioniformes
- Banchinae
- Ctenopelmatinae
- Lycorininae
- Mesochorinae
- Metopiinae
- Neorhacodinae
- Oxytorinae
- Sisyrostolinae
- Stilbopinae
- Tatogastrinae
- Tersilochinae
- Tryphoninae

'higher' ophioniformes
- Anomaloninae
- Campopleginae
- Cremastinae
- Hybrizontinae
- Nesomesochorinae
- Ophioninae

orthopelmatiformes
- Orthopelmatinae

pimpliformes
- Acaenitinae
- Collyriinae
- Cylloceriinae
- Diplazontinae
- Orthocentrinae
- Diacritinae
- Pimplinae
- Poemeniinae
- Rhyssinae

ichneumoniformes
- Adelognathinae
- Agriotypinae
- Alomyinae
- Ateleutinae
- Cryptinae
- Ichneumoninae
- Phygadeuontinae

Microleptinae

brachycyrtiformes
- Brachycyrtinae
- Claseinae
- Pedunculinae

labeniformes
- Labeninae

xoridiformes
- Xoridinae

Eucerotinae

Figure 2. Phylogeny of Ichneumonidae, derived from Quicke *et al.* (2009) and Santos (2017). Major groupings are shown but most finer details of inter-subfamily relationships are shown as unresolved pending more evidence. Subfamilies of uncertain affinities are depicted on dashed branches.

probably does not pertain to the Chalcidoidea, nested within a large clade of endoparasitoids (Heraty *et al.*, 2013; Klopfstein *et al.*, 2013). Two major ichneumonid clades, the ichneumoniformes and pimpliformes, may be sister-groups; with this larger assemblage in turn the sister group to the ophioniformes (Quicke *et al.*, 2009; and Fig. 2). There are substantial differences in host utilization trends between these clades. An overview of trends in host utilization is given here; details of host associations and more complete references can be found in the subfamily accounts.

The ichneumoniformes includes several rather small subfamilies, comprising mostly idiobiont ectoparasitoids, together with the very species-rich Cryptinae, Phygadeuontinae and Ichneumoninae, which are clearly closely related, although the precise pattern of relationships is clouded by the small subfamilies Ateleutinae, Adelognathinae, Alomyine and possibly Microleptinae. The vast majority of cryptines and phygadeuontines are idiobiont ectoparasitoids of more or less concealed holometabolous larvae and pupae. From this basic biology, which is presumably ancestral to the whole ichneumoniformes clade, predation on spider egg sacs has evolved on several independent occasions and a few phygadeuontines are known to attack other egg sacs, including those of a pseudoscorpion (Morley, 1907) and even egg-masses of a weevil in Kenya, ovipositing through leaves into the gelatinous egg mass (Benjamin & Demba, 1969; Kerrich, 1969). One cryptine (a Japanese *Thrybius*) is a predator of successive phytophagous Eurytomidae (Chalcidoidea) larvae (Matsumoto & Saigusa, 2001). The general trend towards predatory behaviour (consuming more than one host during development) rather than as strictly parasitoids is more pronounced in Cryptinae and Phygadeuontinae than in other subfamilies, and is probably originally led by an affinity for silk. In Phygadeuontinae there is also a pronounced trend towards pseudohyperparasitism, i.e. attacking the cocoons of ichneumonoids. A small clade of Phygadeuontinae (the stilpnine genera) are koinobiont larva-pupal endoparasitoids of cyclorrhaphous Diptera, with perhaps similar habits in a few species of *Phygadeuon*, which seem to be closely related. Some Phygadeuontinae are endoparasitoids in Lepidoptera pupae (e.g. Shaw et al., 2015) and this may be widespread in some tropical Cryptinae (e.g. in *Polycyrtus* Spinola; Zúñiga Ramírez, 2004) but there has been very little published information. The small subfamily Ateleutinae seems to also be associated with silk, with a couple of species having been reared from the cases of psychids (bag-worms; Lepidoptera: Psychidae): one species of *Ateleute* has been demonstrated to be a solitary idiobiont ectoparasitoid of psychid larvae. Another small subfamily, the Microleptinae, are possibly closely related to Phygadeuontinae (Santos, 2017). *Microleptes* may be endoparasitoids of Stratiomyidae (Diptera) but beyond this host taxon association nothing is known with certainty.

The Adelognathinae, which seem to be closely related to Cryptinae and Phygadeuontinae (and may be nested within the latter subfamily according to molecular data: Laurenne et al., 2006; Santos, 2017), are ectoparasitoids of sawfly larvae, some as idiobionts and others as koinobionts. Another small subfamily, Alomyinae, are koinobiont endoparasitoids of Hepialidae (Lepidoptera) larvae, eventually mummifying the host larva, which (although a general feature of rogadine Braconidae) is otherwise known only from one other genus of Ichneumonidae, *Colpognathus* (Ichneumoninae), and is not to be confused with superficially similar cocoon formation by some genera of Campopleginae inside the otherwise unmodified host larval skin. The Alomyinae have often been included in the Ichneumoninae, which are, as far as is known, entirely endoparasitoids of Lepidoptera. The hosts of Cryptinae encompass Arachnida, Coleoptera, Hymenoptera, Lepidoptera and Neuroptera whilst Phygadeuontinae have an even wider host spectrum, encompassing Arachnida, Coleoptera, Diptera, Hymenoptera, Lepidoptera, Neuroptera, Raphidioptera and Trichoptera. In Ichneumoninae emergence is always (except for *Colpognathus*) from the pupa but oviposition is into either the larva (acting as a koinobiont) or pupa (as an idiobiont), varying according to tribe or in the larger tribes varying between some clades (even varying across the species of the genus *Ichneumon*). The aberrant *Agriotypus* (the sole genus of the Agriotypinae) are unusual in being aquatic, as they are idiobiont ectoparasitoids of Trichoptera prepupae or pupae. Parasitism of aquatic hosts has evolved convergently in at least four other ichneumonid subfamilies (Campopleginae, Cremastinae, Cryptinae and Metopiinae) as well as four subfamilies of Braconidae (Alysiinae, Braconinae, Microgastrinae and Opiinae) (Bennett, 2011).

Another aberrant subfamily, of uncertain affinities, is the Eucerotinae; *Euceros* species are unique in the Ichneumonidae in laying stalked eggs on the vegetation from which hatch planidial larvae that attach to passing sawfly or Lepidoptera larvae. *Euceros* are true hyperparasitoids, completing development in or on an ichneumonoid primary parasitoid. They are difficult to characterise as ecto- or endoparasitoids although they are clearly koinobionts. There are three small extralimital subfamilies in the brachycyrtiformes clade. Brachycyrtinae are (presumed) idiobiont ectoparasitoids of Neuroptera pupae whilst one species of the closely related Pedunculinae has been recorded as reared from a spider egg sac. Claseinae attack cocooned Hymenoptera and deeply concealed Coleoptera, although very little is known of their biology.

The pimpliformes assemblage again includes several subfamilies that comprise mainly idiobiont ectoparasitoids of concealed hosts, as represented by the Poemeniinae, Rhyssinae and many of the Pimplinae. Although Coleoptera are often attacked there are many parasitoids of Hymenoptera and this association with Hymenoptera seems to have been important in the radiation of Ichneumonidae, in contrast to Braconidae. Possibly sister-group to these subfamilies (although there is little certainty in any phylogenies of the pimpliformes) is a group of koinobiont endoparasitoid subfamilies: the Acaenitinae, which specialise on Coleoptera; Collyriinae, egg-larval parasitoids of Cephidae (Hymenoptera); and a clade of parasitoids of Diptera, the Cylloceriinae, Diplazontinae and Orthocentrinae each using different groups. The Diacritinae have never been reared and may or may not be part of this endoparasitoid clade (Wahl & Gauld, 1998). The greatest variety in biology lies in the Pimplinae, where one major clade (the tribe Pimplini) are idiobiont endoparasitoids of Lepidoptera and in some cases (as a pseudohyperparasitoid) Hymenoptera pupae, and another clade (the *Polysphincta* genus-group of the Ephialtini) are koinobiont ectoparasitoids of spiders, which both presumably arose from idiobiont ectoparasitism, a strategy retained by the Delomeristini and most of the Ephialtini. Parasitism of spiders in the Ephialtini includes genera specialising on concealed egg-nesting spiders, and predatory behaviour on successive eggs within sacs, but otherwise hosts of the ectoparasitoids are concentrated on concealed Hymenoptera, Coleoptera and Lepidoptera, with some ephialtines also regularly parasitizing Diptera.

The Orthopelmatinae was given its own grouping (orthopelmatiformes) as it seems only distantly related to other major groupings of subfamilies, although it may be closest to the ophioniformes (Quicke *et al*, 2009), in line with its biology, or very basal within Ichneumonidae (Bennett *et al*., in prep.). *Orthopelma* are koinobiont endoparasitoids of gall wasps (Hymenoptera: Cynipidae).

All known ophioniformes are koinobiont parasitoids of holometabolous insect larvae, although none are known to attack Diptera. Basal to this clade is the Tryphoninae, which may (Bennett, 2015) or may not (Quicke *et al*., 2009) be monophyletic but are all koinobiont ectoparasitoids. The Lycorininae, another apparently basal subfamily, comprises only the genus *Lycorina*, which functionally seem to be ectoparasitoids too, albeit within the hind gut. Other ophioniformes are endoparasitoids. Most Tryphoninae and the vast majority of Ctenopelmatinae, another species-rich component of the 'lower' ophioniformes, are parasitoids of sawflies and this association is a feature of enough lineages of ophioniformes (as well as pimpliformes) to suggest that initial diversification of the clade may have depended upon Hymenoptera as hosts, or at least Hymenoptera and Lepidoptera. Many other 'lower' ophioniformes attack Lepidoptera, including the rather large subfamilies Banchinae and Metopiinae. The Mesochorinae are probably exclusively true hyperparasitoids of (almost always) ichneumonoid larvae within a range of insect secondary hosts. The small subfamily Neorhacodinae (at least based on the only British species *Neorhacodes enslini* (Ruschka)) are parasitoids of stem-nesting Crabronidae (Hymenoptera). This association with Hymenoptera is not continued in the 'higher' ophioniformes, an assemblage of subfamilies that were mostly lumped together as the Ophioninae in old classifications, on the basis of a distinctly petiolate and often laterally compressed metasoma. The Tersilochinae was included in this grouping too but recent phylogenies place them basal to the 'higher' ophioniformes. Tersilochinae and the small (extralimital) subfamily Sisyrostolinae have radiated as parasitoids of Coleoptera larvae, although a few tersilochines have taken to parasitizing sawflies (and the sisyrostolines are very poorly known). The 'higher' ophioniformes, comprising Anomaloninae, Campopleginae, Cremastinae, Ophioninae and the extralimital Nesomesochorinae, represent a large radiation of basically Lepidoptera parasitoids (although there are no reliable host records for Nesomesochorinae). The small numbers of species attacking Coleoptera (in Anomaloninae, Campopleginae, one species of Ophioninae), Raphidioptera, terrestrial Trichoptera and more significantly Hymenoptera (all in Campopleginae) seem to have adopted these wider hosts from a background of Lepidoptera parasitism. The small and highly aberrant subfamily Hybrizontinae is possibly part of this assemblage too. Hybrizontines are koinobiont endoparasitoids of ant (Formicidae) larvae but with most details of their development (and affinities) still obscure. Whilst the majority of ophioniformes emerge from the host when it is still a larva (or prepupa), the Metopiinae and Anomaloninae are larva-pupal parasitoids, as are a few Campopleginae and Ophioninae (although in a less specialised way). Egg-larval parasitism is only known within the Campopleginae (a few species), Ctenopelmatinae (some Ctenopelmatini and Pionini) and Stilbopinae (*Silbops*) but may occur in some other ophioniformes with very slender ovipositors.

Just a few small subfamilies are entirely unknown biologically; in Britain these are the Diacritinae and Oxytorinae, respectively possibly part of an endoparasitoid clade of pimpliformes, and probably nested within the Ctenopelmatinae. Clearer phylogenetic affinities should give us some clues as to their hosts; or, conversely, accurate host and developmental data might help to provide a more robust classification for them.

Collecting, rearing and preservation

Collecting

Standard collecting techniques such as sweep netting, hand netting and yellow pan trapping all produce plenty of ichneumonids. However, because most ichneumonids seem to occur at low densities, the most productive way of both collecting and assessing the size of a local fauna is by Malaise trapping. A Malaise trap set by the first author in Monks Wood NNR (Cambridgeshire) caught a little over 2,000 ichneumonid specimens over 12 months. The main drawback of Malaise trapping is that significant time is required for sorting the large catches (particularly in the summer). Ideally, willing entomologists should be identified who will put the rest of the catch (Diptera invariably comprise the bulk) to good use. Light trapping samples a suite of genuinely nocturnal species that only turn up in small number in Malaise traps, and many diurnal, or possibly crepuscular, species can also be found at light, probably individuals still active in the evening or disturbed from nearby vegetation. The nocturnal Ichneumonoidea recording scheme, run by the first author, is collating data on distribution, phenology and hosts of these nocturnal species. Mercury vapour, blended UV and tungsten (as used in Rothamsted light traps) bulbs are all effective at attracting ichneumonids, especially when background light levels are low. LED lights have not yet been tested for their attractiveness to ichneumonids but there is no reason to think they will not be effective. It is striking how some species can be readily collected using a particular method but almost never by other means, including rather a lot of species that can be easily reared once the host is known but are hardly ever found otherwise. Thus it is best to employ several collecting techniques in building up a diverse collection.

Rearing ichneumonids and reliability of data

Fitton *et al.* (1988) and Shaw (1997) cover much useful ground on collecting and preserving parasitoids. Rather than repeating much of this information we emphasise techniques most relevant to Ichneumonidae. Rearing from host species is the best way to add substantially to our knowledge of ichneumonids. Within the Ichneumonidae we have the most detailed knowledge of host relations for species that parasitise relatively easily-collected Lepidoptera. We probably know least about parasitoids of Diptera and Coleoptera.

Shaw's (1997) handbook covers all of the essential details of basic equipment, hygiene, environment, specimen preparation etc., and pays special attention to data standards. While reared specimens are of particular value, unless it is absolutely certain that the recorded host determination is correct they can be a source of damaging misinformation. Isolation from other potential hosts during rearing, accounting for all individuals reared together, and recovery of host remains are all important safeguards. As well as host accuracy, it is of course also vital, if any publication results, that the identification of the parasitoid is correct and that both expressions of doubt and the depository from which it could be recalled for verification are stated. Throughout this volume there are references to the dubious nature of many host records: these are all too frequent and continue to bedevil the literature (Noyes, 1994; Shaw, 1994). It cannot be stressed enough that the repetition by re-publication of host or distribution records from catalogues, which are so fraught with problems, simply adds to the miasma of misinformation surrounding parasitoids. These are, after all, poorly known insects and all literature records should be treated with caution. Recent work on British *Netelia* and *Ophion* (Broad, in prep.) has shown that well used names such as *Ophion luteus* (L.) and *Netelia testacea* (Gravenhorst) have been widely misapplied, to the extent that the information in catalogues, field-guides, online (e.g. the National Biodiversity Network) and identifications in collections are mostly incorrect. Data are becoming ever more accessible so it is tempting to take information at face value (data providers rarely issue strong health warnings), but this should be resisted and it is particularly damaging to reiterate unchecked rearing records in print, thereby giving an apparent endorsement of dubious information. The host range of a parasitoid is one of its most interesting and important properties. Understanding the realised (quantitative) host ranges of particular parasitoid species (Shaw, 1994; Shaw *et al.*, 2016) allows parasitoids reared in studies of the parasitoid complex of a given host to be understood in the wider context of their reach to other parasitoid complexes; to get an idea of how specialised a particular parasitoid might be, and the context of its specialisation (Shaw, 2017b).

As a final point on identification, it is important to bear in mind, when trying to find images of an ichneumonid species online, that a very high proportion of images are mislabelled, even on websites that seem authoritative.

Preparing specimens

The 'best' methods of pinning and carding specimens are to some extent subjective but the two key considerations are that all aspects of the specimen should be visible and that the specimen needs to be as protected as much as possible from damage. Ichneumonids have long antennae, a narrow connection of the metasoma to the mesosoma and often long legs and ovipositor sheaths. These render specimens rather easily damaged. From our experience, the best results are achieved from direct pinning and, for smaller specimens, card pointing, as shown in Figs 3 & 4 respectively. Whether for direct pinning or for carrying a card, the pin should have a head (for ease of handling) and a good sharp point (to enhance grip). Direct pinning should be of specimens large enough to take a size 1 or stouter pin, ideally of the readily available 38 mm length. The pin should enter the mesoscutum off-centre (to leave intact the centre and one side of the mesoscutum, where there can be useful sculptural features) then emerge from the mesosternum well behind the fore legs. It is all too easy to angle the pin too far forwards and then detach the prothorax (with fore legs) and head. The specimen should be at about 2/3 of the height of the pin, leaving enough room for labels. It is unnecessary – and usually impractical given the numbers of specimens being processed – to arrange specimens in any particular way, and the wings being spread in Lepidoptera fashion just takes up more room in a collection. Specimens with long appendages should be pinned into a block of plastazote of the right height so that antennae, legs and sometimes ovipositor sheaths can be cross-pinned to dry in the correct position; the long antennae of ophionines and some other groups can easily curl up to rest near the pin head and then are easily damaged. There is no need to open the mandibles and this is difficult to achieve anyway.

Figure 3. Specimen of *Achaius oratorius* (Fabricius) direct-pinned.

Figure 4. Specimen of *Zaglyptus multicolor* (Gravenhorst) glued to a card point.

Staging via micropins (usually into a plastazote or polyporus stage) is popular for some groups of insects but we do not recommend this method for ichneumonids as an extra pin increases the danger of the specimen swinging loosely from one or other pin or being knocked against objects. Those delicate antennae are easily lost. Anything that obscures good views of all parts should be avoided, as critical identification features can be anywhere on the body. Many older specimens in collections are mounted face down on card with their wings and appendages carefully outspread: very neat-looking but with many of the useful features (especially face and mouthparts, also mesopleuron and mesosternum) hidden from view. Card pointing, where the specimen is glued to one mesopleuron (and can include the very base of the wings on that side, for better grip), allows any part of the body to be seen. Gluing to the mesosternum is not recommended as there are useful features that are then obscured, such as the completeness of the posterior transverse mesosternal carina or the sculpture of the mesosternal furrow (particularly useful in Banchinae).

When pinning certain groups, particularly Ophioninae and *Netelia*, the number of antennal segments should be counted, which can be a very useful aid to identification. The earlier the better, in case of damage, with the number then recorded in the labelling. 'F' or 'Ant' are suitable abbreviations for the number of flagellar segments or the full antenna (flagellum plus scape and pedicel), respectively.

Dry specimens can be relaxed most easily in a tube with finely chopped young leaves of cherry laurel *(Prunus laurocerasus)*. Specimens should be glued to card points with water-soluble glue such as seccotine so that they can be soaked off if necessary. Gluing to the side of a pin with shellac is a time-saving method of rapidly preparing many specimens but it has its problems. The shellac needs to have the correct waxiness; if it is too brittle specimens are easily dislodged from their pins.

Ichneumonids can generally be dried and mounted directly from alcohol (prior brief immersion into clean 96% ethanol helps) but the metasomas of smaller specimens can shrivel a little. Small orthocentrines are particularly delicate and the entire body can collapse (as in Fig. 5). For these, critical point or chemical drying is recommended, which keeps the body structure intact and the setae erect, allowing surface sculpture to be seen more clearly. Our preferred method is that of van Achterberg (2009), in which specimens are first dehydrated in a 3:2 ethanol:xylene mixture for one to three days, then transferred to amyl acetate for at least a day. There is minimal collapse of delicate specimens and setae stay erect. However, both xylene and amyl acetate are potentially dangerous chemicals and should be used in a fume cupboard. Ethyl acetate is more widely used by entomologists (for killing or anaesthetising) and can be used instead of amyl acetate, although the results are not as good in our experience.

Figure 5. *Neurateles papyraceus* Ratzeburg (Ichneumonidae: Orthocentrinae) air-dried and partly collapsed.

Checklist of the British genera of Ichneumonidae

Following Broad (2016), with changes following Santos (2017); genera are arranged in subfamilies and tribes (subtribes are not recognised).

ACAENITINAE
Acaenitus Latreille, 1809
Arotes Gravenhorst, 1829
Coleocentrus Gravenhorst, 1829
Leptacoenites Strobl, 1902
Phaenolobus Förster, 1869

ADELOGNATHINAE
Adelognathus Holmgren, 1857

AGRIOTYPINAE
Agriotypus Curtis, 1832

ALOMYINAE
Alomya Panzer, 1806

ANOMALONINAE
ANOMALONINI
Anomalon Panzer, 1804
GRAVENHORSTIINI
Agrypon Förster, 1860
Aphanistes Förster, 1869
Atrometus Förster, 1869
Barylypa Förster, 1869
Erigorgus Förster, 1869
Gravenhorstia Boie, 1836
Habrocampulum Gauld, 1976
Habronyx Förster, 1869
Heteropelma Wesmael, 1849
Parania Morley, 1913
Therion Curtis, 1829
Trichomma Wesmael, 1849

ATELEUTINAE
Ateleute Förster, 1869

BANCHINAE
ATROPHINI
Alloplasta Förster, 1869
Arenetra Holmgren, 1859
Cryptopimpla Taschenberg, 1863
Lissonota Gravenhorst, 1829
Syzeuctus Förster, 1869
BANCHINI
Banchus Fabricius, 1798
Exetastes Gravenhorst, 1829
Rhynchobanchus Kriechbaumer, 1894
GLYPTINI
Apophua Morley, 1913
Diblastomorpha Förster, 1869

Glypta Gravenhorst, 1829
Teleutaea Förster, 1869

CAMPOPLEGINAE
Alcima Förster, 1869
Bathyplectes Förster, 1869
Callidora Förster, 1869
Campoletis Förster, 1869
Campoplex Gravenhorst, 1829
Casinaria Holmgren, 1859
Charops Holmgren, 1859
Clypeoplex Horstmann, 1987
Cymodusa Holmgren, 1859
Diadegma Förster, 1869
Dolophron Förster, 1869
Dusona Cameron, 1901
Echthronomas Förster, 1869
Enytus Cameron, 1905
Eriborus Förster, 1869
Gonotypus Förster, 1869
Hyposoter Förster, 1869
Lathroplex Förster, 1869
Lathrostizus Förster, 1869
Lemophagus Townes, 1965
Leptocampoplex Horstmann, 1970
Macrus Gravenhorst, 1829
Melanoplex Horstmann, 1987
Meloboris Holmgren, 1859
Nemeritis Holmgren, 1860
Nepiesta Förster, 1869
Olesicampe Förster, 1869
Phobocampe Förster, 1869
Porizon Fallén, 1813
Pyracmon Holmgren, 1859
Rhimphoctona Förster, 1869
Scirtetes Hartig, 1838
Sinophorus Förster, 1869
Synetaeris Förster, 1869
Tranosema Förster, 1869
Tranosemella Horstmann, 1978
Venturia Schrottky, 1902

COLLYRIINAE
Collyria Schiødte, 1839

CREMASTINAE
Cremastus Gravenhorst, 1829
Dimophora Förster, 1869
Pristomerus Curtis, 1836
Temelucha Förster, 1869

CRYPTINAE
APTESINI
Aconias Cameron, 1904
Aptesis Förster, 1850
Colocnema Förster, 1869
Cratocryptus Thomson, 1873
Cubocephalus Ratzeburg, 1848
Demopheles Förster, 1869
Giraudia Förster, 1869
Hemiphanes Förster, 1869
Javra Cameron, 1903
Listrocryptus Brauns, 1905
Megaplectes Förster, 1869
Oresbius Marshall, 1867
Parmortha Townes, 1962
Plectocryptus Thomson, 1873
Pleolophus Townes, 1962
Polytribax Förster, 1869
Rhembobius Förster, 1869
Schenkia Förster, 1869
CRYPTINI
Acroricnus Ratzeburg, 1852
Agrothereutes Förster, 1850
Apsilops Förster, 1869
Aritranis Förster, 1869
Buathra Cameron, 1903
Caenocryptus Thomson, 1873
Cryptus Fabricius, 1804
Echthrus Gravenhorst, 1829
Enclisis Townes, 1970
Gambrus Förster, 1869
Helcostizus Förster, 1869
Hidryta Förster, 1869
Hoplocryptus Thomson, 1873
Idiolispa Förster, 1869
Ischnus Gravenhorst, 1829
Listrognathus Tschek, 1871
Meringopus Förster, 1869
Mesostenus Gravenhorst, 1829
Nematopodius Gravenhorst, 1829
Picardiella Lichtenstein, 1920
Sphecophaga Westwood, 1840
Thrybius Townes, 1965
Trychosis Förster, 1869
Xylophrurus Förster, 1869

CTENOPELMATINAE
CHRIONOTINI
Olethrodotis Förster, 1869
CTENOPELMATINI
Ctenopelma Holmgren, 1857
Homaspis Förster, 1869
Notopygus Holmgren, 1857
Xenoschesis Förster, 1869
EURYPROCTINI
Anisotacrus Schmiedeknecht, 1913

Euryproctus Holmgren, 1857
Gunomeria Schmiedeknecht, 1907
Hadrodactylus Förster, 1869
Hypamblys Förster, 1869
Hypsantyx Pfankuch, 1906
Mesoleptidea Viereck, 1912
Occapes Townes, 1970
Pantorhaestes Förster, 1869
Phobetes Förster, 1869
Syndipnus Förster, 1869
Synodites Förster, 1869
Synomelix Förster, 1869
Zemiophora Förster, 1869
MESOLEIINI
Alexeter Förster, 1869
Anoncus Townes, 1970
Arbelus Townes, 1970
Azelus Förster, 1869
Barytarbes Förster, 1869
Campodorus Förster, 1869
Himerta Förster, 1869
Hyperbatus Förster, 1869
Lagarotis Förster, 1869
Lamachus Förster, 1869
Mesoleius Holmgren, 1856
Otlophorus Förster, 1869
Perispuda Förster, 1869
Protarchus Förster, 1869
Rhinotorus Förster, 1869
Saotis Förster, 1869
Scopesis Förster, 1869
Semimesoleius Ozols, 1963
Smicrolius Thomson, 1893
PERILISSINI
Absyrtus Holmgren, 1859
Lathiponus Förster, 1869
Lathrolestes Förster, 1869
Lophyroplectus Thomson, 1883
Oetophorus Förster, 1869
Opheltes Holmgren, 1859
Perilissus Holmgren, 1857
Priopoda Holmgren, 1856
Synoecetes Förster, 1869
Trematopygodes Aubert, 1968
Zaplethocornia Schmiedeknecht, 1912
PIONINI
Asthenara Förster, 1869
Glyptorhaestus Thomson, 1894
Labrossyta Förster, 1869
Lethades Davis, 1897
Phaestus Förster, 1869
Pion Schiødte, 1839
Rhaestus Thomson, 1883
Rhorus Förster, 1869
Sympherta Förster, 1869
Syntactus Förster, 1869
Trematopygus Holmgren, 1857

SCOLOBATINI
Scolobates Gravenhorst, 1829

CYLLOCERIINAE
Allomacrus Förster, 1869
Cylloceria Schiødte, 1838
Hyperacmus Holmgren, 1858

DIACRITINAE
Diacritus Förster, 1869

DIPLAZONTINAE
Bioblapsis Förster, 1869
Campocraspedon Uchida, 1957
Diplazon Nees, 1819
Enizemum Förster, 1869
Eurytyloides Nakanishi 1978
Fossatyloides Klopfstein, Quicke, Kropf & Frick, 2011
Homotropus Förster, 1869
Phthorima Förster, 1869
Promethes Förster, 1869
Sussaba Cameron, 1909
Syrphoctonus Förster, 1869
Syrphophilus Dasch, 1964
Tymmophorus Schmiedeknecht, 1913
Woldstedtius Carlson, 1979
Xestopelta Dasch, 1964

EUCEROTINAE
Euceros Gravenhorst, 1829

HYBRIZONTINAE
Ghilaromma Tobias, 1988
Hybrizon Fallén, 1813

ICHNEUMONINAE
EURYLABINI
Eurylabus Wesmael, 1845
GOEDARTIINI
Goedartia Boie, 1841
HERESIARCHINI
Amblyjoppa Cameron, 1902
Callajoppa Cameron, 1903
Coelichneumon Thomson, 1893
Coelichneumonops Heinrich, 1958
Heresiarches Wesmael, 1859
Lymantrichneumon Heinrich, 1978
Protichneumon Thomson, 1893
Psilomastax Tischbein, 1868
Syspasis Townes, 1965
Trogus Panzer, 1806
ICHNEUMONINI
Achaius Cameron, 1903
Acolobus Wesmael, 1845
Amblyteles Wesmael, 1845
Aoplus Tischbein, 1874

Baranisobas Heinrich, 1972
Barichneumon Thomson, 1893
Chasmias Ashmead, 1900
Cratichneumon Thomson, 1893
Crypteffigies Heinrich, 1961
Crytea Cameron, 1906
Ctenichneumon Thomson, 1894
Ctenochares Förster, 1869
Deuterolabops Heinrich, 1975
Diphyus Kriechbaumer, 1890
Eristicus Wesmael, 1845
Eupalamus Wesmael, 1845
Eutanyacra Cameron, 1903
Exephanes Wesmael, 1845
Gareila Heinrich, 1980
Hepiopelmus Wesmael, 1845
Homotherus Förster, 1869
Hoplismenus Gravenhorst, 1829
Ichneumon Linnaeus, 1758
Limerodes Wesmael, 1845
Limerodops Heinrich, 1949
Melanichneumon Thomson, 1893
Obtusodonta Heinrich, 1962
Platylabops Heinrich, 1950
Probolus Wesmael, 1845
Rictichneumon Heinrich, 1961
Spilichneumon Thomson, 1894
Spilothyrateles Heinrich, 1967
Stenaoplus Heinrich, 1938
Stenichneumon Thomson, 1893
Stenobarichneumon Heinrich, 1961
Sycaonia Cameron, 1903
Thyrateles Perkins, 1953
Tricholabus Thomson, 1894
Triptognathus Berthoumieu, 1904
Virgichneumon Heinrich, 1977
Vulgichneumon Heinrich, 1961
LISTRODROMINI
Anisobas Wesmael, 1845
Listrodromus Wesmael, 1845
Neotypus Förster, 1869
OEDICEPHALINI
Notosemus Förster, 1869
PHAEOGENINI
Aethecerus Wesmael, 1845
Baeosemus Förster, 1869
Centeterus Wesmael, 1845
Colpognathus Wesmael, 1845
Diadromus Wesmael, 1845
Dicaelotus Wesmael, 1845
Dilleritomus Aubert, 1979
Dirophanes Förster, 1869
Eparces Förster, 1869
Epitomus Förster, 1869
Eriplatys Förster, 1869
Hemichneumon Wesmael, 1857

Herpestomus Wesmael, 1845
Heterischnus Wesmael, 1859
Mevesia Holmgren, 1890
Misetus Wesmael, 1845
Nematomicrus Wesmael, 1845
Oiorhinus Wesmael, 1845
Oronotus Wesmael, 1845
Orotylus Holmgren, 1890
Paraethecerus Perkins, 1953
Phaeogenes Wesmael, 1845
Stenodontus Berthoumieu, 1897
Trachyarus Thomson, 1891
Tycherus Förster, 1869
PLATYLABINI
Apaeleticus Wesmael, 1845
Asthenolabus Heinrich, 1951
Cyclolabus Heinrich, 1936
Dentilabus Heinrich, 1974
Ectopius Wesmael, 1859
Hypomecus Wesmael, 1845
Linycus Cameron, 1903
Platylabus Wesmael, 1845
Platymischos Tischbein, 1868
Poecilostictus Ratzeburg, 1852
Pristicerops Heinrich, 1961
Pristiceros Gravenhorst, 1829
ZIMMERIINI
Cotiheresiarches Telenga, 1929

LYCORININAE
Lycorina Holmgren, 1859

MESOCHORINAE
Astiphromma Förster, 1869
Cidaphus Förster, 1869
Dolichochorus Strobl, 1904
Mesochorus Gravenhorst, 1829

METOPIINAE
Carria Schmiedeknecht, 1924
Chorinaeus Holmgren, 1858
Colpotrochia Holmgren, 1856
Exochus Gravenhorst, 1829
Hypsicera Latreille, 1829
Ischyrocnemis Holmgren, 1859
Metopius Panzer, 1806
Periope Haliday, 1839
Scolomus Townes, 1969
Stethoncus Townes, 1959
Synosis Townes, 1959
Triclistus Förster, 1869
Trieces Townes, 1946

MICROLEPTINAE
Microleptes Gravenhorst, 1829

NEORHACODINAE
Neorhacodes Hedicke, 1922

OPHIONINAE
Enicospilus Stephens, 1835
Eremotylus Förster, 1869
Ophion Fabricius, 1798
Stauropoctonus Brauns, 1889

ORTHOCENTRINAE
Aniseres Förster, 1871
Aperileptus Förster, 1869
Apoclima Förster, 1869
Batakomacrus Kolarov, 1986
Catastenus Förster, 1871
Dialipsis Förster, 1869
Entypoma Förster, 1869
Eusterinx Förster, 1869
Gnathochorisis Förster, 1869
Helictes Haliday, 1837
Megastylus Schiødte, 1838
Neurateles Ratzeburg, 1848
Orthocentrus Gravenhorst, 1829
Pantisarthrus Förster, 1871
Picrostigeus Förster, 1869
Plectiscidea Viereck, 1914
Plectiscus Gravenhorst, 1829
Proclitus Förster, 1869
Proeliator Rossem, 1982
Stenomacrus Förster, 1869
Symplecis Förster, 1869

ORTHOPELMATINAE
Orthopelma Taschenberg, 1865

OXYTORINAE
Oxytorus Förster, 1869

PHYGADEUONTINAE
Aclastus Förster, 1869
Acrolyta Förster, 1869
Agasthenes Förster, 1869
Amphibulus Kriechbaumer, 1893
Arotrephes Townes, 1970
Atractodes Gravenhorst, 1829
Bathythrix Förster, 1869
Cephalobaris Kryger, 1915
Ceratophygadeuon Viereck, 1924
Charitopes Förster, 1869
Chirotica Förster, 1869
Clypeoteles Horstmann, 1974
Cremnodes Förster, 1850
Diaglyptidea Viereck, 1913
Dichrogaster Doumerc, 1855
Encrateola Strand, 1917

Endasys Förster, 1869
Ethelurgus Förster, 1869
Eudelus Förster, 1869
Fianoniella Horstmann, 1992
Gelis Thunberg, 1827
Glyphicnemis Förster, 1869
Gnotus Förster, 1869
Gnypetomorpha Förster, 1869
Grasseiteles Aubert, 1965
Hemiteles Gravenhorst, 1829
Holcomastrus Horstmann, 2012
Isadelphus Förster, 1869
Leptocryptoides Horstmann, 1976
Lochetica Kriechbaumer, 1892
Lysibia Förster, 1869
Mastrulus Horstmann, 1978
Mastrus Förster, 1869
Medophron Förster, 1869
Megacara Townes, 1970
Mesoleptus Gravenhorst, 1829
Micromonodon Förster, 1869
Neopimpla Ashmead, 1900
Obisiphaga Morley, 1907
Odontoneura Förster, 1869
Oecotelma Townes, 1970
Orthizema Förster, 1869
Phygadeuon Gravenhorst, 1829
Platyrhabdus Townes, 1970
Pleurogyrus Townes, 1970
Polyaulon Förster, 1869
Pygocryptus Roman, 1925
Stibeutes Förster, 1850
Stilpnus Gravenhorst, 1829
Sulcarius Townes, 1970
Thaumatogelis Schwarz, 1995
Theroscopus Förster, 1850
Tricholinum Förster, 1869
Tropistes Gravenhorst, 1829
Uchidella Townes, 1957
Xenolytus Förster, 1869
Xiphulcus Townes, 1970
Zoophthorus Förster, 1869

PIMPLINAE
DELOMERISTINI
Delomerista Förster, 1869
Perithous Holmgren, 1859
Pseudorhyssa Merrill, 1915
EPHIALTINI
Acrodactyla Haliday, 1839
Acropimpla Townes, 1960
Clistopyga Gravenhorst, 1829
Dolichomitus Smith, 1877
Endromopoda Hellén, 1939
Ephialtes Gravenhorst, 1829

Exeristes Förster, 1869
Flavopimpla Betrem, 1932
Fredegunda Fitton, Shaw & Gauld, 1988
Gregopimpla Momoi, 1965
Iania Matsumoto, 2016
Iseropus Förster, 1869
Liotryphon Ashmead, 1900
Megaetaira Gauld & Dubois, 2006
Oxyrrhexis Förster, 1869
Paraperithous Haupt, 1954
Piogaster Perkins, 1958
Polysphincta Gravenhorst, 1829
Reclinervellus He & Ye, 1998
Scambus Hartig, 1838
Schizopyga Gravenhorst, 1829
Sinarachna Townes, 1960
Townesia Ozols, 1962
Tromatobia Förster, 1869
Zaglyptus Förster, 1869
Zatypota Förster, 1869
PIMPLINI
Apechthis Förster, 1869
Itoplectis Förster, 1869
Pimpla Fabricius, 1804
THERONIINI
Theronia Holmgren, 1859

POEMENIINAE
POEMENIINI
Deuteroxorides Viereck, 1914
Neoxorides Clément 1938
Podoschistus Townes, 1957
Poemenia Holmgren, 1859

RHYSSINAE
Rhyssa Gravenhorst, 1829
Rhyssella Rohwer, 1920

STILBOPINAE
Panteles Förster, 1869
Stilbops Förster, 1869

TERSILOCHINAE
Allophroides Horstmann, 1971
Aneuclis Förster, 1869
Astrenis Förster, 1869
Barycnemis Förster, 1869
Diaparsis Förster, 1869
Epistathmus Förster, 1869
Gelanes Horstmann, 1981
Heterocola Förster, 1869
Phradis Förster, 1869
Phrudus Förster, 1869
Probles Förster, 1869
Pygmaeolus Hellén, 1958
Sathropterus Förster, 1869

Spinolochus Horstmann, 1971
Tersilochus Holmgren, 1859

TRYPHONINAE
ECLYTINI
Eclytus Holmgren, 1857
IDIOGRAMMATINI
Idiogramma Förster, 1869
OEDEMOPSINI
Cladeutes Townes, 1969
Hercus Townes, 1969
Neliopisthus Thomson, 1883
Oedemopsis Tschek, 1869
Thymaris Förster, 1869
PHYTODIETINI
Netelia Gray, 1860
Phytodietus Gravenhorst, 1829
SPHINCTINI
Sphinctus Gravenhorst, 1829
TRYPHONINI
Acrotomus Holmgren, 1857
Cosmoconus Förster, 1869
Cteniscus Haliday, 1832
Ctenochira Förster, 1869
Cycasis Townes, 1965
Dyspetes Förster, 1869
Eridolius Förster, 1869
Erromenus Holmgren, 1857
Excavarus Davis, 1897
Exenterus Hartig, 1837
Exyston Schiødte, 1839
Grypocentrus Ruthe, 1855
Kristotomus Mason, 1962
Monoblastus Hartig, 1837
Neleges Förster, 1869
Orthomiscus Mason, 1955
Otoblastus Förster, 1869
Polyblastus Hartig, 1837
Smicroplectus Thomson, 1883
Tryphon Fallén 1813

XORIDINAE
Ischnoceros Gravenhorst, 1829
Odontocolon Cushman, 1942
Xorides Latreille, 1809

Morphological terminology

The principal aim of this section is to enable the non-specialist user of this *Handbook*, and other RES *Handbooks* on ichneumonids, to locate and identify the main features mentioned in the keys and text. More specialist terms and character states are illustrated or explained at appropriate points and/or in the glossary. The glossary is also the place to refer to when trying to find the meaning of various commonly used terms that we do not employ here, particularly some of those used by Townes (1969-1971).

In taxonomy there is a tendency to persist with incorrect morphological interpretations and outdated terminology. This may provide continuity for specialists, but it can be confusing and off-putting for those whose concern is not primarily with the Ichneumonidae. However, morphological terminology can be complicated and we retain simpler traditional terms for some structures where confusion will not arise. Richards (1956 (second edition 1977)) treated the morphology of adult Hymenoptera in some detail and many of his findings and terms are adopted here. Further advances in interpretation applicable to the Ichneumonidae relate to wing venation (Hamilton, 1972; Wootton, 1978), genitalia (Smith, 1970), mesosoma anatomy (Gibson, 1985), ovipositor structure (Quicke *et al.*, 1992a, 1994) and the nature of the sternaulus (Wharton, 2006). For those keen to investigate Hymenoptera morphology in more depth, some very useful compendia of morphological terminology for the wider Hymenoptera include Karlsson & Ronquist's (2012) summary of ichneumonoid morphology, Schulmeister's (2003) summary of hymenopteran male genitalia terms (an area of complicated homology but peripheral to this handbook) and the online Hymenoptera Anatomy Ontology (http://glossary.hymao.org/projects/32/public/ontology/) (Yoder *et al.*, 2010; Seltmann *et al.*, 2012). Snodgrass's (1956) anatomy of the honey bee (*Apis mellifera* L.) is a meticulously compiled source of much valuable information on Hymenoptera morphology. A diverse set of character systems has been explored for phylogenetic and life history correlates. This work is largely outside of the scope of this handbook but includes studies on head capsule anatomy (particularly of the sawfly and woodwasp groups; Vilhelmsen 1996, 1997b, 1999), antennal cleaner structures (Basibuyuk & Quicke, 1994, 1995); hamuli (Basibuyuk & Quicke, 1997), the orbicula (Basibuyuk *et al.*, 2000), grooming behaviour (Basibuyuk & Quicke, 1999), sperm structure (Quicke *et al.*, 1992b), ovipositor sheaths (Vilhelmsen, 2003) and utilisation of trace metals (Quicke *et al.*, 1998).

Recommended terminology is shown here mainly by means of annotated figures. Alternative terms found in the ichneumonid literature, especially in this series of RES *Handbooks* (Perkins 1959, 1960; Gauld & Mitchell, 1977b; Fitton *et al.*, 1988; Brock, 2017), are included in the glossary. The nomenclature used for cuticular microsculpture follows Eady (1968) and Eady's illustrations are reproduced here in Fig. 24. Measurements are made as shown in Fig. 23. It is important to use an eyepiece micrometer to make measurements and both endpoints need to be in focus simultaneously to ensure consistency. For purposes of orientation, legs and wings are treated as if they were extended horizontally at right angles to the main axis of the body, thus having dorsal, ventral, anterior and posterior aspects. Terms such as basal and apical are frequently misinterpreted and preference is given to anterior, posterior, and for appendages proximal and distal. Aguiar & Gibson (2010) described the difficulties in attempting to refer by homologous terms to surfaces on legs and propose some novel solutions; however, terms such as 'kickface' and 'gripface' have yet to gain general acceptance.

Descriptions of colour are difficult to standardise, although there have been several attempts, all more or less unsuccessful. Aguiar (2005) describes a method for precisely expressing colours using illustration software but we stick here to using only broad descriptions of colour that should be unambiguous.

The main subdivisions of the ichneumonid body are shown in Figs 6 and 7.

HEAD (Figs 8-10). The main taxonomic features of the head are shown in Fig. 8, with different mandible types shown in Fig. 10.

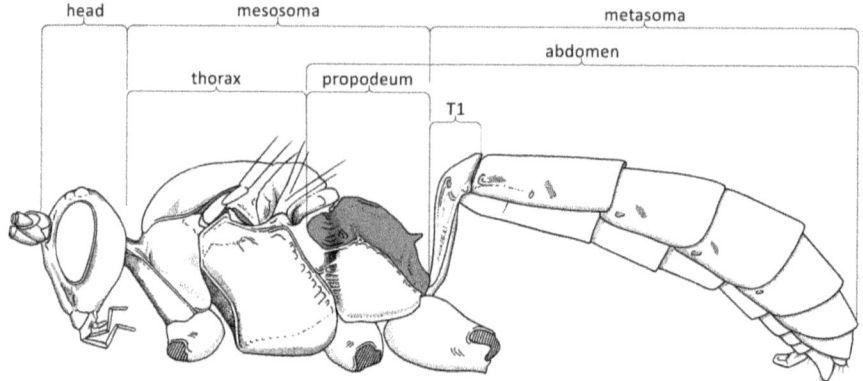

Figure 6. Major subdivisions of the body, illustrated by male *Amblyteles armatorius* (Forster): T1 - first metasomal tergite.

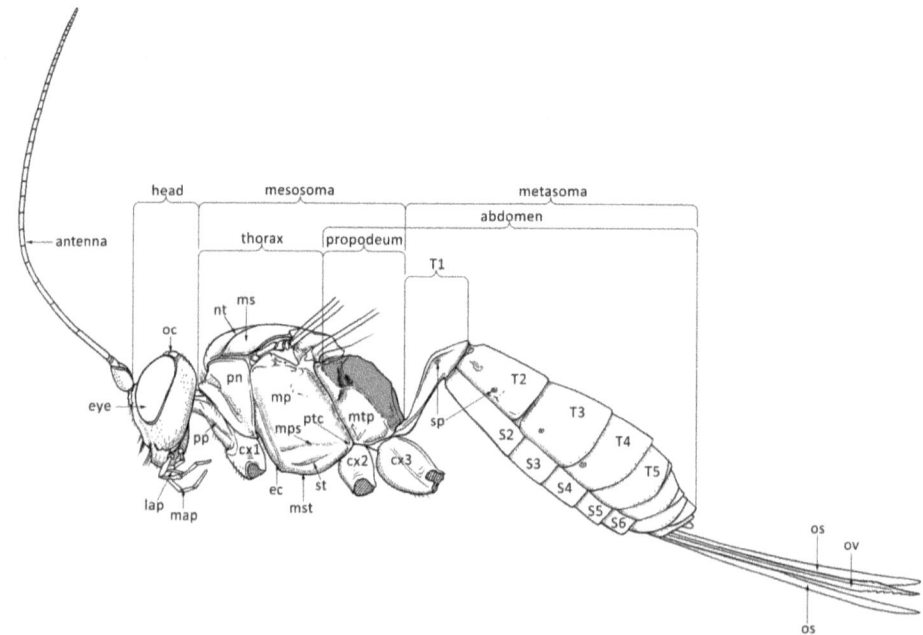

Figure 7. Finer subdivisions of the body, illustrated by (partly stylised) female *Buathra laborator* (Thunberg): cx1 - fore coxa; cx2 - mid coxa; cx3 - hind coxa; ec - epicnemial carina; lap - labial palp; map - maxillary palp; mp - mesopleuron; mps - mesopleural (precoxal) sulcus; ms - mesoscutum; mst - mesosternum; mtp - metapleuron; nt - notaulus; oc - ocellus; os - ovipositor sheath; ov - ovipositor; pn - pronotum; pp - propleuron; ptc - posterior transverse carina; sp - spiracle; st - sternaulus; S2, S3, etc. - second metasomal sternite, third metasomal sternite, etc.; T1, T2, etc. - first metasomal tergite, second metasomal tergite, etc.

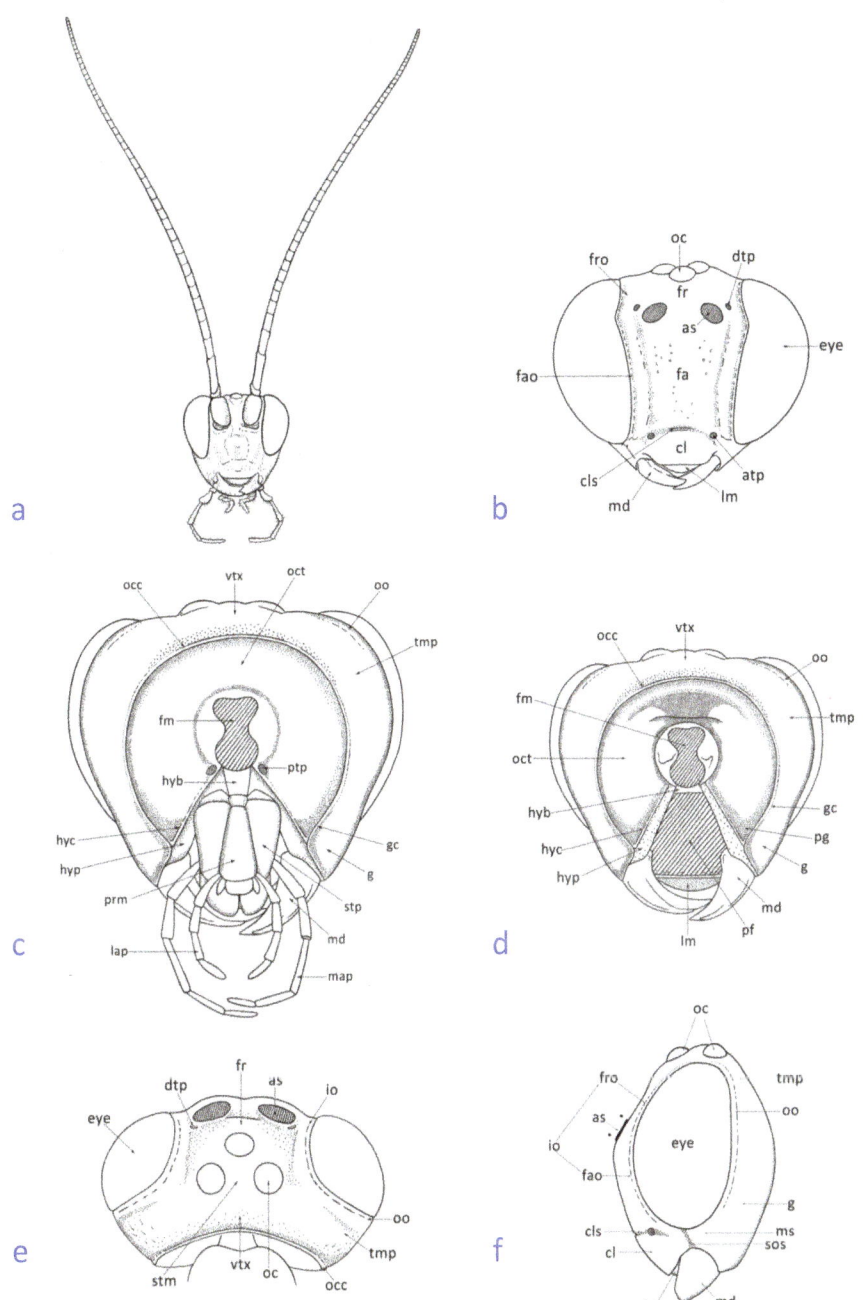

Figure 8. Ichneumonid head (a: based on *Buathra laborator* (Thunberg), remaining stylised to show suite of potential characters): (a) whole head, with antennae; (b) anterior view of head; (c) posterior vew of head; (d) posterior view of head with the proboscidial fossa (and foramen magnum) shaded (labio-maxillary complex removed); (e) dorsal view of head; (f) lateral view of head. Abbreviations: as - antennal socket; atp - anterior tentorial pit; cl - clypeus; cls - clypeal sulcus; dtp - dorsal tentorial pit; eye - compound eye; fa - face; fao - inner (or facial) orbit; fm - foramen magnum; fr - frons; fro - frontal orbit; g - gena; gc - genal carina (continuous with occipital carina); hyb - hypostomal bridge; hyc - hypostomal carina; hyp - hypostoma; io - inner orbit; lap - labial palp; lm - labrum; map - maxillary palp; md - mandible; ms - malar space; oc - ocellus; occ - occipital carina; oct - occiput; map - maxillary palp; oo - outer (or temporal) orbit; pf - proboscidial fossa; pg - postgena; prm - prementum; ptp - posterior tentorial pit; sos - subocular sulcus; stm - stemmaticum (area bounded by the ocelli); stp - stipes; tmp - temple (part of the gena); vtx - vertex.

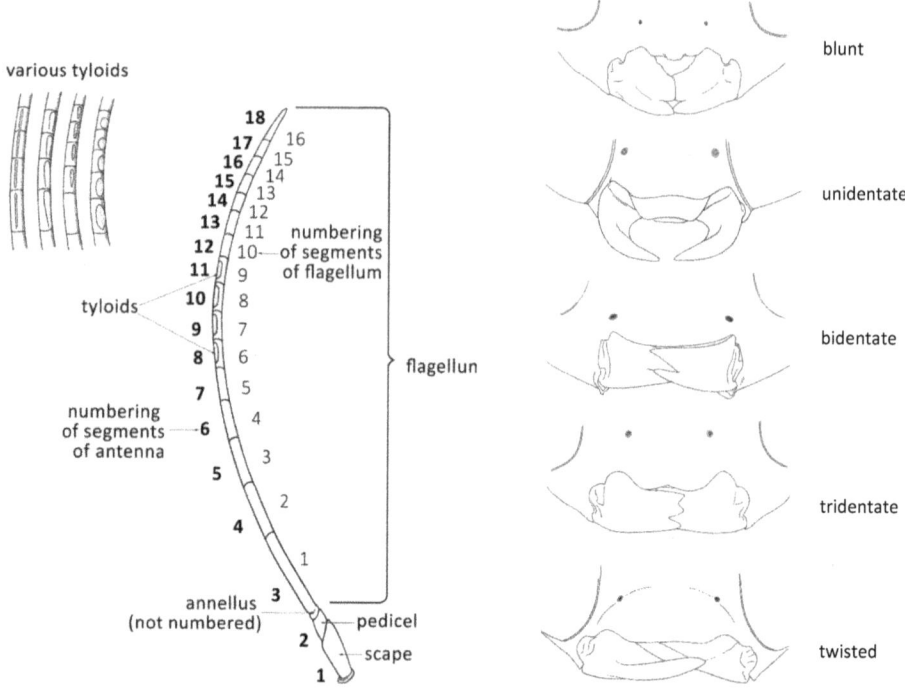

Figure 9. Stylised ichneumonid antenna showing conventions for numbering segments and forms of tyloids.

Figure 10. Ichneumonid mandible types.

ANTENNAE (Fig. 9). The antennae typically have a large number of segments, usually 20 or more, sometimes as many as 90, but in some taxa the number is reduced to 16 or fewer. The two proximal segments of the antenna, the scape and pedicel, are differentiated from the remainder, which constitute the flagellum. The first segment of the flagellum has the proximal section, the annellus, separated by a distinct groove. The annellus is not counted separately in the numbering of the segments. It should be noted that some workers number the segments of the flagellum, rather than the antenna as a whole. As well as various sensilla, tyloids may be present on a limited number of flagellar segments (of males only) and may take the form of longitudinal ridges (not to be confused with placoid sensilla), raised polished areas, or notches. It is not thought that all tyloids are homologous structures.

MESOSOMA (Fig. 11). The mesosoma is so-called because it is the combined thorax and first segment of the abdomen (the latter being the propodeum). Thus the functional joint found between the thorax and abdomen in other insects is between the first and second abdominal segments in apocritan Hymenoptera. The general shape of the ichneumonid mesosoma varies somewhat, for instance in wingless forms such as some *Gelis* the volume of the mesothorax is considerably reduced, corresponding to loss of flight muscles, and sutures also become obsolete (Salt, 1952). The form and development of various carinae and grooves on the thorax, such as the epomia, epicnemial carina, sternaulus and notaulus vary considerably between groups. The surface of the propodeum is divided into areae by a regular system of carinae. These carinae are often reduced in number and in some groups are almost entirely absent.

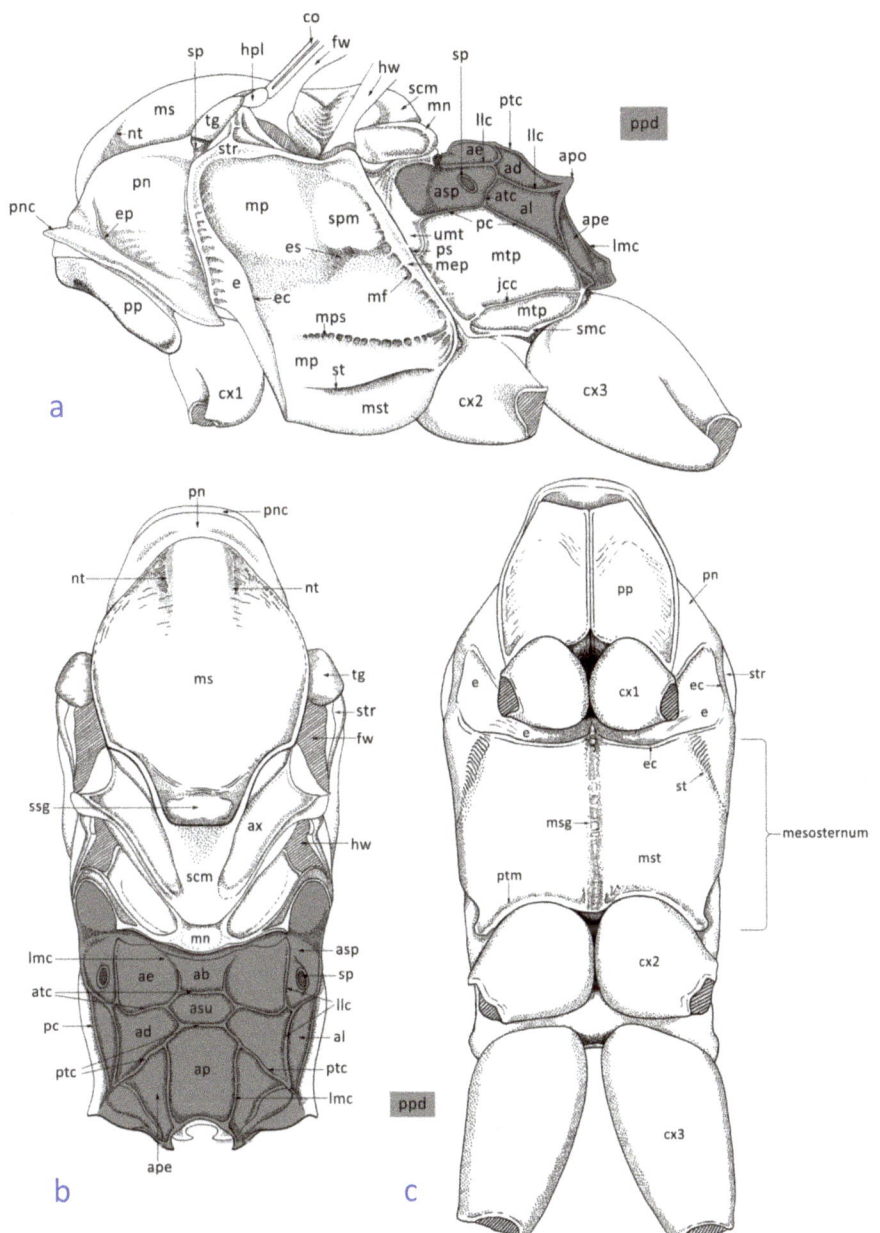

Figure 11. Stylised ichneumonid mesosoma (thorax plus propodeum), with propodeum shaded: (a) lateral view, (b) dorsal view, (c) ventral view. Abbreviations: ab - area basalis; ad - area dentipara; ae - area externa; al - area lateralis; ap - area petiolaris; ape - area postero-externa; apo - apophysis; asp - area spiracularis; asu - area superomedia; atc - anterior transverse carina; ax - axilla; co - costa; cx1 - fore coxa; cx2 - mid coxa; cx3 - hind coxa; e - epicnemium; ec - epicnemial carina; ep - epomia; es - episternal scrobe; fw - fore wing; hpl - humeral plate; hw - hind wing; jcc - juxtacoxal carina; llc - lateral longitudinal carina; lmc - lateromedian longitudinal carina; mep - mesepisternum; mf - mesopleural furrow; mn - metanotum (postscutellum); mp - mesopleuron; mps - mesopleural (precoxal) sulcus; ms - mesoscutum; msg - medial sternal groove; mst - mesosternum; mtp - metapleuron; nt - notaulus; pc - pleural carina (separating propodeum from metapleuron); pn - pronotum; pnc - pronotal collar; pp - propleuron; ppd - propodeum; ps - pleural sulcus; ptc - posterior transverse carina; ptm - posterior transverse carina of the mesosternum; scm - scutellum (mesoscutellum); smc - submetapleural carina; sp - spiracle; spm - speculum; ssg - scuto-scutellar groove; st - sternaulus; str - subtegular ridge; tg - tegula; umt - upper division of the metapleuron.

LEGS (Figs 7 & 12). The three pairs of legs are known as fore, mid and hind to avoid confusion with segment numbering (of the tarsi, for instance) and ambiguous terms such as metatarsus. In ichneumonids the hind legs are invariably very much larger than the fore and mid legs. The hind coxae are usually particularly massive. In most subfamilies the proximal part of the femur is differentiated as a trochantellus. The tarsal claws are often simple but may bear a variety of accessory lobes or teeth, sometimes developed into a conspicuous pecten.

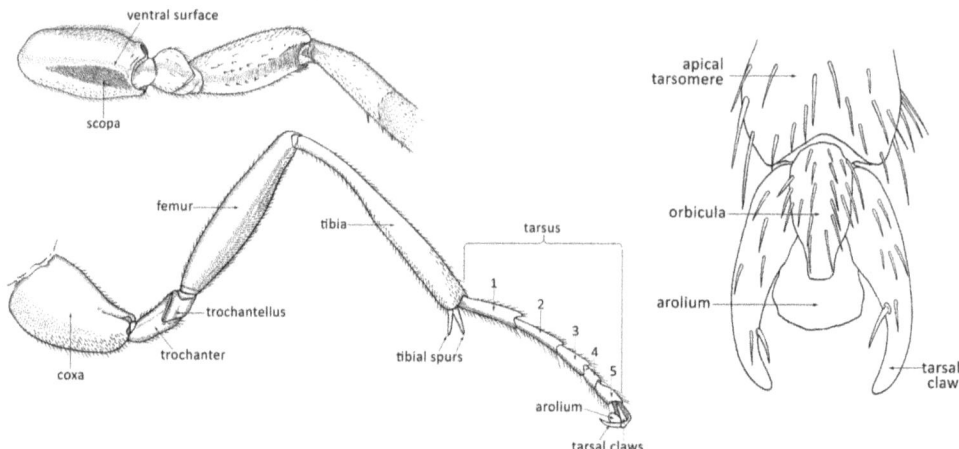

Figure 12. Ichneumonid hind leg and details of the tarsal claws. Upper view of hind leg showing the scopa present on the underside of the hind coxa of some Ichneumoninae.

METASOMA (Fig. 13). The metasoma comprises the second and following segments of the abdomen. Richards (1956, 1977) advocated the term gaster for this, the third, major subdivision of the body. However, gaster continues to be used in more restricted senses in some groups of Hymenoptera (notably ants and Chalcidoidea), where the petiole (first or second metasomal segments) is frequently referred to separately, and in the meantime the unambiguous term metasoma has gained currency. It is usual to number its segments beginning with 1, even though the first segment of the metasoma is in fact the second abdominal segment.

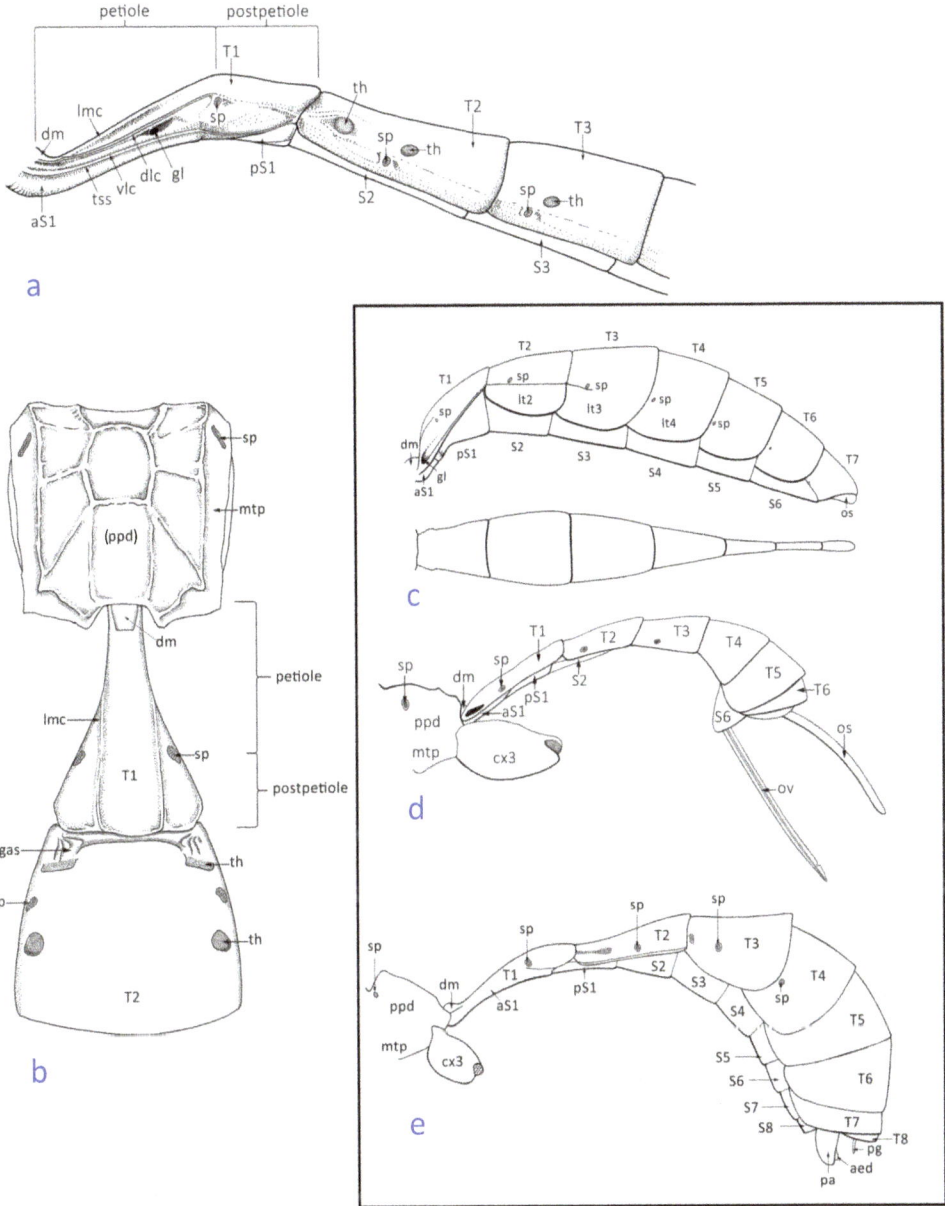

Figure 13. Ichneumonid metasoma: (a) lateral view of anterior section of stylised ichneumonid metasoma; (b) dorsal view of anterior section of metasoma and propodeum of Ichneumoninae; (c) lateral (above) and dorsal (below) views of laterally compressed female metasoma (*Banchus* sp.); (d) lateral view of dorso-ventrally compressed female metasoma (*Cryptopimpla* sp.); (e) lateral view of male metasoma (*Eremotylus marginatus* (Jurine)). Abbreviations: aed - aedeagus; aS1 - anterior sclerotized section of first metasomal sternite; cx3 - hind coxa; dlc - dorso-lateral carina of first metasomal tergite; dm - dorsal muscle; gas - gastrocoelus; gl - glymma; lmc - latero-median carina of first metasomal tergite; lt2, lt3, etc. - laterotergite of second metasomal tergite, laterotergite of third metasomal tergite, etc.; mtp - metapleuron; os - ovipositor sheath; ov - ovipositor; pa - paramere; pg - pygidium; ppd - propodeum; pS1 - posterior, 'membranous' section of first metasomal sternite; S1, S2, etc. - first metasomal sternite, second metasomal sternite, etc.; sp - spiracle; S6 [female] - hypopygium (sixth metasomal sternite); S8 [male] - hypopygium (eighth metasomal sternite); T1, T2, etc. - first metasomal tergite, second metasomal tergite, etc.; th - thyridium; tss - tergal-sternal suture of first metasomal segment; vlc - ventro-lateral carina of first metasomal tergite.

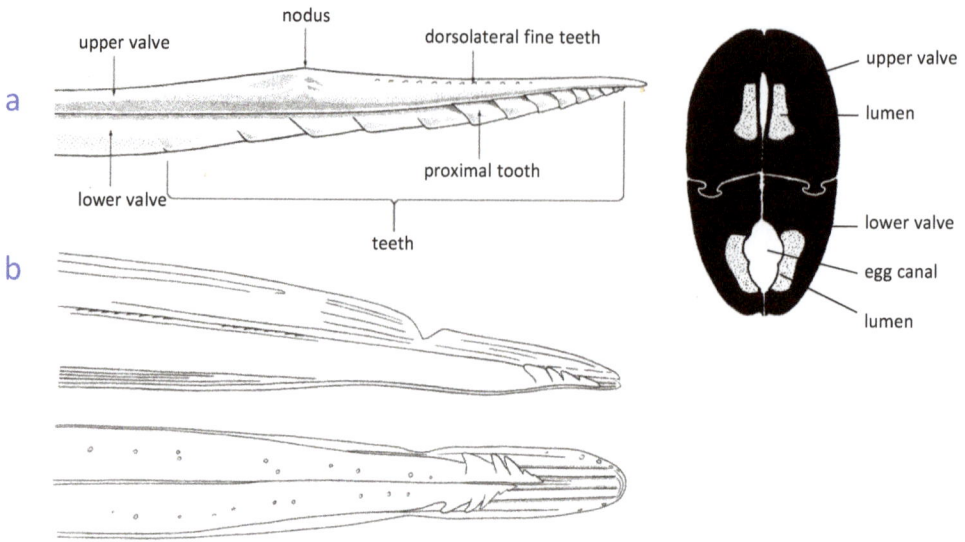

Figure 14. Detail of ovipositor tips: (a) stylised ovipositor with pronounced teeth and nodus, based on *Agrothereutes saturniae* (Boie) (Cryptinae) together with a larger scale, stylised cross section at about the mid-length of the ovipositor showing the tongue (rhachis) and groove (aulax) interlocking mechanism (olistheter); the lumina contain haemolymph, tracheae, nerves, etc., but no muscles; (b) ovipositor with a dorsal subapical notch, based on *Diblastomorpha cylindrator* (Fabricius) (Banchinae): lateral and ventral views showing independent longitudinal movement of the three valves in relation to each other.

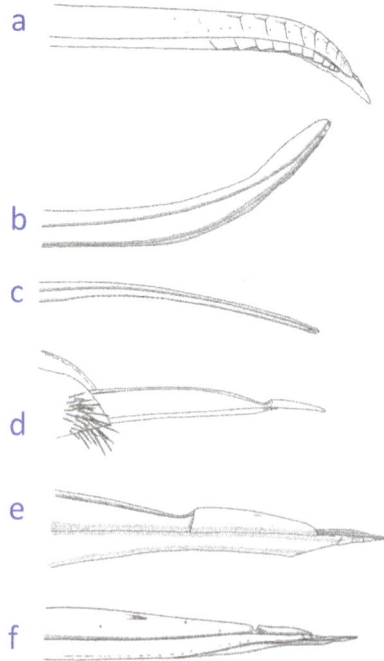

Figure 15. Diversity of ichneumonid ovipositors: (a) *Apechthis quadridentata* (Thomson) (Pimplinae); (b) *Barycnemis harpura* (Schrank) (Tersilochinae); (c) *Collyria coxator* (Villers) (Collyriinae); (d) *Xenoschesis fulvipes* (Gravenhorst) (Ctenopelmatinae); (e) *Lamachus eques* (Hartig) (Ctenopelmatinae); *Exetastes* sp. (Banchinae).

GENITALIA (Figs 14-16). The female ovipositor has three intimately connected components: one upper and two lower valves (Fig. 14). These terms and consequent references to dorsal and ventral assume the ovipositor is directed posteriorly from the end of the metasoma. The ovipositor is protected by a pair of ovipositor sheaths. In males the main visible components are the parameres, which enclose the aedeagus (Fig. 16). In males and females the last visible sternite (metasomal sternite 6 in females and 8 in males), protecting the genitalia, is known as the hypopygium.

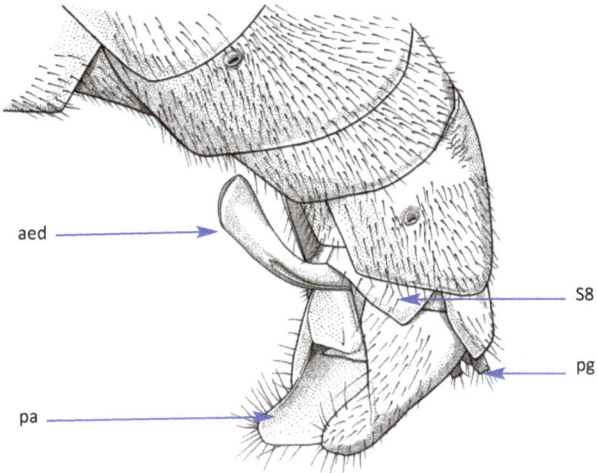

Figure 16. Aedeagus of male ichneumonid (*Xenoschesis fulvipes*). Abbreviations: aed - aedeagus; pa - paramere; pg - pygidium; S8 - hypopygium (8th metasomal sternite).

WINGS (Figs 17-22). The principal feature of the wings is the venation. The Comstock-Needham system of vein nomenclature was first successfully applied to Hymenoptera by Ross (1936) and was widely adopted and developed. There are relatively few outstanding problems of interpretation but nomenclature has not yet fully stabilised. Also, many workers continue to use a modified Jurinean system (as adopted by Townes, 1969) of naming veins which is not easily related to other groups of insects and often conflicts with morphological interpretation. For comparison with the Comstock-Needham system we include an illustration (Fig. 17c) of the Jurinean nomenclature (as applied by Townes) because of this widespread usage. A somewhat different system is used for Braconidae and this needs to be allowed for when keying that family. In the Comstock-Needham system, conventionally the veins are referred to by abbreviations (see below). Primitively each main vein has two elements, for example the radius (R) is made up of radius anterior (RA) and radius posterior (RP), but only a limited subset remains in Hymenoptera and thus the abbreviations are simplified. Abbreviations of main veins are in capitals and of cross veins in lower case. Composite veins with the elements joined in parallel are shown with a plus sign, +, for example M+CU. Composite veins with the elements joined end to end are shown with an ampersand, &, for example M&RS. Cross veins between main veins are hyphenated, and are numbered if there is more than one between a pair of main veins, for example 2rs-m and 3rs-m. The numbers reflect the presumed plesiomorphic condition. Cross veins between the two elements of a main vein simply have the abbreviation of the main vein, hence 2r (rather than 2ra-rs). An abscissa of a vein is a section of the vein between two intersections with other veins (or the last section of a vein from an intersection to the wing margin). Two veins are not shown in the figures because they are not present as separate veins in ichneumonids. The subcosta (abbreviation SC) may be combined with the costa or the radius. The cubitus posterior

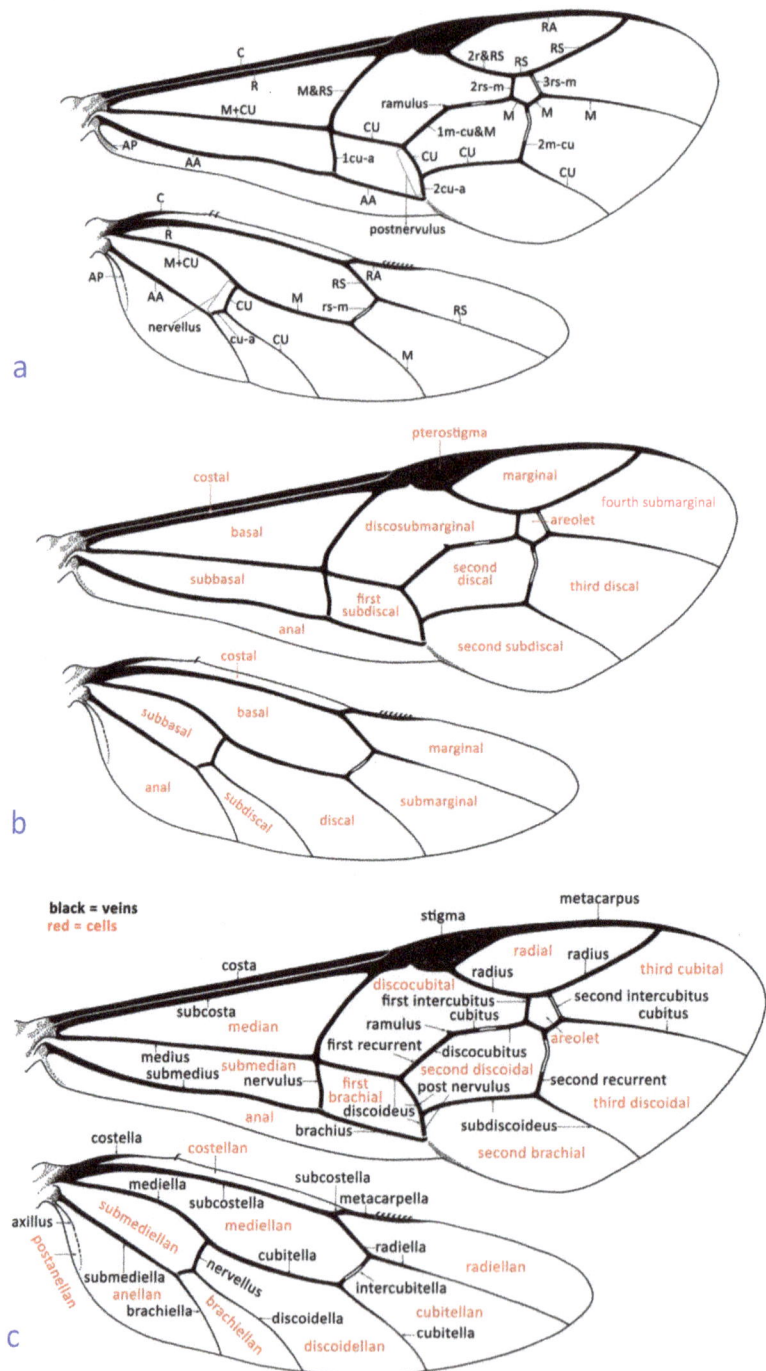

Figure 17. Ichneumonid wings (fore wing uppermost, hind wing below): (a) with wing veins named; (b) with cells named; (c) with Townes's (1969) interpretation of veins (black) and cells (red).

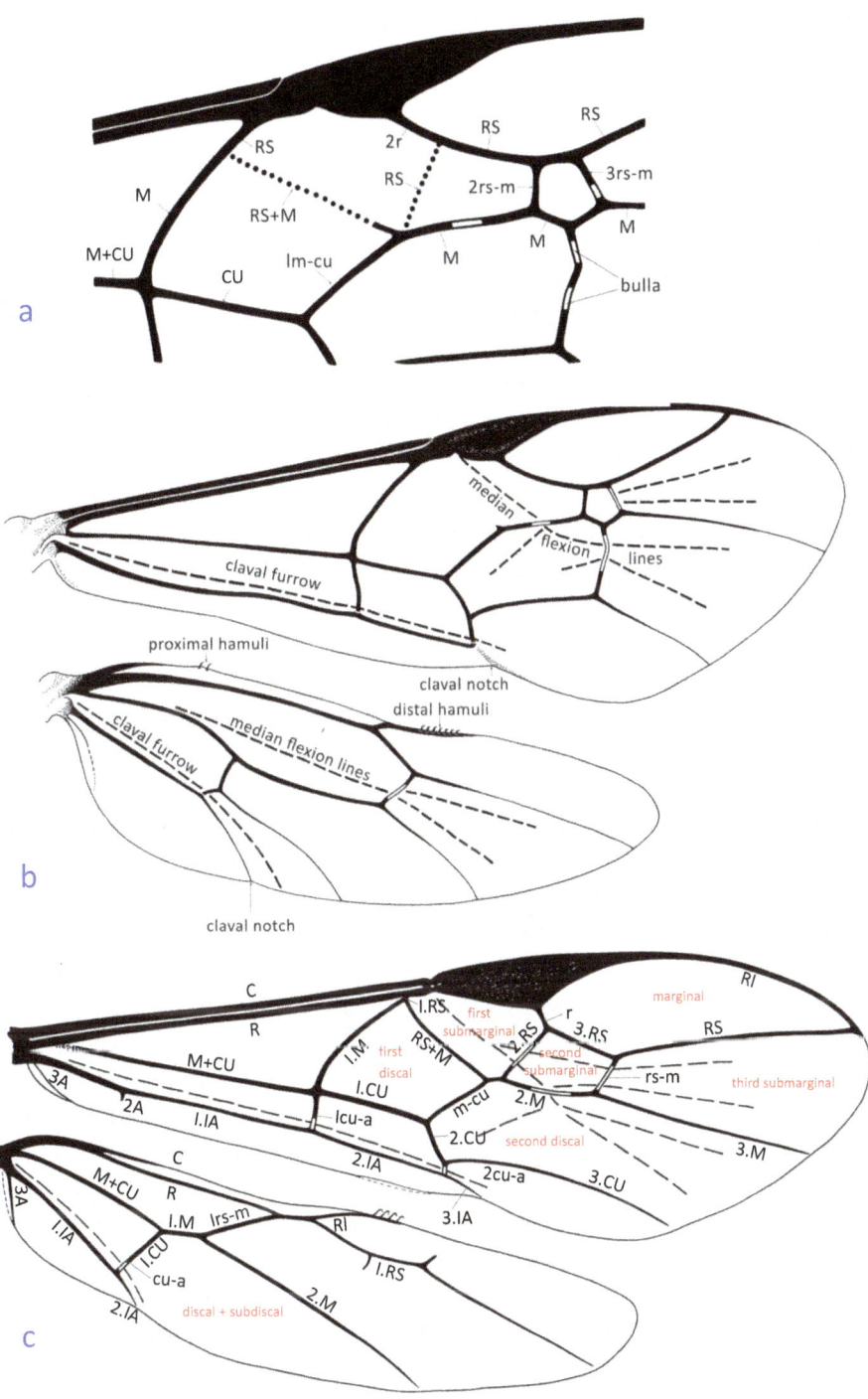

Figure 18. Ichneumonid wings (fore wing uppermost, hind wing below): (a) detail of discosubmarginal cell, with hypotheses of wing vein homology; (b) with folds and hamuli named; (c) wings of Braconidae for comparison.

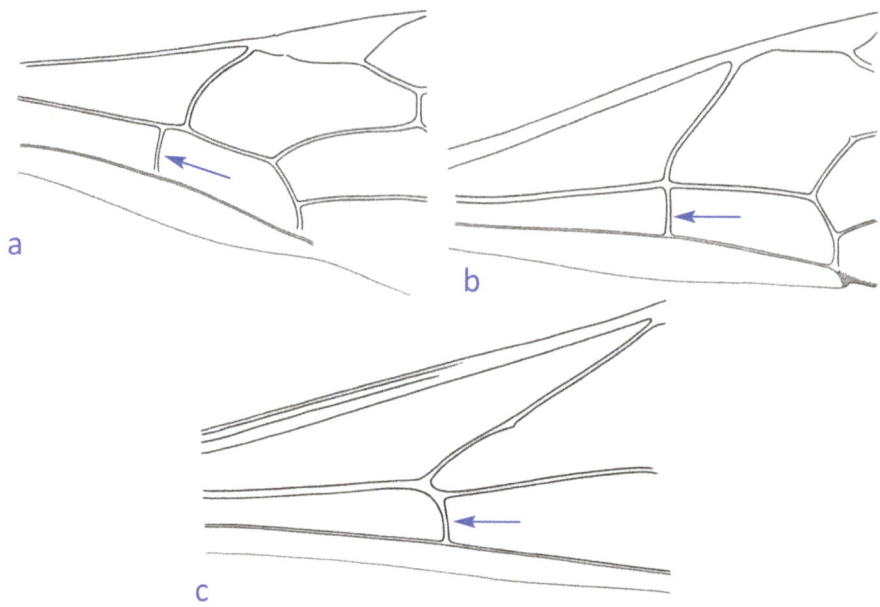

Figure 19. Examples of wing vein positional adjectives. Fore wing vein *1cu-a* (arrowed) relative to *M&RS*: (a) antefurcal, (b) interstitial, (c) postfurcal.

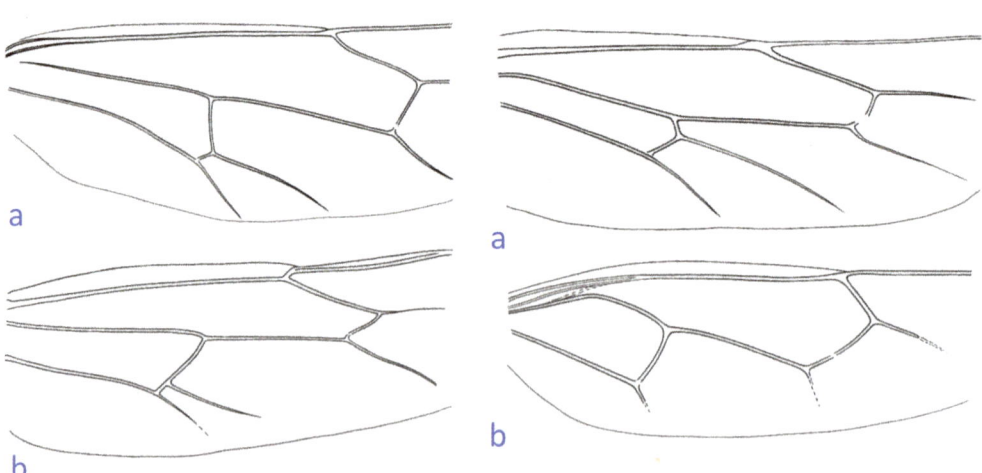

Figure 20. Examples of wing vein positional adjectives. Hind wing nervellus (first abscissa of *CU* plus *cu-a*), in both, intercepted below the middle: (a) inclivous, (b) reclivous.

Figure 21. Hind wing nervellus (first abscissa of *CU* plus *cu-a*): (a) intercepted (above the middle) by second abscissa of CU, (b) not intercepted (second abscissa of CU wholly absent, as figured, or not joining nervellus).

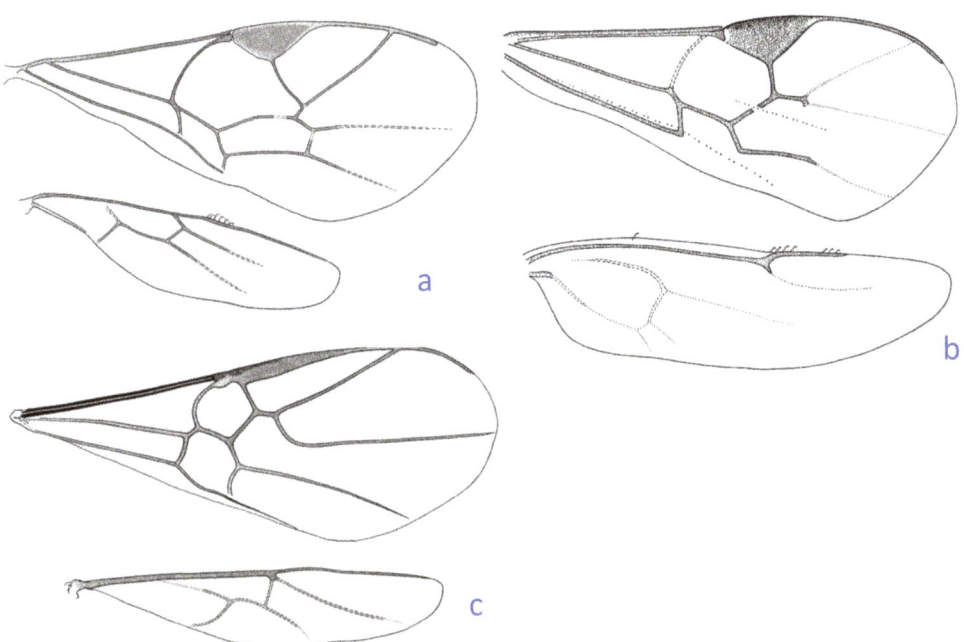

Figure 22. Examples of reduced wing venation: (a) *Phrudus defectus* Stelfox (Tersilochinae); (b) *Neorhacodes enslini* (Ruschka) (Neorhacodinae); (c) *Hybrizon buccatus* (de Brébisson) (Hybrizontinae).

(abbreviation CUP) in the fore wing may follow (or have followed) the distal part of the claval furrow, having diverged from M+CU near the base of or part way along the wing. As far as we have detected this only occurs among extant Hymenoptera in Pamphiliidae (e.g. *Pamphilius* and *Pseudocephaleia*, in which it is often labelled as m-cu-a). Linked to this is the necessity in ichneumonids of recognising what have been known as cu-a as 1cu-a and CU1b as 2cu-a (and CU1a as plain CU). It also raises the possibility of the spurious vein in ophionines being a remnant of CUP. For more detail see Mason (1990).

For wing cells, Eady's (1974) nomenclature is used (Fig. 17b). The name areolet is given to the taxonomically important third submarginal cell in the fore wing (it should not be cxonfused with areola, a term often used for the area superomedia of the propodeum) (Figs 17b). Sometimes vein 3rs-m is absent, in which case the areolet may be described as 'open'. In practice few cells are referred to in keys and descriptions of ichneumonids. The areolet (third submarginal) in ichneumonids is not the same as the similarly located cell (second submarginal) in braconids. The discosubmarginal cell in the fore wing is a large and distinctive feature and is derived through the loss of two abscissae of RS+M and RS, as shown in Fig. 18a). The ramulus is a small stub of RS+M marking the junction of 1m-cu and M in the composite vein forming the posterior margin of the discosubmarginal cell. The postnervulus in the fore wing is the composite vein CU&2cu-a. In the fore wing the relationship of M+CU, M&RS, CU and 1cu-a at their junction is often of taxonomic importance (see Fig. 19).

In the hind wing, nervellus is used for the composite vein first abscissa of CU&cu-a, which can be described as intercepted above, at or below its mid-point by the final abscissa of CU (or not intercepted when the final abscissa of CU is absent) (Figs 20, 21). The overall direction of the nervellus is also important (Fig. 20).

Some ratios are very frequently used in ichneumonid identification and are explained here (Fig. 23), expressing the relative sizes of the eyes (and consequently the shape of the face inbetween) and the proportions of the metasoma. Note that some authors express the length of the ovipositor as its total length, from its base to apex, but that this is more difficult to measure than the length of the ovipositor sheaths due to variation in the contortion of the metasoma in the specimen to hand, often obscuring the base of the ovipositor. Cuticular microsculptures most frequently found in ichneumonids are illustrated in Fig. 24.

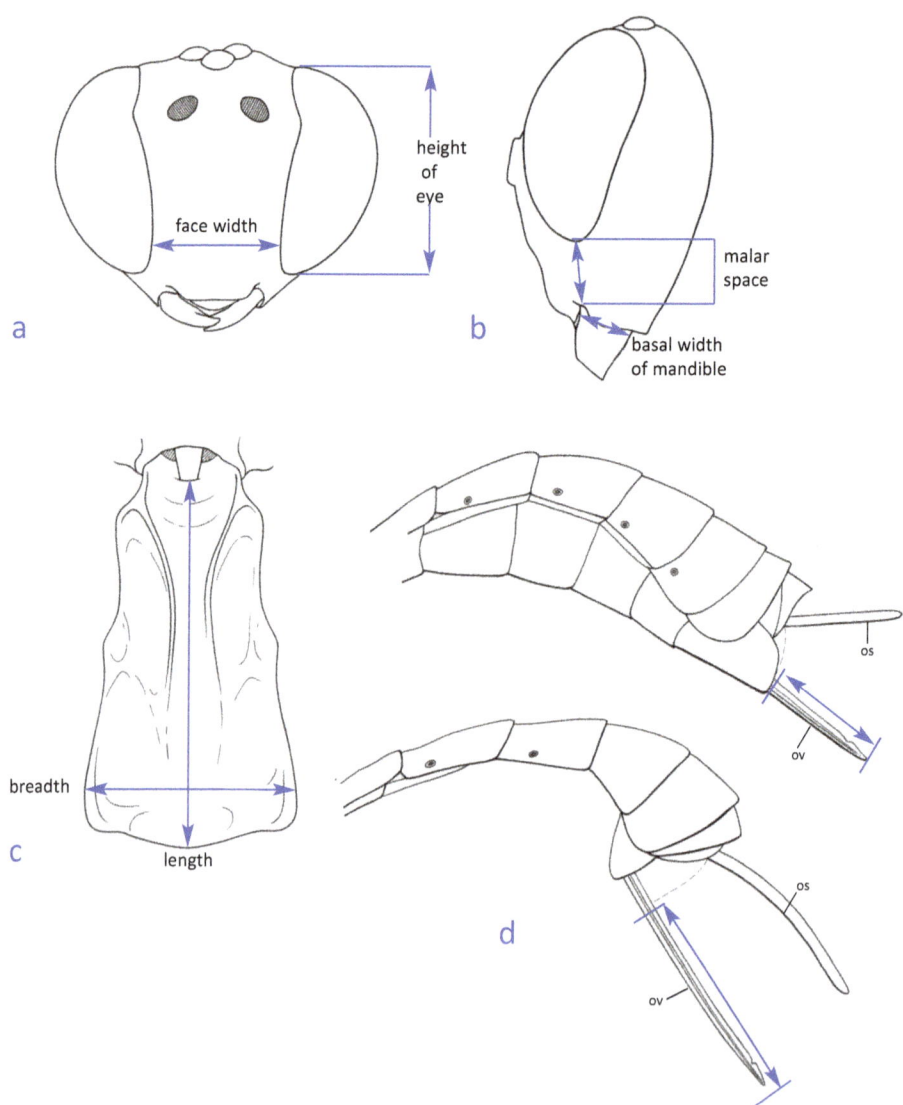

Figure 23. Methods of measuring ratios of different structures: (a) minimum face width to height of eye; (b) malar space to basal width of mandible; (c) width to length of first metasomal tergite; (d) length of ovipositor (ov), as the portion extending beyond the metasomal apex, equivalent to length of the ovipositor sheath (os).

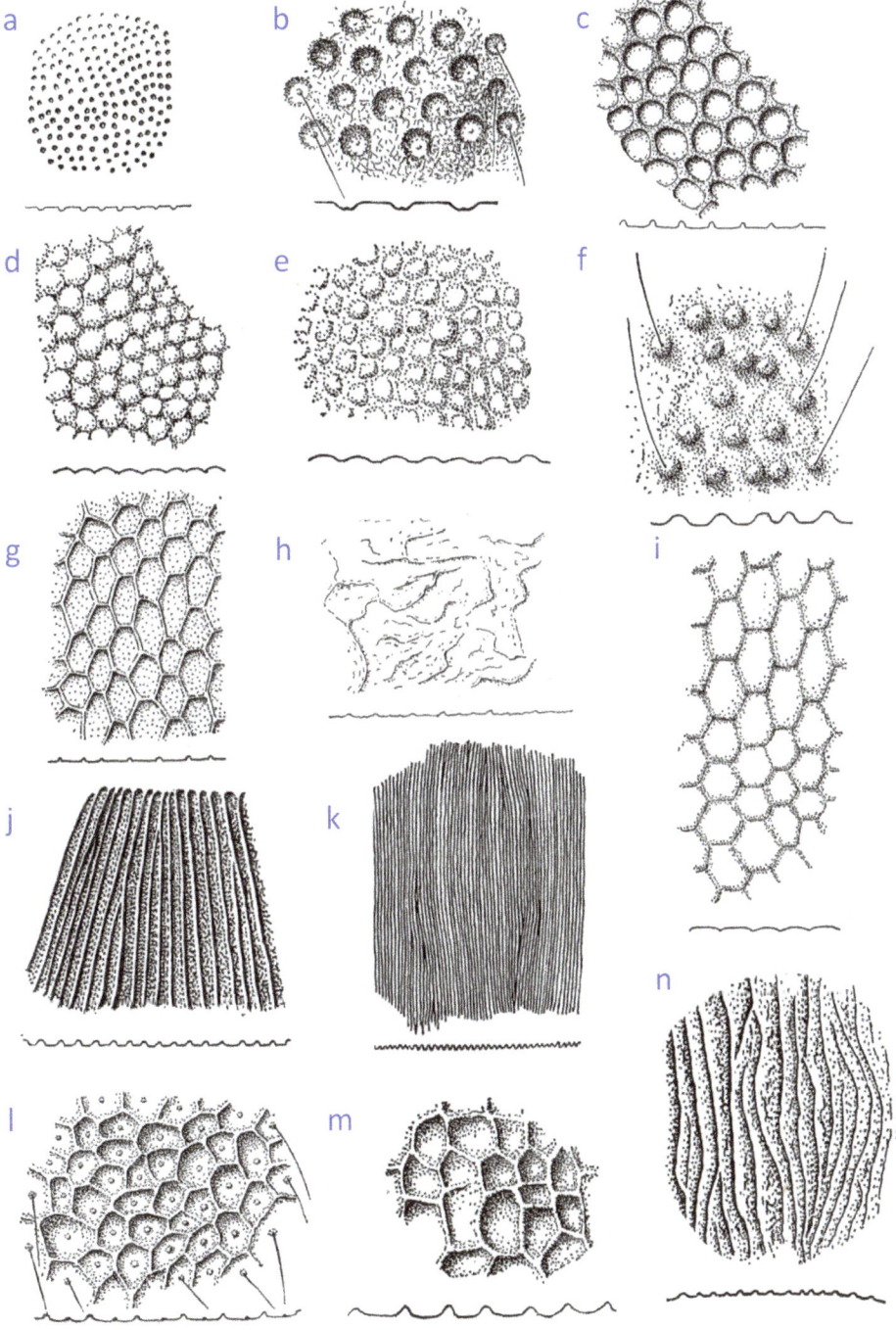

Figure 24. Cuticular microsculpure, selected illustrations from Eady (1968), with the line below each figure representing a section through the cuticle: (a) punctulate; (b) punctate, with setae; (c) punctate reticulate; (d) granulate; (e) pustulate; (f) papillate, with setae; (g) reticulate; (h) rugulose; (i) reticulate coriaceous; (j) striate; (k) aciculate; (l) rugose punctate; (m) reticulate rugose; (n) strigose.

Gavin R. Broad, Mark R. Shaw and Michael G. Fitton

Identification of the subfamilies of British Ichneumonidae

Identifying ichneumonids

Rather few ichneumonid subfamilies are diagnosable on the basis of simple morphological character states; rather, they tend to be recognisable by combinations of characters that occur convergently and in different permutations across various groups of ichneumonids. This is not to say that ichneumonid subfamilies are unrecognisable. Members of most subfamilies are easily recognised by their overall appearance, once a little experience is gained, but this lack of discrete characters for each subfamily results in a large key. Previous keys, such as those of Perkins (1959) and Townes (1969), have tried to key out subfamilies at single couplets and produced rather unworkable couplets with many 'ifs' and 'buts'. Wahl's (1993c) key to world subfamilies was a great improvement but will still be found to contain grey areas where it is difficult to know if you have chosen the correct half of a couplet. With this key we have tried to rely on rather simple characters (with a restricted geographical remit) with the result that most of the larger subfamilies key out in several places. The alternative would be long and unwieldy key couplets that attempt to cover all exceptions.

Notes on use of the keys

Each key couplet number (except the first!) has the couplet number that led to the present couplet in brackets, to allow backtracking, i.e. working your way backwards if you think you have taken a wrong turning. In illustrations, the anterior (head) end of the wasp is to the left or is uppermost. Illustrations referred to in the first half of a couplet are labelled in the format 'a', 'b', 'c', etc., with 'aa', 'bb', 'cc', etc., for the second half of the couplet. Figure numbering for the key is separate from figure numbering for the rest of the book. Each letter is not always represented in each half of the couplet; for example, a scutellar spine can be illustrated for one half of the couplet whereas a scutellum lacking a spine is not illustrated, because the absence of a spine should be obvious. Where multiple characters are included in a key couplet, the specimen needs to agree with all of them, unless stated otherwise. Photographs were taken using specimens in BMNH, using a Canon SLR EOS 5DSR with 65 mm macro lens mounted on a copy stand with an automated Z-stepper; images were aligned using Helicon Focus software version 6.6.1.

Recognition of Ichneumonoidea

Gauld & Bolton (1988) and Goulet & Huber (1993) provide good keys to superfamilies and families of Hymenoptera. Ichneumonoidea, comprising the families Ichneumonidae and Braconidae, can generally be recognised by the wing venation (costal cell of the leading edge of the fore wing more-or-less obliterated by the close apposition of veins *C* and *R*) and the long, simple antennae (usually with more than 16 segments). Another particularly useful character for recognising ichneumonoids is the very poorly sclerotized, almost membranous, sternites, forming the mid venter of the metasoma. Other parasitoid and aculeate groups usually have the sternites as well sclerotized as the dorsal tergites. Note that there is an exception within Ichneumonidae: *Agriotypus* has the sternites as sclerotized as the tergites.

The first key below separates the two families Ichneumonidae and Braconidae, before the main key to ichneumonid subfamilies.

Handbooks for the Identification of British Insects: Ichneumonid Wasps

Key for the separation of British and Irish Braconidae and Ichneumonidae

1. Wings present and not reduced (apparently capable of sustained flight) (macropterous) (**a**) 2

–   Wings absent (**aa**) or reduced (incapable of flight) (**bb**), not projecting beyond tergite 1 of metasoma (apterous, micropterous or brachypterous) [One species (*Sphecophaga vesparum* Curtis) sometimes brachypterous with wings extending to half the length of the metasoma.] 5

2(1). Fore wing vein *2m-cu* present (**a**), sometimes spectral (**b**) .... **Ichneumonidae** (most) (p. 41)

–   Fore wing vein *2m-cu* absent (**aa**, **bb**, **cc**) 3

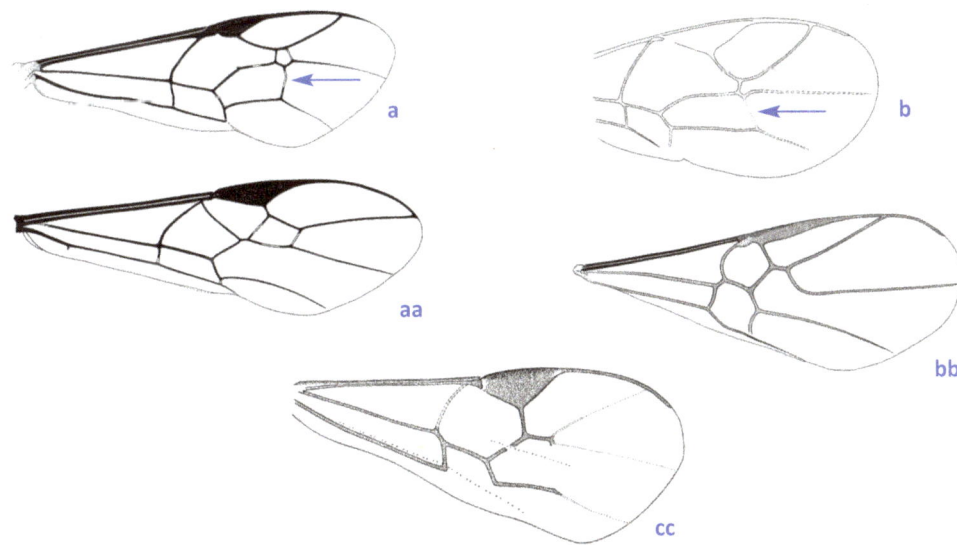

3(2). Metasomal tergites 2 and 3 fused, forming a syntergite (the apparent second tergite) which thus bears two pairs of spiracles (may be on laterotergites) (**a**). Fore wing usually with vein *RS+M* present, separating the first discal and first submarginal cells (**b**) ... **Braconidae** (most)

– Metasomal tergites 2 and 3 separated by a distinct suture, with a pair of spiracles on each tergite (**aa**). Fore wing with vein *RS+M* absent, the contiguous first discal and first submarginal cells forming part of the large discosubmarginal cell (**bb, cc, dd**) 4

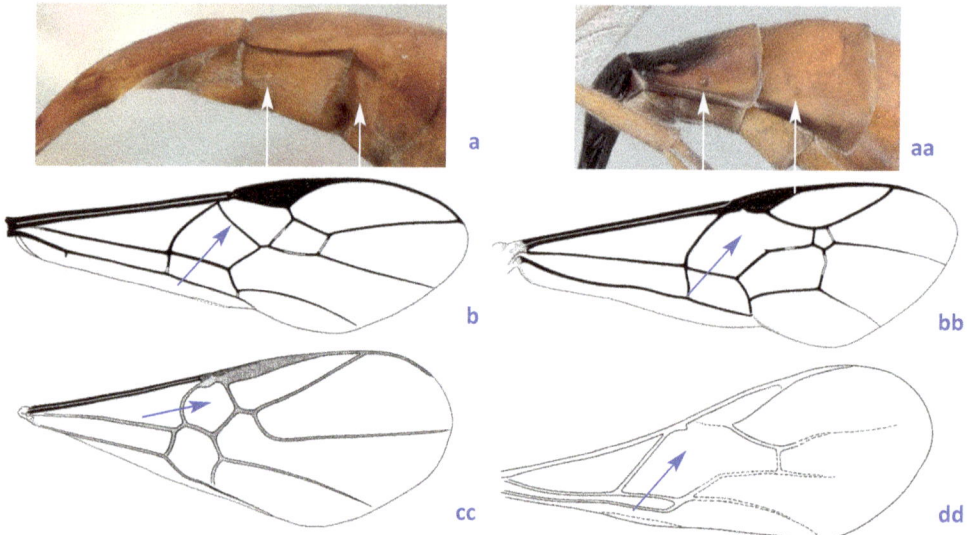

4(3). Hind wing venation reduced, subbasal cell not closed (nervellus absent) (**a**) ....
.... **Braconidae** (some Aphidiinae)

– Hind wing with closed subbasal cell (nervellus present) (**aa**) ... **Ichneumonidae** (a few) (p. 41)

 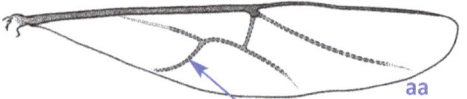

5(1). Metasomal tergite 1 with spiracles behind the middle; with the tergite and sternite fused anteriorly and the sternite extending beyond the middle of the tergite (**a**) ....
.... **Ichneumonidae** (a few, but including some frequently collected species of *Gelis* (Phygadeuontinae)) (p. 41)

– Metasomal tergite 1 with spiracles at or in front of the middle (**aa, bb**); with the tergite and sternite fused into a long petiole (**bb**) or with the sternite separated from the tergite by a flexible suture and not reaching the middle of the tergite (**aa**) 6

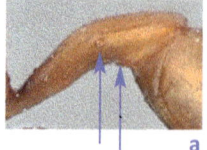 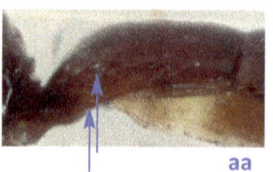 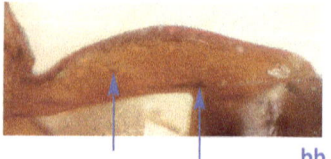

6(5). Clypeal sulcus absent, face and clypeus forming uniform, convex surface (**a**). Mandibles not exodont. Not cyclostome (face without hypoclypeal depression). Metasomal tergites 2 and 3 separated by a distinct suture, with a pair of spiracles on each tergite .... **Ichneumonidae** (a few) (p. 41)

– Clypeal sulcus distinct. Mandibles sometimes exodont (splayed outwards and not meeting when closed, with three or more teeth) (**aa**). Sometimes cyclostome (face with hypoclypeal depression above mandibles) (**bb**). Metasomal tergites 2 and 3 usually fused, forming a syntergite (the apparent second tergite), which thus bears two pairs of spiracles, but tergites separate in Aphidiinae .... **Braconidae** (a few)
[Flightless braconids in Britain and Ireland can be found in the subfamilies Alysiinae, Aphidiinae, Brachistinae (Blacini), Doryctinae, Orgilinae and Pambolinae.]

a

aa

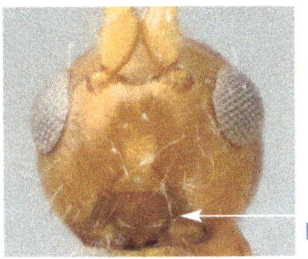
bb

Key for the identification of British and Irish subfamilies of Ichneumonidae

1. Wings present and not reduced (apparently capable of sustained flight) (macropterous) (**a**) 2

– Wings absent (**aa**) or reduced (incapable of flight) (**bb**), usually not projecting beyond tergite 1 of metasoma (apterous, micropterous or brachypterous) 98
[One species (*Sphecophaga vesparum* Curtis) sometimes brachypterous with wings extending to half the length of the metasoma, *Orthizema graviceps* with wings sometimes extending to apex of 2nd tergite; wing cells look shortened in these species and they are incapable of flight. A few macropterous ichneumonids (particularly Anomaloninae) have rather short wings (mainly an illusion, because of the very long metasoma) but the venation is normal-looking.]

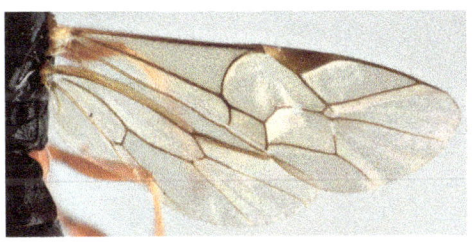
a

bb

aa

2(1). Fore wing vein *2m-cu* absent (**a**, **b**) or spectral (**c**) 3

– Fore wing vein *2m-cu* present (**aa**) 6

3(2). Fore wing with veins *RS* and *M* not fused in the position of the areolet, cross vein *2rs-m* present (**a**), but may be very short (**b**); discosubmarginal cell large, longer than wide 4

– Fore wing with veins *RS* and *M* fused over a short distance so that there is no areolet and no *rs-m* cross veins (**aa**, **bb**); discosubmarginal cell smaller and rather quadrate 5

4(3). Clypeus wide with row of regular strong setae along apical margin (**a**). Fore wing veins thickened around 2*rs-m*, this cross vein very short (**b**) .... **Tersilochinae** (a few) (p. 298)

– Clypeus barely wider than high, margin without row of setae (**aa**). Fore wing veins around 2*rs-m* not thickened, *rs-m* cross veins distinct (**bb**) .... **Phygadeuontinae** (a few) (p. 260)
[Only three species (one of *Aclastus*, two of *Gnypetomorpha*) should key out here.]

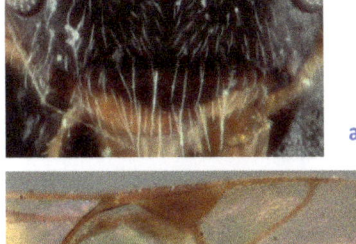
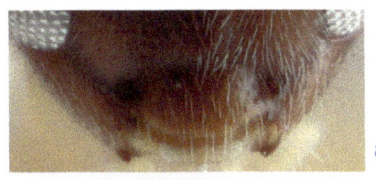

5(3). Metasomal tergites 1-3 with granular sculpture, some longitudinal striae and transverse impressions just behind the middle (**a**); sclerotized part of sternite 1 not reaching spiracle (**b**). Clypeus wider than high; mandible fully formed but twisted, with two teeth (**c**) ....
.... **Neorhacodinae** (p. 239)
[Just one British species (*Neorhacodes enslini* (Ruschka)).]

– Metasomal tergites 1-2 with longitudinal striation, lacking transverse impressions and granular sculpture (**aa**); sclerotized part of sternite 1 reaching beyond spiracle (**bb**). Clypeus higher than wide; mandibles vestigial, lacking teeth (**cc**) .... **Hybrizontinae** (p. 201)
[Often referred to as Paxylommatinae]

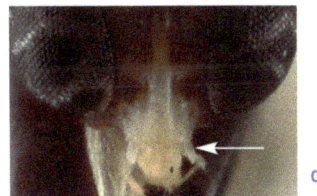

6(2). Mesoscutum with conspicuous transverse rugae across much of surface (**a**) 7

– Mesoscutum lacking transverse rugae (**aa**) 8

7(6). Occipital carina medio-dorsally absent (**a**). Fore wing vein *1cu-a* distal to *M&RS* (**b**) ....
.... **Rhyssinae** (p. 288)

– Occipital carina medio-dorsally complete (**aa**). Fore wing vein *1cu-a* opposite *M&RS* (**bb**) ......
.... **Pimplinae** (*Pseudorhyssa*) (p. 268)
[Just one British species (*Pseudorhyssa alpestris* (Holmgren)).]

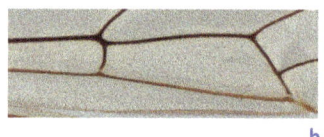

8(6). In ventral view, genae meeting or almost meeting medially (**a**); in lateral view, gena long, wider than width of eye, and widening ventrally (**b**). Lower mandibular tooth much wider and longer than upper tooth (**a**). Clypeus with ventral edge straight and coverered in erect, golden setae (**c**). Fore leg lacking trochantellus (**d**) .... **Alomyinae** (females) (p. 122)

– Genae not so massive, never nearly meeting medially; in lateral view, gena not nearly so long
.... 9

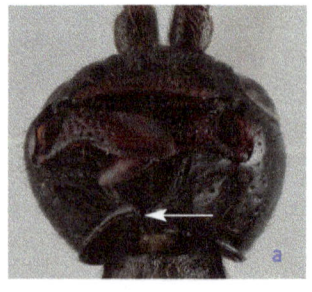
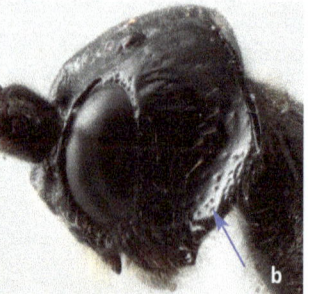
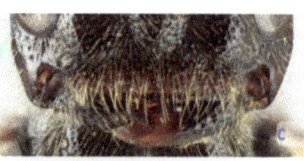

9(8). Metasomal tergite 1 with spiracle behind the middle, usually in the posterior third of the tergite (**a**, **b**, **c**), tergite lacking deep anterior glymmae (may have pits around or posterior to mid-length (**a**)); sclerotized part of sternite 1 extending to the posterior third of the segment, sometimes suture between sternite and tergite obsolete (**b**); metasomal segment 1, in dorsal view, narrow anteriorly and widened posteriorly (**c**, **d**) 10

– Metasomal tergite 1 with spiracle around the middle of the tergite or in the anterior half (**aa**, **bb**), *if* a little beyond the middle *then* tergite with deep glymmae (elongate, pit-like structures) laterally (**cc**); sclerotized part of sternite 1 *usually* not extending beyond the middle of the tergite (**aa**, **bb**, **cc**), *if* extending beyond the middle *then* the spiracle is around or anterior to the mid-length of the tergite (**dd**); metasomal segment 1, in dorsal view, either gradually widened anterio-posteriorly or more parallel-sided (**ee**, **ff**) 32

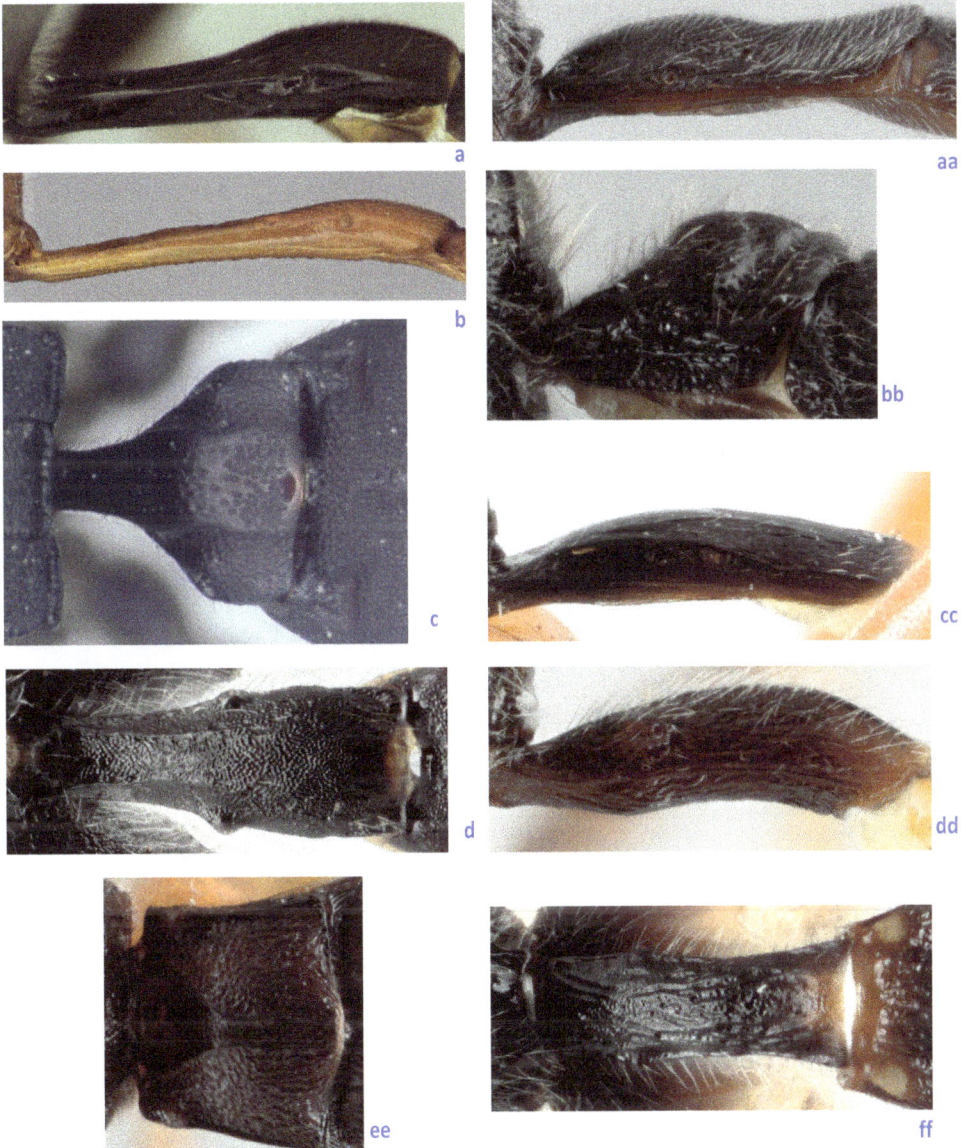

10(9). Fore wing without an areolet *and* the remaining *rs-m* vein distal to vein *2m-cu*, thus discosubmarginal cell extending beyond *2m-cu* (**a, b, c**) 11

– Fore wing with or without an areolet, *if* without *then* the remaining *rs-m* vein is proximal to or opposite vein *2m-cu*, thus discosubmarginal cell not extending beyond *2m-cu* (**aa**) ... 13

11(10). Fore wing with a dark brown adventitious vein in second subdiscal cell, parallel to wing margin (**a**). Body usually entirely pale orange-brown or pale orange-brown but sometimes with copious black markings or pale stripes (**b, c**). Ovipositor barely extends beyond the end of the metasoma .... **Ophioninae** (p. 242)

– Fore wing lacking a dark brown adventitious vein in second subdiscal cell (**aa**). Head and mesosoma with ground colour black or brown, sometimes with paler markings (**bb**). Ovipositor obviously projecting beyond the end of the metasoma 12

12(11). Propodeum lacking regular areas defined by carinae, covered in reticulate macrosculpture (**a**) (one species with anterior transverse carina and parts of median longitudinal carinae (**b**), mid tibia with one spur (**c**)). Fore wing veins around *rs-m* of normal thickness (**d**). Clypeus without a row of regular strong setae along apical margin (**e, f**) .... **Anomaloninae** (*Anomalon, Gravenhorstia*) (p. 125)

– Propodeum with some regular areas defined by carinae, lacking reticulate macrosculpture (**aa**) (and mid tibia always with two spurs). Fore wing veins around 2*rs-m* thickened (2*rs-m* very short) (**dd**). Clypeus wide with row of regular strong setae along apical margin (**ee**) .... **Tersilochinae** (some *Heterocola, Phradis*) (p. 298)

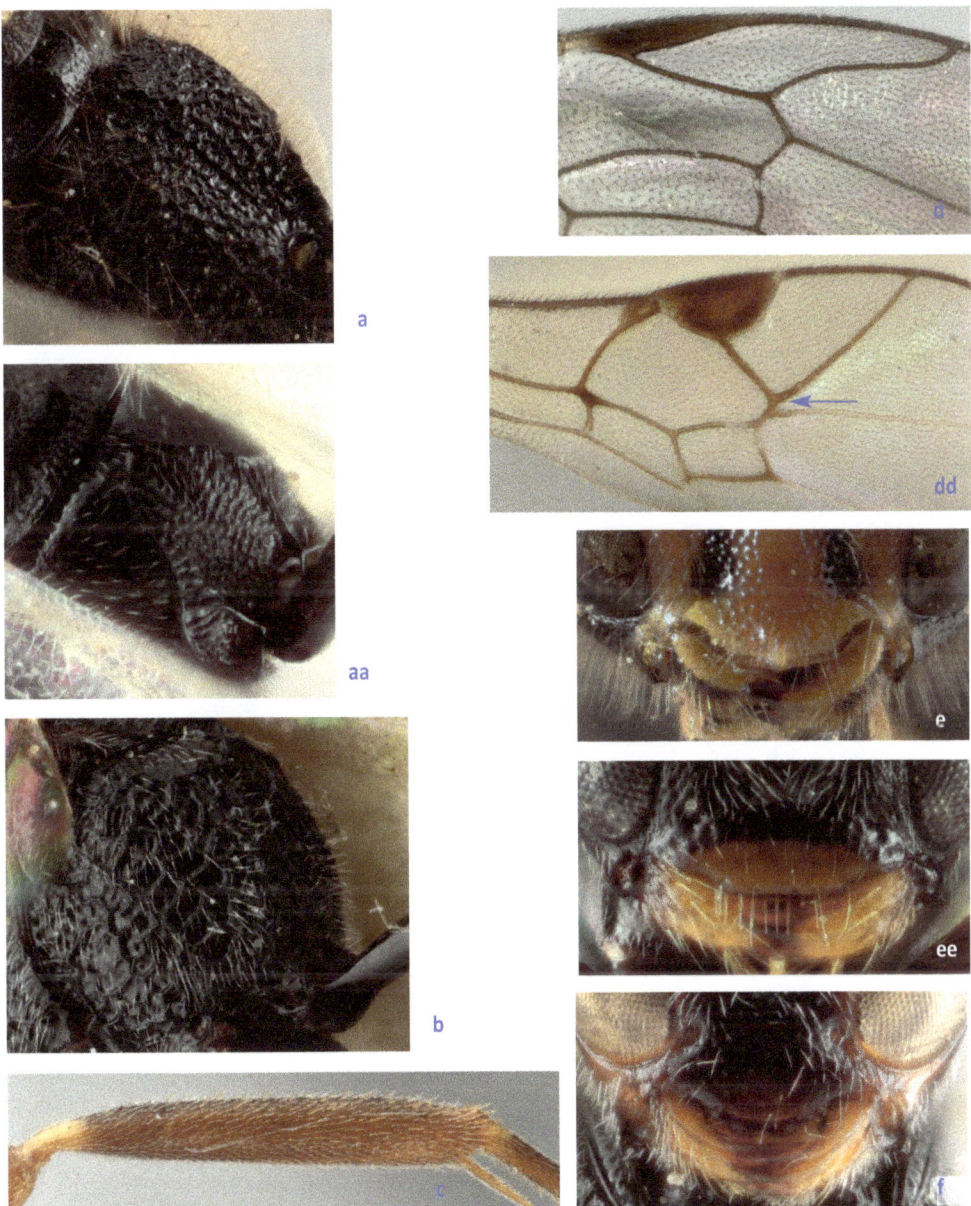

13(10). Propodeum lacking regular areas defined by carinae but instead with reticulate or areolate macrosculpture (**a**). Ovipositor 'pinched' apically, producing an abruptly finer point (**b**) .... **Anomaloninae** (most) (p. 125)

– Propodeum with carinae delimiting regular areas or *if* lacking carinae *then* not with reticulate or areolate macrosculpture (**aa**). Ovipositor with a nodus (**bb**), notch (**cc**) or plain apically, narrowed but not 'pinched' 14

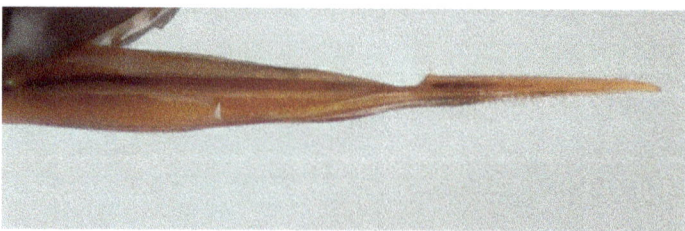

14(13). Mesosoma short, 'hunched', almost round in profile (**a**). Fore wing vein *1cu-a* distal to *M&RS* by more than half the length of *1cu-a* (**c**). Hind wing with vein *rs-m* conspicuously longer than abscissa of *RS* proximal to *rs-m* (**d**). First metasomal tergite and sternite fused, no trace of a suture visible (**e**). [Antenna of female clubbed, distal flagellar segments transverse] .... **Brachycyrtinae** (p. 143)
[Not currently known from Britain or Ireland]

– Mesosoma longer, not 'hunched' (**aa**) [*if* mesosoma short *then* sternaulus conspicuous (**bb**)]. Fore wing vein *1cu-a* much less distal (**cc**), or proximal to *M&RS*. Hind wing vein *rs-m* shorter than abscissa of *RS* proximal to *rs-m* (**dd**). First metasomal tergite and sternite with at least a suture between them (**ee**) 15

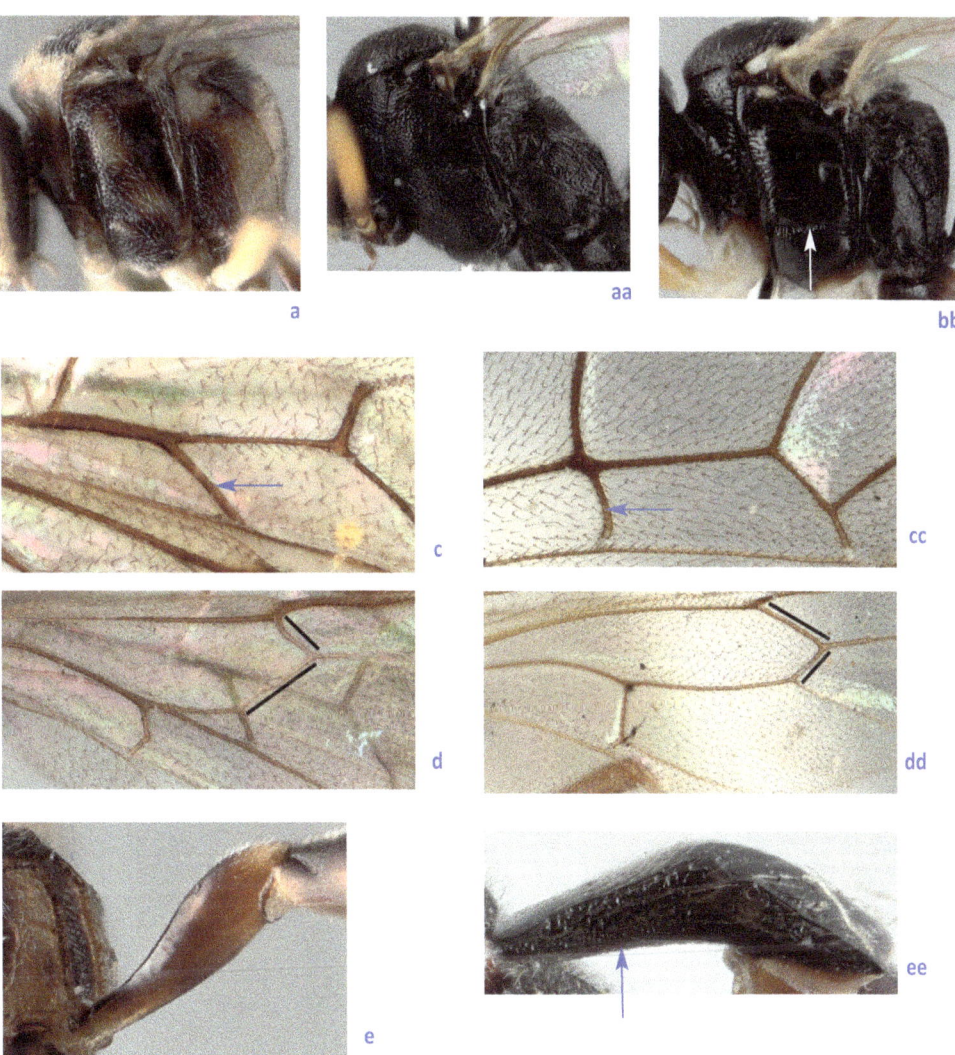

15(14). Mesopleuron with a sternaulus running usually for at least half of its length (**a, b**) 16

– Mesopleuron without sternaulus (**aa**) but sometimes (some Tersilochinae) with a mesopleural sulcus anteriorly in a higher position (**bb**) 19

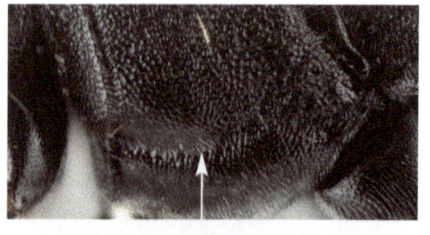

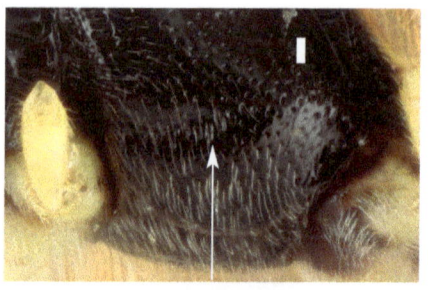

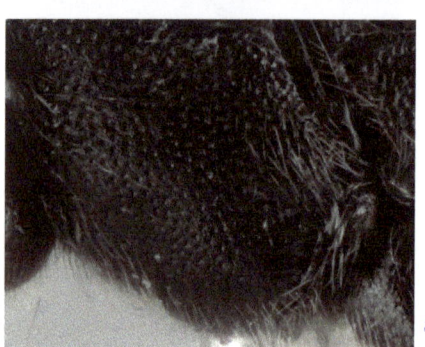

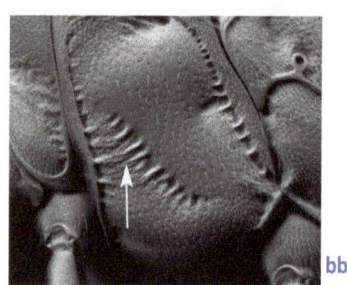

16(15). Mandible very large with two prominent teeth *and* face and clypeus forming uniformly convex surface (**a**) .... **Ichneumoninae** (Listrodromini) (p. 204)
[Sternaulus weak and only present anteriorly in *Anisobas* and *Nycthemerus* but these will key to Ichneumoninae if the sternaulus is overlooked.]

– Mandible never massive (**aa**). Clypeal sulcus sometimes weak but face and clypeus not forming such a convex surface (**bb**) 17

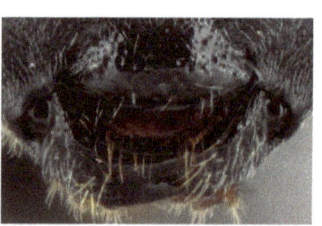

17(16). Mandible with lower tooth much shorter than upper tooth (**a**) *and* area superomedia of propodeum large (c. 1/3 of propodeum width), indented posteriorly, roughly heart-shaped viewed anteriorly (**b**). [Female with ovipositor sheaths short, stiff and straight (**d**)]
.... **Ichneumoninae** (*Dicaelotus*) (p. 204)

– Lower tooth of mandible usually as long as upper tooth (**aa**), sometimes shorter, sometimes longer, but rarely as short. Area superomedia not heart-shaped (**bb**), often open (**cc**), sometimes absent. [Female with ovipositor sheaths usually longer, more flexible-looking (**dd**)] 18

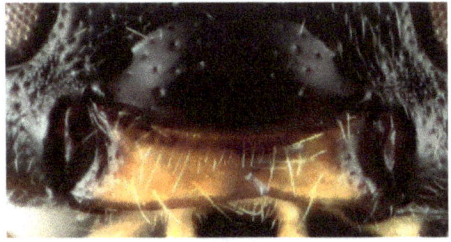

a

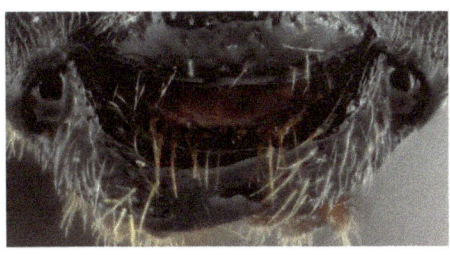

aa

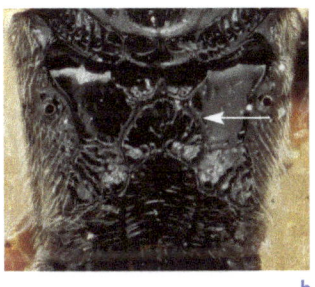

b

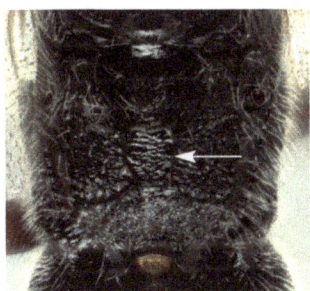

bb

cc

d

dd

18(17). Sternaulus posteriorly ending (or pointing towards and fading out) above the posterior ventral corner of mesopleuron (**a**, **b**); fore wing vein *2m-cu* frequently with two bullae (**c**); propodeum usually with longitudinal carinae as well as transverse carinae (**d**) .... **Phygadeuontinae** (most) (p. 260)
[*Thymaris* species should key out to the other part of the key at couplet 9 but if not they could be confused with Pygadeuontinae because of the long sternaulus and complete propodeal carinae. Tergite 1 of *Thymaris* has deep glymmae laterally.]

– Sternaulus posteriorly ending (or pointing towards and fading out) below the posterior ventral corner of mesopleuron (**aa**); fore wing vein *2m-cu* always with one bulla (**cc**); propodeum usually lacking longitudinal carinae, often with just transverse carina(e) (**dd**) .... **Cryptinae** (most) (p. 165)

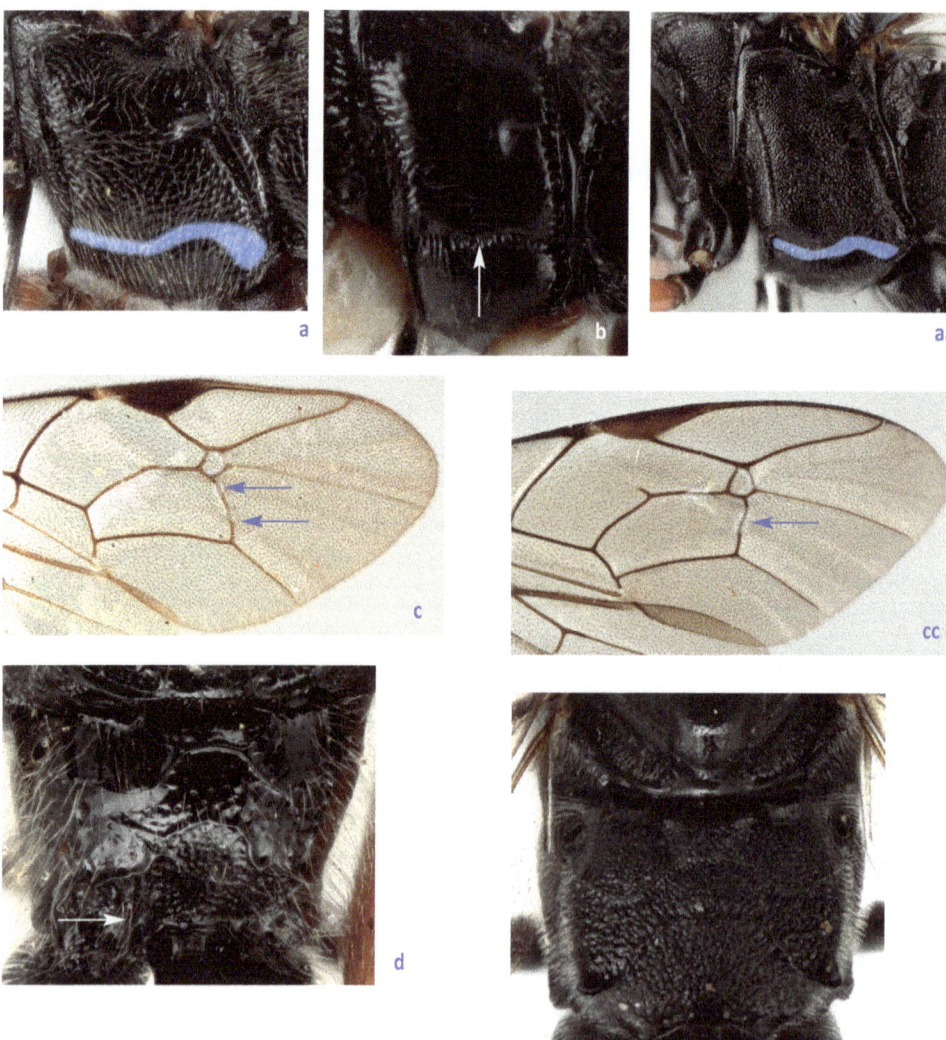

19(15). Hind wing with vein *M+CU* strongly curved and spectral or absent proximally, nervellus not intercepted (**a**). Laterotergite of metasomal tergite 3 pendant and not separated by a sharp crease (**b, c**).. 20

– Hind wing with vein *M+CU* usually fully sclerotized, often weakly curved or straight (**aa**), if strongly curved and spectral proximally then nervellus intercepted. Laterotergite of metasomal tergite 3 usually separated by a sharp crease (**bb**) 21

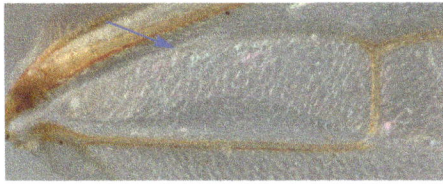
a

aa

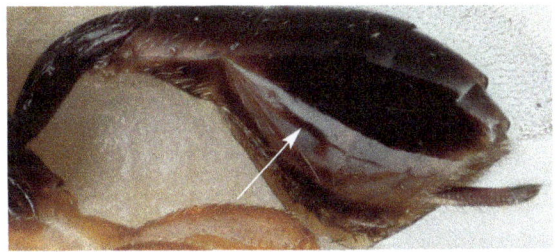
b

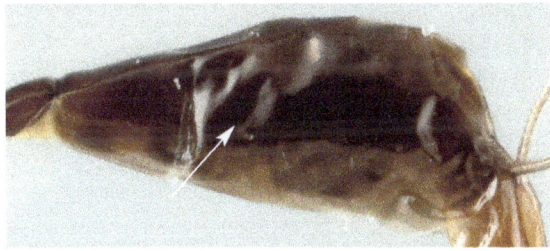
c

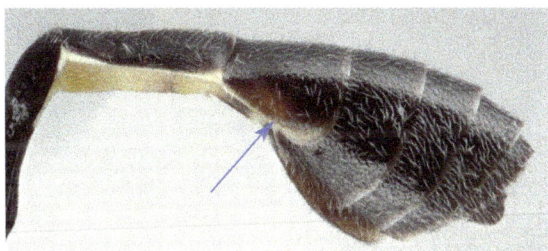
bb

20(19). Fore wing either with areolet present or, if absent, veins around vein 2*rs-m* of normal thickness (**a**). Clypeus without a row of regular strong setae along apical margin (**b**). Tarsal claws pectinate, at least proximally (**c**) .... **Tersilochinae** (*Astrenis* and *Phrudus*) (p. 298)
[Keyed out separately here as the *Phrudus* group of genera have, until very recently, been treated as belonging to the subfamily Phrudinae.]

– Fore wing with veins thickened around vein 2*rs-m*, 2*rs-m* almost obliterated (**aa**). Clypeus wide with row of regular strong setae along apical margin (**bb**). Tarsal claws usually simple (**cc**) (occasionally pectinate) .... **Tersilochinae** (most) (p. 298)

a

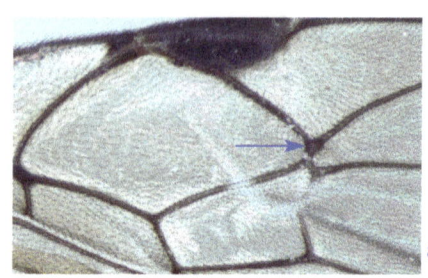
aa

b

bb

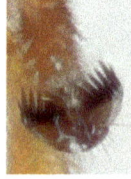

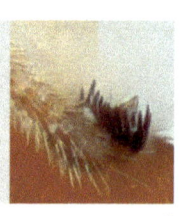
c

cc

21(19). Clypeus apically produced into a strong, median point (**a, b**) 22

– Clypeus not produced into a point, although sometimes with a small tooth on an otherwise truncate clypeus edge (**aa**) 23

a

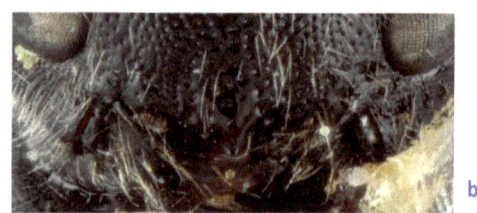
b

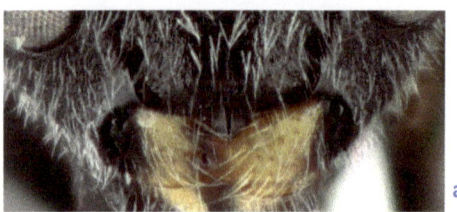
aa

22(21). Hind tibia with one spur (**a**); all tarsal claws pectinate (**b**). Metasoma black and yellow-striped (**c**) .... **Tryphoninae** (Sphinctini) (p. 305)

[Just one possibly extinct species in Britain (*Sphinctus serotinus* Gravenhorst).]

– Hind tibia with two spurs (**aa**); all tarsal claws simple (**bb**). Metasoma black and reddish (**cc**) .... **Metopiinae** (*Ischyrocnemis*) (p. 228)

[One very rare species in Britain (*Ischyrocnemis goesi* Holmgren).]

23(21). Hind tibia with one spur (**a**). Antenna clavate (flagellum club-shaped, apical flagellar segments wider than long) (**b**) .... **Metopiinae** (*Periope*) (p. 228)

[Because *Periope auscultator* (Haliday) (the only British species in the genus) could conceivably be keyed out either way at couplet 9, it has been accommodated in both halves of the key.]

– Hind tibia with two spurs (**aa**). Antenna not clavate (**bb**) 24

24(23). Fore wing vein *2m-cu* with a single bulla (**a**, **b**) (if bullae are hard to distinguish from the rest of the vein, their positions can be located by folds in the wing membrane, but care needs to be taken in ascertaining whether or not there is a section (however small) of fully sclerotized vein separating them if there are two) 25

– Fore wing vein *2m-cu* with two bullae (**aa**) 28

a

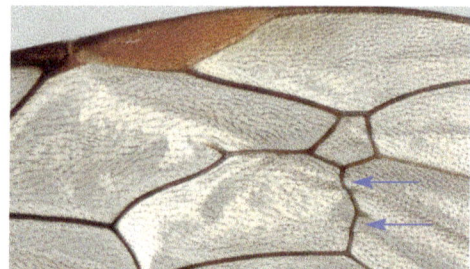

aa

b

25(24). Antenna with 12 flagellar segments (**a**). Labrum conspicuously exposed below clypeus (**b**) .... **Adelognathinae** (in part) (p. 115)
[*Adelognathus dorsalis* (Gravenhorst) will key out here.]

– Antenna with more than 16 flagellar segments. Labrum inconspicuous, a short part exposed below clypeus (**bb**) 26

a

b   bb

26(25). Maxillary palps elongate, able to reach far beyond fore coxae (**a**). Clypeus subbasally raised, apically flattened (**b**). Posterior transverse carina of the mesosternum absent (**c**). Tarsal claws not pectinate (**d**). Ovipositor very short, not extending beyond the end of the metasoma; ovipositor sheaths broad, about as wide as long (**e**) .... **Oxytorinae** (p. 257)

– Maxillary palps not or barely able to reach the fore coxae (**aa**). Clypeus uniformly convex, not apically flattened (**bb**). Posterior transverse carina of the mesosternum usually complete (**cc**). Tarsal claws usually pectinate (**dd**). Ovipositor extending beyond the end of the metasoma; ovipositor sheaths thinner, even on shortest ovipositor at least twice as long as wide (**ee**) **27**

27(26). End of hind tibia with tarsus and spurs inserted in a common membranous area (**a**). Metasomal tergite 2 with varied sculpture but never longitudinal striation (**b**). Clypeal sulcus very weak or absent, the clypeus and face forming a continuous weakly convex surface, with silvery setae conspicuous (**c**). Fore wing pterostigma narrow (**d**). Head usually black [partly yellow in one common species (**e**) and partly or wholly yellow in a few others, *if* so *then* check hind tibial spurs; never with hind femoral tooth] ....
.... **Campopleginae** (p. 145)

– End of hind tibia with a sclerotized bridge between the tarsus and spurs so that they insert into separate sockets (**aa**). Metasomal tergite 2 with fine longitudinal striation (**bb**). Clypeal sulcus distinct and clypeus more convex in lateral view (**cc**). Fore wing pterostigma broad (**dd**). Head often with ground colour pale brown, with darker markings, or at least with conspicuous yellow marks on face (**ee**) [one widespread species with ventral tooth on hind femur (**ff**)] .... **Cremastinae** (p. 161)

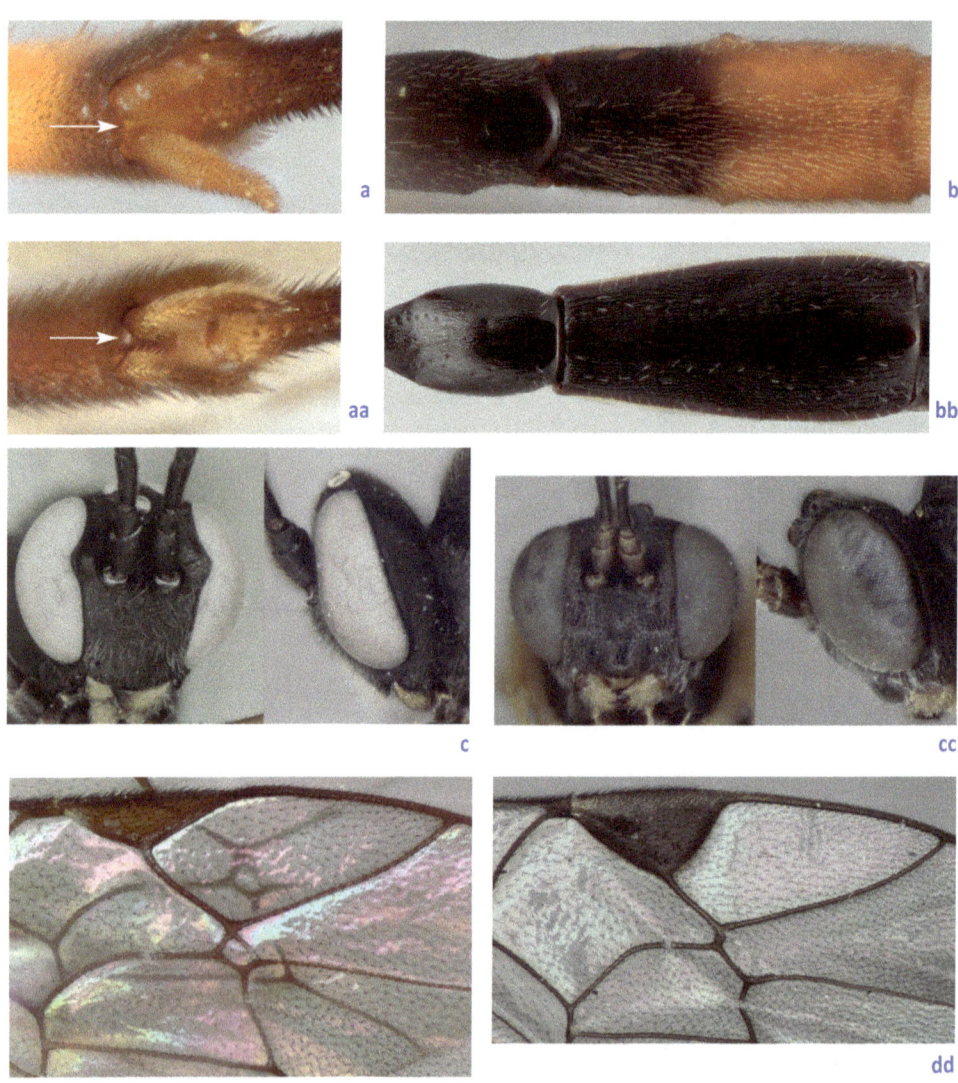

27 figures continued overleaf:

27 figures continued:

e

ee

ff

28(24). Mandibles thin and strongly narrowed towards the apex; clypeus not or hardly wider than deep (**a**). Hind tibia with inner, apical row of dense setae (**c**). Metasoma with shallow thyridia (**d**). Eyes often strongly convergent ventrally (**a**) ....
.... **Orthocentrinae** (*Catastenus, Symplecis*) (p. 248)

– Mandibles usually not thin or strongly narrowed apically but often twisted so that they may look thin in anterior view; clypeus wider than deep, usually markedly so (**aa**); *if* mandibles strongly narrowed and clypeus about as wide as deep (**bb**) *then* hind tibia with inner, apical row of setae much shorter (**cc**) and second metasomal tergite with deeply impressed thyridia and gastrocoeli (**dd**). Eyes not convergent ventrally 29

a

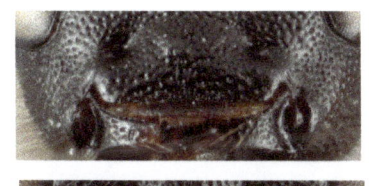
aa

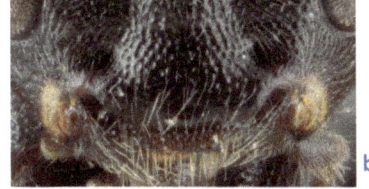
bb

c

cc

dd

29(28). Fore wing with areolet petiolate (2*rs-m* and 3*rs-m* joining before they reach *RS*) (**a**). Fore tibia with an apical, distal tooth (**b**) (never shiny blue-black with white spots on metasoma). Mandible with lower tooth often longer than upper tooth (**c**); clypeus wider than high and strongly convex, often with a blunt transverse ridge (**d**) or apically rounded, blunt (**c**) .... **Ctenopelmatinae** (a few Ctenopelmatini and Euryproctini) (p. 175)

– Fore wing with areolet quadrate or pentagonal (**aa**), sometimes not closed (3*rs-m* missing), never petiolate. Fore tibia usually without a tooth apically (**bb**) (*if* with a tooth *then* shiny blue-black with white spots on metasoma). Mandible often with lower tooth shorter than upper tooth or the same length; clypeus either flat or convex but about as wide as high, lacking a swollen ridge and apically usually sharp (**cc**) 30

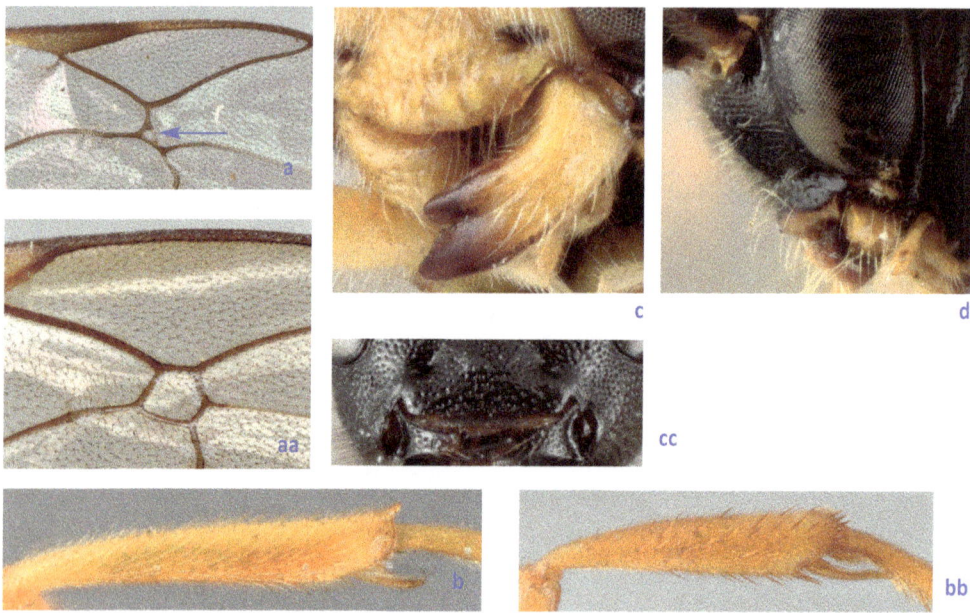

30(29). Fore wing with areolet pentagonal, first abscissa of vein *M* much longer than second abscissa (**a**). Clypeus very wide, abruptly truncate, covered in stiff hairs; mandible widened in apical half, lower tooth larger than upper (**b**) .... **Alomyinae** (p. 122)

– Fore wing areolet rhombic or pentagonal, first abscissa of vein *M* not so obviously longer than second abscissa (**aa**). Clypeus with edge not so abruptly truncate; mandible not widened in apical half, with lower tooth rarely longer than upper tooth (**bb**); *if* longer *then* no wider than upper tooth. 31

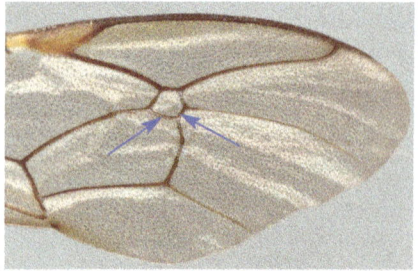
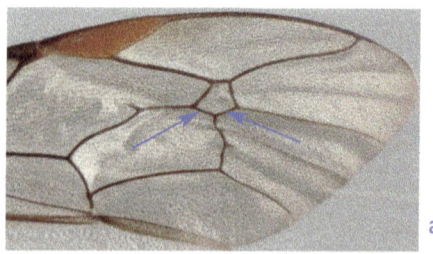

30 figures continued overleaf:

30 figures continued:

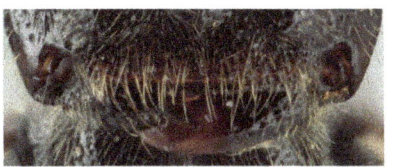 

b   bb

31(30). Clypeus wide and flat, truncate apically, often shiny and punctate; labrum usually exposed as a thin strip with long setae (**a**). Metasomal tergite 2 with anterior thyridia and gastrocoeli often deeply impressed, gastrocoeli usually with rugosity (**b**). Fore wing pterostigma usually uniformly coloured, sometimes paler proximally but not sharply differentiated (**c**). Mandibles usually with the lower tooth shorter than the upper and the mandible slightly twisted (**a**). Ovipositor sheaths stiff and straight, usually barely projecting beyond metasoma apex [long and glabrous in one genus (*Crypteffigies*) (**e**)] .... **Ichneumoninae** (most) (p. 204)

– Clypeus convex with an impressed apical rim, usually coriaceous, almost impunctate; labrum usually concealed, lacking long setae (**aa**). Metasomal tergite 2 with anterior thyridia small and superficial, gastrocoeli unsculptured (**bb**). Fore wing pterostigma often with a paler proximal corner (**cc**). Mandibles with the lower tooth usually the same length as the upper tooth but sometimes a little shorter; mandible not twisted (**dd**). Ovipositor sheaths thinner and flexible (**ee**) .... **Phygadeuontinae** (a few)(p. 260)

[Only males of *Gelis* should key out here but other genera might run to this couplet if a weak sternaulus is overlooked.]

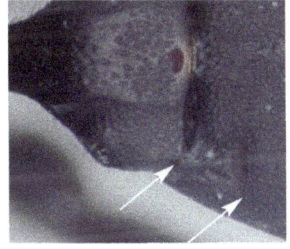 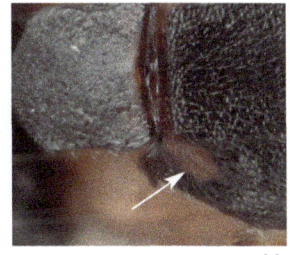

a

aa   b   bb

 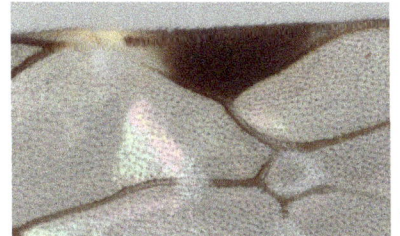

c   cc

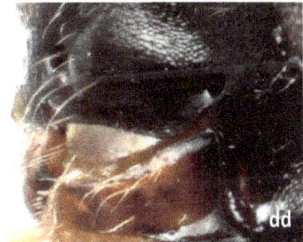  

dd   e   ee

32(9). Female with egg(s) conspicuously hanging from the lower valves of the ovipositor (**a, b**) .... **Tryphoninae** (some) (p. 305)

– Male, or female with no eggs hanging from the ovipositor 33

a

b

33(32). Areolet 1.5x as long as high (if not closed, 3*rs-m* indicated by bend of vein and areolet still discernibly 1.5x as long as high) (**a**). Posterior transverse carina of mesosternum complete (**b**). Distal abscissa of hind wing vein *AA* missing (**c**) .... **Ateleutinae** (p. 131)
[One widespread species in Britain (*Ateleute linearis* Förster).]

– Areolet not longer than high, different shape (**aa**) or open (vein 3*rs-m* entirely absent). Posterior transverse carina of mesosternum usually incomplete (**bb**). Distal abscissa of hind wing vein *AA* usually present (**cc**) 34

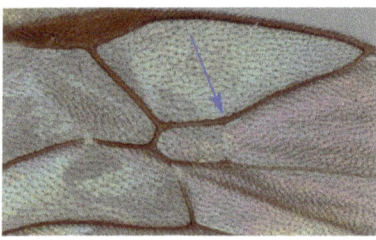
a

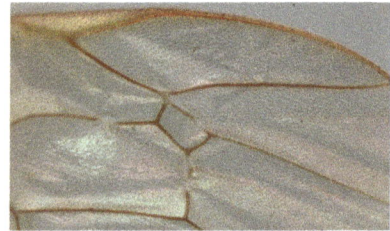
aa

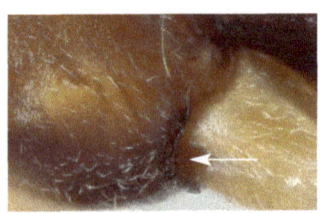
b

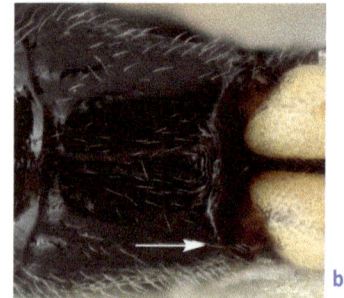
bb

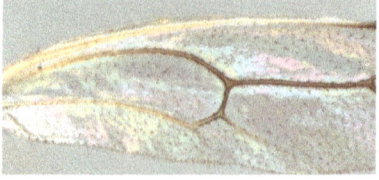
c

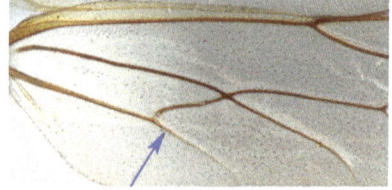
cc

34(33). Antenna with 12 or 13 flagellar segments (**a**). Labrum conspicuously exposed below the clypeus (**b**). Fore wing vein 2*m-cu* with one bulla (**c**) .... **Adelognathinae** (most) (p. 115)

– Antenna with more than 13 flagellar segments, usually with more than 16 (**aa**). *If* labrum conspicuously exposed below the clypeus (**bb**) *then* antenna with more than 16 flagellar segments. Fore wing vein 2*m-cu* with one or two bullae 35

35(34). Either hind or mid tibia (or both) with no spur (**a**) or one spur (**b**) 36

– Hind and mid tibiae each with two spurs (**aa**) 38

36(35). Face with carinae delimiting a shield-shaped area (**a**) [mid tibia with one spur, hind tibia with two] .... **Metopiinae** (*Metopius*) (p. 228)

– Face without carinae delimiting a shield-shaped area (**aa**) 37

37(36). Hind tibia lacking spurs, mid tibia with one spur (**a**). Antennae not clavate, all flagellar segments longer than wide (**b**).. **Tryphoninae** (Tryphonini: *Exenterus* genus-group) (p. 305)

– Hind tibia with one spur, mid tibia with two (**aa**). Antennae clavate, penultimate and several preceding flagellar segments wider than long (**bb**) .... **Metopiinae** (*Periope*) (p. 228)

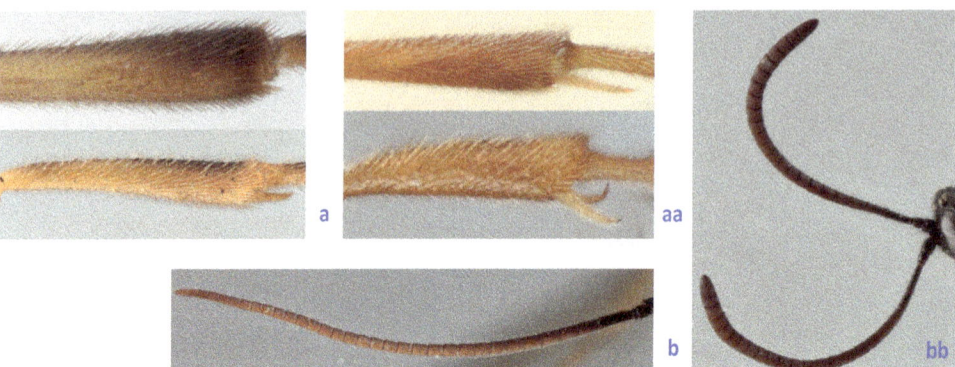

38(35). Female with long, posteriorly narrow, pointed hypopygium, extending slightly beyond the metasomal apex (**a**) or level with the apex of the elongate posterior tergite (**b**). Male (and most females) with fore tarsal claws bifurcate (**c**). Mandible bidentate; clypeus usually with preapical sharp transverse ridge (**e**) and usually with longitudinal ridge between antennal sockets (**f**) .... **Acaenitinae** (in part) (p. 111)

– Female with hypopygium much shorter, never posteriorly narrow and pointed (**aa, bb**). Fore tarsal claws simple, lobed or pectinate (**cc**) but rarely bifurcate; *if* bifurcate *then* mandible unidentate (**dd**). Clypeus lacking preapical sharp transverse ridge, although often convex subbasally (**ee**). Rarely with weak longitudinal ridge between antennal sockets (**ff**) 39

38 figures continued overleaf:

38 figures continued:

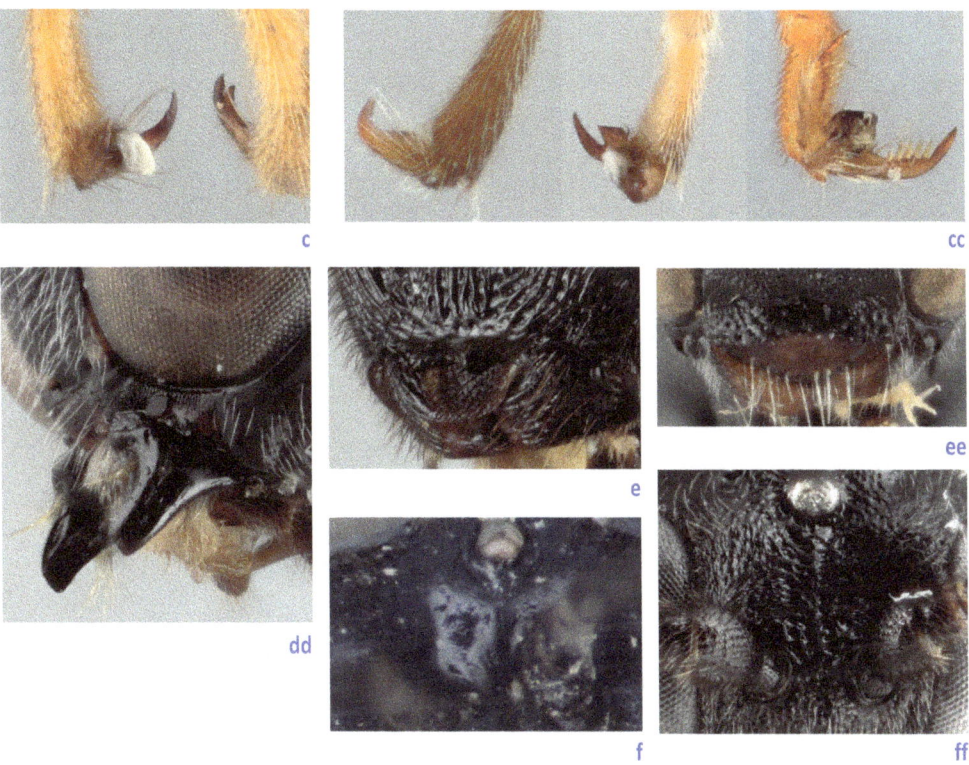

39(38). Mesopleuron with sternaulus extending across at least half of its length (**a, b**)............. 40

– Mesopleuron without a sternaulus (**aa**) or with a short sternaulus extending less than a third of the length of the mesopleuron (or, if with strong sternaulus, this difficult to see amongst dense setae and scutellum has a spine (*Agriotypus armatus* Curtis)) 41

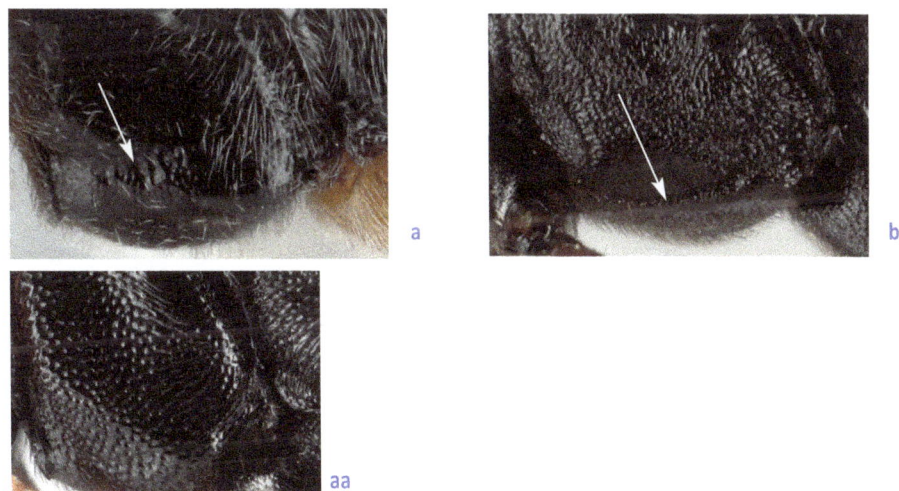

40(39). Metasomal tergite 1 with deep glymmae laterally (**a**). Clypeus with apical row of regularly spaced setae, without a median tooth (**b**) ...... **Tryphoninae** (some Oedemopsini) (p. 305)

– Metasomal tergite 1 without glymmae (**aa**). Clypeus without apical row of regularly spaced setae, with a median tooth (**bb**) .... **Cryptinae** (*Echthrus*) (p. 165)

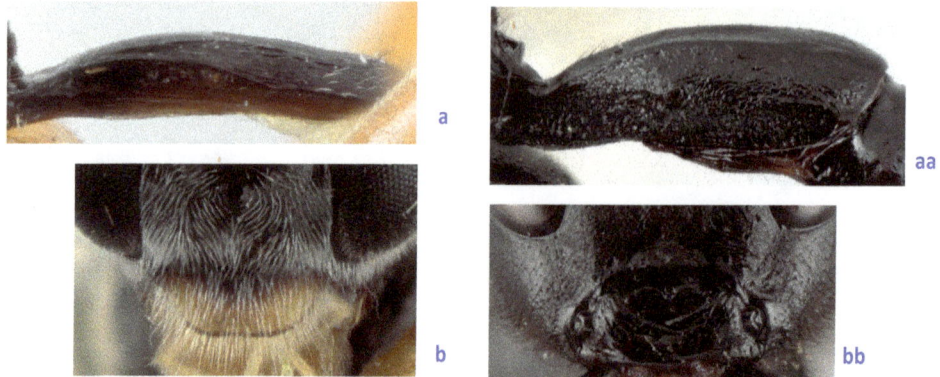

41(39). Pronotum dorsally with a forwards-projecting, bilobed flange (**a**). Male with central flagellar segments expanded, much wider than long (**b**). Ovipositor and sheaths tiny, vestigial (**c**) .... **Eucerotinae** (*Euceros*) (p. 197)
[Confirmatory characters, in combination: fore wing vein 2*m-cu* with one bulla; clypeal sulcus barely discernible; pronotal epomia absent; submetapleural carina expanded into an anterior flange.]

– Pronotum lacking bilobed flange (**aa**). Antennae not medially expanded (**bb**). Ovipositor longer, sheaths fully formed (**cc**), but occasionally concealed by the large hypopygium (**dd**) ... 42

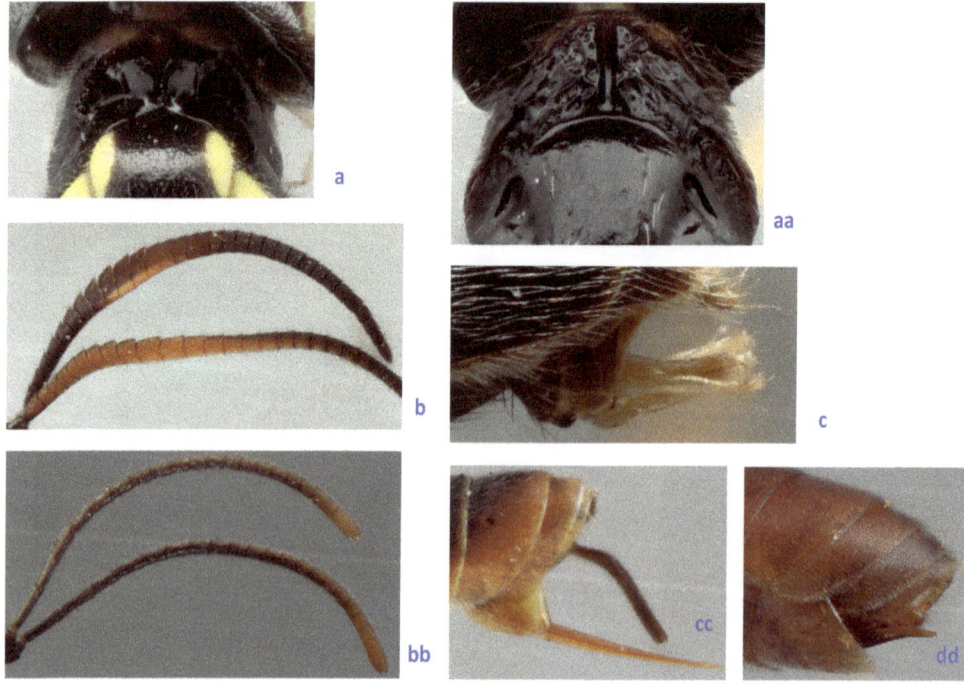

42(41). Female: ovipositor sheaths projecting stiffly, much longer than high, cuticle shiny; ovipositor thin, needle-like (**a**, **b**). Male: parameres elongate, rod-like (**c**). Both sexes: metasomal tergite 1 with deep glymmae, separated centrally only by thin, transparent cuticle (**d**); fore wing with regularly rhombic (diamond-shaped), relatively large areolet (**e**) *or*, if areolet obliquely rhombic (**f**), large (7+ mm fore wing length) and mostly testaceous species with densely pectinate claws (**g**) .... **Mesochorinae** (p. 220)

– Female: ovipositor sheaths not so stiffly projecting, with microsculpture (**aa**), *if* shiny and stiff *then* no longer than high (**bb**). Male: parameres not elongate or rod-like. Both sexes: metasomal tergite 1 with or without deep glymmae; fore wing areolet never so large and diamond-shaped; *if* obliquely rhombic *then* not predominantly testaceous and claws not so densely pectinate 43

43(42). Clypeus weakly separated from face, face dorsally with no inter-antennal projection; malar space long, about 2 times the basal width of the mandible (**a**). Fore tarsus with segments 2-4 short, obviously foreshortened compared to segment 5 (**b**). Fore wing areolet rhombic, rather diamond-shaped (**c**). Metasomal tergite 1 with glymmae deep, separated medially by thin, translucent cuticle, situated distinctly behind anterior end of tergite (**d**). Female hyopygium large, extending to about apex of metasoma, ovipositor sheaths very short, concealed, ovipositor slender, simple (**e**) ....
.... **Metopiinae** (*Scolomus*) (p. 228)
[One very rarely collected species in Britain (*Scolomus borealis* (Townes)).]

– Clypeus separated from face by clypeal sulcus (**aa**); *or* if clypeus not separated from face (**bb**) *then* not with same combination of characters. *If* with foreshortened tarsal segments *then* glymmae not so deep and situated anteriorly on segment (**dd**), often with upper edge of face produced as a triangular inter-antennal point or ridge (**ee**) 44

44(43). Clypeal sulcus absent, the face and clypeus forming a single slightly convex or bulging surface (**a, b, c, d**) 45

– Clypeal sulcus present as a groove or transverse impression, the whole surface not strongly bulging (**aa, bb**) 49

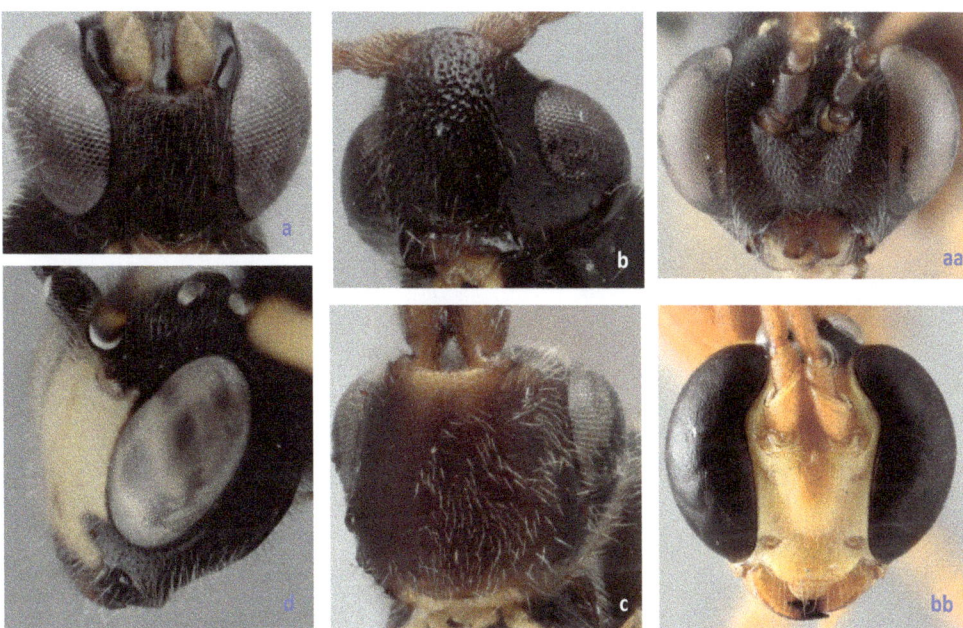

45(44). Eyes with conspicuous, long setae over entire surface (**a**). Female tarsal claws with a lobe and arolium large, extending well beyond claws (**b**) .... **Pimplinae** (*Schizopyga*) (p. 268)

– Eyes bare, or with very inconspicuous setae (**aa**). Female tarsal claws lacking a lobe, sometimes pectinate, otherwise bare; arolium not extending far beyond claws (**bb**) 46

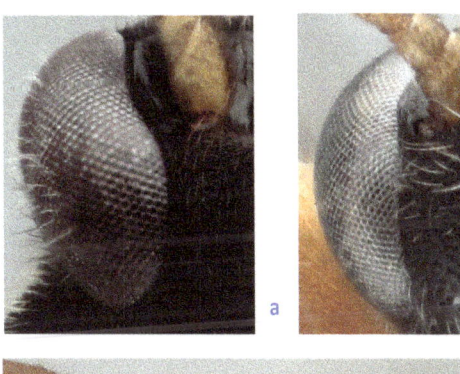
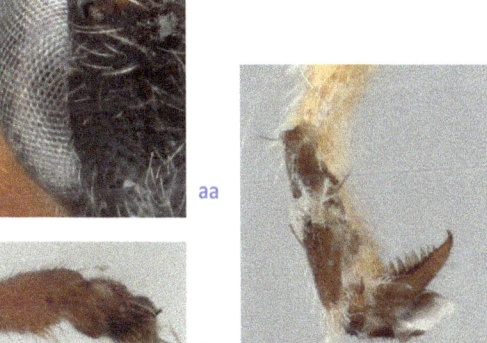

46(45). Fore tarsus with segments 2-4 short, obviously foreshortened compared to segment 5, often as wide as or wider than long (**a**); fore trochantellus often not differentiated from femur (**b**). Top of face in the form of a triangular projection or ridge between the antennal sockets (**c**) or occasionally a transverse ridge (**d**). Metasomal tergite 1 with a pair of strong median longitudinal carinae (**e, f**) .... **Metopiinae** (p. 228)

– Fore tarsus with all segments longer than wide, not foreshortened (**aa**); fore trochantellus differentiated from femur (**bb**). Face with upper edge simple (**cc**). Metasomal tergite 1 often without a pair of strong median longitudinal carinae (**ee**) 47

47(46). Antenna with scape rather cylindrical, about 3 times as long as wide (**a**). Malar space long (space between eye and mandibles 3-4 times as long as the width of the mandible base) and sometimes with a well-defined sulcus (**b**). Mandible small, thin, narrowed apically, lower tooth shorter than upper tooth (**b**). Tarsal claws simple (**d**) .... **Orthocentrinae** (*Orthocentrus* group) (p. 248)

– Antenna with scape more globose, from 1-2 times as long as wide (**aa**). Malar space shorter (not more than 1.5 times as long as basal breadth of mandible) and lacking a sulcus (**bb**). Mandible robust, not strongly narrowed apically (**cc**). Tarsal claws often pectinate (**dd**) 48

48(47). Mesoscutum shining, unsculptured, with notauli distinct and narrow (**a**). Metasomal tergite 1 heavily sculptured (rugose) (**b**). Tarsal claws simple (**c**). Female: antenna with all flagellar segments wider than long and face strongly protruding in lateral view (**d**); male: antenna with deeply excavate tyloids on 4th and 5th segments (**e**) .... **Cylloceriinae** (*Hyperacmus*) (p. 186)

[One uncommon species in Britain (*Hyperacmus crassicornis* (Gravenhorst)).]

– Mesoscutum sculptured, notauli weak (**aa**). Metasomal tergite 1 polished or weakly sculptured (**bb**). Tarsal claws pectinate (**cc**). Both sexes: antenna with all flagellar segments longer than wide and face not protruding (**dd**); male: antenna lacking tyloids .... **Ctenopelmatinae** (*Rhorus*) (p. 175)

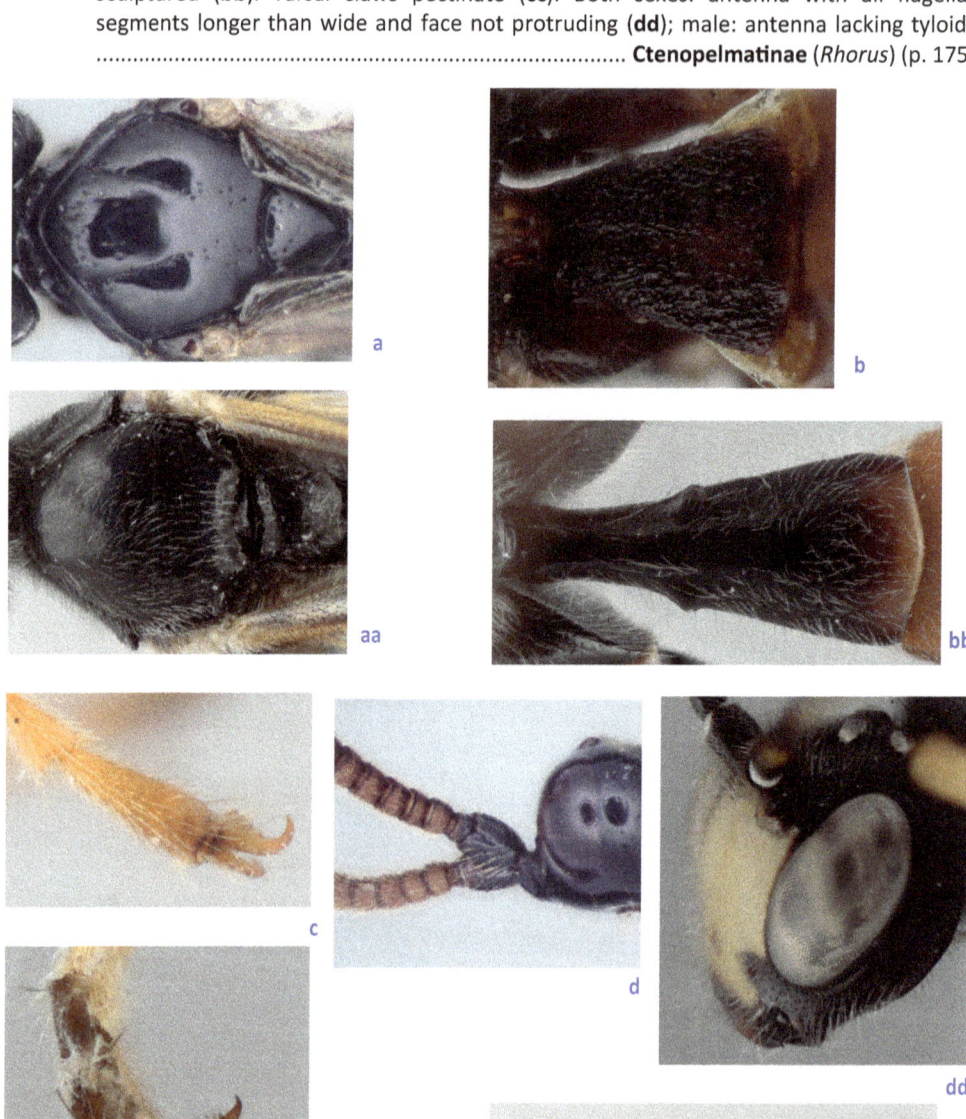

49(44). Metasomal tergites 2-4 (at least) with grooves delimiting a triangular or rhombic pattern (**a**, **b**, **c**). Submetapleural carina often expanded anteriorly into a deep lobe (**d**) 50

– Metasomal tergites 2-4 without grooves delimiting a triangular or rhombic pattern, at most tergites 2 and 3 with weak grooves, with a wide space between them anteriorly, cutting off the anterior corners (**aa**) and in these cases with submetapleural carina not expanded anteriorly into a deep lobe (**dd**) 52

a

aa

b

c

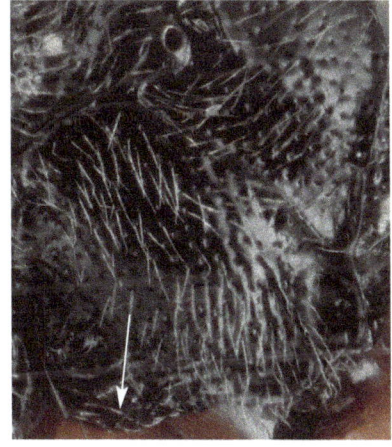

d

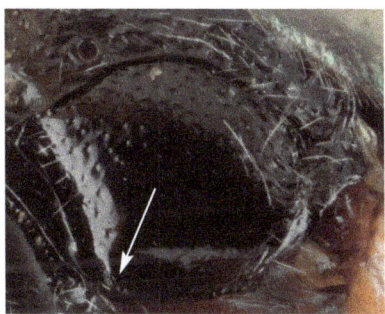

dd

50(49). Metasomal tergites 2-4 with grooves delimiting a rhombic pattern (**a**). Submetapleural carina not expanded anteriorly (**b**). Fore tarsus with fifth segment conspicuously broadened and with arolium (pad) projecting beyond claws; claws with basal lobe (**c**). Ovipositor lacking obvious teeth or a notch (**d**) .... **Pimplinae** (*Zatypota*) (p. 268)

– Metasomal tergites 2-4 with grooves delimiting a triangular area (**aa**). Submetapleural carina expanded anteriorly into a lobe (**bb**). Fore tarsal claws with fifth segment not conspicuously broadened, arolium not projecting beyond claws, claws simple or pectinate, not lobed (**cc**). Ovipositor with dorsal notch (**dd**) or ventral teeth (**ee**) 51

51(50). Metasomal tergites 2-4 with triangular areas extending from the anterior margin of the tergites, defined by grooves laterally but not posteriorly (**a**). Propodeum with or without posterior transverse carina, sometimes with area superomedia weakly demarked with carinae (**b**). Upper division of metapleuron not produced dorso-posteriorly into a 'catch' (**c**). Ovipositor lacking obvious teeth and with a dorsal apical notch (**d**) ....
.... **Banchinae** (Glyptini) (p. 134)

– Metasomal tergites 2-4 with triangular areas near the centre of the tergite, defined by grooves laterally and posteriorly (**aa**). Propodeum with combined area superomedia and area basalis defined by strong carinae (**bb**). Upper division of metapleuron produced dorso-posteriorly into a 'catch' that overlies the anterior end of the propodeum (**cc**). Ovipositor with obvious teeth (**dd**) .... **Lycorininae** (*Lycorina*) (p. 217)
[One rare species in Britain (*Lycorina triangulifera* Holmgren).]

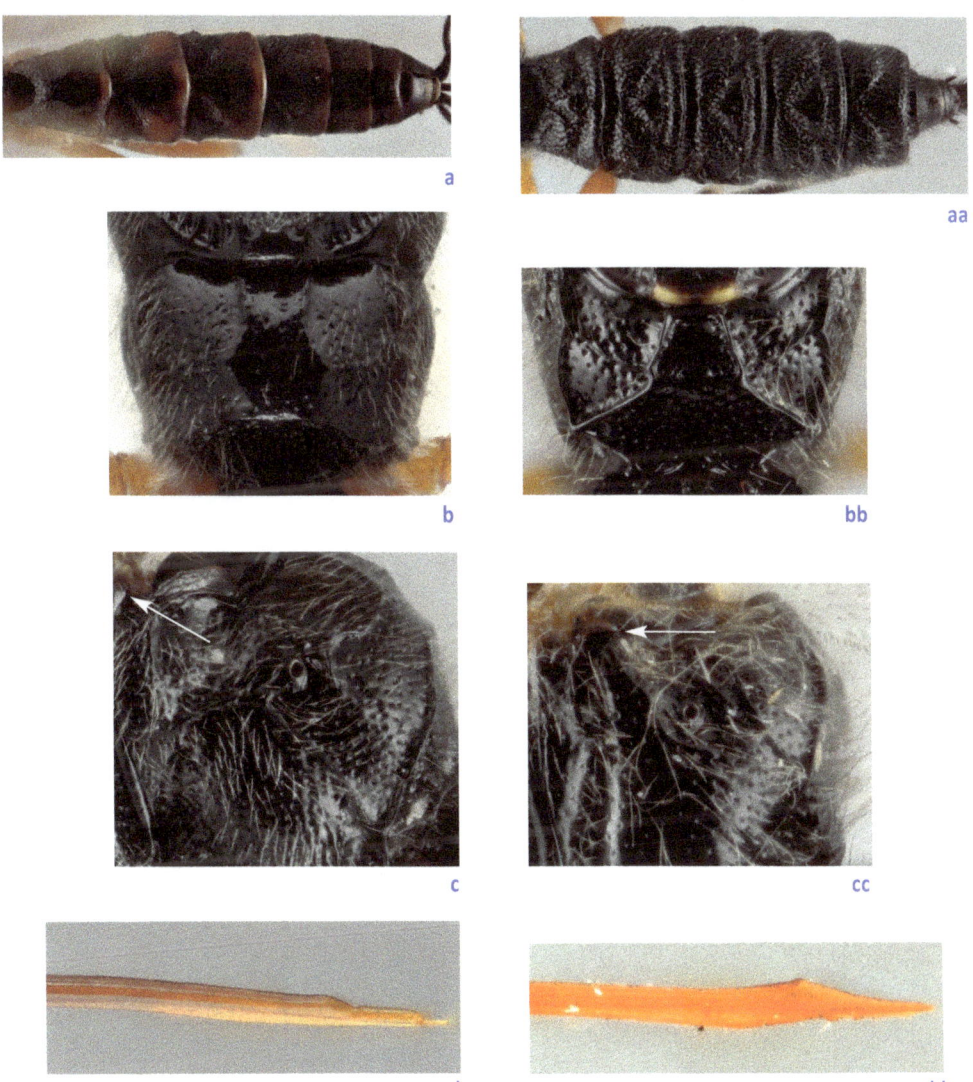

52(49). Metasomal segment 1 with sclerotized part of sternite extending at least 0.75 length of tergite, with spiracles near or anterior to mid-length; tergite roughly parallel-sided, narrow, rather cylindrical (lateral: **a**, **b**, **c**, **d**; dorsal: **e**) 53

– Metasomal segment 1 with sclerotized part of sternite usually not extending to far behind the spiracle; tergite almost always widened posteriorly or stout, dorso-ventrally flattened, never cylindrical (**aa**, **bb**) 56

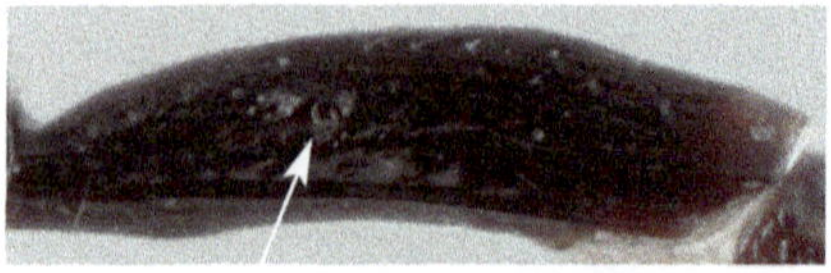

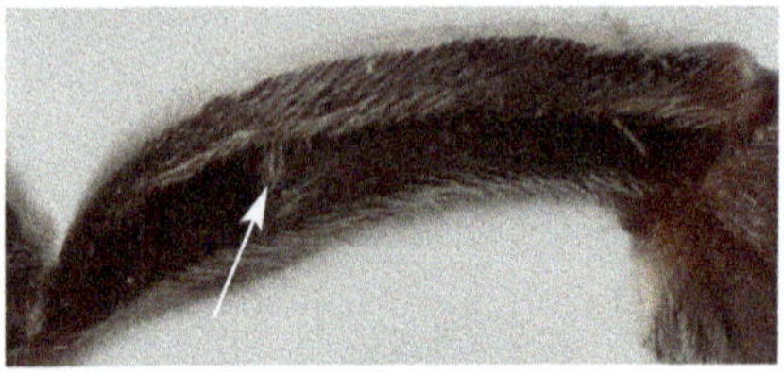

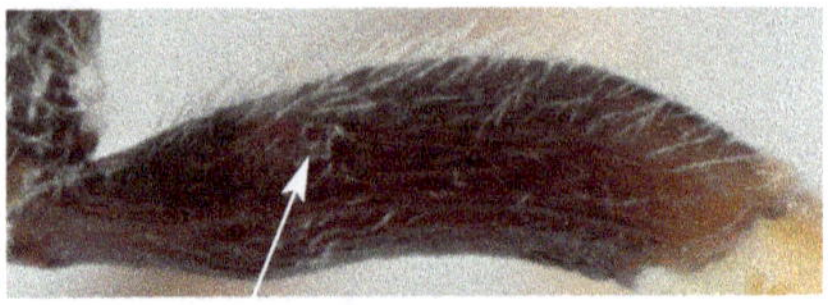

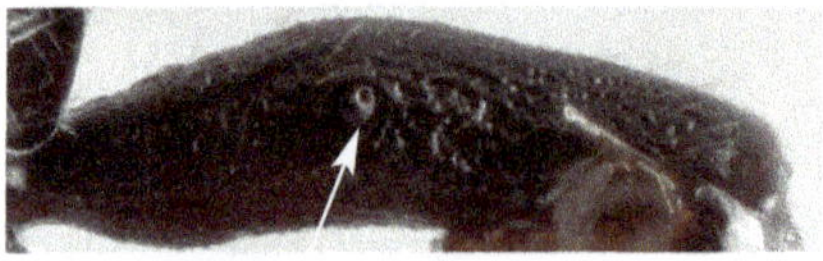

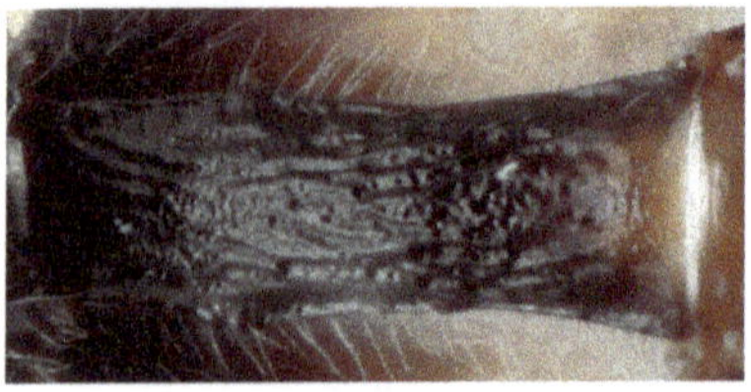

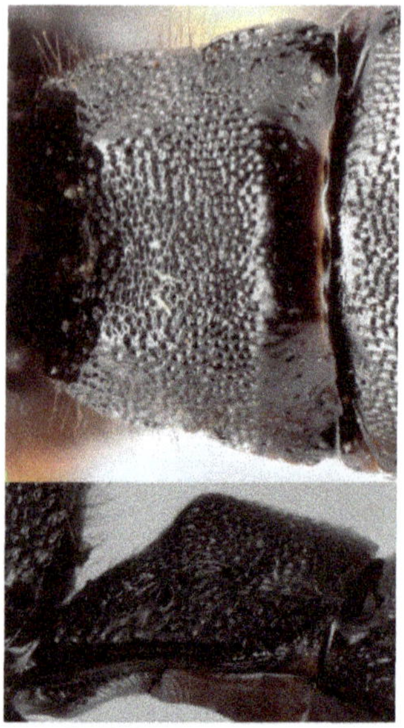

53(52). Metasomal sternites heavily sclerotized, metasoma rather cylindrical (**a**). Scutellum with long backwards-directed spine (**b**). Covered in dense, silvery pubescence (**a**, **c**) .... **Agriotypinae** (p. 119)
[One species in Britain (*Agriotypus armatus* Curtis), associated with flowing water.]

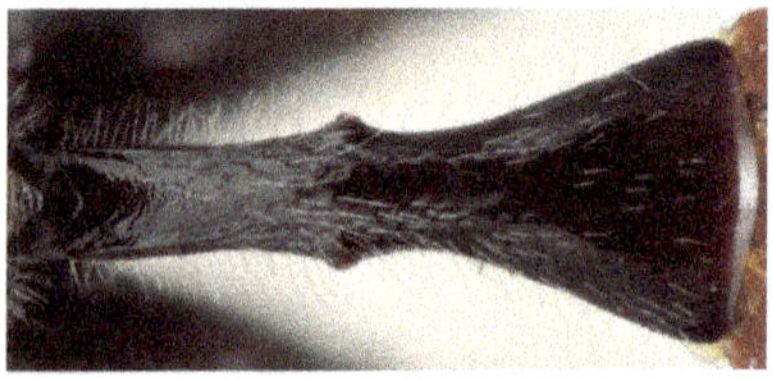

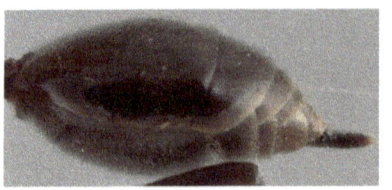

53 figures continued overleaf:

53 figures continued:

c

– Metasomal sternites (except anterior of 1 and the most posterior segments) very weakly sclerotized, almost membranous, and folded in under the tergites in dry pinned specimens, giving the metasoma a rather flattened appearance (**aa**). Scutellum lacking spine. Not densely pubescent, setae less conspicuous (**cc**) 54

aa

cc

54(53). Fore wing with areolet obliquely quadrate (**a**). Metasomal tergites from 2 onwards black or dark brown with contrastingly pale posterior bands (**b**) .... **Diacritinae** (p. 189)
[One, widespread, species in Britain (*Diacritus aciculatus* (Vollenhoven)).]

– Fore wing with areolet absent (**aa**). Metasomal tergites lacking contrastingly pale posterior bands (**bb, cc**) 55

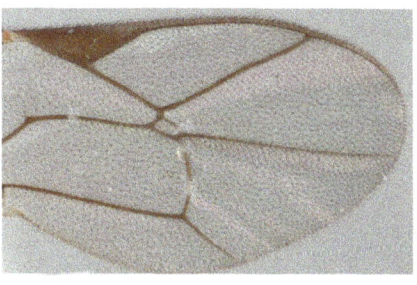
a

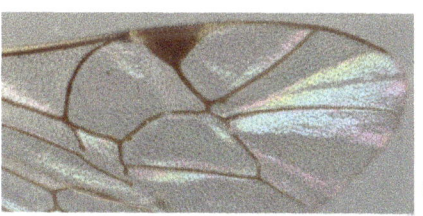
aa

b

bb

cc

55(54). Head, in anterior view, squared-off ventrally owing to sides of head ending ventrally in strong 'corners' at the mandible base (**a**); labrum barely visible and mandibles blunt-ended, weakly divided into two teeth or appearing unidentate (**b**). Hind tibia with apical, slanted row of dense setae (**c**). Mesoscutum with lateral carina ending anterior of scuto-scutellar groove (**d**). Metasomal tergite 1 with spiracles at mid-length (**e**). Ovipositor very short and mostly concealed within a large, triangular hypopygium (**f**) .... **Microleptinae** (p. 236)

– Head, in anterior view, not squared-off ventrally, eyes ventrally convergent (particularly noticeable in females) (**aa**); labrum protruding, semicircular and mandible twisted, strongly narrowed, divided into two sharp teeth (**bb**). Hind tibia without an apical row of dense setae (**cc**). Mesoscutum with lateral carina continuing across scuto-scutellar groove (**dd**). Metasomal tergite 1 with spiracles in anterior third (**ee**). Ovipositor projecting beyond metasomal apex, hypopygium short (**ff**) .... **Orthopelmatinae** (p. 254)

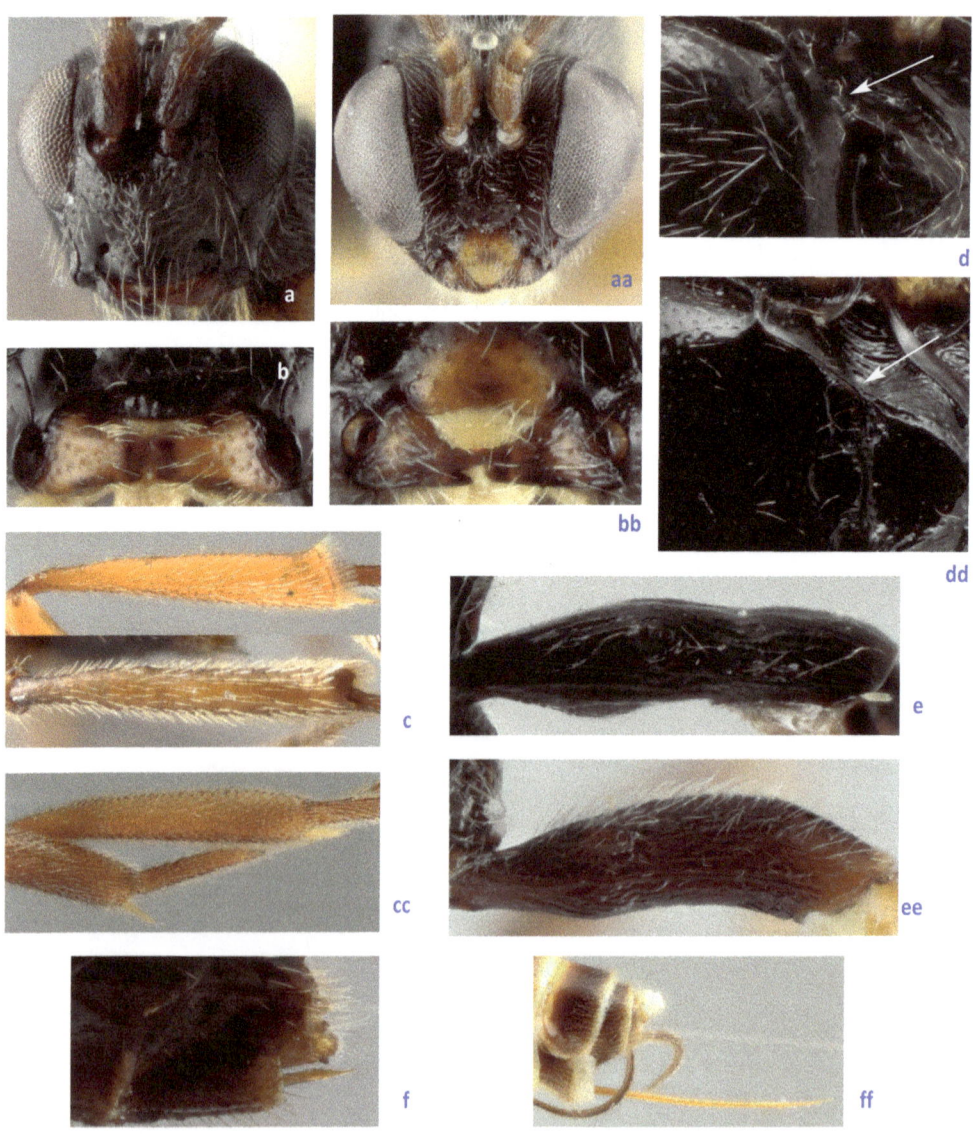

56(52). Female: antenna with first flagellar segment very long and slender, 7-10x as long as wide (**a**). Ovipositor long with an inconspicuous subapical notch (**b**). Male: antenna with tyloids in the form of deep, semi-circular notches on sub-proximal flagellar segments (**c**) .... **Cylloceriinae** (*Cylloceria*) (p. 186)

– Female: antenna with first flagellar segment not so elongate (**aa**) (ovipositor various) *or*, if elongate (**cc**), ovipositor with apical teeth and lacking subapical notch (**bb**). Male: antenna without deep, semi-circular notches on flagellar segments 57

57(56). Hind wing with nervellus intercepted well above the middle (**a**) 58

– Hind wing with nervellus intercepted about or below the middle or not intercepted (**aa**, **bb**).. 65

58(57). Epicnemial carina absent (**a**) 59

– Epicnemial carina present ventrally and usually laterally (**aa**) 60

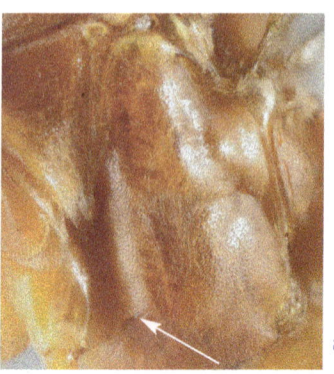

59(58). Tarsal claws pectinate (**a**). Mandible with upper tooth wider than lower tooth and indented so that mandible appears weakly tridentate (**b**). Maxillary palps often with distal segments expanded (**c**). Scutellum usually with a small, apical spine pointing backwards (**d**). Ovipositor very short, not projecting beyond the metasoma apex, with a dorsal, supapical notch (**e**) .... **Banchinae** (*Banchus*, *Rhynchobanchus*) (p.134)

– Tarsal claws simple or with a single accessory tooth (**aa**). Mandible unidentate (**bb**) or bidentate (**cc**). Maxillary palps simple. Scutellum lacking spine. Ovipositor projecting far beyond the metasoma apex, with distal teeth on lower valves (**ee**) .... most **Poemeniinae** (p. 283)

59 figures continued overleaf:

59 figures continued:

d

e

ee

60(58). Fore wing areolet roughly rhombic and strongly petiolate (**a**). Clypeus unsculptured, shiny, either long and flat (*Leptacoenites*) (**b**) or shorter and flattened laterally (*Coleocentrus*) (**c**). Female with ovipositor longer than metasoma ....
.... some **Acaenitinae** (*Coleocentrus* males, *Leptacoenites* females, males) (p. 111)

– Fore wing areolet absent, sessile or with short petiole, not of same shape (**aa**). Clypeus not so shiny, neither long and flat nor laterally flattened (**bb**). Female with ovipositor shorter ... 61

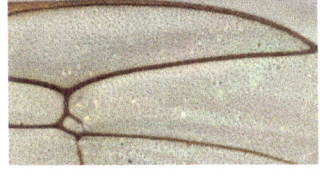
a

b

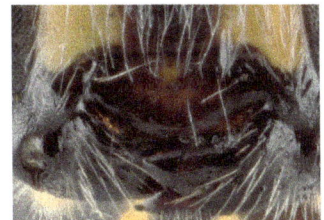
c

aa

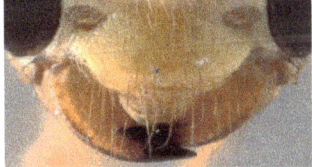
bb

61(60). Propodeum with strong, straight lateromedian and lateral longitudinal carinae but no transverse carinae (**a**). Clypeus with a weak median tooth (**c, d**). Frons with a longitudinal carina (**d**). Ovipositor down-curved and with very weak teeth along much of the ventral surface (**e**) .... **Collyriinae** (p. 82)

– Propodeum without strong lateral longitudinal carinae, with a transverse carina (**aa**), traces of area superomedia (**bb**), or entirely lacking carinae. Clypeus lacking tooth (**cc**). No longitudinal ridge between antennal sockets. Ovipositor straight or kinked downwards at the tip, with or without strong teeth apically, never with weak teeth along much of the ventral surface (**ee**) 62

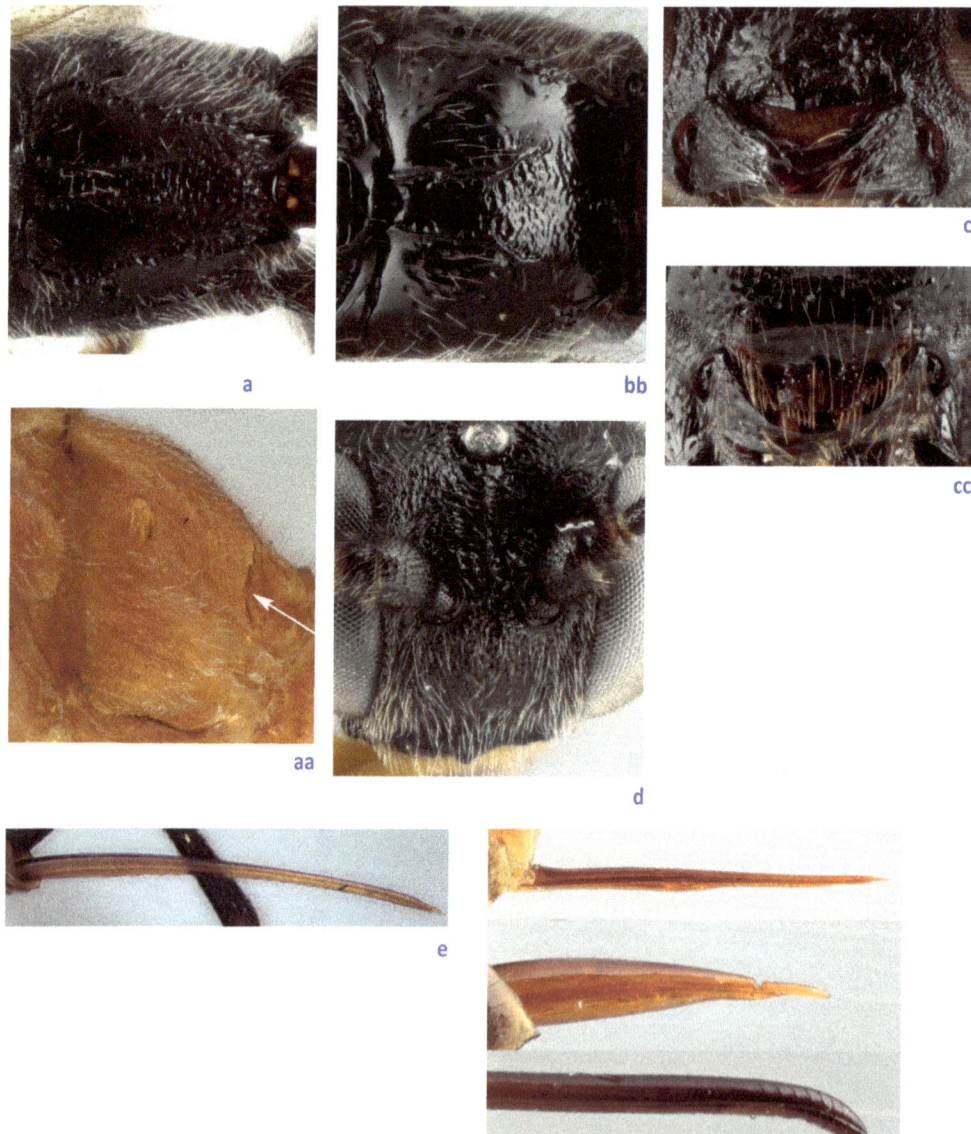

62(61). Mandibles strongly twisted, lower tooth scarcely or not visible when looking face-on (**a**). Metasomal tergite 1 with deep glymmae (**b**). Ovipositor short (not more than half metasoma length), simple, lacking notch or teeth, rigid (**c**). Claws with long, strong pectination (**d**). Male: genitalia large and extruded (**e**). Predominantly pale, orange-brown species .... **Tryphoninae** (Phytodietini: *Netelia*) (p. 305)

– Mandibles not twisted, both teeth visible when looking face-on (**aa**). Metasomal tergite 1 often with shallow glymmae (**bb**) or lacking them. Ovipositor with apical teeth ventrally or notch or nodus dorsally (**cc**). Claws simple or pectinate, *if* pectinate *then* pectination often shorter and sparser (**dd**). Male: genitalia less conspicuous (**ee**). Predominantly black species, with or without paler markings 63

63(62). Fore tibia with a small dorsal apical tooth on the outer margin (**a**). Ovipositor usually short, not longer than apical depth of metasoma, with dorsal notch (**b**). [Some Euryproctini with first metasomal segment lacking glymmae, long and narrow (**c**)] ... some **Ctenopelmatinae** (p. 175)

– Fore tibia without a small, dorsal apical tooth on the outer margin (**aa**). Ovipositor various ... 64

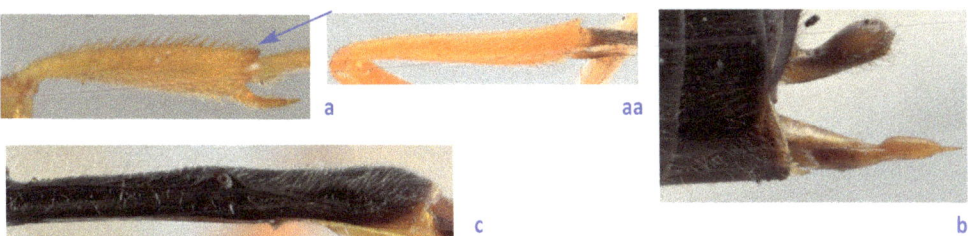

64(63). Metasomal tergites (from 2nd) each heavily sculptured anteriorly and with unsculptured posterior band (**a**). Clypeus flattened, shiny, often notched centrally (**b**). Ovipositor clearly extending beyond the metasomal apex, with ventral, apical teeth (**c**). Fore wing with areolet rather small, with first abscissa of *M* longer than second abscissa (**d**) .... some **Pimplinae** (p. 268)

– Metasomal tergites uniformly, weakly sculptured (**aa**). Clypeus flattened, usually strongly coriaceous, unnotched (**bb**). Ovipositor with subapical notch, usually short, not projecting much beyond metasomal apex (**cc**). Fore wing areolet large, with first abscissa of *M* shorter than second abscissa (**dd**) .... **Banchinae** (*Exetastes*) (p. 134)

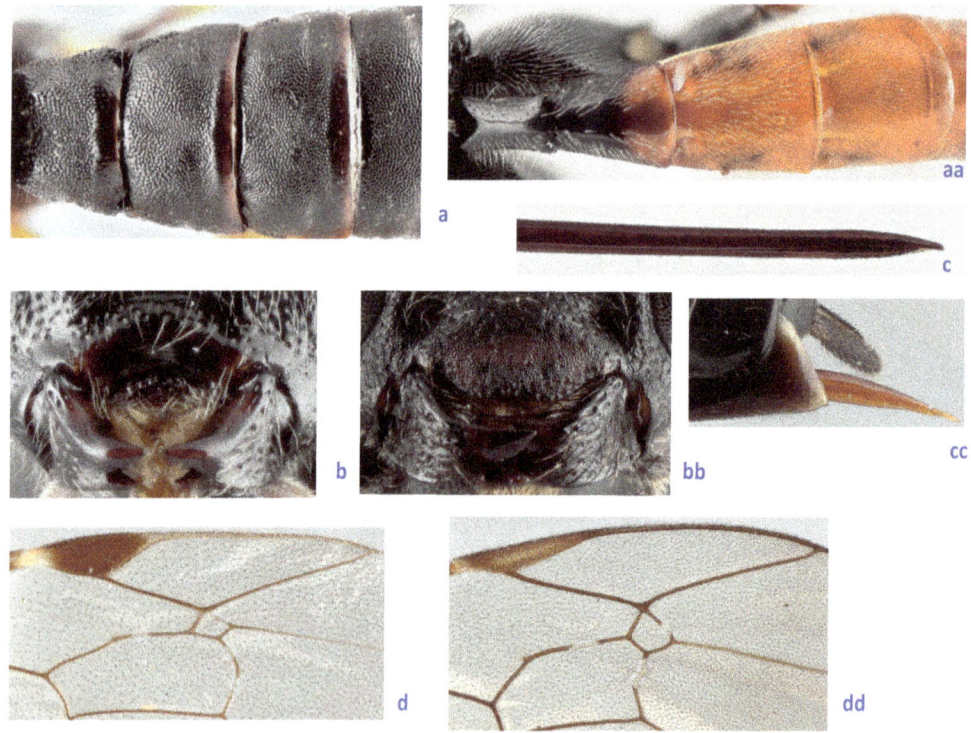

65(57). Mandible with upper tooth divided, thus appearing tridentate (**a**) [fore tibia lacking dorsal apical tooth on the outer margin and pronotal epomia absent]. Metasomal tergite 1 straight-sided from near base to apex, often square (**b**), sometimes rectangular (**c**) .... **Diplazontinae** (p. 191)

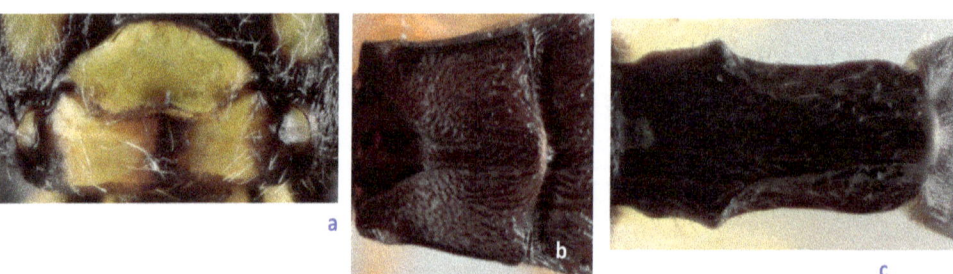

– Mandible with upper tooth undivided, mandible bidentate (**aa**), rarely unidentate [*if* appearing vaguely tridentate *then* fore tibia with a dorsal apical tooth on the outer margin, pronotal epomia present, and metasomal tergite 1 long and narrow]. Metasomal tergite 1 rarely square, often widened apically (**bb, cc, dd**) 66

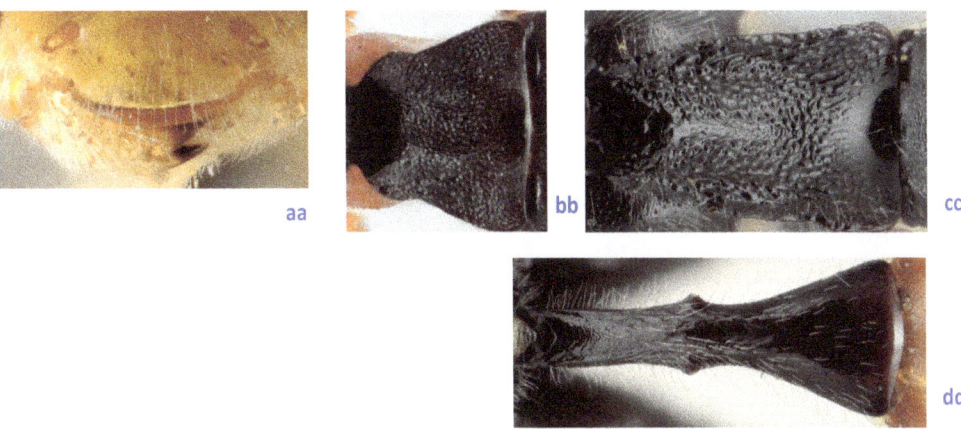

66(65). Propodeum lacking dorsal carinae except for strong, evenly curved posterior transverse carina (**a**). Submetapleural carina strong, usually expanded as a wide lobe anteriorly (**c**). Clypeus strongly convex sub-medially, flatter apically (**d**) .... **Banchinae** (Atrophini) (p. 134)

– Propodeum various but never with just a strong, evenly curved posterior transverse carina (**aa, bb**). Submetapleural carina usually not expanded as a lobe anteriorly, or as much smaller lobe (**cc**). Clypeus various, often more evenly rounded (**dd**) 67

67(66). Tarsal claws pectinate (**a**) 68

– Tarsal claws not pectinate (but may have basal lobes (**aa**)) 71

a

aa

68(67). Propodeum with only sections of the posterior transverse carina or no carinae (**a**). Fore tibial spur long with comb only reaching to half its length (**b**). Female with ovipositor with a conspicuous nodus (**c**). Tibiae with numerous spines, stronger than surrounding setae (**d**). Fore wing with areolet triangular (**e**).... **Tryphoninae** (Phytodietini: *Phytodietus*) (p. 305)

– Propodeum with at least traces of anterior transverse and longitudinal carinae (**aa**). Fore tibial spur with comb reaching more than half its length (**bb**). Female with ovipositor lacking nodus (**cc**). Tibiae usually not spinose (**dd**). Fore wing with areolet usually quadrate (**ee**, **ff**) or 3*rs-m* missing 69

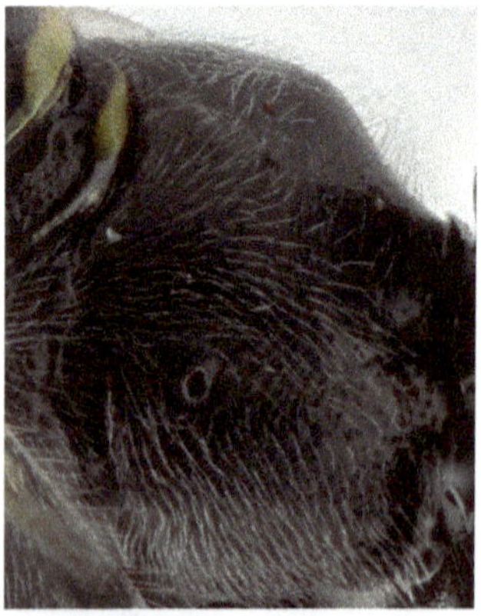

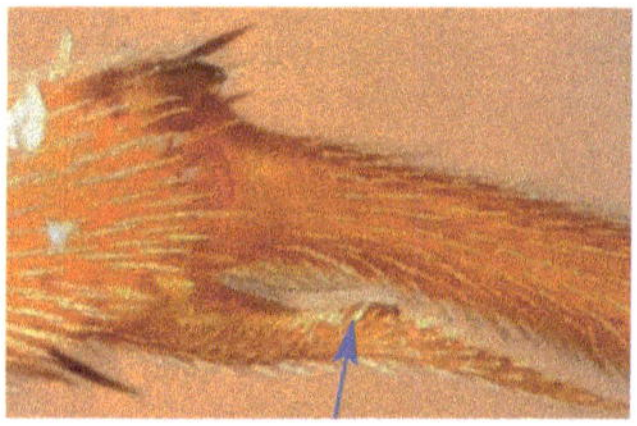

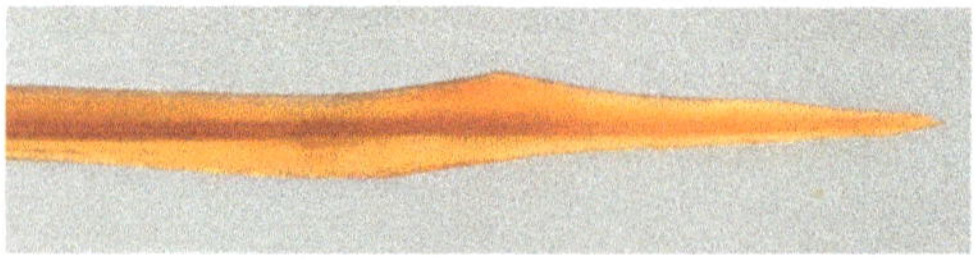

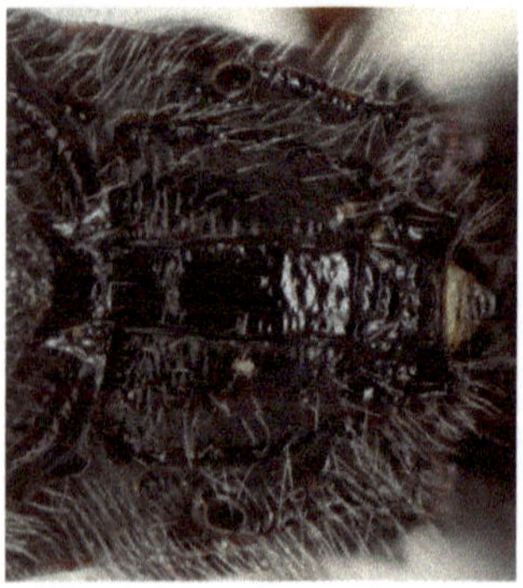

68 figures continued overleaf:

68 figures continued:

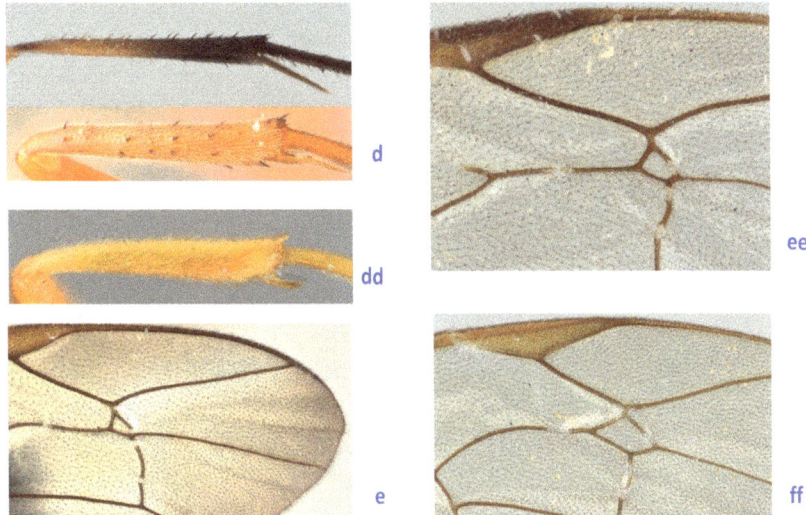

69(68). Mandible thin, narrowed apically (**a**). First metasomal tergite very wide posteriorly (**b**). Clypeus about as wide as deep, apically slightly flattened and shiny (**a**). Ovipositor with *dorsal* subapical teeth (**d**) .... **Stilbopinae** (*Panteles*) (p. 293)
[One rare species in Britain and Ireland (*Panteles schuetzeanus* (Roman)); male with face entirely yellow.]

– Mandible broader (**aa**). First metasomal tergite narrower posteriorly (**bb**). Clypeus often wider (**cc**), various shapes. Ovipositor lacking dorsal teeth (**dd**) 70

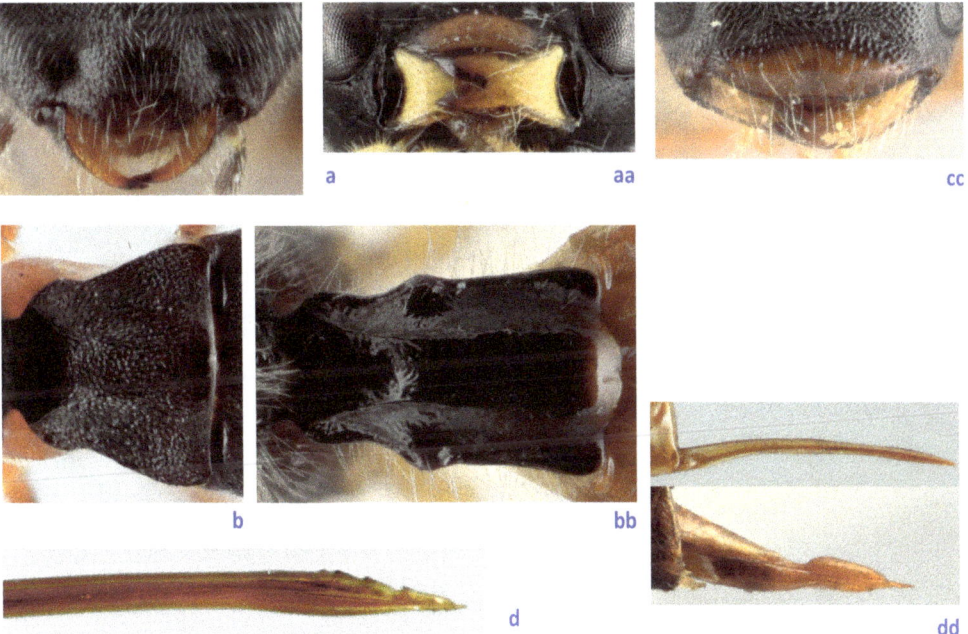

70(69). Fore tibia with a dorsal apical tooth on the outer margin (**a**). Clypeus lacking apical row of evenly spaced setae, clypeus often with flattened lateral lobes (**b**). Fore wing vein *2m-cu* with one bulla (**c**). Ovipositor with dorsal, subapical notch (**d**) or very thin over most of its length (**e**) .... some **Ctenopelmatinae** (p. 175)

– Fore tibia without a dorsal apical tooth on the outer margin (**aa**). Clypeus with apical row of evenly spaced setae, clypeus lacking lateral flattened lobes (**bb**). Fore wing vein *2m-cu* with two bullae (**cc**). Ovipositor lacking notch and/or not so thin (**dd**) ....
.... **Tryphoninae** (some Tryphonini) (p. 305)

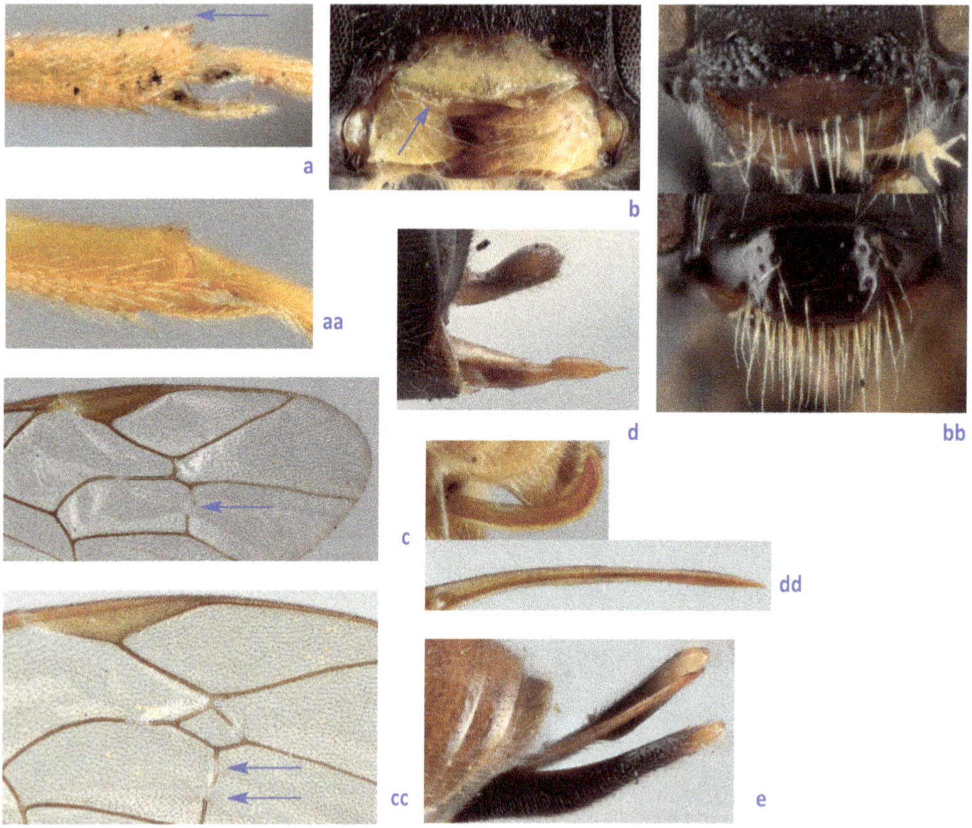

71(67). Fore tibia with a dorsal apical tooth on the outer margin (**a**) 72
[The fore tibial tooth can be difficult to discern or hardly present on small ctenopelmatines, so these specimens can still be keyed out later in the key.]

– Fore tibia without a dorsal apical tooth on the outer margin (**aa**) 73

72(71). Antenna with scape and pedicel same size (**a**); flagellum with 14 segments. Metasomal tergite 2 with laterotergite not separated by a crease, smoothly curved over (**b**). Fore wing vein 2*rs-m* almost obliterated and marginal cell very short (**c**) ... **Tersilochinae** (*Pygmaeolus*)(p. 298)
[One rare species in Britain (*Pygmaeolus nitidus* (Bridgman)).]

– Antenna with scape longer than pedicel (**aa**); flagellum with 16 or more segments. Metasomal tergite 2 with laterotergite separated by a crease (**bb**). Fore wing vein 2*rs-m* longer and marginal cell longer (**cc**) .... most **Ctenopelmatinae** (p. 175)

73(71). Fore wing vein 2*m-cu* with one bulla (**a**) 74

– Fore wing vein 2*m-cu* with two bullae (**aa**), or one long bulla and the vein angled at this point (**bb**) 83

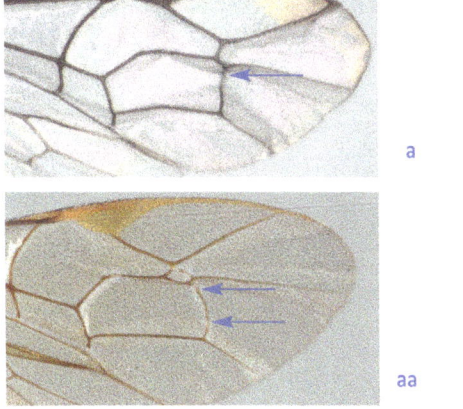
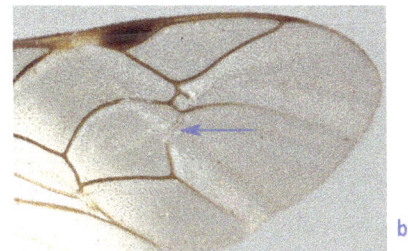

74(73). Female (if ovipositor not visible, hypopygium is obvious in outline and metasomal apex appears to be enclosed in sclerotized tergites and sternites (**a**)) 75

– Male 78

75(74). Hypopygium large, reaching almost to the metasomal apex; ovipositor very short, hardly extending beyond the metasomal apex (**a**, **b**) 76

– Hypopygium small, not reaching metasomal apex (**aa**); *or*, if hypopygium large, roughly triangular in outline, ovipositor extending obviously beyond the metasomal apex (**bb**) ... 78

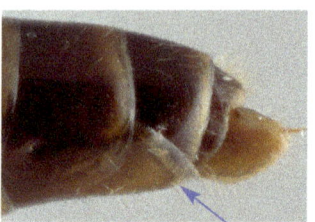

76(75). Maxillary palps elongate, able to reach far beyond fore coxae (**a**). Clypeus subtly flattened apically; labrum not exposed (**b**). Ovipositor sheaths as wide as long (**d**).....................
.... **Oxytorinae** (p. 257)

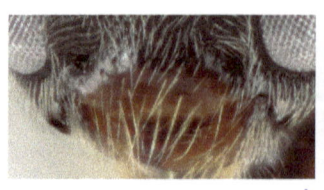

– Maxillary palps shorter, not reaching beyond fore coxae (**aa**). Clypeus rather evenly convex, slightly concave apically with labrum exposed, with (**bb**) or without (**cc**) an impressed edge. Ovipositor sheaths narrower (**dd**) 77

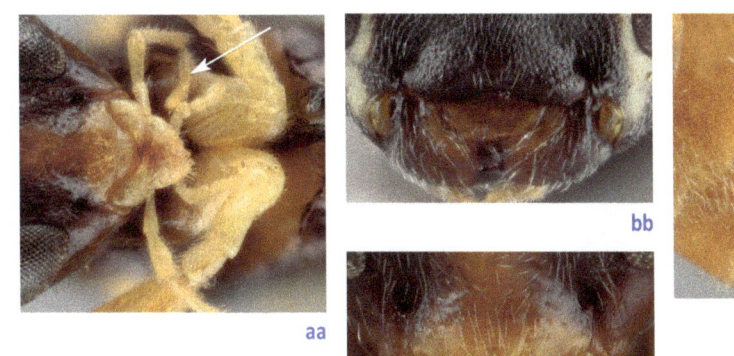

77(76). Apical edge of clypeus more strongly concave, clypeus with narrow apical flange; mandibles not twisted (**a**). Propodeum with wider area petiolaris (lateromedian longitudinal carinae strongly diverging) (**b**) .... **Cryptinae** (*Sphecophaga*)(p. 165)
[One species in Britain (*Sphecophaga vesparum* (Curtis)), a parasitoid of Vespinae pupae in their nests.]

– Apical edge of clypeus very weakly indented medially, without narrow apical flange; mandibles down-curved, slightly twisted with lower tooth 0.5 times length of upper tooth (**aa**). Propodeum with narrower area petiolaris (lateromedian longitudinal carinae weakly diverging) (**bb**) ....**Cryptinae** (*Hemiphanes*) (p. 165)

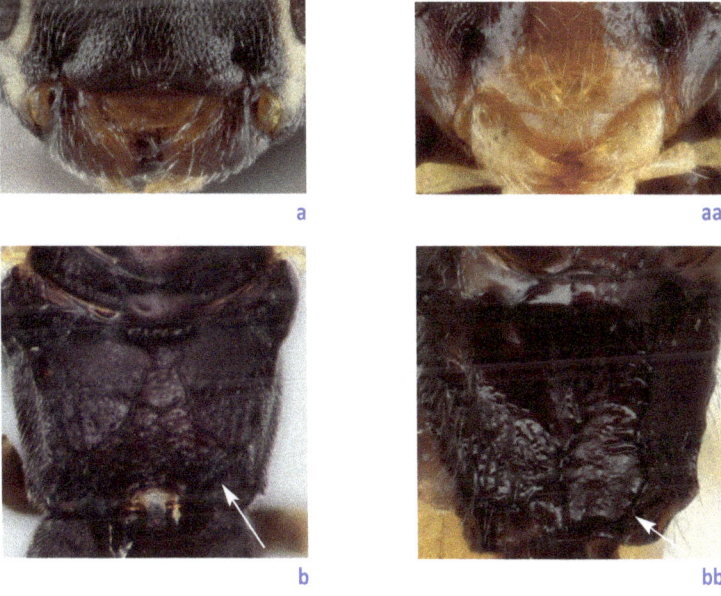

78 (74,75). Propodeum with no longitudinal carinae, only a single transverse carina which is 'V'-shaped (**a**). Female fore tibia inflated (**b**) and ovipositor with conspicuous teeth (**c**). Male antenna with raised, tooth-like tyloids on several flagellar segments (starting at 10th flagellar segment) (**d**) .... **Cryptinae** (*Helcostizus*) (p. 165)

[One species in Britain, *Helcostizus restaurator* (Fabricius), a parasitoid of wood-boring beetles.]

– Propodeum usually with at least traces of longitudinal carinae, never just one 'V'-shaped transverse carina (**aa**). Female fore tibia not inflated (**bb**) and ovipositor with subapical dorsal notch or simple (**cc**). Male antenna lacking raised, tooth-like tyloids (may have flattened areas) 79

79(78). Female. [Fore tibia with an apical, distal tooth that may be difficult to see. Clypeus either uniformly convex or raised sub-apically (**a**)] .... some **Ctenopelmatinae** (p. 175)

– Male 80

80(79). Clypeus uniformly convex, mandible down-curved, slightly twisted and lower tooth 0.5 times length of upper tooth; apical edge of clypeus truncate, slightly indented medially (**a**) .... **Cryptinae** (*Hemiphanes*) (p. 165)

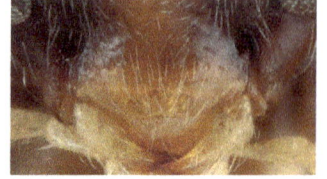

- Clypeus either uniformly convex (**aa**), with subapical swelling (**bb**) or abruptly declivous apically (**cc**); mandible straight and lower tooth approximately same length as upper tooth or slightly shorter or longer; apical edge of clypeus often with narrow (**bb**) or laterally wide (**cc**) flange 81

  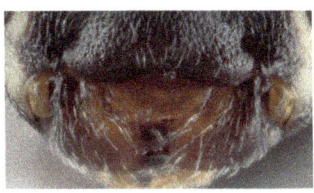

81(80). Apical edge of clypeus weakly concave, clypeus with the apical third abruptly declivous (**a**) ....
.... **Cryptinae** (*Sphecophaga*) (p. 165)

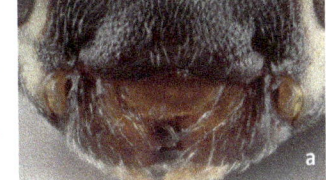

– Apical edge of clypeus straight, clypeus uniformly convex (**aa**) or raised basally (**bb**) or subapically (**cc**) 82

 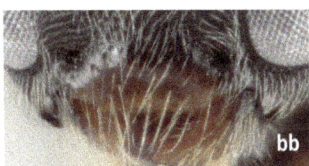 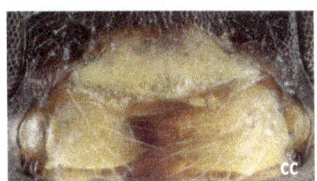

82(81). Maxillary palps elongate, able to reach far beyond fore coxae (**a**). Clypeus convex basally, otherwise flat (**b**). First metasomal tergite without a glymma (**c**) ... **Oxytorinae** (*Oxytorus*) (p. 257)

– Maxillary palps not or barely able to reach beyond fore coxae (**aa**). Clypeus either uniformly convex (**bb**) or raised sub-apically (**cc**). First metasomal tergite with or without a glymma .... some **Ctenopelmatinae** (p. 175)

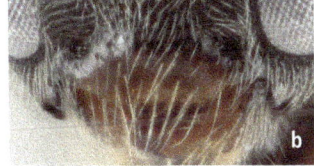  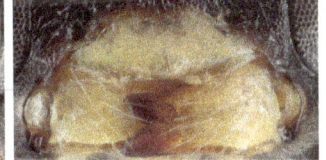

83(73). Mesoscutum with a transverse suture immediately anterior of scuto-scutellar groove and often with median longitudinal carina across scuto-scutellar groove (**a**). Metasomal first tergite and sternite fused, glymmae absent (**b**). With one of the following characters: hind femur with a large ventral tooth (**c**) *or* mandible with a single tooth (**d**) (female also with pegs projecting from subapical flagellar segments (**e**)) *or* frons with a median horn (**f**) [Claws without a basal lobe. Females with long ovipositor bearing apical ventral teeth] ... . **Xoridinae** (p.321)

– Mesoscutum without a transverse suture anterior of scuto-scutellar groove and scuto-scutellar groove without carinae (**aa**). Hind femur never with a ventral tooth. Frons never with a median horn. If, rarely, mandible apparently unidendate, first metasomal tergite with glymmae and not fused with sternite (**bb**) 84

84(83). Metasomal tergites (from 2nd) each heavily sculptured anteriorly and with unsculptured posterior band (**a**, **b**). Clypeus flattened, often notched centrally (**c**). Ovipositor without a notch, with ventral teeth apically, sometimes with a nodus; ovipositor sheaths with long, strong, dense setae, appearing bristly (**d**) ....
.... some **Pimplinae** (some Ephialtini, Delomeristini) (p. 268)

– Metasomal tergites usually uniformly sculptured, lacking posterior band of differentiated sculpture (**aa**). Clypeus usually convex (**bb**), seldom notched centrally (**cc**). Ovipositor lacking ventral teeth apically, usually lacking nodus, sometimes with a notch; ovipositor sheaths often much less 'bristly' (setae shorter and/or sparser) (**dd**) 85

85(84). Clypeus wide and very short; mandibles broad, with carinae along dorsal and ventral edges (**a**). Ovipositor longer than metasoma and with weak dorsal nodus (**b**) .... **Tryphoninae** (Idiogrammatini) (p. 305)

[One rarely collected species, *Idiogramma euryops* Förster.]

– Clypeus much narrower and higher; mandibles narrower and lacking carinae along dorsal and ventral edges (**aa**). Ovipositor often shorter than metasoma, lacking nodus (**bb**) 86

86(85). Female 87

– Male 92

87(86). Ovipositor without a dorsal subapical notch (**a**, **b**) 88

– Ovipositor with a dorsal subapical notch (**aa**) [If the ovipositor cannot be seen as it is very short and concealed by the sheaths, go to 90] 90

88(87). Claws with basal lobe (**a**, **b**); fifth tarsal segment conspicuously broadened, arolium projecting beyond claws (**b**) *or* ovipositor relatively stout, upcurved, ovipositor sheaths with long, dense setae (**c**); 2nd and 3rd metasomal tergites with elevated median areas (**e**) .... some **Pimplinae** (Ephialtini) (p. 268)

– Claws lacking lobe; fifth tarsal segment not conspicuously broadened, arolium not projecting beyond claws (**aa**). Ovipositor straight or, if ovipositor upcurved, slender, ovipositor sheaths with short setae (**cc**) *or* ovipositor stout, no longer than height of metasomal apex, and abruptly upcurved apically (**dd**); 2nd and 3rd metasomal tergites lacking median raised areas (**ee**) 89

a

b

aa

c

cc

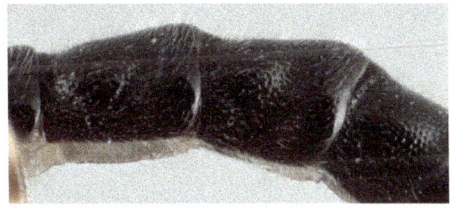
e

ee

dd

89(88). Clypeus small, about as wide as long, convex but apex slightly flattened and shiny, lacking apical row of evenly spaced, robust bristles (**a**). Apical flagellar segment conspicuously longer and a little wider than preceding segments (**b**). Hypostomal and occipital carinae meet at mandibular base (**c**), malar space usually elongate, about as long as mandible width at its base (**a**). Face often covered in long, dense, silvery setae (**c**). Ovipositor evenly sclerotized (**d**) .... **Stilbopinae** (*Stilbops*) (p. 293)

– Clypeus usually wider than long, not apically flattened and shiny, with apical row of evenly spaced, robust bristles (**aa**). Apical flagellar segment not conspicuously longer nor wider than preceding segments (**bb**). Hypostomal and occipital carinae usually meet away from mandibular base, malar space not elongate, shorter than mandible width at its base (**cc**). Face not covered in long, dense, silvery hairs (**cc**). Ovipositor sometimes with weakly sclerotized area on lower valves (tribe Oedemopsini) (**dd**) .... some **Tryphoninae** (p. 305)

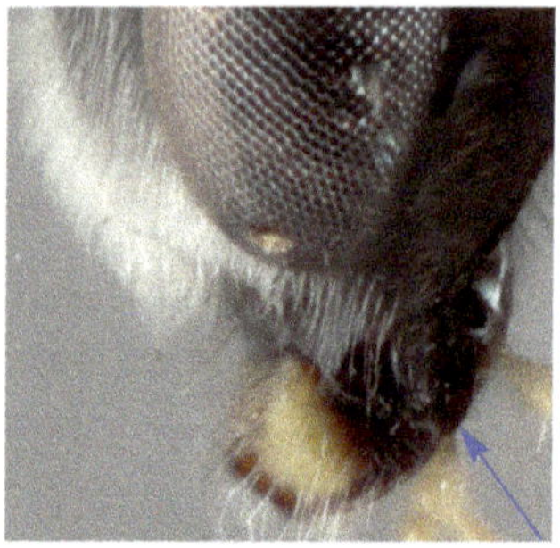

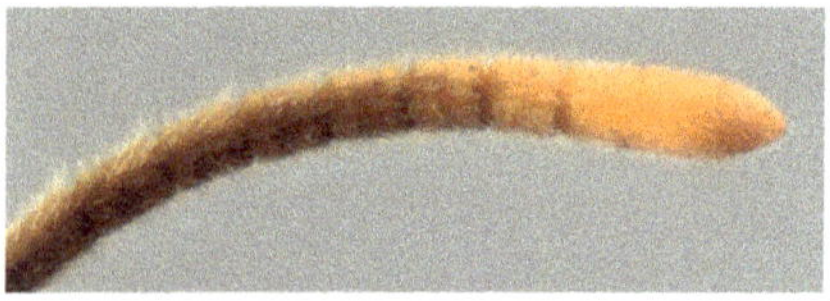

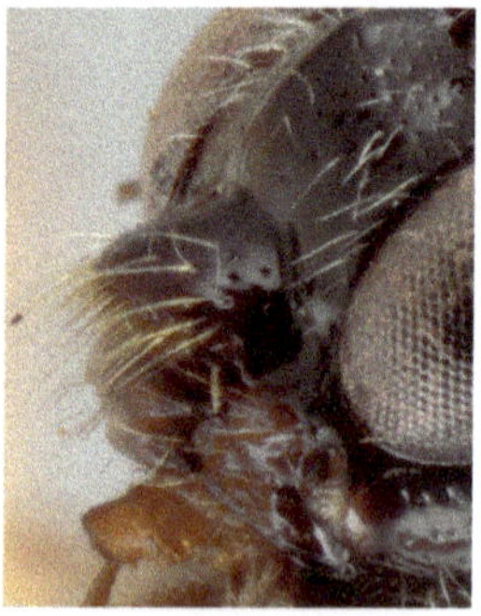

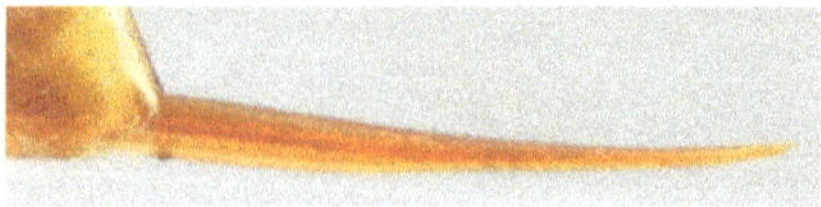

90(87). Mandibles broad, not narrowed, lower tooth not much shorter than upper tooth, same length or longer (**a**). Hind tibia usually without apical, inner fringe of closely spaced setae. [Fore tibia with an apical, distal tooth that may be difficult to discern] ....
.... **Ctenopelmatinae** (p. 175)
[A few Mesoleiini and Euryproctini potentially.]

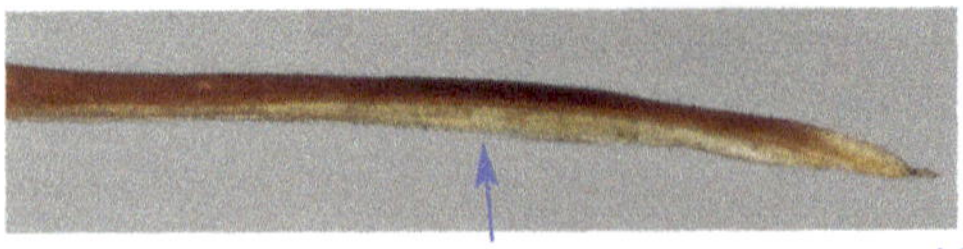

- Mandibles thin and apically narrowed, often with lower tooth shorter than upper (**aa**). Hind tibia with apical, inner fringe of closely spaced setae (**bb**) 91

aa

bb

91(90). Face with yellow marks on inner orbits, adjacent to antennal sockets (**a**). 1st and 2nd metasomal tergites with coriaceous sculpture (**b**). Ovipositor upcurved (**c**). Face lacking subocular sulcus and clypeus wider than high (**e**) .... **Cylloceriinae** (*Allomacrus*)(p. 186)
[One, fairly widespread, species in Britain (*Allomacrus arcticus* (Holmgren)).]

- Face lacking discrete yellow marks on inner orbits, although may have extensively pale face (**aa**). Without combination of coriaceous sculpture on first and second metasomal tergites (**bb**) and upcurved ovipositor (**cc**) *or*, if 1st and 2nd metasomal tergites with coriaceous/striate sculpture (**dd**) *then* malar space with discrete subocular sulcus and clypeus about as wide as high (**ee**) .... some **Orthocentrinae** (p. 248)

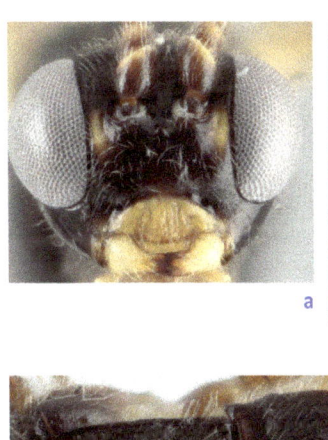
a

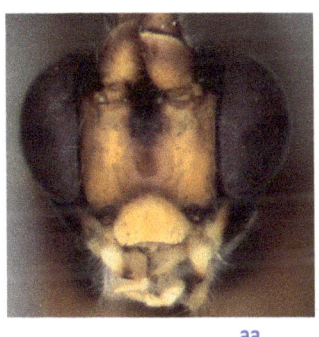
aa

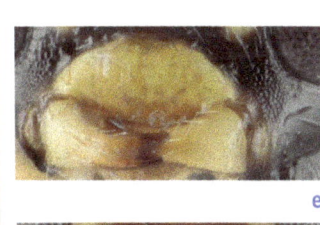
e

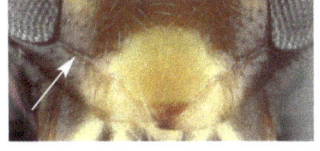
ee

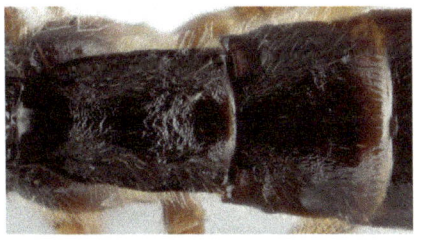
b

bb

c

dd

cc

92(86). Metasomal tergites (from 2nd) each with unsculptured posterior band contrasting with more strongly sculptured anterior area (**a**). Tarsal claws often with basal lobes (**b**) [Aedeagus dorso-ventrally flattened, smoothly curved (**c**)] .... some **Pimplinae** (p. 268)

– Metasomal tergites without posterior unsculptured bands contrasting with more strongly sculptured anterior areas (**aa**). Tarsal claws never with basal lobes (**bb**) 93

93(92). Mandibles thin and apically narrowed, lower tooth often shorter than upper tooth (**a**). Hind tibia with apical, inner fringe of closely spaced setae (**b**). Aedeagus dorso-ventrally flattened, smoothly curved (**c**) 94

– Mandibles not as narrowed apically (**aa**), lower tooth often the same length as upper tooth, sometimes shorter or longer. Hind tibia without apical, inner fringe of closely spaced setae (**bb**). Aedeagus not flattened, with thickened, angled apex (**cc**) 95

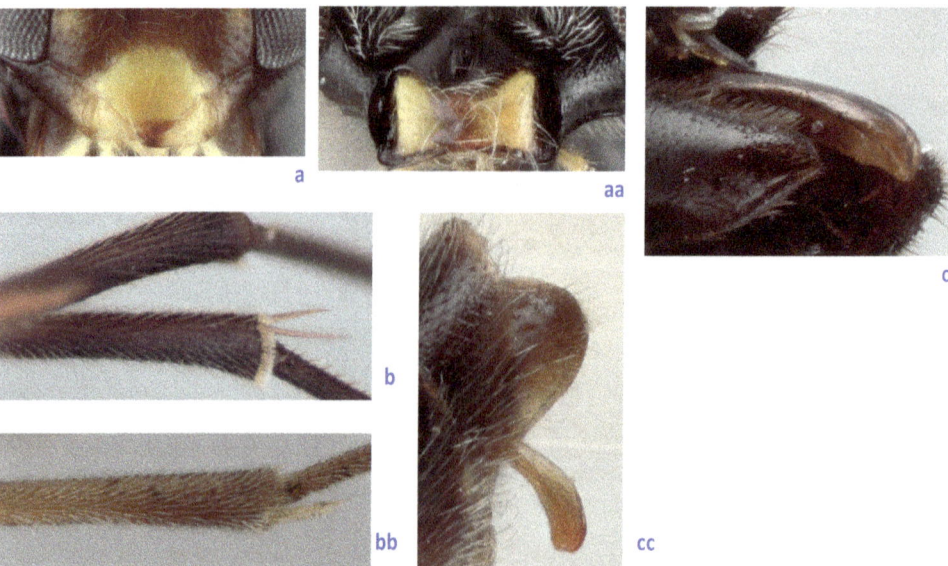

94(93). Clypeus wide and flat (**a**). Propodeum and first and second metasomal tergites with coriaceous sculpture (**b**). Face with yellow marks on inner orbits adjacent to antennal sockets, or mostly yellow (**c**) .... **Cylloceriinae** (*Allomacrus*) (p. 186)

– Clypeus narrower and often strongly convex (**aa**). *If* clypeus wider and flattened (**bb**) *then* propodeum and first and second metasomal tergites lacking coriaceous sculpture (**cc**, **dd**) and face lacking yellow marks (**bb**) (these characters otherwise variable) .... some **Orthocentrinae** (p. 248)

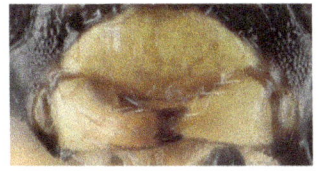

a

aa

b

c

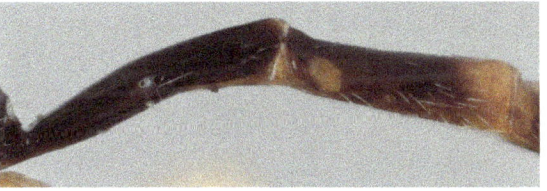

bb

cc

dd

95(93) Clypeus small, about as wide as high, flattened apically and shiny (**a**). Fore wing vein 1*cu-a* strongly inclivous and separated from *M&RS* by 1/3 or more of the length of 1*cu-a* (**b**). Metasoma with at least 1st tergite predominantly punctate (**c**). Face often covered in dense, silvery setae (**d**) .... **Stilbopinae** (*Stilbops*) (p. 293)

– Clypeus wider, not with the same combination of being shiny and flattened apically (**aa**). Fore win vein 1*cu-a* vertical or weakly inclivous and separated from *M&RS* by less than 1/3 of the length of 1*cu-a* (**bb**). Metasoma usually unsculptured or weakly coriaceous or punctate, occasionally heavily punctate or rugose-striate (**cc**). Face with sparse setae, not unusually dense (**dd**) 96

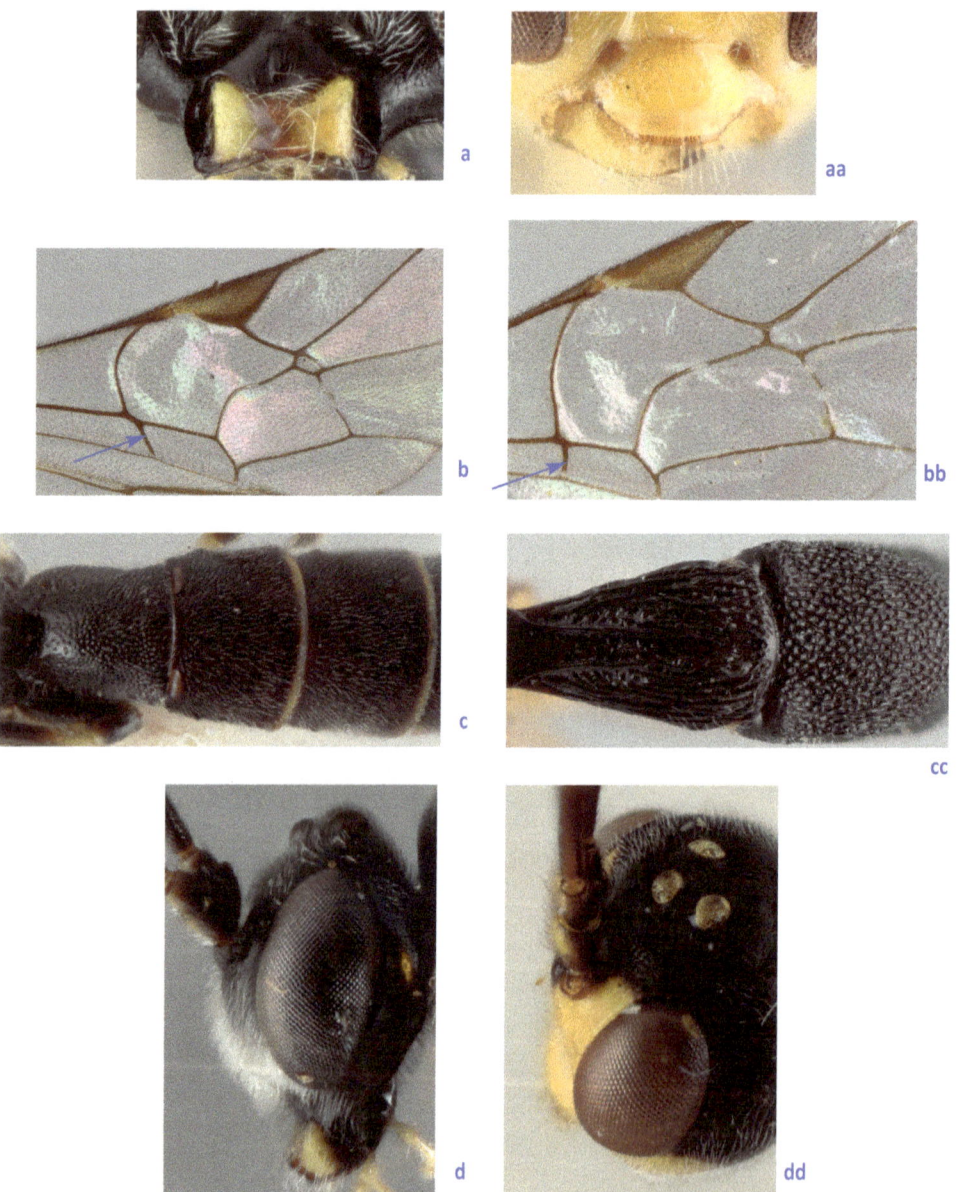

96(95). Notauli strong, usually extending to middle of mesoscutum (**a**). Metasoma sometimes heavily sculptured (**b**) .... **Tryphoninae** (Oedemopsini and Eclytini) (p. 305)

– Notauli weak or absent (**aa**). Metasoma weakly sculptured or unsculptured (**bb**) 97

a

aa

b

bb

97(96). Clypeus with an apical row of closely spaced, erect setae (**a**). Fore wing sometimes with an elongate, angled bulla in the half of 2*m-cu* directly below areolet (**b**) .... some **Tryphoninae** (Tryphonini) (p. 305)

– Clypeus without an apical row of closely spaced setae (**aa**). Fore wing with a shorter, straight bulla in half of 2*m-cu* below areolet (**bb**). [Fore tibia with an apical, distal tooth that may be difficult to discern (**cc**)] .... **Ctenopelmatinae** (p. 175)
[A few Mesoleiini and Euryproctini potentially.]

a

aa

cc

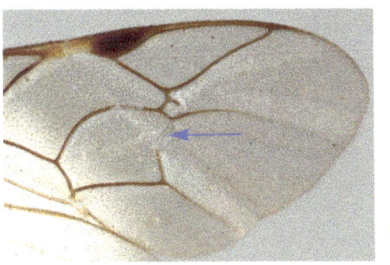
b

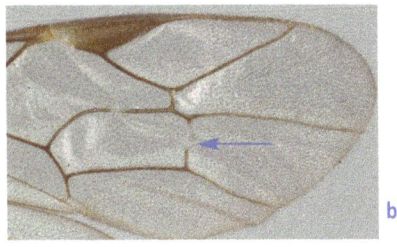
bb

98(1). Clypeal sulcus absent, the whole face and clypeus strongly convex; antennal scape more than twice as long as apically wide (**a**). Spiracle of metasomal tergite 1 anterior to midlength, tergite not narrowed except at very anterior end (**b**). Small species, c.2.5 mm body length .... **Orthocentrinae** (*Stenomacrus*) (p. 248)

– Clypeal sulcus distinct, face in profile flat or slightly convex without or with a broad subocular sulcus (**aa**), scape much shorter. Spiracle of metasomal tergite 1 usually in posterior third and tergite narrowed anteriorly (**bb**). Usually significantly larger species, a few *Gelis* as small as 2.5 mm 99

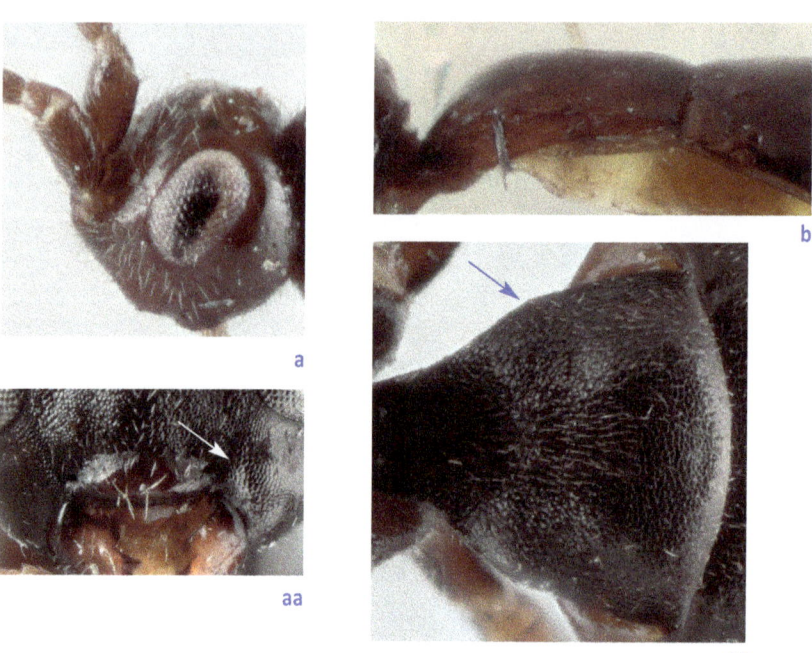

99(98). Clypeus wide and flat, apical edge truncate (**a**). Thyridia of metasomal tergite 2 deeply impressed at anterior edge (**b**). Brachypterous females (**c**) .... **Ichneumoninae** (p. 204)
[Only some ♀♀s of *Ichneumon oblongus* Schrank should key out here.]

– Clypeus not more than 1.5x as wide as deep, convex, apical edge with a narrow flange (**aa**). Thyridia of metasomal tergite 2 small, not deeply impressed (**bb**). Brachypterous or micropterous or apterous (**cc**), male or female 100

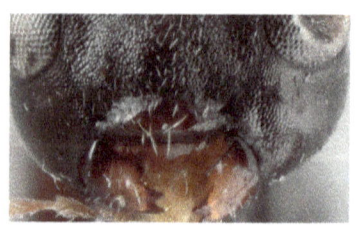

99 figures continued overleaf:

99 figures continued:

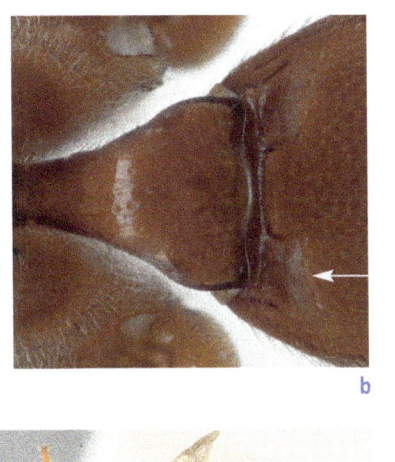

b

bb

c

cc

100(99). Apterous, with no remnants of wings (**a**) .... some **Phygadeuontinae** (p. 260)
[Apterous individuals will always belong to *Gelis*, *Thaumatogelis* or *Polyaulon*, most specimens will be found to be *Gelis* species. Schwarz (2001, 2002b) keys out females of *Thaumatogelis* and *Gelis*, respectively. Schwarz (1995) keys out the genera with apterous females.]

– Brachypterous or micropterous, wings present but reduced (**aa**) ....
.... some **Cryptinae** (p. 165) and **Phygadeuontinae** (p. 260)
[Both cryptine tribes and Phygadeuontinae have brachypterous representatives and the subfamilies are not easy to separate due to simplification of the mesosoma concomitant with wing reduction. Horstmann (1993) provides keys to the genera and species of both subfamilies (not treated as such then) with brachypterous females. Schwarz (1994) provides an updated key to brachypterous females of *Gelis*.]

a

aa

List of the species illustrated in the keys

Key for the Separation of British and Irish Braconidae and Ichneumonidae

1a: *Panteles schuetzeanus* (Roman) (Stilbopinae)
1aa: *Gelis melanocephalus* (Schrank)
  (Phygadeuontinae)
1bb: *Agrothereutes abbreviatus* (Fabricius)
  (Cryptinae: Cryptini)
2a: *Buathra laborator* (Thunberg) (Cryptinae:
  Cryptini)
2b: *Aclastus solutus* (Thomson) (Phygadeuontinae)
2aa: *Zele albiditarsus* Curtis (Braconidae:
  Meteorinae)
2bb: *Hybrizon buccatus* (de Brébisson)
  (Hybrizontinae)
2cc: *Neorhacodes enslini* (Ruschka) (Neorhacodinae)
3a-b: *Zele albiditarsus*
3aa: *Dusona stragifex* (Förster) (Campopleginae)
3bb: *Buathra laborator*
3cc: *Hybrizon buccatus*
3dd: *Aphidius colemani* Viereck (Braconidae:
  Aphidiinae)
4a: *Aphidius colemani*
4aa: *Hybrizon buccatus*
5a: *Gelis melanocephalus*
5aa: *Stenomacrus pedestris* (Holmgren)
  (Orthocentrinae)
5bb: *Spathius pedestris* Wesmael (Braconidae:
  Doryctinae)
6a: *Stenomacrus pedestris*
6aa: *Chasmodon apterus* (Nees) (Braconidae:
  Alysiinae)
6bb: *Spathius pedestris*

Key for the Identification of British and Irish subfamilies of Ichneumonidae

1a: *Panteles schuetzeanus*
1aa: *Gelis melanocephalus*
1bb: *Agrothereutes abbreviatus*
2a: *Hybrizon buccatus*
2b: *Neorhacodes enslini*
2c: *Aclastus solutus*
2aa: *Buathra laborator*
3a: *Aclastus solutus*
3b: *Sathropterus pumilus* (Holmgren) (Tersilochinae)
3aa: *Neorhacodes enslini*
3bb: *Hybrizon buccatus*
4a-b: *Sathropterus pumilus*
4aa-bb: *Aclastus solutus*
5a-c: *Neorhacodes enslini*
5aa-cc: *Hybrizon buccatus*
6a: *Rhyssa persuasoria* (L.) (Rhyssinae)
6aa: *Oedemopsis scabricula* (Gravenhorst)
  (Tryphoninae: Oedemopsini)
7a-b: *Rhyssa persuasoria*
7aa-bb: *Pseudorhyssa alpestris* (Holmgren)
  (Pimplinae: Delomeristini)
8a: *Alomya debellator* (Fabricius) (Alomyinae)
8b-d: *Alomya semiflava* Stephens (Alomyinae)
9a: *Dusona bicoloripes* (Ashmead) (Campopleginae)
9b: *Enicospilus adustus* (Haller) (Ophioninae)
9c: *Virgichneumon albilineatus* (Gravenhorst)
  (Ichneumoninae: Ichneumonini)
9d: *Cubocephalus anatorius* (Gravenhorst)
  (Cryptinae: Aptesini)
9aa: *Dyspetes luteomarginatus* Habermehl
  (Tryphoninae: Tryphonini)
9bb: *Lycorina triangulifera* Holmgren (Lycorininae)
9cc: *Thymaris tener* (Gravenhorst) (Tryphoninae:
  Oedemopsini)
9dd: *Orthopelma mediator* (Thunberg)
  (Orthopelmatinae)
9ee: *Enizemum ornatum* (Gravenhorst)
  (Diplazontinae)
9ff: *Diacritus aciculatus* (Vollenhoven) (Diacritinae)
10a: *Ophion mocsaryi* Brauns (Ophioninae)
10b: *Gravenhorstia picta* Boie (Anomaloninae:
  Gravenhorstiini)
10c: *Heterocola similis* Horstmann (Tersilochinae)
10aa: *Lycorina triangulifera*
11a: *Ophion mocsaryi*
11b: *Ophion longigena* Thomson (Ophioninae)
11c: *Enicospilus combustus* (Gravenhorst)
  (Ophioninae)
11aa: *Gravenhorstia picta*
11bb: *Anomalon cruentatum* (Geoffroy)
  (Anomaloninae: Anomalonini)
12a: *Gravenhorstia picta*
12b-d: *Anomalon cruentatum*
12e: *Gravenhorstia picta*
12f: *Anomalon cruentatum*
12aa-ee: *Heterocola similis*
13a: *Agrypon canaliculatum* (Ratzeburg)
  (Anomaloninae: Gravenhorstiini)
13b: *Agrypon batis* (Ratzeburg) (Anomaloninae:
  Gravenhorstiini)

13aa: *Diadegma semiclausum* (Hellén) (Campopleginae)
13bb: *Buathra laborator*
13cc: *Dusona bicoloripes*
14a-e: *Brachycyrtus ornatus* Kriechbaumer (Brachycyrtinae)
14aa: *Dusona bicoloripes*
14bb: *Dichrogaster liostylus* (Thomson) (Phygadeuontinae)
14cc-dd: *Phygadeuon flavimanus* Gravenhorst (Phygadeuontinae)
15a: *Cryptus dianae* (Gravenhorst) (Cryptinae: Cryptini)
15b: *Dicaelotus pumilus* (Gravenhorst) (Ichneumoninae: Phaeogenini)
15aa: *Dusona bicoloripes*
15bb: *Stethantyx niger* Khalaim & Broad (Tersilochinae) (Scanning Electron Micrograph by A. Khalaim from Costa Rican specimen in BMNH).
16a: *Listrodromus nycthemerus* (Gravenhorst) (Ichneumoninae: Listrodromini)
16aa: *Cryptus armator* Fabricius (Cryptinae: Cryptini)
16bb: *Dicaelotus pumilus*
17a-d: *Dicaelotus pumilus*
17aa: *Cryptus armator*
17bb: *Cubocephalus anatorius*
17cc: *Phygadeuon flavimanus*
17dd: *Cryptus armator*
18a: *Phygadeuon flavimanus*
18b: *Dichrogaster liostylus*
18c-d: *Phygadeuon flavimanus*
18aa-dd: *Cryptus dianae*
19a: *Barycnemis gravipes* (Gravenhorst) (Tersilochinae)
19b: *Phrudus defectus* Stelfox (Tersilochinae)
19c: *Sathropterus pumilus*
19aa-bb: *Diadegma semiclausum*
20a-c: *Phrudus monilicornis* Bridgman (Tersilochinae)
20aa-cc: *Barycnemis gravipes*
21a: *Sphinctus serotinus* Gravenhorst (Tryphoninae: Sphinctini)
21b: *Ischyrocnemis* sp. (Metopiinae)
21aa: *Campoletis varians* (Thomson) (Campopleginae)
22a-c: *Sphinctus serotinus*
22aa-cc: *Ischyrocnemis* sp.
23a-b: *Periope auscultator* Haliday (Metopiinae)
23aa: *Dusona terebrator* (Förster) (Campopleginae)
23bb: *Diadegma semiclausum*
24a: *Adelognathus dorsalis* (Gravenhorst) (Adelognathinae)
24b: *Pristomerus vulnerator* (Panzer) (Cremastinae)
24aa: *Diphyus quadripunctorius* (Müller)

(Ichneumoninae: Ichneumonini)
25a-b: *Adelognathus dorsalis*
25bb: *Diphyus quadripunctorius*
26a-e: *Oxytorus luridator* (Gravenhorst) (Oxytorinae)
26aa: *Dusona erythrogaster* (Förster)
26bb: *Pristomerus vulnerator*
26cc: *Dusona erythrogaster*
26dd: *Dusona confusa* (Förster) (Campopleginae)
26ee: *Dusona bicoloripes*
27a-b: *Dusona terebrator*
27c: *Dusona bicoloripes*
27d: *Diadegma semiclausum*
27e: *Alcima orbitale* (Gravenhorst) (Campopleginae)
27aa-bb: *Cremastus spectator* Gravenhorst (Cremastinae)
27cc, dd, ff: *Pristomerus vulnerator*
27ee: *Cremastus geminus* Gravenhorst (Cremastinae)
28a-d: *Symplecis breviuscula* Roman (Orthocentrinae)
28aa: *Diphyus quadripunctorius*
28bb-dd: *Platylabus heteromallus* (Berthoumieu) (Ichneumoninae: Platylabini)
29a-c: *Hadrodactylus indefessus* (Gravenhorst) (Ctenopelmatinae: Euryproctini)
29d: *Homaspis analis* (Holmgren) (Ctenopelmatinae: Ctenopelmatini)
29aa-cc: *Diphyus quadripunctorius*
30a-b: *Alomya debellator*
30aa-bb: *Diphyus quadripunctorius*
31a, c: *Diphyus quadripunctorius*
31b: *Virgichneumon albilineatus*
31e: *Crypteffigies lanius* (Gravenhorst) (Ichneumoninae: Ichneumonini)
31aa-dd: *Gelis* sp. (Phygadeuontinae)
31ee: *Aclastus eugracilis* Horstmann (Phygadeuontinae)
32a: *Tryphon latrator* (Fabricius) (Tryphoninae: Tryphonini)
32b: *Polyblastus wahlbergi* Holmgren (Tryphoninae: Tryphonini)
33a-c: *Ateleute linearis* Förster (Ateleutinae)
33aa: *Cidaphus atricillus* (Haliday) (Mesochorinae)
33bb: *Diacritus aciculatus*
33cc: *Banchus volutatorius* (L.) (Banchinae: Banchini)
34a-b: *Adelognathus pallipes* (Gravenhorst) (Adelognathinae)
34c: *Adelognathus brevicornis* Holmgren (Adelognathinae)
34aa: *Periope auscultator*
34bb: *Leptacoenites notabilis* (Desvignes) (Acaenitinae)
35a: *Excavarus apiarius* (Gravenhorst) (Tryphoninae: Tryphonini)

35b: *Periope auscultator*
35aa: *Netelia cristata* (Thomson) (Tryphoninae: Phytodietini)
36a: *Metopius citratus* (Geoffroy) (Metopiinae)
36aa: *Eridolius pachysoma* (Stephens) (Tryphoninae: Tryphonini)
37a-b: *Excavarus apiarius*
37aa-bb: *Periope auscultator*
38a, f: *Acaenitus dubitator* (Panzer) (Acaenitinae)
38b: *Coleocentrus excitator* (Poda) (Acaenitinae)
38c-e: *Phaenolobus terebrator* (Scopoli) (Acaenitinae)
38aa: *Lissonota impressor* Gravenhorst (Banchinae: Atrophini)
38bb: *Monoblastus brachyacanthus* (Gmelin) (Tryphoninae: Tryphonini)
38cc: *Oxytorus luridator* (left), *Clistopyga incitator* (Fabricius) (Pimplinae: Ephialtini) (middle), *Banchus volutatorius* (right)
38dd: *Podoschistus scutellaris* (Desvignes) (Poemeniinae: Poemeniini)
38ee: *Tryphon latrator*
38ff: *Collyria trichophthalma* (Thomson) (Collyriinae)
39a: *Thymaris tener*
39b: *Echthrus reluctator* (L.) (Cryptinae: Cryptini)
39aa: *Pimpla rufipes* (Miller) (Pimplinae: Pimplini)
40a-b: *Thymaris tener*
40aa-bb: *Echthrus reluctator*
41a: *Euceros serricornis* Haliday (Eucerotinae)
41b: *Euceros albitarsus* Gravenhorst (Eucerotinae)
41c: *Euceros pruinosus* (Gravenhorst) (Eucerotinae)
41aa: *Diacritus aciculatus*
41bb-cc: *Stilbops vetula* (Gravenhorst) (Stilbopinae)
41dd: *Sphecophaga vesparum* (Curtis) (Cryptinae: Cryptini)
42a: *Cidaphus atricillus*
42b: *Mesochorus politus* Gravenhorst (Mesochorinae)
42c: *Mesochorus punctipleuris* Thomson (Mesochorinae)
42d, f, g: *Cidaphus atricillus*
42e: *Mesochorus punctipleuris*
42aa: *Stilbops ruficornis* (Gravenhorst) (Stilbopinae) (upper), *Proclitus praetor* (Haliday) (Orthocentrinae) (lower)
42bb: *Oxytorus luridator*
43a-e: *Scolomus borealis* (Townes) (Metopiinae)
43aa: *Enizemum ornatum*
43bb: *Stethoncus monopicida* Broad & Shaw (Metopiinae)
43dd: *Triclistus lativentris* Thomson (Metopiinae)
43ee: *Triclistus globulipes* (Desvignes) (Metopiinae)

44a: *Schizopyga frigida* Cresson (Pimplinae: Ephialtini)
44b: *Stethoncus monopicida*
44c: *Orthocentrus fulvipes* Gravenhorst (Orthocentrinae)
44d: *Rhorus longicornis* (Holmgren) (Ctenopelmatinae: Pionini)
44aa: *Enizemum ornatum*
44bb: *Netelia infractor* Delrio (Tryphoninae: Phytodietini)
45a-b: *Schizopyga frigida*
45aa: *Triclistus globulipes*
45bb: *Rhorus longicornis*
46a-c, e: *Triclistus globulipes*
46d: *Stethoncus monopicida*
46f: *Trieces tricarinatus* (Holmgren)(Metopiinae)
46aa: *Batakomacrus caudatus* (Holmgren (Orthocentrinae)
46bb-ee: *Rhorus longicornis*
47a: *Orthocentrus fulvipes*
47b, d: *Batakomacrus caudatus*
47aa-dd: *Rhorus longicornis*
48a-e: *Hyperacmus crassicornis* (Gravenhorst) (Cylloceriinae)
48aa-dd: *Rhorus longicornis*
49a, d: *Glypta similis* Bridgman (Banchinae: Glyptini)
49b: *Zatypota albicoxa* (Walker) (Pimplinae: Ephialtini)
49c: *Lycorina triangulifera*
49aa-dd: *Ephialtes manifestator* (L.) (Pimplinae: Ephialtini)
50a-d: *Zatypota albicoxa*
50aa-bb: *Glypta similis*
50cc: *Lycorina triangulifera*
50dd: *Glypta similis*
50ee: *Lycorina triangulifera*
51a-d: *Glypta similis*
51aa-dd: *Lycorina triangulifera*
52a: *Diacritus aciculatus*
52b: *Agriotypus armatus* Curtis (Agriotypinae)
52c: *Orthopelma mediator*
52d: *Microleptes aquisgranensis* (Förster) (Microleptinae)
52e: *Diacritus aciculatus*
52aa: *Pimpla rufipes*
52bb: *Pion fortipes* (Gravenhort) (Ctenopelmatinae: Pionini)
53a-c: *Agriotypus armatus*
53aa, cc: *Diacritus aciculatus*
54a-b: *Diacritus aciculatus*
54aa-bb: *Orthopelma mediator*
54cc: *Microleptes rectangulus* (Thomson) (Microleptinae)
55a, f: *Microleptes aquisgranensis*

55b-e: *Microleptes rectangulus*
55aa-ff: *Orthopelma mediator*
56a-c: *Cylloceria caligata* (Gravenhorst) (Cylloceriinae)
56aa: *Stilbops vetula*
56bb-cc: *Pimpla rufipes*
57a: diagrammatic nervellus, intercepted above the middle
57aa: diagrammatic nervellus, intercepted below the middle
57bb: diagrammatic nervellus, not intercepted
58a: *Banchus volutatorius*
58aa: *Netelia infractor*
59a-e: *Banchus volutatorius*
59aa, bb, ee: *Podoschistus scutellaris*
59cc: *Poemenia hectica* (Gravenhorst) (Poemeniinae: Poemeniini)
60a, c: *Coleocentrus croceicornis* (Gravenhorst) (Acaenitinae)
60b: *Leptacoenites notabilis*
60d: *Coleocentrus excitator*
60aa: *Collyria trichophthalma* (top),
Pimpla turionellae (L.) (Pimplinae: Pimplini) (middle),
Netelia infractor (bottom)
60bb: *Netelia infractor*
61a-e: *Collyria trichophthalma*
61aa: *Netelia infractor*
61bb: *Scambus brevicornis* (Gravenhorst) (Pimplinae: Ephialtini)
61cc: *Pimpla rufipes*
61ee: *Netelia infractor* (top),
Exetastes illusor Gravenhorst (Banchinae: Banchini) (middle),
Apechthis rufata (Gmelin) (Pimplinae: Pimplini) (bottom)
62a-e: *Netelia infractor*
62aa: *Pimpla turionellae*
62bb, ee: *Pimpla rufipes*
62cc: *Scambus brevicornis* (top),
Exetastes illusor (middle),
Pimpla rufipes (bottom)
62dd: *Exetastes illusor*
63a: *Ctenopelma tomentosum* (Desvignes) (Ctenopelmatinae: Ctenopelmatini)
63b: *Perispuda facialis* (Gravenhorst) (Ctenopelmatinae: Mesoleiini)
63c: *Hadrodactylus indefessus*
63aa: *Exetastes illusor*
64a-c: *Pimpla rufipes*
64d: *Pimpla turionellae*
64aa-dd: *Exetastes illusor*
65a: *Syrphophilus tricinctorius* (Thunberg) (Diplazontinae)
65b: *Enizemum ornatum*

65c: *Sussaba cognata* (Holmgren) (Diplazontinae)
65aa: *Eclytus multicolor* (Kriechbaumer) (Tryphoninae: Eclytini)
65bb: *Panteles schuetzeanus*
65cc: *Lissonota impressor*
65dd: *Pion fortipes*
66a-d: *Lissonota impressor*
66aa, cc, dd: *Phytodietus ornatus* Desvignes (Tryphoninae: Phytodietini)
66bb: *Panteles schuetzeanus*
67a: *Phytodietus montanus* Tolkanitz (Tryphoninae: Phytodietini)
67aa: *Clistopyga incitator*
68a: *Phytodietus montanus*
68b, e: *Phytodietus geniculatus* Thomson (Tryphoninae: Phytodietini)
68c: *Phytodietus ornatus*
68d: *Phytodietus montanus* (hind tibia, upper), *Phytodietus geniculatus* (fore tibia, lower)
68aa: *Pion fortipes*
68bb, dd: *Hadrodactylus indefessus*
68cc: *Monoblastus brachyacanthus* (top), *Panteles schuetzeanus* (middle), *Perispuda facialis* (bottom)
68ee: *Protarchus testatorius* (Thunberg) (Ctenopelmatinae: Mesoleiini)
68ff: *Dyspetes luteomarginatus*
69a-d: *Panteles schuetzeanus*
69aa-bb: *Polyblastus varitarsus* (Gravenhorst) (Tryphoninae: Tryphonini)
69cc: *Dyspetes luteomarginatus*
69dd: *Monoblastus brachyacanthus* (top), *Perispuda facialis* (bottom)
70a-c: *Campodorus holmgreni* (Schmiedeknecht) (Ctenopelmatinae: Mesoleiini)
70d: *Perispuda facialis*
70e: *Pion fortipes*
70aa: *Polyblastus varitarsus*
70bb: *Tryphon latrator* (top), *Grypocentrus cinctellus* Ruthe (Tryphoninae: Tryphonini) (bottom)
70cc: *Dyspetes luteomarginatus*
70dd: *Grypocentrus cinctellus* (above), *Monoblastus brachyacanthus* (below)
71a: *Hadrodactylus indefessus*
71aa: *Polyblastus varitarsus*
72a-c: *Pygmaeolus nitidus* (Bridgman) (Tersilochinae)
72aa: *Pion fortipes*
72bb: *Alexeter multicolor* (Gravenhorst) (Ctenopelmatine: Mesoleiini)
72cc: *Campodorus holmgreni*
73a: *Sphecophaga vesparum*
73aa: *Scambus brevicornis*

73bb: *Tryphon latrator*
74a: *Sphecophaga vesparum*
75a: *Oxytorus luridator*
75b: *Sphecophaga vesparum*
75aa: *Helcostizus restaurator* (Fabricius) (Cryptinae: Cryptini)
75bb: *Pion fortipes*
76a-d: *Oxytorus luridator*
76aa, cc, dd: *Hemiphanes erratum* Humala (Cryptinae: Aptesini)
76bb: *Sphecophaga vesparum*
77a-b: *Sphecophaga vesparum*
77aa-bb: *Hemiphanes erratum*
78a-d: *Helcostizus restaurator*
78aa: *Sphecophaga vesparum*
78bb: *Polyblastus varitarsus*
78cc: *Pion fortipes* (top), *Perispuda facialis* (bottom)
79a: *Campodorus holmgreni*
80a: *Hemiphanes erratum*
80aa: *Pion fortipes*
80bb: *Campodorus holmgreni*
80cc: *Sphecophaga vesparum*
81a: *Sphecophaga vesparum*
81aa: *Pion fortipes*
81bb: *Oxytorus luridator*
81cc: *Campodorus holmgreni*
82a-c: *Oxytorus luridator*
82aa *Homaspis analis*
82bb: *Pion fortipes*
82cc: *Campodorus holmgreni*
83a, b, f: *Ischnoceros rusticus* (Geoffroy) (Xoridinae)
83c: *Odontocolon dentipesm* (Gmelin) (Xoridinae)
83d-e: *Xorides fuligator* (Thunberg) (Xoridinae)
83aa: *Ephialtes manifestator*
83bb: *Monoblastus brachyacanthus*
84a, c: *Delomerista borealis* Walkley (Pimplinae: Delomeristini)
84b, d: *Scambus brevicornis*
84aa: *Stilbops vetula*
84bb: *Dyspetes luteomarginatus*
84cc: *Stilbops ruficornis*
84dd: *Allomacrus arcticus* (Holmgren) (Cylloceriinae) (top),
Idiogramma euryops Förster (Tryphoninae: Idiogrammatini) (middle),
Proclitus praetor (bottom)
85a-b: *Idiogramma euryops*
85aa: *Dyspetes luteomarginatus*
85bb: *Clistopyga incitator*
87a: *Clistopyga incitator*
87b: *Stilbops vetula*
87aa: *Allomacrus arcticus*
88a, c, e: *Clistopyga incitator*
88b: *Polysphincta longa* Kasparyan (Pimplinae: Ephialtini)
88aa: *Stilbops vetula*

88cc, ee: *Stilbops ruficornis*
88dd: *Grypocentrus cinctellus*
89a-d: *Stilbops vetula*
89aa-cc: *Grypocentrus cinctellus*
89dd: *Oedemopsis scabricula*
90a: *Hadrodactylus indefessus*
90aa: *Allomacrus arcticus*
90bb: *Megastylus cruentator* Schiødte (Orthocentrinae)
91a-e: *Allomacrus arcticus*
91aa: *Megastylus cruentator*
91bb-cc: *Proclitus praetor*
91dd-ee: *Gnathochorisis crassula* (Thomson) (Orthocentrinae)
92a: *Scambus brevicornis*
92b: *Clistopyga incitator*
92c: *Pimpla rufipes*
92aa: *Allomacrus arcticus*
92bb: *Gnathochorisis dentifer* (Thomson) (Orthocentrinae)
93a: *Gnathochorisis dentifer*
93b: *Megastylus cruentator*
93c: *Pimpla rufipes*
93aa-cc: *Stilbops vetula*
94a-c: *Allomacrus arcticus*
94aa: *Gnathochorisis dentifer*
94bb-dd: *Proclitus praetor*
95a-d: *Stilbops vetula*
95aa, bb, dd: *Hercus fontinalis* (Holmgren) (Tryphoninae: Oedemopsini)
95cc: *Oedemopsis scabricula*
95dd: *Hercus fontinalis*
96a-b: *Oedemopsis scabricula*
96aa: *Tryphon latrator*
96bb: *Tryphon relator* (Thunberg) (Tryphoninae: Tryphonini)
97a-b: *Tryphon latrator*
97aa: *Perispuda facialis*
97bb: *Campodorus holmgreni*
97cc: *Hadrodactylus indefessus*
98a-b: *Stenomacrus pedestris*
98aa-bb: *Gelis exareolatus* Förster (Phygadeuontinae)
99a-c: *Ichneumon oblongus* Schrank (Ichneumoninae: Ichneumonini)
99aa-bb: *Gelis exareolatus*
99cc: *Gelis melanocephalus*
100a: *Gelis melanocephalus*
100aa: *Agrothereutes abbreviatus* (Fabricius)

Accounts of British subfamilies

Subfamily ACAENITINAE

This relatively small subfamily comprises about 281 species in 29, mainly small, genera (updated from Yu *et al.*, 2012) found in the Holarctic and the Old World tropics, with one genus (*Arotes*) reaching South America (Castillo *et al.*, 2011). In Britain only six species in five genera have been found, all rarely and three of them not for many years. The wider European fauna comprises about 30 species in eight genera.

Figure 25. *Acaenitus dubitator* (Panzer).

Recognition. Acaenitinae females are easily recognised by the large, elongate hypopygium, although this is not so pronounced in *Leptacoenites* and the non-British *Procinetus* Förster (which has erroneously been recorded as British; Fitton, 1981). Other useful characters to help with recognition (including of males) are: the clypeus with a median or subapical declivity, or if flat then the labrum widely exposed; the clypeus often with a median tooth; the hind wing with the nervellus intercepted above the centre; the propodeum rather flat, without distinct dorsal and posterior slopes; the fore and mid claws sometimes with an accessory tooth; and the long ovipositor, lacking a notch but with weak teeth on the lower valves. Even within the limited British fauna the species are diverse morphologically, making it difficult to characterise the subfamily, though most are readily recognisable as acaenitines.

Systematics. Similarities between the larvae of Orthocentrinae (in the modern sense) and Acaenitinae have been recognised since Townes *et al.* (1960) separated the Acaenitinae from the Pimplinae *s. l.* This pattern of relationships has been confirmed by subsequent studies (Wahl, 1986; 1990; Wahl & Gauld, 1998; Quicke *et al.*, 2009), with Acaenitinae always being recovered within the pimpliformes group of subfamilies. The Diacritinae (biology completely unknown) and Acaenitinae are likely to be basal taxa in the clade of koinobiont endoparasitoid pimpliformes (Wahl & Gauld, 1998; Quicke *et al.*, 2009). Many authors recognise two tribes within Acaenitinae, Acaenitini and

Coleocentrini, but Wahl & Gauld (1998) found that the Coleocentrini are defined only by the absence of apomorphies relative to Acaenitini and argued that tribes should not be recognised, and we follow that here. No study has adequately addressed the issue of the monophyly of the subfamily, although in Quicke *et al.*'s (2009) results, the Acaenitinae were never monophyletic as *Procinetus* always fell outside of the clade comprising the other genera (the *Coleocentrus*-group were sparsely sampled). Also, there are some distinct similarities between acaenitines and collyriines (Sheng *et al.*, 2012) and various authors had included *Collyria* with the acaenitines in the past.

The constituent genera are very narrowly defined, resulting in a relatively large number of species-poor genera and a trickle of newly described genera for species that do not fit current generic definitions (e.g. Sheng & Sun, 2010a); several undescribed taxa in BMNH do not fit within any genera as currently defined.

Biology. The few reliable rearing records for Acaenitinae are from Coleoptera. Records from xylophagous Hymenoptera and Lepidoptera have been repeated frequently in the literature but there seem to be no carefully isolated host records to support these assertions. Aubert (1969) lists various cerambycids as hosts for several species of acaenitine but his primary sources provide little information. Wahl (1990) records two separate rearings of the North American *Arotes amoenus* Cresson from a species of Melandryidae. Some other records are consistent, for example those of *Phaenolobus* species reared from cerambycid larvae in plant stems, with several records (including European specimens in BMNH) from *Phytoecia* and *Oberea* species (e.g. Fiori, 1947; Haeselbarth, 1983; Scaramozzino, 1982; Wahl, 1986). There are, intriguingly, several records of *Phaenolobus* species being reared from Sesiidae (Lepidoptera), the most credible of which is that of the non-British *P. saltans* (Gravenhorst) being reared from a sesiid mining *Euphorbia* stems in Iran (Karimpour *et al.*, 2007), but parasitoids seem to have been reared in all instances from infested roots rather than from isolated hosts and several species of Cerambycidae are known to infest *Euphorbia* roots. Records of *Procinetus decimator* (Gravenhorst) being reared from the noctuid moth *Gortyna flavago* (Denis & Schiffermüller) relate to misidentifications of *Lissonota digestor* (Thunberg) (Banchinae). Previously published statements suggested variously that acaenitines are ectoparasitoids or endoparasitoids, although Wahl (1986) surmised that they were endoparasitoid, based on the larval head capsule.

Figure 26. *Coleocentrus excitator* (Poda).

It has been confirmed that one British species, *Acaenitus dubitator* (Panzer), is a koinobiont endoparasitoid, suggesting a similar mode of development for the whole subfamily (Shaw & Wahl, 1989). This has also been demonstrated for *Leptacoenites notabilis* (Desvignes) by Schwarz (2002a), who published a brief biological summary stating that oviposition was into early instar larvae of the weevil *Otiorhynchus coecus* Germar (as *O. niger* (Fabricius)) in their earthen cells, with the parasitoid developing as a koinobiont endoparasitoid and pupating in the hosts' earthen cells. Schwarz (2002a) studied *L. notabilis* in the Austrian Alps; although it was described from a purportedly British specimen, there have been no further British specimens of *L. notabilis* and the Austrian host species has not been reliably recorded in Britain (Duff, 2012). The life history of *A. dubitator* in Scotland, including details of the egg and first, second and final larval instars, is described by Shaw & Wahl (1989). Oviposition, in early summer, is into the first and second larval instars of the weevil *Cleonis pigra* (Scopoli) (Coleoptera: Curculionidae) galling in the root of *Cirsium arvense* in sand dunes. Eggs and larvae were found free in the haemocoel of the hosts. The egg (dissected from the host) is elongate and tapered at one end. There is evidence that host development is retarded, the full grown *Acaenitus* larva emerging from the host larva during late July or August, when most unparasitized hosts had already pupated. The cocoon is spun within the host's pupation chamber in the thistle root. The cocoon is similar to those of several other species which are known: cylindrical, rounded at the ends, smooth and paper-like in texture and brownish in colour. *Acaenitus dubitator* can overwinter in its cocoon in one of two ways. Some individuals remain as mature larvae and do not become adult the following summer, while others overwinter as morphologically distinct prepupae and are committed to pupate early in the next year and become adult. In those individuals which will become adult in the following year, the change to the prepupal condition takes place soon after cocoon formation. The mature larvae which do not become prepupae are committed to at least a one year delay in their development and do not change to prepupae until late in the following (or perhaps a further subsequent) summer. The habit of overwintering more than once as a cocooned stage is very unusual in temperate ichneumonids and in this species appears to be an adaptation to life in a particularly harsh and uncertain sand dune environment.

In Britain, all acaenitines are either very localised or quite possibly extinct. The only species with recent records are *Acaenitus dubitator* from eastern Scottish sand dunes (Shaw & Wahl, 1989), *Coleocentrus excitator* (Poda) from a native Scottish pine wood (Shaw, 1986) and *Phaenolobus terebrator* (Scopoli) from Wicken Fen (National Museum of Wales). Other records are from fifty years ago or more (Fitton, 1981). Large and spectacular species such as *Arotes albicinctus* Gravenhorst and *Coleocentrus croceicornis* (Gravenhorst) are not likely to have been overlooked for so long.

Identification. The Acaenitinae will be included in the *Handbook* to British Pimplinae and related subfamilies (Shaw *et al.*, in prep.). A key to five of the British species is given by Fitton (1981) and the sixth is dealt with by Shaw (1986). Kolarov (1997) gives a useful key to the Bulgarian fauna, which includes most of the European species.

Figure 27. Morphology of Acaenitinae: (a) *Acaenitus dubitator* (Panzer) face; (b) *Coleocentrus croceicornis* (Gravenhorst) face; (c) *Leptacoenites notabilis* (Desvignes) face; (d) *Phaenolobus terebrator* (Gravenhorst) head; (e) *P. terebrator* fore tarsus claws; (f) *A. dubitator* hypopygium; (g) *Coleocentrus excitator* (Poda) hypopygium; (h) *P. terebrator* aedeagus.

Subfamily ADELOGNATHINAE

Adelognathinae comprises only the genus *Adelognathus*, which is Holarctic (extending in east Asia as far south as Taiwan), and includes 46 named species. Adelognathines are generally poorly represented in collections, although they occur regularly in samples from Malaise traps. Nineteen species have been found in Britain.

Figure 28. *Adelognathus pallipes* (Gravenhorst).

Recognition. Adelognathines are small ichneumonids, fairly easily recognised by the conspicuously exposed labrum and the short antennae, with usually 14 segments (15 in only one British species, *A. thomsoni* Schmiedeknecht). Other useful characters are the pentagonal areolet (when present) and the spiracles of the first tergite slightly posterior to the middle. They may be confused with small phygadeuontines; this applies especially to *Adelognathus dorsalis* (Gravenhorst), which has a rather petiolate first metasomal segment with the sclerotized part of the sternite long. The exposed labrum, short antennae and lack of a sternaulus should distinguish *Adelognathus*. The ovipositor is simply pointed, lacking a notch or teeth.

Systematics. At various times *Adelognathus* has been considered to be a ctenopelmatine, an orthocentrine (in the modern sense), a cryptine *s.l.* or a tryphonine. Perkins (1943) was the first to suggest that it formed a distinct subfamily. Laurenne *et al.* (2006) and Quicke *et al.* (2009) found that *Adelognathus* either formed the sister group to the Cryptinae *s.l.* or the Cryptinae+Ichneumoninae, or was nested within the basal phygadeuontines, and Santos (2017) also found an association with the ichneumoniformes, with *Adelognathus* nested within the Phygadeuontinae in combined morphological and molecular analyses of ichneumonid phylogeny. While surprising in view of the biology of *Adelognathus*, this is not so surprising in terms of overall morphology.

Biology. Fitton *et al.* (1982) summarised our knowledge of the biology of the British Adelognathinae but additional information has been recorded for a few species since then. It is probable that all species develop as ectoparasitoids of well-grown sawfly larvae (Hymenoptera: Pamphilioidea and Tenthredinoidea). Although Fitton *et al.* (1982) found that the species they had studied (for example, *A. laevicollis* Thomson) were koinobiont (the host larvae continuing to feed after oviposition, until overwhelmed by the *Adelognathus* larvae), the development of *Adelognathus* larvae was extremely rapid. Other authors have reported idiobiont development (permanent paralysis and no subsequent feeding by the host larva) in *A. chrysopygus* (Gravenhorst) (Rahoo & Luff, 1987; confirmed by F.D. Bennett, pers. comm.), *A. difformis* Holmgren (Heitland & Pschorn-Walcher, 2005) and the non-British *A. cubiceps* Roman (Kopelke, 1987). Gregarious parasitism has been most often observed but Rahoo & Luff (1987) record that, for *A. chrysopygus* at least, development on small larvae is solitary, but clutches of eggs are laid on larger host larvae. Small larvae of its host, *Pristiphora pallipes* (Lepeletier) (Tenthredinidae), were chosen by *A. chrysopygus* preferentially for host feeding, with larger larvae preferred for oviposition (Rahoo & Luff, 1987). As development is so rapid in this species, the first generation of *A. chrysopygus* can complete development solitarily on small hosts and the second generation is then gregarious on larger host larvae of the same host generation. Kopelke (1987) found the same for *A. cubiceps* developing on *Euura* (as *Pontania*) spp. (Tenthredinidae) and also noted that the first generation parasitoids did not make cocoons but simply pupated in the host gall. Host feeding in *A. chrysopygus* was destructive and caused much higher mortality of sawfly larvae than parasitism; occasionally the same host was used for oviposition and feeding,

Figure 29. Three larvae of *Adelognathus leucotrochi* Shaw & Wahl on a final instar *Nematus leucotrochus* Hartig larva (M.R. Shaw).

sometimes involving the *Adelognathus* female eating her own eggs. Shaw & Wahl (2014) have studied and illustrated the biology of *A. leucotrochi* Shaw & Wahl, a univoltine koinobiont parasitoid of *Nematus leucotrochus* Hartig, in some detail. They found that the egg issues directly from the genital opening without involvement of the ovipositor and the eggs are laid on the host's dorsum in the position at which the female had applied adhesive spread by the ovipositor. Host-feeding was noted only from wounds inflicted by the mandibles.

The broad obovoid eggs are laid in central dorsal positions on the abdominal segments of the host and are firmly stuck by their flat bases, as opposed to being anchored by a pedicel as in the Tryphoninae (Fitton *et al.*, 1982; Shaw & Wahl, 2014). In gregarious species the eggs are strongly grouped on two or three adjacent segments, where the larvae subsequently feed (Fig. 29). Eggs are white, centrally yellowish, with fine surface sculpture. Hatching takes place within a few hours of oviposition and larval development is extremely rapid (four days at ambient temperature in *A. laevicollis*). The larva remains anchored by its posterior end to the collapsed egg shell until feeding is nearly completed. The rapid development may help reduce the vulnerability of these ectoparasitoids. Interestingly, eggs of the idiobiont *A. chrysopygus* and *A. difformis* are laid on the ventral surface of the host, in contrast to some other species of *Adelognathus*, possibly to reduce predation risk (Rahoo & Luff, 1987; Heitland & Pschorn-Walcher, 2005), although lateral egg placement seems to be usual in the idiobiont *A. cubiceps*, which is concealed within the host gall (Kopelke, 1987). The fully-grown larvae leave the host and spin cocoons in cavities in ground debris or, if there is one, within the host retreat. Shaw & Wahl (2014) record five larval instars for *A. leucotrochi*. Some species are plurivoltine, others univoltine, and all are thought to overwinter in the cocoon, probably as a final instar larva. The summer cocoons of plurivoltine species are relatively frail and pale brown while the overwintering cocoons are stout, dark brown, and in effect double-skinned. The adult chews an emergence hole at one side well below the apparent apex. A specialist parasitoid, *Telepsogina adelognathi* Hedqvist (Hymenoptera: Pteromalidae), develops as a koinobiont endoparasitoid of *Adelognathus* larvae, emerging from the host larva after it has spun its cocoon (Bennett *et al.*, 2002).

There is more information on hosts since the summary given by Fitton *et al.* (1982), and host ranges are rather narrow and well-defined in known cases. Overall, adelognathines appear to be particularly associated with Tenthredinidae and Pamphiliidae feeding on trees and bushes, involving both fully exposed hosts and those concealed in leaf rolls and galls. However, the somewhat aberrant *A. dorsalis* is a parasitoid of dolerines (Tenthredinidae) feeding on grasses and sedges. Kopelke (2003), summarising his studies on the parasitoid complexes of gall-forming sawflies on *Salix*, found that *Adelognathus pusillus* Holmgren attacked a range of *Euura* (as *Phyllocolpa*) species that cause swellings and rolls on leaf edges, and another species, *A. cubiceps*, was the only *Adelognathus* found in the larger swelling galls of *Euura* (as *Pontania*). *Adelognathus cubiceps* is unusual in that the female gains access to the concealed host larva by chewing a hole through the wall of the gall and entering the gall to access the host (Kopelke, 1987).

Identification. A key to the British species is included in Fitton *et al.* (1982) but the nomenclature of several species has been revised and additional species have been recorded since, so Kasparyan's (1986, 1990) keys to the genus and the latest checklist should be consulted. Shaw & Wahl (2014) have subsequently described a further species.

Figure 30. Morphology of Adelognathinae: (a) *Adelognathus dorsalis* (Gravenhorst) clypeus; (b) *A. pallipes* (Gravenhorst) head in anterior view; (c) *A. pallipes* antennae; (d) *A. pallipes* first metasomal tergite (lateral); (e) *A. dorsalis* first metasomal tergite (lateral); (f) *A. chrysopygus* (Gravenhorst) ovipositor; (g) *A. brevicornis* (Holmgren) fore wing.

Subfamily AGRIOTYPINAE

This subfamily, which has been revised by Bennett (2001), comprises only sixteen species in one genus, *Agriotypus*. One species occurs in Britain and Europe and the others are found in the Himalayas and eastern Asia.

Figure 31. *Agriotypus armatus* Curtis.

Recognition. *Agriotypus armatus* Curtis can be readily identified in the British fauna by the following combination of characters: fully sclerotised metasomal sternites (unique in Ichneumonoidea); the second and third metasomal tergites (and sternites) fused, forming a syntergite (and synsternite); a tubular first metasomal segment (sternite and tergite fused, spiracles anterior to mid-length); a long spine on the scutellum; and dark ribboning on the wings.

Systematics. Some features of *Agriotypus*, including its unusual biology, have led some workers to recognize it as a separate family. Certain attributes of the last larval instar have been used by Short (1978) to place it in the Ichneumonidae, whereas others had been used by Mason (1971) to justify its exclusion from the Ichneumonoidea. Mason, mainly on the basis of the structure of the metasoma, went so far as to suggest that it is more closely related to the superfamily Proctotrupoidea than to the Ichneumonoidea. However, *Agriotypus* has wing-venation and larval head capsule autapomorphies of the Ichneumonidae. Recent phylogenetic work has placed *Agriotypus* clearly within the Ichneumonidae, with a possible relationship to the ichneumoniformes group of subfamilies (Quicke *et al.*, 2009).

Biology. *Agriotypus* species parasitize case-bearing caddisflies (Trichoptera) in well-aerated water, such as fast-flowing streams, and are among the relatively few species of Hymenoptera adapted to an aquatic existence. *Agriotypus armatus* has been the subject of several quite detailed life-history investigations since its host association was discovered more than 150 years ago. The most important British studies have been by Fisher (1932) and Elliott (1982, 1983). In Britain *A. armatus* has three main hosts, *Silo pallipes* (Fabricius), *S. nigricornis* (Pictet) and *Goera pilosa* (Fabricius). These and most of its other recorded hosts all belong to the family Goeridae, although it also frequently parasitises *Odontocerum albicorne* (Scopoli) (Odontoceridae) (Wallace, 2003; and Fig. 32). Recorded hosts of other *Agriotypus* species include Odontoceridae and Uenoidae (Bennett, 2001). *Agriotypus armatus* is an idiobiont ectoparasitoid, and the other species are likely to have similar biology. Adults are active in late April and May in Southern England, rather later in the north. Females descend beneath the surface of the water and when submerged they are covered by a film of air, and the antennae and wings are laid back over the body. They periodically come up to the surface to replenish their air supply. Under water they crawl about on stones and water plants and can adjust their buoyancy by scraping off some of the air film with the hind legs. They can take flight directly from the water surface (Clausen, 1931) or they can skim along it like pond skaters but with the wings providing propulsion. Sometimes they are seen making short flights between stones projecting from shallow water.

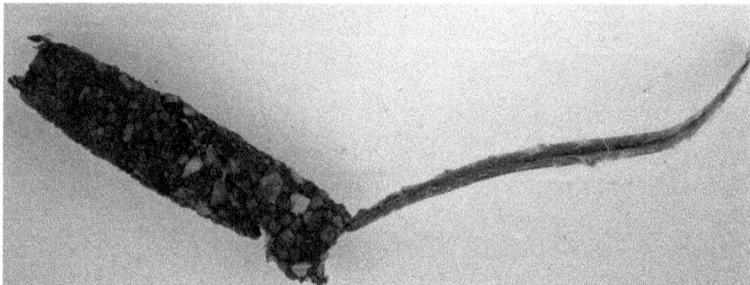

Figure 32. Case of *Odontocerum albicorne* (Scopoli) (Trichoptera: Odontoceridae) with ribbon of silk produced by the larval *Agriotypus armatus*.

Mating takes place soon after emergence; there is no discernible courtship and copulation is brief (Fisher, 1932). About a week after emergence females begin to search for hosts. At this time of the year host cases may contain an active final instar larva, an inactive prepupa or a pupa. The female *Agriotypus* enters the water on or near large stones which are particularly suitable sites for the caddis pupal cases, but once in the water searching appears to be at random (Elliott, 1983). When a case has been found the female inserts her ovipositor between the pebbles of the case or into the small aperture at its posterior end and stabs at the occupant. If the latter is active she leaves immediately. If it is a quiescent prepupa or a pupa she pierces the case with her ovipositor and lays an egg. (None of the published accounts say so explicitly, but presumably the host is paralysed at this time). The egg is anchored in the dermis of the host by a distinct stalk, about a quarter of the length of its main ovoid part. The stalk differs from that of other ectoparasitoid ichneumonids which attach their eggs to the host in a similar manner, in that it is not an extension of the chorion but seems to be formed from material deposited on the anterior pole of the egg at oviposition (Clausen, 1931). The stalk and its enlarged tip (the anchor) become black and shrivelled soon after the egg is laid. The egg chorion itself is very tough. The egg may be attached to the host in a variety of places but is often near the front of the abdomen.

It is not clear whether there are three larval instars (Fisher, 1932) or five (Grenier, 1970; Elliott, 1982). The first instar is of short duration and at this time the larva has transverse rows of bristles on its body segments and a pair of long slender appendages at its posterior end. These adaptations enable it to move between the body of the host and its case to a position where it can feed on the underside of the thorax of a host prepupa or under the wing pads of a pupa. The first instar larva has mandibles that have been described as 'heavy' and simple (Clausen, 1931), a possible adaptation for defending possession of the host in the event of superparasitism. The middle instar(s) is (are) quite different in appearance, lacking the bristles and with the posterior appendages reduced, and is (are) spent entirely in feeding. Elliott (1982) states that the final instar larva moves to the rear end of the host to enable its feeding. The fully grown larva pushes the scant host remains to the posterior end of the case and constructs a stout cocoon of coarse reddish brown silk. At the anterior end a ribbon-like appendage of silk is spun and pushed out from the case as it is made. It consists of an outer layer of wet, closely woven silk covering a mass of tangled dry silk that communicates with the interior of the cocoon, which is also gas-filled. The ribbon, when complete, may be several times the length of the caddis case (Fig. 32) and functions as a plastron for gas exchange, thereby providing an oxygen supply for the parasitoid's metamorphosis (Messner & Taschenberger, 1981); removal of the ribbon results in the death of the *Agriotypus* (Elliott, 1982). Finally the larva lines the cocoon with a layer of fine white silk and pupation takes place about seven to ten days later, at about the end of August. The adult develops in September or October but remains, with its wings expanded, within the cocoon over the winter. In the following year the adult leaves the case when the water temperature reaches about 10°C (Elliott, 1982) and escapes by cutting off the anterior end of the case and floating to the surface in a bubble of gas.

Agriotypus armatus and its host *S. pallipes*, as studied by Elliott (1982), were both univoltine. Potential hosts were present over more than three months in the summer, but adult parasitoids were active for only about five weeks and were thus able to attack only a fraction of the host population. Elliott (1982) showed that females had a potential fecundity of about 45 eggs. He inferred that mortality was very high in the egg stage and found that it was also high in the larval stages, with less than 7% surviving to adulthood.

Identification. The single British species can be identified using the key to subfamilies and is dealt with by Perkins (1960). Bennett (2001) keys the world fauna of *Agriotypus* but no species other than *A. armatus* is likely to be found in Europe.

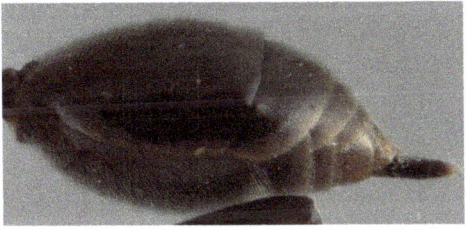 

Figure 33. Morphology of Agriotypinae (*Agriotypus armatus* Curtis): (a) first metasomal segment (lateral); (b) metasoma (lateral); (c) mesonotum, showing posterior spine of scutellum.

Subfamily ALOMYINAE

This Palaearctic and Oriental subfamily comprises about a dozen species of the genera *Alomya* and *Megalomya* Uchida. Two species of *Alomya* occur in Britain.

Figure 34. *Alomya debellator* (Fabricius).

Recognition. *Alomya* species are readily recognised by having large mandibles with blunt teeth, the lower tooth much longer in females (often unidentate in males); golden erect setae on the sharply truncate clypeus; narrowed proboscidial fossa, with the genae almost meeting in females; the fore trochantellus not (females) or weakly (males) separated from the femur; a rather petiolate first metasomal segment, but with spiracles at about the mid-length; and strongly sclerotized metasomal sternites with the hypopygium very large and almost flat. Females are also distinctive in having a very long vertex and pronotum. Males are more ichneumonine-like than females but can be distinguished by the lack of thyridia and gastrocoeli (missing in a few Ichneumoninae: Phaeogenini but these are much smaller) and by the wing venation, with the reclivous nervellus, and the section of fore wing vein *M* on the distal side of the areolet much shorter than the proximal section. The British species are moderately large insects, 12-19 mm long, and strongly sexually dimorphic.

Systematics. The morphology and biology of *Alomya* (the Eastern Palaearctic *Megalomya* is only poorly known) suggest that it is closely related to the Ichneumoninae. At various times *Alomya* has been treated as belonging to the ichneumonine tribe Phaeogenini (which then takes the name Alomyini; Wahl & Mason, 1995), as comprising a separate tribe within Ichneumoninae (e.g. Hinz &

Short, 1983) or as a separate subfamily (e.g. Perkins, 1960). Recent phylogenetic work found *Alomya* to be either the sister group to Cryptinae *s.l.* + Ichneumoninae or the sister group to Ichneumoninae (Laurenne *et al.*, 2006; Quicke *et al.*, 2009). In the light of these findings we retain Alomyinae as a separate subfamily within the ichneumoniformes assemblage.

Biology. Despite Perkins' (1952, 1960) statements to the contrary, it has been known for a long time that the hosts of *Alomya* are swift moths (Lepidoptera: Hepialidae) (Stenton, 1926). However, it is only relatively recently that details of the life history (of *Alomya semiflava* Stephens, a koinobiont larval endoparasitoid of *Hepialus lupulinus* (L.)) have been elucidated (Hinz in Hinz & Short, 1983). The female *A. semiflava* enters the tunnel of the final instar larva of *H. lupulinus*, approaching from the front and making threatening movements with her jaws. The host larva retreats to the end of its tunnel followed by the parasitoid, at which point *Alomya* brings forward her metasoma, between her legs, and oviposits into the host from below, just behind its mouthparts. The host larva can respond aggressively, and *Alomya* females sometimes lose their antennae in these encounters. The parasitized host larvae behave similarly to unparasitized ones but fail to pupate at the expected time. Instead the larva shrinks and darkens and its cuticle hardens, to form a structure known as a 'mummy'. The external appearance then remains unchanged for several months until the adult *Alomya* emerges; it has not been recorded from which end of the mummy this takes place.

Males of *Alomya* are often common and conspicuous, feeding on the flowers of umbellifers or flying over low vegetation. Males of an *Alomya* sp. (probably *A. semiflava*) have been reported as pollinating an orchid, *Coeloglossum viride*, in Germany (Hoffmann, 2014). The more compact females are seen much less often, spending most of their time at soil level searching for hosts, although in hot weather they can sometimes be found on flowers and they have also been captured by netting at night, when they seem to be most active (Hinz in Hinz & Short, 1983). The flight period of *A. semiflava* is in late summer and autumn when *H. lupulinus* is in its final larval instar, and Hinz (in Hinz & Short, 1983) concluded that *A. semiflava* is univoltine and monophagous. Females offered larvae of two other species of *Hepialus* failed to attack them. Hinz speculated that the other common European species of *Alomya*, *A. debellator* (Fabricius), is possibly associated with *H. sylvinus* (L.) or with several *Hepialus* species. In most of continental Europe its flight period is in May but in Britain, the Alps and the Pyrenees, it extends to late summer. This might be because some *Hepialus* species have a two-year life cycle in these areas and spend a prolonged period in the last larval instar (Hinz in Hinz & Short, 1983) or, perhaps less likely, that *A. debellator* has more than one generation per year (Perkins, 1960).

Identification. The two British species are keyed by Perkins (1960).

Figure 35. Morphology of Alomyinae: (a) *Alomya debellator* (Fabricius) female head (lateral); (b) *A. semiflava* Stephens female head (ventral); (c) *A. semiflava* female clypeus; (d) *A. debellator* fore wing; (e) *A. semiflava* female fore leg; (f) *A. semiflava* female first metasomal segment (lateral); (g) *A. debellator* male propodeum and anterior metasoma.

Subfamily ANOMALONINAE

Tribes
Anomalonini
Gravenhorstiini

This is a medium-sized subfamily with 41 species in Britain in 13 genera, out of a world fauna of 750 described species assigned to 45 genera (updated from Yu *et al.*, 2012). Moderate numbers of anomalonine species can be found across the world and in various types of habitat, although the dominant genera in the tropics are not those that are commonest in temperate climes. Anomalonines are distinctive in flight and the larger species tend to be noticed as they patrol hedgerows and the lower foliage of trees, but many of the British species are rarely collected.

Figure 36. *Anomalon cruentatum* (Geoffroy), tribe Anomalonini.

Recognition. This is one of the easier ichneumonid subfamilies to characterise. Anomalonines have a laterally compressed metasoma, a petiolate first metasomal segment with the spiracle far behind the middle and the sclerotized part of the sternite very long, conspicuously long hind legs (often with the tarsi stout and strikingly pale yellow), and long trochanters. The tribe Gravenhorstiini, which contains all but one of the British species, have a characteristic propodeum, usually extending far beyond the hind coxal insertions and with the surface coarsely reticulate, imparting an irregular honeycombed appearance. The Anomalonini, with one British species, have the propodeum usually with a series of delimited areas but can be readily recognised by the mid tibia having only one apical spur (also the case in *Metopius* and *Periope* of Metopiinae, and the *Exenterus* group of genera of Tryphoninae, which are otherwise very different in appearance) and the fore wing with the discosubmarginal cell extending far beyond 2*m-cu* (as in Ophioninae).

Systematics. Gauld (1976a) revised the tribal and generic classification of Anomaloninae and his groupings seem to have gained general acceptance, except that some of his subgenera are more usually treated as valid genera. The two tribes, Anomalonini and Gravenhorstiini, are readily separated but there is evidence that they form a monophyletic group (e.g. Gauld, 1976a; Quicke *et al.*, 2009) and there seems to be no obvious justification for separating them as distinct subfamilies, as was done, without explanation, by Dasch (1984). The Anomalonini comprises the single genus *Anomalon*. Townes (1971) and others had treated the Gravenhorstiini, as now understood, as three tribes, but Gauld (1976a) did not recover these as monophyletic groupings, although his methodology was phenetic rather than cladistic. Although some authors had suggested a close relationship between Anomaloninae and Metopiinae, on the basis of similarities in the larval head capsule (e.g. Townes, 1971), this is now thought to be the result of convergence in larval morphology between two subfamilies that are, unusually within the ophioniformes group of subfamilies, larva-pupal endoparasitoids that spin only a flimsy cocoon within the host pupa (Viktorov, 1968; Gauld, 1976a; Short, 1978). Although the reduction in silk-spinning has been suggested as the reason for reductions in certain sclerites of the larval head capsule (see Metopiinae chapter for more details), it also seems entirely probable that the development of the anomalonine (or metopiine) larva on a histolysed host could explain the reduction in head capsule musculature and associated sclerotization. Phylogenetic reconstructions have generally shown the Anomaloninae to be part of a 'higher' ophioniformes clade of subfamilies with (mostly) petiolate and laterally compressed metasomas, along with Campopleginae, Cremastinae, Ophioninae and Nesomesochorinae (Gauld, 1985; Wahl, 1991; Quicke *et al.*, 2009). The Tersilochinae do not seem to be part of this clade (Quicke *et al.*, 2009).

Biology. Species of *Anomalon* are often reported to be parasitoids of Tenebrionidae (Coleoptera), in marked contrast to the Gravenhorstiini, which are parasitoids of Lepidoptera. However, there is very little evidence behind these host records, which have been magnified by repetition. The European *Anomalon cruentatum* (Geoffroy) was reported by Bogush (1959) to be a parasitoid of *Gonocephalum rusticum* (Olivier) (Tenebrionidae) and Gauld (1978) says that an Australian *Anomalon* species has been reared from two tenebrionid species. Gauld (1978) also reports a series of a North American species reared from lepidopterous hosts and Pair *et al.* (1986) list *Anomalon ejuncidum* Say as a parasitoid of *Spodoptera frugiperda* (Smith) (Lepidoptera: Noctuidae). Dasch (1984) lists a series of hosts of North American species on the basis of label data but there is no reason to accept any of these host records uncritically. In the absence of any published life history studies it remains unresolved as to whether *Anomalon* are parasitoids of either Coleoptera, Lepidoptera or both. *Anomalon cruentatum* was thought by Gauld & Mitchell (1977b) only to be adventitious in southern England but the species has recently been found to be

Figure 37. *Agrypon batis* (Ratzeburg), tribe Gravenhorstiini.

Figure 38. *Agrypon flaveolatum* (Gravenhorst) ovipositing in a larva of *Operophtera brumata* (L.) (Geometridae) in culture (M.R. Shaw).

established in at least one area of Kent (Clemons, 2009). In France small aggregations of males hovering against head-high foliage of mature *Populus* trees have been seen on several occasions (M. R. Shaw, unpublished), but quite what they were doing was not clear.

As far as is known all Gravenhorstiini attack larval Lepidoptera, emerging from the host pupa. They are all solitary koinobiont endoparasitoids and the recorded host range is wide, in terms of both host taxonomy and size. Gauld & Mitchell (1977b) summarize data known to them on the hosts of British species but some of their host records require verification. One feature of the British anomalonines and their hosts is that the majority appear to be univoltine (M.R. Shaw, unpublished) and this may reflect constraints imposed by their biology. An exception is *Therion circumflexum* (L.), some of whose hosts are plurivoltine in which cases the parasitoid sometimes emerges soon after host pupation.

Little has been published on the details of anomalonine biology. Females searching for hosts have a characteristic flight pattern, hovering and making gentle up-and-down movements close to the vegetation, during which they can be very conspicuous due to the often bright yellow hind tarsi. The antennae and hind legs are held outstretched and the metasoma is raised (Gauld, 1976b). Gauld (1976b) notes that *Heteropelma megarthrum* (Ratzeburg) (as *calcator*) cling tightly to conifer needles when disturbed and raise their metasomas, presumably to blend in with their surroundings. It is not known whether there is host-feeding by any anomalonines. The most complete account of the biology of an anomalonine is that of Tothill (1922), who studied *Therion morio* (Fabricius) as a univoltine parasitoid of the forestry pest, *Hyphantria cunea* (Drury) (Lepidoptera: Erebidae, Arctiinae). M.R. Shaw (unpublished) has studied some aspects of the biology of *Agrypon flaveolatum* (Gravenhorst) when presented with *Operophtera brumata* (L.) (Geometridae) larvae, and it is these two studies that form the basis of this account. The host range of *A. flaveolatum* is wide, as in a number of British anomalonines, but concentrated on just a few spring-feeding arboreal geometrids with only occasional rearings from other hosts including noctuids. Searching females are interested only in hosts that are moving, and oviposition seems to be very rapid (Fig. 38), without any paralysis of the host, usually aimed at the head region of the caterpillar, although dissections failed to reveal any eggs to confirm that oviposition had taken place. One rapid oviposition (into a host raised in culture) did, however, subsequently result in an

A. flaveolatum emerging. Oviposition by *T. morio* is mainly targeted at the caudal end of the host, and that of *Heteropelma megarthrum* is apparently targeted at caudal and head regions (Plotnikov, 1914). Tothill (1922) observed that *Therion morio* females could oviposit in several host caterpillars in rapid succession and that oviposition could commence a few hours following emergence of the adult female. Gauld (1976b) describes how the egg of *Heteropelma megarthrum* is moved into position towards the apex of the ovipositor just prior to oviposition and H. Schnee (pers. comm.) reports that in *Erigorgus* species oviposition is achieved from a distance, presumably by extending the long metasoma, to avoid the defensive thrashing of the host caterpillar. Penultimate or final instar hosts were preferred by *A. flaveolatum*. The ovaries of *A. flaveolatum* produce a large number of eggs, with a large potential fecundity. The very small eggs are distinctive, pear-shaped with a roughened protrusion of the surface, and this is probably a characteristic of many anomalonine eggs, judging by Tothill's (1922), Iwata's (1960) and Gauld's (1976a) illustrations and descriptions of the eggs of several genera of Anomaloninae. Tothill (1922) found that the egg of *T. morio* is attached to the body wall by means of this small holdfast, which Gauld (1976a) supposes to act as an anchor or 'barb' but which Tothill describes as adhering via a sticky substance on the 'cushion', which would be more in line with the lack of any protrusions that look capable of penetrating cuticle. However, in *Erigorgus*, at least, the egg and then the first instar larva are free within the haemocoel (H. Schnee, pers. comm.), and Nuzhna (2013) records that the ovarian eggs of several anomalonines, including two species of *Erigorgus*, lack protrusions. According to Tothill (1922) the first instar larva remains enclosed within a trophamnion, which is a membrane of embryonic origin, and Rosenberg (1934) similarly reports that what he assumed to be young larvae of *Trichomma enecator* (Rossi) were found within trophamnions of about 2.5 mm in length. Within the Ichneumonidae a trophamnion has otherwise only been observed in the tribe Euryproctini (Ctenopelmatinae). The larva increases in size during the first instar, within the trophamnion, but there is no ecdysis until the host has pupated. The first larval instar is the usual overwintering stage and thus it is also the stage in which the anomalonine spends the greater part of the year. First instar larvae of anomalonines have a distinct caudal appendage (Plotnikov, 1914; Tothill, 1922), as is present in most ophioniform larvae, and large mandibles. Tothill (1922) distinguished three larval 'stages', although the precise number of instars has not been established. Development of *Therion morio* in the host pupa is rapid, from two to three weeks, following

Figure 39. Mating pair of *Aphanistes ruficornis* (Gravenhorst) (L. Doherty).

ecdysis to the second instar (Tothill, 1922). The fully grown larva spins at most only a flimsy cocoon within the host pupal shell and spends two to three weeks as a pupa before the adult parasitoid escapes by biting a roughly circular hole at the anterior end. There is some evidence (M. R. Shaw, unpublished) to suggest that the pupa is formed isolated from the host's bodily remains, sometimes (but not always) in the ecdysal space between the host's pupal cuticle and its body. Alongside the findings by Tothill (1922) and Rosenberg (1934) that at least the first instar larva exists within a trophamnion and is thus isolated from the host, this raises the intriguing possibility that larval development after the host's pupation might conceivably take place effectively outside the host's body. However, careful investigation of the developmental biology of anomalonines would be needed to test this idea.

For several of the rarer species the rearing records are so few that there is an illusion, whether real or not, of extreme specialisation; for example, *Agrypon anomelas* (Gravenhorst) seems to be a parasitoid only of lycaenid butterflies, in Britain having been reared only from *Neozephyrus quercus* (L.) (George, 1978; Shaw et al., 2009). At least three British anomalonines have host ranges centred on low-feeding Satyrinae (Nymphalidae), but also encompassing Hesperiidae and Noctuidae in the same habitat (Shaw et al., 2009). Whilst smaller anomalonines, such as *Habronyx* (*Camposcopus*) species, are parasitoids of Tortricidae, larger species, such as *Barylypa delictor* (Thunberg), attack noctuid larvae such as *Acronicta* species. *Trichomma* species have rather long ovipositors relative to other Anomaloninae and some have been reared from semi-concealed 'microlepidopteran' larvae such as leaf-rollers (Gauld & Mitchell, 1977b), although *T. fulvidens* Wesmael parasitizes low-feeding noctuids. *Heteropelma megarthrum* is a frequently cited and often common (Gauld, 1976b) parasitoid of *Bupalus piniarius* (L.) (Geometridae) and other conifer-feeding Lepidoptera so it is surprising that there has not been more published on the interactions of this species with its economically important hosts. Gauld (1976b) notes that oviposition is into early instar hosts, and this has been stated regarding Anomaloninae in general (e.g. Gauld, 1976a) although there is very limited evidence for it (mainly from Tothill's (1922) study, but also confirmed for the congeneric *T. circumflexum* (M. R. Shaw, unpublished)), and there is some evidence that at least *Agrypon flaveolatum* oviposits into penultimate or final instar larvae (see above). This lack of detail extends to most of our knowledge of anomalonine biology and seems rather remiss, given that several species can be reared fairly commonly from frequently collected Lepidoptera. Studies on their biology would be very rewarding. Malaise traps tend to produce only small numbers of anomalonines in Britain; sweeping and hand-netting are probably more productive general collecting methods. At least one species, *Agrypon batis* (Ratzeburg), is light-trapped frequently enough to suggest that it is either regularly nocturnal or crepuscular. There are few known British specimens of *Gravenhorstia picta* Boie, and certainly none collected at all recently. This is a large and conspicuously yellow-striped species and is unlikely to have been overlooked. Gauld & Mitchell (1977b) suggest that this mainly Mediterranean species has been an occasional immigrant, although it is also possible that it is a now extinct former resident.

Identification. There is a handbook to the British species (Gauld & Mitchell, 1977b); although still useful, this is out of date, with numerous subsequent name changes, a number of reidentifications and several species recognised as British since then, especially in the genus *Agrypon* in which '*A. anxium*' of Gauld & Mitchell is now known to be an aggregate of some half dozen British species. Taxonomic work on the European fauna includes a paper on *Perisphincter* Townes (Schnee, 1978) and resolution of various names (Schnee, 1989, 2008, 2018) but the fauna is in need of revision and modern identification keys. Atanasov (1975, 1977, 1978) provides keys to Western Palaearctic *Erigorgus*, *Habronyx* and *Aphanistes*, respectively, and compilatory keys to the Anomaloninae of European Russia (Atanasov, 1981), as well as other regional guides, which, whilst not representing thorough revisions, may be useful, although written almost entirely in Russian. Schnee (2018) includes a key to Western Palaearctic *Aphanistes*. Gauld (1976, modified in 1976b) provides a key to the world genera which supersedes that of Townes (1971).

Figure 40. Morphology of Anomaloninae: (a) *Agrypon batis* (Ratzeburg) (Gravenhorstiini) head; (b) *Agrypon anomelas* (Gravenhorst) (Gravenhorstiini) head; (c) *Gravenhorstia picta* Boie (Gravenhorstiini) head; (d) *Anomalon cruentatum* (Geoffroy) (Anomalonini) head; (e) *G. picta* fore wing; (f) *A. batis* ovipositor tip; (g) *A. cruentatum* fore wing; (h) *A. cruentatum* mid tibia; (i) *G. picta* propodeum; (j) *A. cruentatum* propodeum.

Subfamily ATELEUTINAE

Ateleutinae is a small subfamily, recently separated from Cryptinae (Santos, 2017) on the basis of phylogenetic analyses of several genes and morphology. There are two included genera, *Ateleute* Förster and *Tamaulipeca* Kasparyan, the former cosmopolitan with 41 described species (and many more undescribed) and the latter with five described species restricted to the Neotropics (Bordera & Sääksjärvi, 2012; Sheng *et al.*, 2013). There is a single European species, *Ateleute linearis* Förster, which is found in England (Schwarz & Shaw, 1998).

Figure 41. *Ateleute linearis* Förster female.

Recognition. Most Ateleutinae are readily recognised by the shape of the fore wing areolet alone: elongate pentagonal, although with 3*rs-m* generally weakly defined, and with 2*m-cu* intercepting the areolet proximally (in *Tamaulipeca*, however, 3*rs-m* is entirely absent, and the clypeus is produced medially). Additionally, the first metasomal tergite is petiolate (first sternite and tergite fused) with the spiracle at about the middle, and the posterior transverse carina of the mesosternum is complete. The hind tibia is conspicuously spiny. Unlike most Cryptinae and Phygadeuontinae, the sternaulus is often only weakly indicated anteriorly (especially in males). Ateleutines are strongly sexually dimorphic and small males are frequently confused with orthocentrines as they tend to be more uniformly brown, the mandibles are small and the distinctive areolet shape can be difficult to discern, although the spiny hind tibia and complete posterior transverse carina of the mesosternum are useful recognition features.

Systematics. The Ateleutinae belong in the ichneumoniformes group of subfamilies and have generally been classified as rather aberrant genera of Cryptinae (e.g. Townes, 1970, before the description of *Tamaulipeca*). The distinctive morphology was found by Santos (2017) to reflect a distinct evolutionary lineage, with ateleutines being recovered as the sister group to the Cryptinae or the Cryptinae+Ichneumoninae under different analysis parameters. Santos (2017) therefore took the logical step of elevating Ateleutinae to the status of a separate subfamily (see further discussion under Cryptinae).

Biology. The ecology and biology of ateleutines are poorly known; the only detailed study is that of Momoi *et al.* (1965) on the Japanese *Ateleute minusculae* (Uchida). This was shown to be a solitary, idiobiont ectoparasitoid of larvae of a species of Psychidae (Lepidoptera), *Eumeta minuscula* Butler (as *Clania minuscula*). Oviposition is through the wall of the case and, after the host larva is paralysed, the egg is deposited on the host or the inner wall of the case. As in at least many Cryptinae, the egg is not attached to the host. Larval development is rapid and *A. minusculae* spends the winter as a prepupa, with two generations per year. The cocoon is thin, composed of a few sheets of silk, and strongly tapered caudally, with the ovipositor of the parasitoid outstretched in the cocoon. Momoi *et al.* (1965) proposed that *A. minusculae* is absolutely host specific and, by transferring host larvae between cases, they demonstrated that *A. minusculae* would accept only cases made by this one species of psychid, but would oviposit on hosts of different species if they were transferred to the case of *C. minuscula*. The North American *Ateleute carolina* Townes has also been reared from a psychid (Townes, 1967) but *A. linearis* has never been reared. It is found in scrub and woodland (Schwarz & Shaw, 1998) and is evidently quite widespread in England.

Identification. The single British (and single European) species, *A. linearis*, can be identified through the key to subfamilies and is illustrated in Figs 41-43.

Figure 42. *Ateleute linearis* male.

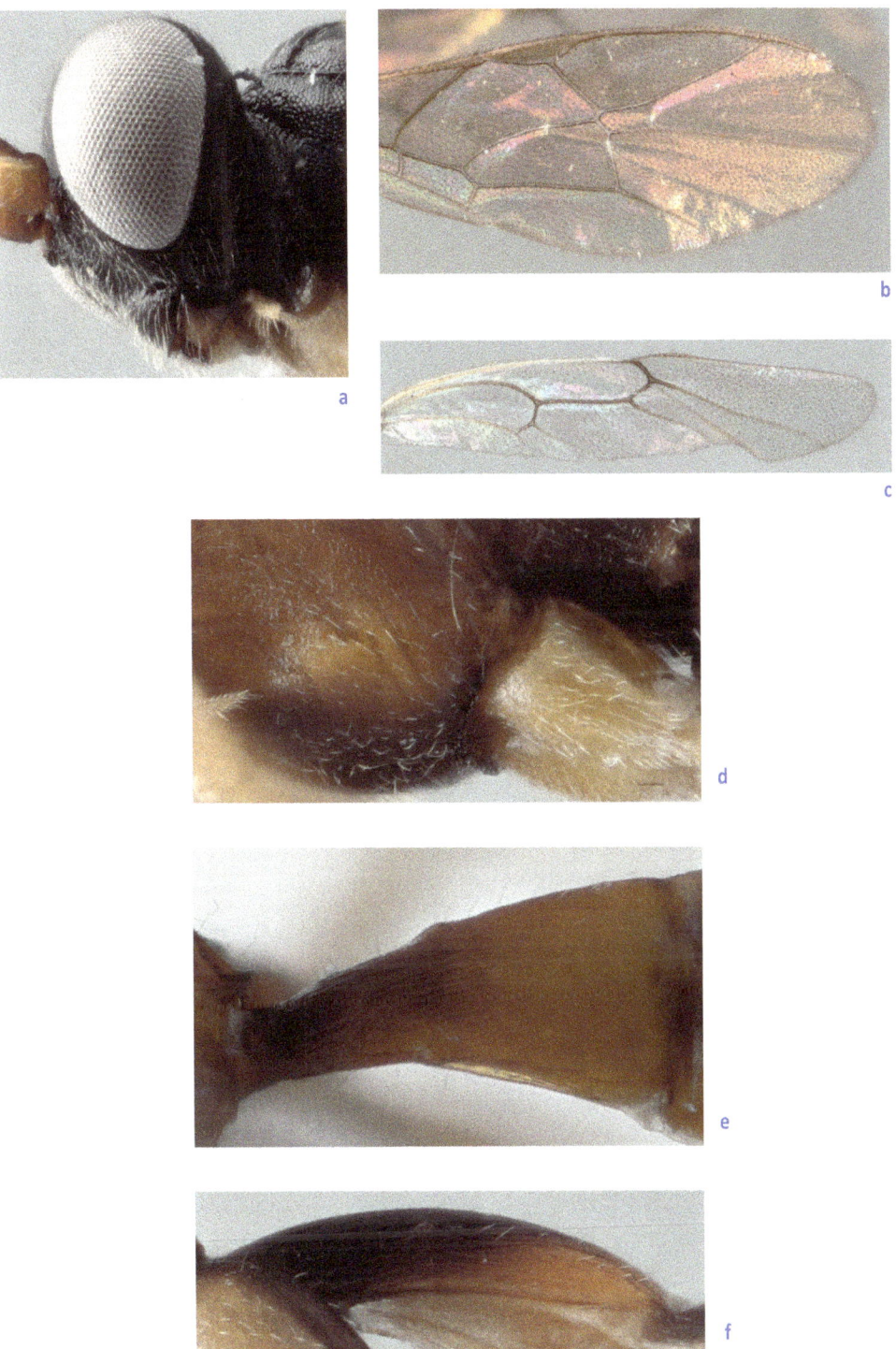

Figure 43. Morphology of Ateleutinae (*Ateleute linearis* Förster): (a): head, lateral; (b) fore wing; (c) hind wing; (d) mesosternum; (e) first metasomal tergite (dorsal); (f) first metasomal segment (ventro-lateral).

Subfamily BANCHINAE

Tribes
Atrophini
Banchini
Glyptini

Banchinae is a species-rich subfamily found throughout the world, with currently a little over 1,760 described species in 64 genera classified in three tribes. Numerous new genera have been described from Central and South America in recent years (Gauld *et al.*, 2002a; Broad *et al.*, 2011) and collections hold several undescribed genera from various parts of the world. There are currently 138 species in 12 genera on the British list (Brock, 2017). There are a few more genera in the western Palaearctic that are not present in Britain but these are found mainly around the Mediterranean and on the Atlantic islands. Banchines are species-rich in both temperate and tropical regions (e.g. Gauld *et al.*, 2002a; Veijalainen *et al.*, 2012a), but in the Holarctic region the fauna is dominated by two genera, *Lissonota* and *Glypta*; the latter is particularly species-rich in North America with 311 described species (Dasch, 1988), far higher than the European total, which may be a reflection of the greater size of the Nearctic tortricid moth fauna.

Figure 44. *Lissonota luffiator* Aubert, tribe Atrophini.

Figure 45. *Banchus volutatorius* (L.) (female), tribe Banchini.

Recognition. Banchines are most frequently basically black, or black with an extensively red metasoma, often with long ovipositors, and superficially resembling some Pimplinae. A few banchines (some species of *Banchus* and *Syzeuctus*) are black and yellow-striped, and several species have yellow marks on the mesosoma and/or the head, especially in the male. Unlike pimplines, Banchinae have notched ovipositors, often pectinate tarsal claws, and a characteristic propodeum and metapleuron: the submetapleural carina is strongly expanded anteriorly and on the propodeum dorsal (longitudinal) carinae are mostly missing with the posterior transverse carina prominent and strongly arched (some Glyptini have remnants of the anterior transverse and median longitudinal carinae and *Rhynchobanchus* species entirely lack dorsal carinae). Additional useful recognition features are the rather large, sub-triangular, apically notched female hypopygium, and the strongly convex clypeus (except in the *Banchus*-group). Species of the tribe Glyptini all have characteristic chevron-shaped grooves on the metasomal tergites, delimiting an anteriorly-pointing triangular area on at least the second and third tergites; the Banchini have the nervellus intercepted very high up and have relatively short or very short ovipositors (some exceptions in *Exetastes*, particularly in two European but not British species); most *Banchus* species have a scutellar spine, which among the British Ichneumonidae is otherwise found only in *Agriotypus* (Agriotypinae), in which the spine is longer.

Figure 46. *Glypta elongator* Holmgren, tribe Glyptini.

Systematics. Because of their non-petiolate first metasomal segment, with a short sclerotized section of the sternite and the spiracles anterior to the middle, banchines were traditionally classified in the Pimplinae assemblage (e.g. Morley, 1908). It has long been recognised, however, that the Banchinae are more closely allied to the ophioniformes group of koinobiont endoparasitoids (Pampel, 1914). Gauld (1985) included the Banchinae as a basal subfamily within the ophioniformes, on the basis of their biology, notched ovipositor, female reproductive tract anatomy and larval morphology (lacking a labral sclerite and with a Y-shaped prelabial sclerite). This rather basal position within the ophioniformes has also been recovered in the cladistic analyses of Wahl (1991) and Quicke *et al*. (2009). Exact relationships in the latter study varied with analysis but Banchinae was usually recovered as a clade (the non-British *Townesion* did not group with the remaining banchines under some of the analysis parameters, although its 28S D2 rRNA sequence is aberrant). Banchinae formed part of a grade, along with Lycorininae, Tersilochinae and the non-British Sisyrostolinae, of ophioniform subfamilies between the 'lower' ophioniformes comprising Tryphoninae and allies, and the 'higher' ophioniformes consisting of Anomaloninae, Campopleginae, Cremastinae, Nesomesochorinae and Ophioninae (and possibly some aberrant genera).

The tribal composition of Banchinae has varied somewhat; because of the rather plesiomorphic nature of the subfamily relative to other ophioniformes groups, various genera with a stout first metasomal segment have tended to be lumped in. Townes (1970b) included five tribes in the Banchinae although that number had been reduced to three by 1978 (Townes & Townes, 1978) with the elevation of Neorhacodinae and Stilbopinae to subfamily status, following Short (1957) on the larval morphology of *Stilbops* and Horstmann (1968) on the biology and morphology of *Neorhacodes*. Townes & Townes (1978) still retained *Panteles* in the Atrophini, although Wahl (1988) concluded that the Stilbopinae should also include *Panteles*. Subsequently, Quicke *et al.* (2009) transferred *Neorhacodes* and related genera to the Tersilochinae and found that the Stilbopinae may or may not comprise a monophyletic group but are more closely related to elements of a paraphyletic Tryphoninae than to Banchinae. Note that until Gauld (1984), Atrophini was generally referred to as Lissonotini, unfortunately a preoccupied name.

There has been no real test of the monophyly or otherwise of the three remaining banchine tribes that remain, although there is molecular evidence that together they do form a clade. The Atrophini and Glyptini each include a very species-rich, cosmopolitan genus, *Lissonota* and *Glypta*, respectively. The Banchini are not so dominant, although *Exetastes* species are fairly numerous across the Holarctic and the Old and New World tropics. Banchini and Glyptini can each be defined by a discrete character relative to Atrophini, namely the position of the hind wing nervellus in Banchini and the metasomal grooves in Glyptini, whereas there are no obvious characters of the adult morphology that define Atrophini. However, Wahl (1988) found that, on the basis of larval morphology, it is the Glyptini that are not supported by any apomorphies. A possibly plesiomorphic character of Glyptini adult morphology is the tendency towards a more fully areolated propodeum. There has been much difference of opinion as to whether or not the *Banchus*-group and *Exetastes*-group are each other's closest relatives, as Townes (1944; 1970b) advocated, or whether *Exetastes* and relatives are more closely related to the Atrophini (e.g. Cushman, 1937b) or form a separate tribe (Beirne, 1941a; Aubert, 1978b). Quicke *et al.* (2009) found that the Banchini is probably not monophyletic, although the phylogenetic results were not consistent across analyses. One finding was that the Nearctic genus *Agathilla* Westwood was more closely related to the *Banchus*-group than to the *Exetastes*-group, in which it has traditionally been classified, which partly accords with the results of Wahl (1988) that there are no larval synapomorphies linking *Agathilla* and *Exetastes*.

Although most banchines are readily recognisable as such, and in Europe all fall into one of only a few body forms, extralimitally there are a few very aberrant genera that are difficult to recognise as banchines (Gauld & Wahl, 2000a; Broad *et al.*, 2011). Indeed, Kasparyan (1993a) described a new subfamily, Townesioninae, to accomodate *Sachtlebenia* and the newly described *Townesion*, which Gauld & Wahl (2000a) concluded are actually highly derived glyptines. Although Gauld *et al.* (2002a) attempted to rigorously define putatively monophyletic genera in Costa Rica, over much of the world atrophines with long ovipositors which lack the distinctive characters of other genera have tended to be classified in *Lissonota*, which is now a hopelessly polythetic assemblage (*Lissonota* species are scattered across the Atrophini in the topology of Quicke *et al.*, 2009), concealing significant evolutionary diversification. Brock (2017) employs several subgenera and species-groups to subdivide *Glypta* and *Lissonota* but there is no evidence that these divisions can be applied to taxa beyond Europe; indeed the subgeneric divisions could be positively misleading in misclassifying extralimital taxa. Nevertheless, reference is made here to the subgenera, for consistency with Brock's (2017) handbook.

Host associations and life histories. All banchines with known host associations are koinobiont endoparasitoids of Lepidoptera larvae. Whilst the Banchini, with their mainly short ovipositors, are parasitoids of more or less exposed 'macro-lepidoptera', the Atrophini and Glyptini, most of which have long ovipositors, attack variably concealed Lepidoptera larvae, from those in superficial concealment in leaf-rolls, flower heads, etc., to xylophagous Sesiidae and Cossidae larvae. Most banchines are solitary but *Lissonota mutator* Aubert is known to be gregarious, developing in broods of about four to nine in Tortricidae larvae (Shaw, 1999). As broods seem invariably to consist of one sex only, Shaw (1999) speculates that this could be a case of polyembryony, which would certainly be worth investigating. Polyembryony, whilst known in several Hymenoptera families including the Braconidae (subfamily Macrocentrinae), has not been demonstrated in Ichneumonidae. Two Costa Rican species of *Diradops* Townes (tribe Atrophini) are gregarious parasitoids of Erebidae (Arctiinae) and Lasiocampidae (Gauld *et al.*, 2002a).

Species of Glyptini are remarkably consistent in their known overall host range, with the vast majority of reliable host records being from Tortricidae that feed in leaf-rolls, galls, flower-heads, roots and other sites of concealment. This is certainly true of *Glypta* and close relatives (e.g. *Apophua* and the non-British *Cephaloglypta* Obrtel, which may just represent specialised off-shoots of *Glypta*), although some putatively more distantly related genera have been reared from other families of Lepidoptera elsewhere in the world. Some species-groups of *Glypta* share particular host associations, such as the *G. mensurator*-group, parasitoids of tortricids in Apiaceae seed heads and stems, whereas species of *Diblastomorpha* (recently separated from *Glypta* by Watanabe & Maeto, 2013) attack foliage-feeding, leaf-rolling hosts. *Teleutaea brischkei* (Holmgren) has not been reared from an identified host in Britain but there are rearing records from Tortricidae for this (e.g. Aubert, 1978b) and other *Teleutaea* species.

Hosts of British Banchini are all Noctuidae which pupate in the soil, typically species of the subfamilies Hadeninae and Noctuinae but with some host records from Cuculliinae (*Exetastes*) and Xyleninae (*Banchus*). Although *Rhynchobanchus flavopictus* Heinrich has been reared from a cocoon collected in soil (Fitton, 1987), there are no host records for the genus. The extralimital host range of the *Banchus*-group as a whole is slightly wider, with two species of the Australian *Philogalleria* Cameron having been reared from Geometridae (Fitton, 1987).

The Atrophini comprises a more morphologically diverse assemblage of genera than the other two tribes and this is reflected in a wider range of hosts. Larger species in genera such as *Alloplasta*, *Lissonota s.s.* and *Lissonota* ('*Meniscus*'), tend to be parasitoids of noctuid larvae, and *Lissonota* ('*Lampronota*') species apparently specialise on wood-boring and stem-boring Cossidae and Sesiidae. The largest British banchine, the impressive *Lissonota setosa* (Geoffroy), is a specialist parasitoid of *Cossus cossus* (L.) and tends to be illustrated in field guides as a 'typical' ichneumonid, despite being very rare; indeed, almost all online images purporting to be of *L. setosa* are misidentified. Hosts of British *Lissonota* span a wide variety of families; Amphypyrinae (Noctuidae) which feed as young larvae in grass heads (moving down to stems and roots) are targeted by several species of *L.* ('*Lissonota*') while *L.* ('*Campocineta*') species have been reared from smaller hosts of the families Adelidae, Choreutidae, Crambidae (Scopariinae), Incurvariidae, Psychidae, Tortricidae and Yponomeutidae (Brock, 2017). Some species-groups which seem to consist of closely related species have distinct and narrow host associations, for example several species of the *Lissonota saturator*-group parasitise Psychidae and species of the *L. gracilenta*-group attack Scopariinae in mosses. One bivoltine species, *Lissonota biguttata* Holmgren, is a well-known parasitoid of the foliage-feeding Winter Moth, *Operophtera brumata* (L.) (Geometridae), but the alternating generation seems to be associated with the very different host, *Esperia sulphurella* (Fabricius) (Oecophoridae), which feeds on decaying wood (Brock, 2017). Species of *Cryptopimpla*, which morphologically could probably be accommodated within *Lissonota*, seem, on the basis of rather few rearings, to be parasitoids of Geometridae larvae, predominantly in low vegetation. *Arenetra pilosella* (Gravenhorst) is a parasitoid of noctuid larvae and both British *Alloplasta* species have been at least tentatively associated with *Orthosia* (Noctuidae) species: the more common *A. piceator* (Thunberg) is a regular parasitoid of *Orthosia gothica* (L.) and *O. gracilis* (Denis & Schiffermüller) (Noctuidae). The distinctive genus *Syzeuctus* seems to comprise mainly parasitoids of Pyralidae. Some extralimital genera with short ovipositors (*Diradops* and *Hylesicida* Ugalde & Gauld) have been reared in Costa Rica from exposed Erebidae (Arctiinae), Lasiocampidae and

Saturniidae larvae (Gauld *et al.*, 2002a). A few species of *Quillonota* Ugalde & Gauld and *Wahlamia* Ugalde & Gauld have been reared in Costa Rica from lepidopterous stem-galls and, no doubt for this reason, have distinct denticles on their ovipositors (Gauld *et al.*, 2002a), but in Britain the hosts of most atrophines are usually concealed in far less tough substrates, with the exception of some *L.* ('*Lampronota*') species which presumably use cracks in the wood to access their cossid or sesiid hosts.

Figure 47. *Lissonota lineolaris* (Gmelin) probing for *Apamea* (Lepidoptera: Noctuidae) larva (P. Adams).

Figure 48. *Glypta* sp. probing for tortricid (Lepidoptera: Tortricidae) larva (L. Counter).

Various authors report particular ecological preferences of banchines. At a broad habitat-scale, *Cryptopimpla* and *Banchus* species seem predominantly to inhabit open areas of low vegetation (Townes & Townes, 1978; Brock, 2017), although *Banchus hastator* (Fabricius) is a parasitoid of the pine-feeding *Panolis flammea* (Denis & Schiffermüller) (Noctuidae) (van Veen, 1982). There are several *Glypta* and *Lissonota* species which inhabit water margins, including ponds and fenlands, and *Teleutaea brischkei* seems to be confined to fens in Britain. At a finer level, *Cephaloglypta murinanae* (Bauer) shows a preference for hosts on the crowns of firs (*Abies*) rather than in dense thickets (Zwölfer, 1961). Felland (1990) found that the North American *Lissonota brunnea* (Cresson) would not be effective at controlling populations of the stalk-boring noctuid *Papaipema nebris* (Guenée) as *L. brunnea* seems to be a habitat specialist, attacking hosts in perennial ragweed but not in crops, where the moth causes damage.

Several species of Banchinae are known in Britain only from a few very old specimens, but some species previously considered very rare have turned out to be widespread when their hosts are reared, or have proved to be frequent in Malaise traps (Brock, 2017). However, there is some evidence for declines of a few banchines; for example, at least three of the eight British *Banchus* species have not been collected for many years, although all *Banchus* are fairly conspicuous, and *Lissonota setosa* has been collected in the last thirty years at only a couple of sites; unsurprisingly, given the much reduced range of its sole host, *Cossus cossus*. A few species may have become more numerous in recent years, such as the distinctive, attractively patterned *Lissonota luffiator* Aubert, which was only described in 1969 and is poorly represented in older collections, yet now seems to be a frequent parasitoid of *Luffia ferchaultella* (Stephens) (Psychidae) in southern England. Several

Exetastes species seem very prone to major swings in abundance, and some have been common in gardens at various times.

Two additional banchine genera are found in northern Europe but not yet in Britain. *Helotorus capitator* Townes (Atrophini) has been reared from an *Argyresthia* sp. (Yponomeutidae) in Sweden (Townes & Townes, 1978); females have a distinctive ovipositor tip but males would probably be passed over as *Lissonota*. *Cephaloglypta murinanae* is a specialist parasitoid of the *Abies*-feeding *Choristoneura murinana* (Hübner) (Tortricidae) (Zwölfer, 1961), a moth which has not yet been found in Britain.

Developmental biology. There are reliable host records for a fair number of Banchinae, unsurprising given that their hosts are Lepidoptera, but details of host-parasitoid interactions are generally sketchy. As is usually the case, the most detailed studies of the biology of Banchinae have been of parasitoids of pest Lepidoptera species, which for banchines have generally been defoliators (e.g. Zwölfer, 1961; Münster-Swendsen, 1979; van Veen, 1982; Rappaport & Page, 1985). *Lissonota dubia* Homgren was numerically the most significant parasitoid of *Epinotia tedella* Clerck (Tortricidae) in the study of Münster-Swendsen (1979) and, in combination with some braconid parasitoids, exerted significant effects on the host population. Conveniently, the most comprehensive life-history studies have concerned species from each of the three tribes, although the glyptines concerned are not British species. Oviposition by banchines is into early, usually first or second, instar hosts. There seems to be no information on how banchines orientate towards their hosts but there are some observations on short-range detection and acceptance of hosts. Faecal piles at the entrance of the larval mine of *E. tedella* elicit probing by *Lissonota dubia* (Münster-Swendsen, 1979), and there is a period of only 3-5 days in which the host is accessible before its mine becomes too long. *Lissonota* species frequently parasitize their early instar hosts before they bore into stems or other more deeply concealed locations; a few species can commonly be found probing in grass seed heads for early instar larvae of *Apamea* (Noctuidae) before they descend to the roots. *Banchus hastator* (Fabricius) is stimulated to oviposit on contact with the cuticle of its exposed-feeding host, *Panolis flammea* (van Veen, 1982). The ovipositor of *B. hastator* is inserted up to the notch, which is sharply defined and seems to hold the host's cuticle in place during oviposition (van Veen, 1982). Eggs are slightly elliptical and narrower at one end. The parasitoid can develop best following oviposition into second to fourth instar larvae, although second instars are preferred, and first instar larvae frequently die when oviposited into (Bledowski & Krianska, 1926) whereas fifth instar hosts are completely ignored (van Veen, 1982). When the host is a fourth instar larva, its development is retarded relative to unparasitised hosts, and it descends to the forest floor later (van Veen, 1982). *Glypta fumiferanae* (Viereck) utilises second instar *Choristoneura fumiferana* (Clemens) (Tortricidae) as a host (Rappaport & Page, 1985) and will oviposit in either exposed larvae or those concealed in their hibernacula. The various studies do not report any preference for oviposition site although larvae tend to migrate to the posterior of the host for development, at least in *L. dubia* (Jørgensen, 1975; Münster-Swendsen, 1979), but this is only after an initial migration to the anterior end where supernumeraries compete. No banchines studied have been found to avoid superparasitism but supernumeraries are suppressed physiologically with just the older larva developing. Parasitism of *Epinotia tedella* by *Lissonota dubia* results in physiological suppression of the host's wing buds and sexual organs (Münster-Swendsen, 1979). In *L. dubia*, the winter is passed as a second instar in the cocooned prepupal host on the forest floor, with development completed the following April; there are five larval instars, which is probably true throughout the subfamily (e.g. Slovák, 1984). The first instar larva is caudate, as in many ichneumonids (Slovák, 1984). The non-British, but European, *Cephaloglypta murinanae* overwinters as a first instar larva, but *Lissonota stigmator* Aubert, which is a common univoltine parasitoid of just the overwintering larval generation of the plurivoltine choreutid *Anthophila fabriciana* (L.) in Britain, overwinters as an unenclosed pharate adult (with full adult cuticular development, but wings unexpanded) in its cocoon, the adult emerging in mid to late February (Shaw, 2017b). There is less information on exactly how it is done by other species which overwinter in their cocoons (which is usual in the tribe Banchini, common in Atrophini, but unusual – possibly never occurring – in Glyptini which have very insubstantial cocoons), though some other species flying early in spring (for example, several rather rarely collected *Banchus* species, and *Arenetra pilosella*) seem likely to adopt the same means of becoming active ahead of real warmth, as has been surmised by Fitton (1985b). In many species of Glyptini, and Atrophini with generations that emerge as adults in the year of cocoon formation, the pupal stage lasts for two to three weeks.

Most banchines are univoltine and overwinter in cocoons or as early instar larvae in their hosts. Maintaining a culture of *G. fumiferanae* in the laboratory meant that the larval diapause was circumvented and the parasitoid could be continuously brooded (Rappaport & Page, 1985). Where known, all Banchini pass the winter in densely woven, black, elongate cocoons in the soil; many other banchines have similar cocoons, though in some *Lissonota* species dark cocoon colour results only under conditions of high humidity at the time of cocoon formation (Shaw, 2017b). The pupation site is more diverse in the Atrophini and Glyptini (though normally still concealed, at least weakly) and some bivoltine Atrophini are known. *Cephaloglypta murinanae* pupates in a thin, transparent cocoon which is exposed on a needle or twig, incurring high levels of pseudohyperparasitism (Zwölfer, 1961). Because many banchines, especially in the largest tribes Atrophini and Glyptini, use hosts that can only be reached by long ovipositors, or which rapidly conceal themselves to feed in secluded sites, hyperparasitism rates of banchines are usually low. However, this is not a true property of the subfamily, as banchines parasitising hosts whose accessibility to parasitoids continues through their larval feeding periods (e.g. most Banchini, some Atrophini) are no less hyperparasitised than other koinobionts with similar development.

Females of *Banchus hastator* emerge with some mature eggs. Female longevity is significantly reduced (to an average 11 days) when they are presented with a continuous supply of hosts, compared to females given nutrition but no hosts (an average of 24 days). An average of 40 progeny were produced by females given continuous access to hosts (van Veen, 1982). As with most ichneumonids, mating in *Exetastes atrator* (Forster) and *Glypta fumiferanae* is brief (up to three and a half minutes duration) and there is no discernible courtship (Rappaport & Page, 1985; Slovák, 1986). Females are receptive for only a brief period, during which time they may mate with multiple males. Females preferred slightly older males, which is presumably common in ichneumonids, with males typically emerging earlier than females. In contrast, courtship in *Banchus hastator* can extend for a few minutes and involves some distinctive behaviours, though actual mating is brief at around 55 seconds. The male faces the female and fans his wings, presumably producing an airstream over the female, and raises his abdomen (van Veen, 1982), which, in contrast to the female, is conspicuously banded with yellow. Sexual dimorphism is pronounced in most *Banchus* species, with the males more extravagantly marked than females, with at least the face more yellow. These colour differences (reflected to a less striking extent in much of the subfamily) may be associated with courtship. Conspicuous colour patterns in *Banchus* species are supposedly also aposematic and, when handled, at least some *Banchus* species release a distinct, pungent odour (Townes & Townes, 1978).

Polydnaviruses have been isolated from species of *Apophua*, *Glypta* and *Lissonota* (Stoltz et al., 1981; Stoltz & Whitfield, 1992; Djoumad et al., 2013) and it would be of interest to know whether they are found in Banchini too. Based on the site of origin of the particles, in the calyx epithelium, and characteristics of the DNA, these were presumed to be polydnaviruses, which has been confirmed for the virus carried by *Glypta fumiferanae* and *Apophua simplicipes* (Cresson) characterised by Lapointe et al. (2007) and Djounad et al. (2013), who concluded that banchine polydnaviruses may be an independent acquisition from other ichnoviruses, although they are more closely related to campoplegine polydnaviruses than to the bracoviruses found in Braconidae. Nothing is known of the function of these polydnaviruses in Banchinae but, by analogy with what is known of their function in Campopleginae, they may be assumed to help overcome the immune responses of their hosts. Some other parasitoids take advantage of the host's compromised immune response – for example, *Exochus tibialis* Holmgren (Ichneumonidae: Metopiinae) habitually cleptoparasitises *Lissonota dubia* in its tortricid host, *Epinotia tedella* (Münster-Swendsen, 1979).

Identification. Brock's (2017) handbook covers all of the British Banchinae. Aubert's (1978b) and Kuslitzky's (1981) keys to the European Banchinae may be found to be useful; Kuzlitsky's Atrophini and Banchini keys are based mainly on Aubert's. However, these works should be used with caution as the keys are rather simplified and there have been numerous misidentifications of European banchines, especially of *Lissonota* (Brock, 2017). Fitton (1985b) revised the Palaearctic *Banchus* species and (Fitton, 1987) provided a key to the genera of the *Banchus*-group, separating the two European species of *Rhynchobanchus* (one has been found in Britain).

Figure 49. Morphology of Banchinae: (a) *Lissonota impressor* Gravenhorst (Atrophini) head; (b) *Banchus volutatorius* (L.) (Banchini) male head; (c) *Exetastes illusor* Gravenhorst (Banchini) head; (d) *B. volutatorius* male palps and fore tarsus; (e) *B. volutatorius* male mesopleuron; (f) *B. volutatorius* scutellum; (g) *B. volutatorius* ovipositor: (continued overleaf).

Figure 49 continued: Morphology of Banchinae: (h) *E. illusor* propodeum and metasoma (lateral); (i) *Glypta similis* Bridgman (Glyptini) metasoma dorsal; (j) *B. volutatorius* male metasoma dorsal; (k) *E. illusor* propodeum and metasoma (dorsal); (l) *B. volutatorius* wings; (m) *E. illusor* fore wing; (n) *L. impressor* propodeum; (o) *G. similis* propodeum; (p) *B. volutatorius* male scutellum and propodeum (dorsal); (q) *L. impressor* mesosoma (lateral).

Subfamily **BRACHYCYRTINAE**

The Brachycyrtinae contains one genus, *Brachycyrtus* Kriechbaumer, with 22 described species (Yu *et al.*, 2012) and having an almost worldwide distribution, although none has yet been described from continental Africa (species have been described from Madagascar; Seyrig, 1952). Previously included as a tribe in the subfamily Labeninae (e.g. Townes, 1969), the Brachycyrtini were elevated to subfamily rank by Wahl (1993a). Genera other than *Brachycyrtus* that were previously included in Brachycyrtini *sensu* Townes (1969) have since been removed to the Pedunculinae (Porter, 1998; Gauld & Ward, 2000).

Figure 50. *Brachycyrtus ornatus* Kriechbaumer.

The single European species, *Brachycyrtus ornatus* Kriechbaumer, has not yet been found in Britain but is a potential colonist, being widely distributed in Europe and having recently been found as far north as Southern Sweden (P. Magnusson, pers. comm.). *Brachycyrtus ornatus* is readily identified, with a very hunched (short and high) appearance to the mesosoma, a long first metasomal segment with tergite and sternite fused, distinctive wing venation (with the areolet elongate triangular, though not closed by vein *3rs-m*, and the abscissa of hind wing vein *RS* between *RA* and *rs-m* conspicuously shorter than *rs-m*) and conspicuous yellow markings. It is described briefly and/or figured by Ceballos (1942), Constantineanu (1961), Townes (1969) and Kasparyan (1981).

There is scant information on the biology of *Brachycyrtus*, and no reliable host records for *B. ornatus*. Some species have been reared from cocoons of green lacewings (Neuroptera: Chrysopidae) (e.g. Townes *et al.*, 1960), with adult emergence from the host cocoon (Kusigemati, 1981). Based on their phylogenetic position and toothed ovipositor, *Brachycyrtus* species are presumed to be idiobiont ectoparasitoids although this has not yet been demonstrated.

Figure 51. Morphology of Brachycyrtinae (*Brachycyrtus ornatus* Kriechbaumer): (a) wings; (b) mesosoma; (c) propodeum and first metasomal segment.

Subfamily CAMPOPLEGINAE

This is a big subfamily that comprises one of the two largest ichneumonid radiations of parasitoids of Lepidoptera (the other being the Ichneumoninae), although in the case of Campopleginae hosts in other orders are also involved. The approximately 2,100 described species in 66 genera certainly represent only a small proportion of the true global diversity of a group that seems to be species-rich in both temperate and tropical areas. The British fauna, of 336 species in 37 genera, makes this the third most species-rich subfamily, but considering the number of poorly revised genera and undescribed species, it is probably the second largest behind Phygadeuontinae.

Figure 52. *Campoplex eudoniae* Horstmann & Yu.

Recognition. The Campopleginae is a morphologically rather homogeneous group with the majority of species recognisable as such by the petiolate first metasomal segment; subapically notched dorsal valve of the ovipositor; usually complete posterior transverse carina of the mesosternum (narrowly interrupted in a few small genera); the face with fairly dense, silvery setae; the face and clypeus only faintly separated by a very shallow sulcus; the predominantly dull, coriaceous mesosomal sculpture; and the usually uniformly black mesosoma, at least in British species (a very few species have yellow-marked faces). Confusion with Cremastinae is likely, but cremastines have the hind tibial spurs inserted in separate sockets (uniquely in Ichneumonidae) and usually have some light colouring on the head and mesosoma, with distinct differences in the shape of the pterostigma and the clypeus: the pterostigma of cremastines is large and triangular, slenderer in campoplegines; and the clypeus of cremastines is distinctly separated from the face and not conspicuously wider than high, whereas the clypeus of campoplegines is poorly differentiated from the face and apparently wider than high. Small male phygadeuontines and campoplegines can be confused but campoplegines lack a distinct sternaulus and most phygadeuontines have only small sections of the posterior transverse carina of the mesosternum. The areolet shape of phygadeuontines, even if the position of 3*rs-m* is only indicated, is distinctively pentagonal compared to the more rhombic or petiolate shape of campoplegines (and various other groups), but this can be a subtle character.

Systematics. The Campopleginae is probably the sister group to Cremastinae within a clade of 'higher' ophioniformes (Quicke *et al.* 2009). Most of this clade comprises parasitoids of Lepidoptera and all possess a petiolate first metasomal segment. Townes (1970b) called this subfamily the Porizontinae and used tribes unsatisfactorily to divide the group. Tribes were sensibly abandoned by Wahl (1991) and two of Townes's small (non-British) tribes have been removed to the Ophioninae (*Hellwigia* Gravenhorst and *Skiapus* Morley) and to a new subfamily (Nesomesochorinae, comprising *Chriodes* Förster, *Klutiana* Betrem and *Nonnus* Cresson) (Quicke *et al.*, 2005; 2009). Taxonomically, the Campopleginae is a difficult and relatively poorly studied group. There are several large, cosmopolitan genera surrounded by satellites of small genera that tend to be defined on the basis of Palaearctic species, which do not take into account the large and diverse Nearctic and Neotropical faunas. The generic classification is certainly due a global revision.

Biology.
Host associations
Species of Campopleginae develop as koinobiont endoparasitoids, the vast majority as solitary parasitoids of Lepidoptera. Campoplegines, as far as is known, always attack the host in its larval stage, very frequently the early instars and it is known that one non-British species of *Hyposoter* oviposits into host larvae just before they actually hatch from their batched eggs (Nouhuys & Kaartinen, 2008). Although several species are known to be able to parasitize quite well grown host larvae, there is probably considerable flexibility in these cases, and often a higher success rate when the host is attacked young. There have been multiple colonisations of hosts in holometabolous insect orders other than Lepidoptera, but subsequent radiations have been limited, resulting in relatively small numbers of species. Several rather small genera use Coleoptera larvae as hosts: three genera are parasitoids of exposed phytophagous beetles, namely *Bathyplectes* on Curculionidae (Horstmann, 1974) and *Lemophagus* and *Nepiesta* on Chrysomelidae (Haye & Kenis, 2004; Jolivet, 1950; Jolivet & Théodoridès, 1952); *Lathroplex* are parasitoids of Dermestidae (Horstmann, 1978a); *Pyracmon* of subterranean phytophagous Elateridae and Artematopodidae (Barron & Walley, 1983); and *Rhimphoctona* of wood-boring Cerambycidae and Curculionidae (Horstmann, 1980c; Sanborne, 1986). Species of *Dolophron*, *Lathrostizus* and *Olesicampe* (the largest of these three genera) are parasitoids of larval sawflies (Hymenoptera: Pamphilioidea and Tenthredinoidea). *Dolophron* attack only *Heterarthrus* leaf-miners (Horstmann, 1978a), *Lathrostizus*

Figure 53. *Dusona stragifex* (Förster).

are parasitoids of gall-forming Nematinae (Tenthredinidae) (e.g. Carleton, 1939; Kasparyan & Kopelke, 2009) and *Olesicampe* parasitise exposed Tenthredinoidea (e.g. Morris *et al.*, 1937). Some *Nemeritis* are parasitoids of Raphidioptera as well as Lepidoptera (Tortricidae) and predatory Coleoptera (Cleroidea) (Horstmann, 1994a); and one species of *Diadegma* has been recorded from a terrestrial caddis (Trichoptera) (van Achterberg, 2002; Horstmann, 2004a), though confirmation is desirable as the validity has been questioned (Shaw, 2017a).

The evolution of host utilisation pathways in Campopleginae is potentially very interesting but any analysis of these pathways is severely hampered by our lack of understanding of the internal phylogeny of this species-rich group. Wahl (1991) proposed some generic groupings and identified an apparently monophyletic group comprising *Bathyplectes*, *Nepiesta*, *Pyracmon*, *Rhimphoctona*, which are parasitoids of Coleoptera, and the non-British *Leptoperilissus* Schmiedeknecht (supposedly a parasitoid of Lepidoptera; Horstmann, 1993d), but in Wahl's (1991) scenario most of the campoplegine genera, and host utilisation transitions, were accommodated within a very large '*Dusona* genus-group'. There seem to have been multiple independent transitions away from Lepidoptera as hosts, probably from within larger genera; for example, Wahl (1991) notes that *Olesicampe* is probably derived from within (Lepidoptera-parasitizing) *Hyposoter*. Compared to the classification of Townes (1970b), Finlayson (1975), on the basis of final instar larval morphology, suggested some novel generic groupings. Quicke *et al.* (2009) included numerous campoplegines in their molecular phylogenetic reconstruction of the Ichneumonidae, which demonstrated that there was some useful phylogenetic signal in the 28S D2 gene, and provided some tentative support to some of Finlayson's (1975) hypotheses (e.g. the close relationship between *Sinophorus* and *Venturia*), but the outcome was very far from indicating a robust phylogeny. These studies have all suggested that the recognition of tribes by Townes (1970b) (Campoplegini and Porizontini) is untenable, which will come as no surprise to all those who have struggled with the key couplet separating the two.

Life histories
It is known or presumed that the vast majority of campoplegines overwinter as a young larva in an overwintering host larva, or in the cocoon stage (Shaw *et al.*, 2016) either as a prepupa or in some cases as a pharate or eclosed adult (e.g. Haye & Kenis, 2004). However, a Holarctic but non-British species (*Enytus montanus* (Ashmead)) has been found by Baltensweiler (1958, as *Horogenes exareolatus*) to overwinter as an adult, in soil, and there are several common British species that emerge as adults late in the year, with no evident 'next host' available, that might do something similar, although direct evidence is lacking (Shaw *et al.*, 2016). In many cases, probably the majority,

Figure 54. Mating pair of *Diadegma fabricianae* Horstmann & Shaw (M.R. Shaw).

hosts are parasitised while still young. Some species attack older host larvae, although in these cases there is probably considerable flexibility and sometimes constraint (e.g. Corbet & Rotherham, 1965; Smilowitz & Iwantsch, 1973, 1975; Haye & Kenis, 2004). In many genera (e.g. *Campoplex*, *Diadegma*, *Dusona* and *Olesicampe* of the large genera in Britain) the host is normally killed as a prepupa in its pupation site, the final and destructive feeding phase of the parasitoid possibly (though see below) being triggered by the host becoming prepupal. *Campoplex* and *Diadegma* have moderately long ovipositors and both predominantly parasitize more or less concealed 'microlepidoptera', with a few species instead specializing on 'macrolepidoptera' having a similar feeding habit. Examples are *Diadegma aculeata* (Bridgman) which is a regular parasitoid of the lycaenid *Cupido minimus* (Fuessly) feeding in *Anthyllis* flowerheads, and *Campoplex brevicornis* (Szépligeti) which parasitises the geometrid *Eupithecia venosata* (Fabricius) in *Silene vulgaris* and *S. uniflora* flowers. A few species of *Campoplex*, and also of *Enytus*, that normally parasitize small 'microlepidoptera' and kill the host as a prepupa, occasionally parasitise young semi-concealed butterfly larvae, such as *Vanessa atalanta* (L.) (Shaw et al., 2009), and in these cases the host is killed while still small, suggesting that prepupal hormonal changes in the host are not the only trigger for final larval development by the parasitoid, but rather that a host size threshold may be important. *Dusona*, the other large genus in which the host is normally killed as a prepupa, is associated with 'macrolepidoptera', especially in the families Geometridae, Noctuidae and Notodontidae. Known hosts (Horstmann, 2011a) are particularly those having smooth as opposed to densely long-setose skins, though taxa with caterpillars that live semi-concealed as well as fully exposed are used. *Dusona* species have a laterally compressed, blade-like metasoma, and this might be an important aid in reaching some of its lightly-concealed hosts as the ovipositor, though variable and sometimes robust, is generally short in this genus. Odd species in a few genera (e.g. *Dusona admontina* (Speiser) and *D. leptogaster* (Holmgren), certain species of *Campoplex* (Carlson, 1979; also *C. brevicornis* in Britain), and *Diadegma scotiae* (Bridgman) (Shaw & Horstmann, 1997)), habitually spin their cocoon within the pupal shell of the host, and in some other species this just happens occasionally (e.g. in *Diadegma fabricianae* Horstmann & Shaw parasitizing the choreutid moth *Anthophila fabriciana* (L.); Shaw, 2017b). In several genera, however, and especially in those that attack exposed 'macrolepidoptera', the host is normally killed well before it has attained full growth, and the parasitoid cocoon is therefore not constructed in a selected or specially prepared relatively safe site, such as a pupation chamber, but rather in the open. The large genera involved are *Casinaria*, *Campoletis*, *Hyposoter* and *Phobocampe*, and rather often the form or colour of the cocoons made by particular species is so distinctive as to be diagnostic. These campoplegines seem on the whole not to influence host behaviour just prior to its death to the extent seen in many braconids (although host manipulation leading to the parasitoid cocoon dangling from an isolated leaf mid-rib has been shown in at least one Costa Rican campoplegine; I.D. Gauld, pers. comm.), so cocoon construction tends to be more or less where the host would normally be found – although

Figure 55. Cocoon of *Casinaria mesozosta* (Gravenhorst).

Figure 56. Cocoon of *Hyposoter tricolor* (Ratzeburg) (Campopleginae) alongside the cadaver of its host, *Abraxas grossulariata* (L.) (Lepidoptera: Geometridae) (M.R. Shaw).

Figure 57 (above). Cocoon of *Hyposoter rhodocerae* (Rondani) (Campopleginae) within the larval cuticle of its host, *Gonepteryx rhamni* (L.) (Lepidoptera: Pieridae) (M.R. Shaw).

Figure 58 (right). Larva of *Aricia artaxerxes* (Fabricius) (Lepidoptera: Lycaenidae) parasitised by *Hyposoter notatus* (Gravenhorst) (Campopleginae). Top: quiescent as the parasitoid larva completes its feeding. Bottom: after the parasitoid has constructed its cocoon within the host's skin (M. R. Shaw).

it seems that there is often at least some movement to a final resting place, such as a twig or an exposed position, that has some advantage to the parasitoid. There are two alternative strategies for protecting the parasitoid cocoon, which is vulnerable both to predators and pseudohyperparasitoids, seen in these cases. First, the cocoon may be made in an exposed situation where the warmth of insolation will accelerate development so that the adult hatches quickly. In these cases (especially species of *Campoletis*, *Casinaria*, *Hyposoter* and one species usually classified as *Phobocampe*: *P. bicingulata* (Gravenhorst)) the parasitoids usually incorporate dark-coloured meconium into the cocoon, so that it resembles a bird's dropping (Figs 55-57) – completely unlike the unicolorous brownish cocoons typical of species that kill the host in its pupation chamber. In a few cases, an alternative is to construct an outer, fluffier, false cocoon, giving the appearance of bird down, or leave a protuding empty sac as a decoy beyond the parasitoid's actual pupation chamber (e.g. as in *Hyposoter carbonarius* (Ratzeburg), a parasitoid of the erebids (Lymantriinae) *Dicallomeria fascelina* (L.) and *Orgyia* species; see also Finlayson, 1966). Several species of *Hyposoter*, including *H. carbonarius*, parasitize setose, spiny or toughly-skinned granulose caterpillars, and in these cases (but generally not when the host's skin is smooth, unless it has aposematic markings) they make their cocoons inside the skin of the dead host, breaking through either not at all or to a minimal extent. This clearly affords some additional protection, whether it be physical or through its appearance. The second strategy, in Britain employed by the remainder of *Phobocampe* and the small genera *Bathyplectes*, *Callidora* and *Scirtetes*, is to have a motile cocoon. This is achieved by spinning a stout, smooth, ovoid cocoon initially held to the substrate by a minimum of attachment. Once the tough cocoon is completed, the parasitoid larva is able to cause the cocoon first to break loose of its moorings, and then to jump or twitch its way to a crevice or some other sheltered spot at ground level. The exposed weevil hosts of *Bathyplectes*, species of which are by far the most active jumpers, are usually prepupal when killed, but in host cocoons that are not durable through the cocooned life of the parasitoid. Finlayson (1964) ascribed this ability of the larva within to produce the necessary violent jerks to a caudal appendage she observed in the final instar larva of *Scirtetes*, and it is well known that the behaviour is stimulated by light and heat. It is significant that many of these 'jumping cocoons' overwinter, in some cases (e.g. in the univoltine *Scirtetes robusta* (Woldstedt)) spending about 10 months in the leaf litter. In some species (especially of *Bathyplectes*, and in *Scirtetes robusta*) the cocoons are distinctively marked by a reinforced band; in several other cases the species concerned are at least partially plurivoltine and there is then often a difference in the strength, and hence resource input, for overwintering and non-overwintering cocoons (cf. Cross & Simpson, 1972), a recurrent phenomenon in cocoons, particularly if more or less exposed, of many plurivoltine ichneumonoids. Other strategies include spinning a patterned cocoon suspended permanently on a stout thread; this is common in several exotic *Casinaria* (Jerman & Gauld, 1988) and species of *Charops*, although our only British repesentative of the latter, *C. cantator* (Degeer), makes its cocoon within that of its host *Zygaena* (Zygaenidae) species.

Within many genera, individual species appear to have rather narrow host ranges and to be physiologically well adapted to their hosts but the host ranges of some genera are overall very wide, for example *Diadegma*, species of which attack many 'microlepidopteran' families, as well as some 'macrolepidoptera', including some butterflies (Lycaenidae) and even a terrestrial caddis (Trichoptera), although this has been questioned (Shaw, 2017a). There is a strong association in *Diadegma* with mining hosts, at least hosts that mine in their early stages (Shaw & Horstmann, 1997). In contrast, *Olesicampe* species are all parasitoids of larval sawflies (Pamphilioidea and Tenthredinoidea) and individual species are thought to be essentially host species-specific (Pschorn-Walcher & Altenhofer, 1999). As in many subfamilies the most obvious outward signs of adaptation to host taxa are in overall size and in ovipositor structure. Genera such as *Campoplex*, with long ovipositors, tend to be parasitoids of larvae in leaf-rolls or spinnings. Species of *Rhimphoctona* have particularly robust ovipositors and an overall morphology reminiscent of poemeniines, xoridines and other parasitoids of wood-boring beetles.

The genus *Diadegma* has, across its many species, a wide spectrum of hosts, so has proved a useful subject for studies of host specificity and the evolution of host range (Horstmann & Shaw, 1984; Shaw & Horstmann, 1997). In particular, Shaw & Horstmann (1997) summarised a large body of rearings of the *Diadegma nanus* species-group and found some evidence for the hypotheses of Shaw (1994), that speciation can occur in the manner of specialists budding from generalists and that taxonomically isolated species are often 'old specialists' (isolated for long periods of evolutionary time on taxonomically or ecologically isolated hosts, without producing near relatives through a process of host recruitment then speciation). In the absence of a phylogeny much of this remains supposition but host ranges within the *nanus* group are illuminating and certain patterns can be teased apart that provide much support to hypotheses of host range evolution. Across the whole *Diadegma nanus* species-group (which may or may not be monophyletic) parasitising essentially leaf-miners, several groups of Ditrysian Lepidoptera serve as hosts, but the only Monotrysian family used seems to be Tischeriidae, with the species-rich and exclusively leaf-mining Nepticulidae apparently escaping attack. The reasons for this restriction are unknown, although it is noteworthy that the several groups of braconid parasitoids of nepticulids are highly specialised to this host group (Shaw & Huddleston, 1991).

Developmental biology
A few species of Campopleginae have been studied in great detail and much of what we know about the biology of the subfamily derives from studies of these species. Whilst several campoplegines have been extensively studied for their potential in biological control of both pest Lepidoptera and sawflies (Hymenoptera), others have been studied because they are easy to rear on commonly cultured lepidopteran hosts and thus serve as model laboratory organisms. The most thorough studies of the developmental biology of Campopleginae are those of Fisher (1959) on the life history of *Diadegma chrysostictos* (Gmelin) and, especially, the many publications by Salt on *Venturia canescens* (Gravenhorst), both being convenient laboratory animals capable of continuous culture on 'flour moths' such as *Ephestia* species (Lepidoptera: Pyralidae). Salt's work was particularly broad, and is summarised towards the end of this chapter. However, some of the attributes that make species such as *Venturia canescens* such attractive experimental subjects may be unusual within Campopleginae as a whole.

Many laboratory studies of North American species in genera such as *Hyposoter* and *Campoletis* parasitizing pest Noctuidae (and sometimes Sphingidae) have revealed a range of complex interactions through which the parasitoid larva, alone and/or mediated by substances injected during oviposition, synchronises with or regulates the development of the host. Much of the pioneer work is summarized by review articles through which the primary literature relevant to Campopleginae can be traced (Vinson & Iwantsch, 1980a,b; Stoltz, 1986; see also Beckage & Templeton, 1985).

Although solitary development is by the far the most common strategy, at least one species, *Olesicampe clandestina* (Holmgren), develops in gregarious broods in the sawfly, *Cimbex femorata* (L.) (Hymenoptera: Cimbicidae) (Shaw, 1999). Although oviposition by campoplegines is, where known, into the haemocoel, larvae of some species will move to specific tissues. Heitland & Pschorn-Walcher (2005) describe how the first instar larva of *Olesicampe vexata* (Holmgren) enters the head capsule of the host larva, the sawfly *Platycampus luridiventris* (Fallén) (Hymenoptera: Tenthredinidae), possibly to evade hyperparasitoid species of *Mesochorus*, which are frequent parasitoids of *Olesicampe*. In *D. chrysostictos* the egg often comes to rest at the posterior end of the host's body; the first instar larva floats free in the haemocoel. Superparasitism is avoided by *D. chrysostictos*, at least under normal host densities, but later in the female's life, under high parasitoid density, superparasitism is frequent. Only one larva can develop in a host and competitors are eliminated by the first instar larva in all species that have been studied in detail (Fisher, 1959; Corbet & Rotherham, 1965; Bartell & Pass, 1978). The older larva punctures supernumerary eggs or larvae with its mandibles, which are then encapsulated at the sites of wounds, or older larvae physiologically inhibit development of younger larvae, as the first instar larvae of at least some *Hyposoter* species lack mandibles (Bahena et al., 1999). The head capsule is well-developed, in contrast to subsequent instars, and early instar campoplegine larvae are conspicuously caudate, with the caudal appendage shrinking in successive instars until it has disappeared in the fourth, at least in *D. chrysostictos*. However, in at least three genera, *Meloboris*, *Scirtetes* (although apparently not the American *S. canadensis* (Walley); Finlayson, 1975) and *Phobocampe*, the final instar larvae retain a conspicuous caudal appendage (Finlayson, 1964). The first instar larva of *Bathyplectes curculionis* (Thomson) does not emerge completely from the egg for the first one to four days, only the head and later the tail emerging at first (Bartell & Pass, 1978). Campoplegine larvae feed initially on the host's haemocoel but switch to feeding on the fat body when they reach a certain size. Eventually, the host organs and tissues, except for the integument, are consumed. As is usual in endoparasitoid larvae, the tracheal system is closed initially. There are five instars in total; Mazanec (1990), studying an exotic *Enytus* sp., found that the tracheal system opens up to the exterior via spiracles in the third instar, when feeding switches to the consumption of the fat body and internal organs, whereas Corbet & Rotherham (1965) report that the spiracles appear in the fifth instar in *Venturia canescens*. In *V. canescens*, and presumably other species, uric acid is deposited in the fat body overlying the gut, appearing as white granules and persisting until the end of the pupal stage (Corbet & Rotherham, 1965). The larva of *D. chrysostictos* spends about six days in its first instar but then only one or two days per subsequent instar as it rapidly feeds and grows. Larvae of *D. chrysostictos* and *Venturia canescens*, when dissected from the same species of host, can be readily distinguished on the shape and position of the head, in the first instar, and differences in the sclerotization of the labial sclerites. By the final instar, there are nine pairs of spiracles connected to the tracheal system. The larva of *D. chrysostictos* pupates in a cocoon within the ruptured skin of its host. *Diadegma* species reared by Shaw & Horstmann (1997) killed the host as a prepupa (or pupa, in one case) and made their cocoon within the host's pupation site. Uniquely amongst insects, most adult parasitoids lack lipogenesis (Jervis et al., 2008; Visser et al., 2010), and therefore require all of their lipids (essential for egg production) to be synthesised by the time of eclosion, which in *V. canescens* means that lipids comprise 94% of teneral resources (Casas et al., 2003). In a range of observed species, oviposition takes place soon after emergence, regardless of whether or not the female has mated, indicating at least partial pro-ovigeny. Mating has been recorded to last from one to 20 minutes, which is longer than in ichneumonids of many other subfamilies, and male sperm of *D. chrysostictos* is depleted after five or six days (Fisher, 1959). In *Bathyplectes anurus* (Thomson), courtship is almost non-existent, other than that males vibrate wings and antennae when females are present, and females mate only once (Gordh & Hendrickson, 1976). When seeking hosts, female *D. chrysostictos* orientate towards the flour and oatmeal food of their hosts and then the odours of their host larvae. Even when hosts are no longer present, the allomones left behind elicit stabbing with the ovipositor into the host medium, as is known also in *V. canescens* (Thorpe & Jones, 1937). When a host is contacted, Fisher (1959) regards the convulsive recoiling of the larva as the stimulus to oviposition, the time taken for which varies considerably from two or three seconds to more than a minute.

Spatial memory has been demonstrated to be an important part of host location in one species of Campopleginae. Nouhuys & Kaartinen (2008) found in Finland that the specialist *Hyposoter horticola* (Gravenhorst) can remember the locations, based on landmarks, of egg masses of its host, the butterfly *Melitaea cinxia* (L.) (Lepidoptera: Nymphalidae) (which is very localised in Britain, with *H. horticola* apparently absent). Many, widely dispersed egg clusters are located and monitored by an individual female until the still unhatched larvae are in a suitable state for parasitism. The high level of spatial memory involved in keeping track of the development of the egg clusters is thought to be an adaptation to hosts being widely scattered and remaining in the appropriate state for oviposition for only a short period of time, and *Hyposoter horticola* found and successfully parasitised a very high percentage of host egg clusters in the study.

Biological control
European species of *Olesicampe* have proved effective in biological control of sawfly pests of forestry introduced to North America, for example *Olesicampe melanogaster* (Thomson) against *Pristiphora erichsonii* (Hartig) and *O. geniculatae* Quednau & Lim against *Pristiphora geniculata* (Hartig) (Kelleher & Hulme, 1984), although neither has been recorded from Britain (Kelleher & Hulme, 1984). Heitland & Pschorn-Walcher (2005) argue that parasitoids such as campoplegines are more successful in biocontrol programmes if only a very limited parasitoid fauna is introduced. If the entire complex is introduced, control is less effective due to competition between parasitoids, which is particularly detrimental to *Olesicampe* if Tryphoninae are included in the guild of biocontrol agents, which, being ectoparasitoids of later host instars, outcompete *Olesicampe*. Muldrew (1967) discusses the biology and efficacy of *O. melanogaster* (as *Olesicampe* sp., and often subsequently referred to by the junior synonym *O. benefactor* Hinz) as a parasitoid of the larch sawfly, *Pristiphora erichsonii*, an economically significant defoliator with a parasitoid complex that has been intensively studied for biocontrol potential. Oviposition is only ever into the host's first instar and development beyond the first instar only occurs when the host has spun a cocoon, with the winter spent in the second and third larval instars; parasitized hosts have retarded growth, which is probably a widespread phenomenon, as in the extralimital *Enytus* sp. studied by Mazanec (1990). Superparasitism is actively avoided, unlike some Ctenopelmatinae attacking the same host. There was never any encapsulation by the host, unlike in *Mesoleius tenthredinis* Morley.

Some *Bathyplectes* species have been studied in detail because of their importance as parasitoids of the alfalfa weevil, *Hypera postica* (Gyllenhal), which was inadvertently introduced to North America from Europe. *Bathyplectes anurus*, *B. curculionis* and *B. stenostigma* (Thomson) (the first two are British species) were introduced to the U.S. in the early 20th Century to control *H. postica*; together with a suite of other parasitoids and a fungal pathogen, they have achieved biological control to a great extent (reviewed by Radcliffe & Flanders, 1998). Release and establishment, particularly of *B. curculionis*, were accompanied by studies on life history parameters and behaviour.

Time for development from oviposition to cocoon formation is 13-18 days for non-diapausing larvae and about three days longer for diapausing larvae (Bartell & Pass, 1978). In contrast to *Diadegma chrysostictos*, *B. curculionis* females frequently superparasitise, including laying more than one egg per oviposition event, and it has been reported that encapsulation rates are lower in cases of superparasitism (Berberet, 1986). *Bathyplectes curculionis* oviposits in the host during its first three instars, when it is concealed in plant buds, and the host is killed as a cocooned prepupa. The *Hypera* cocoon is necessary for the *Bathyplectes* larva to spin its cocoon as it uses the host cocoon as a brace for initial silk spinning (Cross & Simpson, 1972). Non-diapausing larvae took from 24 to 30 hours to spin a cocoon; those destined for diapause took 32-42 hours to spin a thicker cocoon. The *Bathyplectes* larvae were much attacked by hyperparasitoids, especially pteromalids and the ichneumonid *Mesochorus agilis* Cresson. Although *B. anurus* is univoltine and *B. curculionis* plurivoltine, *B. anurus* has ended up displacing *B. curculionis* in many areas where the two have become established. Bartell & Pass (1980) attribute this greater success to the higher fecundity, more rapid handling of hosts and efficiency in eliminating competitors (partly through preferentially attacking slightly older larvae), but greater synchronicity with the host's life cycle and much less frequent encapsulation of its eggs (Puttler, 1967) must surely play important roles too in the dominance of *B. anurus*, which seems to be more closely adapted to *H. postica*. As in *Scirtetes*, and most *Phobocampe*, *Bathyplectes anurus* larvae can cause their cocoon to jump, up to 3 cm (Day, 1970).

Two species of the small genus *Lemophagus* are both parasitoids of *Lilioceris lilii* Scopoli (Coleoptera: Chrysomelidae) (and other *Lilioceris* species), that can be a major pest of ornamental lilies. The two species, *Lemophagus errabundus* (Gravenhorst) and *L. pulcher* (Szépligeti), differ geographically in their relative dominance and only *L. errabundus* has been found in Britain (Salisbury, 2003). Whereas *L. errabundus* is strictly univoltine, the more southerly *L. pulcher* is partly bivoltine (Haye & Kenis, 2004), with the greater part of the population entering larval diapause in the summer, with a partial second generation, although Haye & Kenis (2004) could not establish whether *L. pulcher* was using different host species in this second generation, as *L. lilii* larvae were generally not available in the wild. Host location in *Lemophagus* depends on cues associated with the faecal shield that hides the host chrysomelid larva (Schaffner & Müller, 2001). Oviposition in *L. pulcher* is into first to fourth instar larvae but development is mostly in the host prepupa (Haye & Kenis, 2004). Only three larval instars were recorded in *L. pulcher*, but the possibility that rapid parasitoid development resulted in middle instars being overlooked cannot be ruled out. Five larval instars seems to be more usual in Campopleginae. Mating lasts about a minute and oviposition is very rapid, taking less than a second. Development is reported to be very similar in *L. errabundus* (Haye & Kenis, 2004) but the first instar larva of *L. errabundus* differs in its much greater head capsule sclerotization.

The American *Campoletis sonorensis* (Cameron) is an important parasitoid of another lepidopteran agricultural pest, *Spodoptera frugiperda* (Smith) (Noctuidae), and has been much-studied as a result. The general features of its development agree with what we know about other campoplegines that have been studied in detail (no British *Campoletis* species has been studied in detail). *Campoletis sonorensis* has high fecundity, with a degree of temperature dependence, attacking up to 40+ host larvae per day in the laboratory (Isenhour, 1986). As *C. sonorensis* is easily reared in the laboratory it has become a model organism for research into polydnaviruses. Several species of campoplegines are known to have symbiotic relationships with polydnaviruses which the female injects into the host in the venom at oviposition. These viruses are integrated into the wasp genome but seem to be most closely related to viruses that attack Lepidoptera larvae (Whitfield & Asgari, 2003), which may have been co-opted by the parasitoids to overcome the encapsulation responses of host larvae (Federici & Bigot, 2003). They have been found in various well-studied campoplegines, such as *Diadegma semiclausum* (Hellén), *Hyposoter didymator* (Thunberg) and *Tranosema rostrale* (Brischke). The virus particles replicate in the cells of a thickened part of the lateral oviduct, the calyx, of the ichneumonid. It has been shown that the virus is essential for the successful development of the campoplegine larva but its exact role is unknown. If the virus, the other components of the venom, or both are not injected into the host then the parasitoid egg or larva is encapsulated and dies. In *Tranosema rostrale*, the virions are associated with spiky projections on the egg chorion, which anchor it to basement membranes of the fat body and muscles, whence they migrate to the haemocoel (Cusson *et al.*, 1998), but polydnaviruses are not necessarily associated with the egg chorion (e.g. Huang *et al.*, 2008). Polydnaviruses found in ichneumonids have been classified as ichnoviruses and are thought to have a separate origin from bracoviruses, expressed in Braconidae (Federici & Bigot, 2003). Polydnaviruses have also been found in Banchinae, but that may well represent a separate origin of these viruses (Lapointe *et al.*, 2007), and have been tentatively identified in a ctenopelmatine (Stoltz, 1981). Polydnaviruses ('poly-DNA' because they are double-stranded DNA viruses) are involved in suppressing host immune defences and/or inhibiting metamorphosis and are presumed to be very widespread in Campopleginae. The very well-studied *Venturia canescens* (see below) differs in the presence of virus-like particles which, unlike polydnaviruses, are not integrated into the parasitoid's genome (see reviews by Stoltz & Vinson, 1979; Whitfield & Asgari, 2003).

Biological studies on Venturia canescens
A huge body of literature has built up relating to *Venturia canescens* (usually referred to as *Nemeritis canescens* in the earlier literature), a fairly common and widespread parasitoid of phycitine and galleriine (Pyralidae) moth larvae, particularly well-known as a parasitoid of hosts in flour mills and other such locations. As an easily cultured species in the laboratory, *V. canescens* has served to illuminate topics such as host location, encapsulation and melanisation responses, cocoon spinning, learning ability, foraging strategies, energy budgets, resource allocation, sex determination, sex ratio in relation to latitude, population dynamics, etc. *Venturia canescens* has been a popular laboratory subject for many years, with detailed studies on its development and the encapsulation response of potential hosts going back to the early 1930s (e.g. Rietra, 1932). In a series of papers, George Salt meticulously investigated the nature of the insect immune response, also dealing with host range, initiation of embryogenesis, etc., along the way. As a result there are good accounts of the life history of this species (e.g. Salt, 1964). Salt (1976) analysed in detail the natural and laboratory host range of *V. canescens*. Although its host range is potentially broad he concluded that its realised host range is restricted to certain, mostly pyralid, Lepidoptera larvae. The host range in warehouses, its principal habitat in temperate regions, is artificial to some extent as there is a mix of host species not found in the wild. Salt (1976) found that in the laboratory *V. canescens* can successfully develop in a range of lepidopteran larvae that it would not encounter in the wild and is very occasionally reared, apparently reliably, from wild hosts outside the usual spectrum. There are also, of course, all the usual misidentifications and mistakes in the literature, all of which adds up to a long list of hosts on paper but a much narrower regular host range in nature. Nevertheless, the host range of *V. canescens* includes species in different habitats (e.g. stored grains and fruit orchards) and from rather distantly related families. Much of what we know about *V. canescens* is based on thelytokous, clonal cultures originating from specimens in north temperate bakeries and flour mills, whereas in more southern climes, sexual populations are widespread and both modes of reproduction have apparently been maintained in sympatry for long periods of time. Schneider *et al.* (2003) speculate that certain clones could have done very well in the constant climate of these indoor conditions. In their Mediterranean sampling areas, most thelytokous individuals belonged to one clone, although there were several rarer clones. These were closely related to sexual individuals and it seems that there is occasional sex within otherwise thelytokous populations, as evidenced by Thorpe's (1939) observation of males being produced spontaneously on one occasion in his laboratory culture. Schneider *et al.* (2003) demonstrated that thelytokous females are capable of mating with the male offspring of arrhenotokous individuals and using the sperm to increase genetic diversity, though remaining thelytokous. As Schneider *et al.* (2003) point

out, this has implications for theories as to why sexual reproduction is maintained in nature, given that *V. canescens* clones seem to gain fitness advantages that the sexual lineages do not. This must be borne in mind when interpreting some of the earlier experiments involving *V. canescens*. For example, Salt (1976) and Thorpe & Jones (1937) concluded that *Galleria mellonella* would not be a natural host of northern European *V. canescens*, whereas it seems to be fairly frequently parasitized by *V. canescens* in the wild, particularly in bumblebee, *Bombus* (Apidae), nests. The presence of sexual and thelytokous populations of *V. canescens* has resulted in the species being used to explore the costs of sexual reproduction. Both arrhenotoky and thelytoky are obligatory and not the result of *Wolbachia* infection (Schneider *et al.*, 2003).

Much of what we know about the fate of endoparasitoid larvae in hosts (suitable or not) derives from original experiments using *V. canescens*, including different responses from different life history stages of the host, endopterygotes vs exopterygotes, etc. Salt's work demonstrated that host ranges of endoparasitoids will be strongly constrained by very effective immune responses within insects. In most insects that the *V. canescens* larvae were implanted into, encapsulation and/or melanisation resulted although some hosts, e.g. calliphorid Diptera larvae, that lacked an effective immune response were unsuitable for development of the parasitoid larva in other ways (Salt, 1957). Salt (1964) demonstrated how the size of a host (*Galleria mellonella*) can profoundly alter the success or otherwise of *V. canescens* larvae, with some *G. mellonella* larvae apparently being too large (see also Harvey, 1996). Salt (1965) demonstrated that the female parasitoid's calyx produces a layer around the egg that repels host haemocytes, and also demonstrated that eggs require physical deformation from passage down the ovipositor to be viable. A layer of epithelial cells within the calyx was later identified as the site of production of virus-like particles (Salt, 1973).

Salt's (1977) meticulous unpicking of the cocoon construction behaviour of *V. canescens* illuminates some themes that seem to be common across many Ichneumonidae. The parasitoid larva is under selective pressure to orientate within the cocoon to the easiest escape route from the host's pupation site. To do this, the *Venturia* larva spins rough silk into the smooth lamella layers, forming guide marks orientating the wasp towards the exit. Salt (1977) found such structures in several campoplegines and phygadeuontines, which presumably are used for orientation in these taxa too. Within the host cocoon, the initial orientation of the parasitoid larva seems to be achieved by displacing the host's head capsule to a consistent, posterior position and then commencing silk spinning.

Identification. Despite the frequency with which they are reared and their abundance in many habitats, identification of most campoplegines remains difficult with no major syntheses available and the taxonomic literature scattered (and a good proportion of it in German). However, progress can be made in many genera, especially the smaller ones, although some species-rich and commonly collected genera, such as *Campoplex*, *Hyposoter* and *Olesicampe*, remain almost completely unrevised. Townes's (1970b) key to world genera remains the only generic key but is not entirely reliable and omits various (subsequently recognised) small genera present in the British Isles, namely *Alcima* (Horstmann, 1970), *Clypeoplex* (see Horstmann, 1987a), *Lathroplex* (see Horstmann, 1978a), *Leptocampoplex* (see Horstmann, 1970), *Macrus* (see Horstmann, 1970), *Melanoplex* (see Horstmann, 1987a) and *Tranosemella* (see Horstmann, 1978a). Townes's illustration for *Porizon* is of *Leptocampoplex cremastoides* (Holmgren) (Várkonyi, 1998). References to keys and revisions, where available, are given for each genus in turn (except for those mentioned above with only one species on the British and Irish list, and on the whole we omit descriptions of single or few species not presented with keys). *Bathyplectes* (Horstmann, 1974); *Campoletis*, revision by Riedel (2017), and Horstmann (1979) keys one species group; *Campoplex*, no complete revision, Horstmann (1985) revised the *difformis* group; *Casinaria* (Riedel, 2018b); *Charops*, one British (and European) species; *Cymodusa* (Dbar, 1984, 1985); *Diadegma*, a large genus with a (now incomplete) key to species by Horstmann (1969), a revision of the *nanus* group by Shaw & Horstmann (1997), with a few species described subsequently by Horstmann & Shaw (1984), Horstmann (2004a, 2008a, 2013c), Shaw *et al.* (2016), and the parasitoids of *Plutella xylostella* revised by Azidah *et al.* (2000); *Dolophron*, one described species in Britain, the two European species keyed by Horstmann (1978a); *Dusona* (Horstmann, 2009a; see also Horstmann, 2011a for host data); *Echthronomas* (Horstmann, 1987b); *Enytus* (Horstmann, 1969, as *Diadegma*; see also Horstmann, 1980a); *Eriborus* (Horstmann, 1987b); *Gonotypus*, one British (and European) species; *Lathrostizus* (Horstmann, 1971a; see also Horstmann, 2004a); *Lemophagus* (Horstmann, 2004a); *Meloboris* (Horstmann, 2004a); *Nemeritis* (Horstmann, 1994a); *Nepiesta* (Horstmann, 1973b); *Phobocampe* (Šedivý, 2004; see also Horstmann, 2008a, 2009c); *Pyracmon*, European species revised by Horstmann (1978a), Holarctic species by Barron & Walley (1983); *Rhimphoctona* (Horstmann, 1980c); *Scirtetes*, one British (and European) species; *Sinophorus*, world species revised by Sanborne (1984); *Synetaeris*, one British species, see Horstmann (1987a) for notes on the two European species; *Tranosema* (Horstmann, 1978a); *Tranosemella* (Horstmann, 1978a, 1987a are useful); *Venturia*, one species known from Britain, European species revised by Horstmann (1973a). Shaw *et al.* (2016) report host data for 225 British species. In Britain, campoplegines with yellow on the clypeus or face belong only to *Alcima orbitale* (Gravenhorst) (yellow inner orbits), *Echthronomas* species (yellow face or just the clypeus), *Olesicampe femorella* (Thomson) (yellow clypeus) or *Tranosemella citrofrontalis* (Hedwig) (males with entirely yellow face).

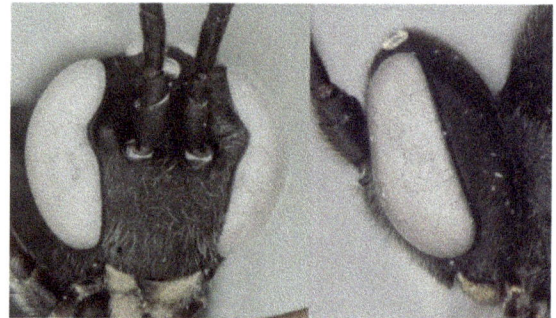

Figure 59. Morphology of Campopleginae: (a) *Dusona bicoloripes* (Ashmead) head; (b) *Diadegma semiclausum* (Hellén) male lateral; (continued overleaf).

Figure 59 continued: Morphology of Campopleginae: (c) *D. semiclausum* wings; (d) *Dusona erythrogaster* (Förster) head and mesosoma, ventral; (e) *D. bicoloripes* metasoma; (f) *Alcima orbitale* (Gravenhorst) head; (g) *Campoletis varians* (Thomson) head; (h) *Dusona terebrator* (Förster) hind tibial spurs; (i) *Dusona confusa* (Förster) mid tarsal claw; (j) *D. semiclausum* propodeum.

Subfamily COLLYRIINAE

The Collyriinae includes only three small, Palaearctic, genera; the monotypic *Aubertiella* Kuslitzky & Kasparyan (Kuslitzky & Kasparyan, 2011), *Bicurta* Sheng, Broad and Sun (Sheng *et al.*, 2012) and the distinctive genus *Collyria*, with nine described species, of which two are found in Britain.

Figure 60. *Collyria trichophthalma* Thomson.

Recognition. *Collyria* can be recognised by the following characters in combination: weak median apical tooth on the clypeus; the vertex not sloping down behind the ocelli, meeting the occiput (at the occipital carina) at almost a right angle; short antennae; the fore and mid tarsal claws with a small tooth at about their midlength (the hind claw is simple); the elongate hind coxa and short, thick hind femur; the hind wing with the nervellus intercepted above the middle; tergite 1 with the spiracles well in front of the midpoint; and the ovipositor compressed, weakly downcurved, tapering evenly from base to apex and with each lower valve with a series of very small teeth, best developed on the proximal half, along its lower edge. The lower valve of the ovipositor is markedly different from that of any other British ichneumonid.

Systematics. The systematic position of *Collyria* has long been unclear, owing to its aberrant morphology and biology. Several authors had placed the genus in the Acaenitinae but Townes (1971) argued that the genus was so dinstinctive that it belonged in a subfamily of its own. Recent phylogenetic analyses (Quicke *et al.*, 2009) have shown that the genus belongs in the pimpliformes group of subfamilies, where the loss of transverse carinae on the propodeum and toothed or lobed claws are found in several groups. The Collyriinae was thought to comprise only one genus until recently, when two, apparently more plesiomorphic, non-British genera were described (Kuslitzky & Kasparyan, 2011; Sheng *et al.*, 2012).

Biology. The hosts of *Collyria* are stem-sawflies (Hymenoptera: Cephidae), although only three species have been reared. A detailed account of the life-history of *C. coxator* (Villers) (as *calcitrator*), a parasitoid of *Cephus pygmeus* (L.) in wheat (*Triticum aestivum*) in Britain, is given by

Salt (1931). Oviposition is into the relatively large egg of the host, which is concealed within the tissue of its food plant. The female *Collyria* moves down the wheat stems searching for the oviposition scars of the sawfly and oviposits facing downwards. As many as four punctures have been found in a host egg containing only one *Collyria* egg. The *Collyria* egg hatches before that of the host, but further development takes place only within the larval host.

The first and particularly the second instar parasitoid larvae have a sclerotized head capsule with well-developed mandibles that are used in cases of superparasitism to eliminate competitors. While mandibulate first instar larvae are a typical feature of Ichneumonoidea, the persistence of the condition into the second instar is very unusual. Superparasitism is common (Salt, 1931, 1932; Wahl *et al.*, 2007) although *Collyria* females evidently avoid it to some extent (Salt, 1932). Although the first instar larva is free in the host haemocoel the transformation to the second stage takes place in a prominent evagination of the skin of the host. From the third instar onwards, the *Collyria* larva is typically hymenopteriform (spindle-shaped, maggot-like, as is the case in most apocritan later instars; see Gauld & Bolton, 1988), and lacking a sclerotized head capsule (Salt, 1931; Wahl *et al.*, 2007). In late summer the host larva is fully grown and constructs a cocoon in the base of the wheat stem, in which it overwinters with the parasitoid within, by then in its penultimate instar. The following spring the *Collyria* larva grows rapidly, finally discarding the remains of its host in May. Salt (1931) suggests that there are five larval instars in total, but points to the difficulty of ascertaining the number of ecdyses. The parasitoid pupates within the host's cocoon, not spinning one of its own, and the adult emerges in June. Emergence of *Collyria* starts about a week before that of *Cephus*. Adult *Collyria* have been observed mating on umbels of *Heracleum*.

Collyria coxator was considered by Salt (1931) to play an important role in regulating the population of *C. pygmeus* in England and was introduced into Canada from England in an attempt to control the native cephid pest of wheat, *Cephus cinctus* Norton. It was released in Saskatchewan (Smith, 1931) but unfortunately was unsuccessful (Carlson, 1979). Further introductions were made between 1935 and 1938 in the eastern United States, where its natural European host, *C. pygmeus*, was becoming a pest. No recoveries were made in succeeding years and that attempt too was judged a failure. However, *C. coxator* has more recently been found in numbers there and in some areas appears to be providing adequate control of *C. pygmeus* (Filipy *et al.*, 1985). The Chinese *Collyria catoptron* Wahl has also been studied for its potential as a biocontrol agent of *Cephus* species (Wahl *et al.*, 2007). Details of its development were similar to *C. coxator*, although *C. catoptron* may be thelytokous as only females were reared, whereas the sex ratio in *C. coxator* is merely slightly female-biased (Salt, 1931). The second British *Collyria*, *C. trichophthalma* (Thomson), is fairly common and widespread but has never been reared.

Identification. Fitton (1984) gives a key to the British species and Collyriinae will be included in a revised handbook covering the British Pimplinae and related subfamilies (Shaw *et al.*, in prep.). The six known Western Palaearctic species are keyed by Gürbüz & Kolarov (2006), who include *Aubertiella nigricator* (Aubert) as a species of *Collyria*.

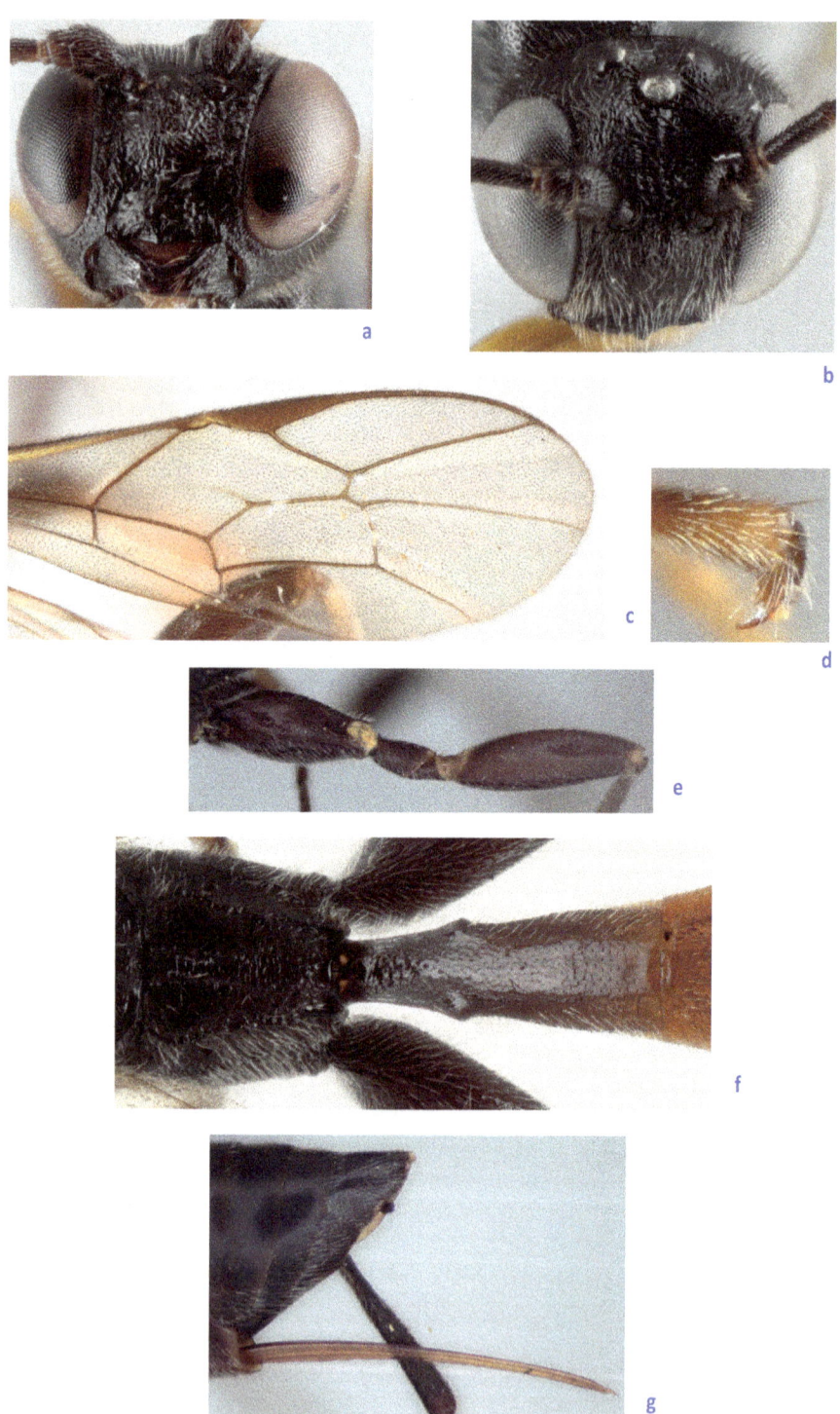

Figure 61. Morphology of Collyriinae: (*Collyria trichophthalma* (Thomson)): (a) face; (b) frons; (c) fore wing; (d) fore tarsal claw; (e) hind coxa and femur; (f) propodeum and first metasomal tergite (dorsal); (g) ovipositor.

Subfamily CREMASTINAE

The Cremastinae is notably more species-rich at lower latitudes, with the northern European fauna comparatively impoverished. Of a world fauna of 790 described species in 35 extant genera, only 15 species in four genera have been found in Britain, although there are some additional, unnamed, species in collections. Although species richness of Cremastinae seems to be highest in tropical regions (e.g. Gauld, 1984, 2000), where they comprise a conspicuous part of the fauna of parasitoids of weakly concealed Lepidoptera larvae, generic diversity is high in the rather dry areas of the southern Palaearctic (e.g. Narolsky, 1990, 1994; Narolsky & Schönitzer, 2001).

Figure 62. *Pristomerus vulnerator* (Panzer).

Recognition. In contrast to most other ichneumonid subfamilies, all cremastines possess one unambiguous morphological character that is not present in any other group, that the hind and mid tibial spurs are inserted in separate sockets so that the spurs are separated by a sclerotized bridge. In all other ichneumonids, the hind and mid tibial spurs arise from a shared socket at the apex of the tibia. This character, though, can be hard to see in small specimens as well as those in which the tarsus flexes toward the tibial spurs. In general habitus, cremastines closely resemble campoplegines, to which they are apparently closely related (Wahl, 1991; Quicke *et al.*, 2009). Both subfamilies share a long, petiolate first metasomal segment, with the sclerotized part of the sternite usually extending beyond the mid-length of the tergite and the spiracle far behind the middle. The posterior transverse carina of the mesosternum is complete or largely complete. In contrast to Campopleginae, the clypeus of Cremastinae is well-defined from the face and obviously convex in lateral view. Many cremastines have pale markings, particularly on the face, which are not usually present in British Campopleginae, almost all of which have uniformly black heads and mesosomas (the exceptions are listed in the Campopleginae chapter). The second metasomal tergite is invariably longitudinally finely striate in cremastines but not in campoplegines. One of the most frequently collected British cremastines is *Pristomerus vulnerator* (Panzer), which has a conspicuous ventral tooth on the hind femur (Fig. 63e), not found in any campoplegines. In the British fauna, *Dimophora* species are aberrant in their robust habitus and large areolet and can be mistaken for Phygadeuontinae; the ovipositor, however, is typically ophioniform, with a dorsal, subapical notch. Some extralimital taxa, namely the African *Belesica* Waterston and *Eurygenys* Townes and the Australian *Gahus* Gauld, are particularly aberrant and difficult to recognise as Cremastinae, although all share the hind tibial spur character.

Systematics. The Cremastinae belongs in the 'higher' ophioniformes clade of subfamilies (in the sense of Quicke *et al.*, 2009), that share a petiolate first metasomal segment and are predominantly parasitoids of Lepidoptera larvae. Gauld (1984) had suggested that the Cremastinae are the sister-group to the Tersilochinae, with the Campopleginae and Ophioninae representing their sister clade. However, these relationships were based on trends in morphology and subsequent studies have agreed that the Cremastinae and Campopleginae are sister groups (Wahl, 1991; Quicke *et al.*, 2009), or, in some scenarios, that the Cremastinae form the sister group to the Campopleginae + Nesomesochorinae (Quicke *et al.*, 2009). Within the Cremastinae, there has been no attempt at a phylogenetic reconstruction other than Quicke *et al.*'s (2009) sparse sampling of 14 genera. Their phylogenetic analyses unsurprisingly recovered a mostly monophyletic Cremastinae; however, *Eurygenys* (representing the aberrant pair of *Belesica*-group genera, which also includes *Belesica*) was never recovered with the remaining cremastines and, in their combined morphological and molecular analyses, occupied a position basal to all of the 'higher' ophioniformes. The African and Malagasy *Belesica* and *Eurygenys* are markedly dissimilar to all other cremastines; their densely pectinate claws, strong fore tibial tooth and narrowed proboscidial fossa are suggestive of their being misplaced in the Cremastinae and that the hind tibial spurs being inserted in separated sockets may have arisen more than once in the Ichneumonidae (*Tersoakus* Narolsky seems to share these characters too, judging by Narolsky's (2002) description). However, *Belesica* and *Eurygenys* also differ in several important respects from each other, morphologically and in terms of host groups, and further studies, including *Belesica*, are necessary before potentially separating off any genera. The unique condition of the hind tibial spurs certainly seems to circumscribe a monophyletic group in much of the world and most cremastines have a very similar habitus.

As pointed out by Gauld (2000), the majority of cremastines are classified in just four genera, *Cremastus*, *Pristomerus*, *Temelucha* and *Trathala* Cameron, although *Eiphosoma* Cresson and *Xiphosomella* Szépligeti dominate the cremastine fauna in the neotropics. Globally, the majority of species are probably undescribed and there is a steady trickle of descriptions of new genera (e.g. Narolsky & Schönitzer, 2001; Narolsky, 2002; Rousse *et al.*, 2011). That 16 genera (46% of the total) are currently monotypic, and that some of the larger genera are vaguely defined, is an indication that the generic classification of the subfamily is in need of re-evaluation.

Biology. The basic biology of cremastines is that of koinobiont endoparasitism of weakly concealed Lepidoptera larvae. All reared species have been solitary. The long ovipositors of many cremastines, such as species in the cosmpolitan genus *Pristomerus*, are adaptations to access host larvae living in such situations as leaf-rolls, silk webbing and inside fruit bodies. Within the British fauna, *Pristomerus vulnerator* is a well-known parasitoid of the codling moth, *Cydia pomonella* (L.), as well as *Gypsonoma dealbana* (Frölich) (Tortricidae) and a few other tortricid species. Species of *Temelucha* have also been reared from weakly concealed tortricid and gelechiid larvae. *Cremastus* species, however, seem to attack more exposed hosts, with a possible connection to silken tubes, on the limited available rearings; *C. kratochvili* Šedivý has been reared from *Scythris empetrella* Karsholt & Nielsen (Scythrididae), which lives in a silken tube on *Calluna* and *Erica*, and *C. cephalotes* Šedivý from a case-bearing psychid, *Pachythelia villosella* (Ochsenheimer) (Fitton & Gauld, 1980). Species of various genera, but especially marked in some species of *Pristomerus* and the neotropical *Eiphosoma* and *Xiphosomella*, have sinuous ovipositor tips, with one or two arches, which presumably allows extra mobility of the ovipositor tip, to contact hosts within their tunnels that are not linearly acccessible (Quicke, 1991). Although rearings of British cremastines have all been from 'microlepidoptera', some extralimital cremastines are parasitoids of Hesperiidae (*Creagrura* Townes: Gauld, 2000), Limacodidae (*Eurygenys*: Rousse *et al.*, 2011) and Noctuidae (*Eiphosoma*: Ashley *et al.*, 1982; *Pristomerus*: Gauld, 2000). Several cremastines are parasitoids of Coleoptera larvae and all reliable host records for these taxa are from the family Chrysomelidae. The aberrant, and possibly misplaced, *Belesica pictipennis* (Tosquinet) has been reared from a toxic chrysomelid, *Diamphidia nigroornata* (Stål) (Waterston, 1929a). One of the two British *Dimophora*, *D. evanialis* (Gravenhorst), has been reared in Europe from *Cryptocephalus moraei* (L.) (Scholler, 1999) and a species of the apparently closely related Neotropical genus *Ptilobaptus* Townes has also been reared from an unidentified chrysomelid (Gauld, 2000). A North American cremastine, *Tanychela pilosa* Dasch, is unusual within Ichneumonidae in that it enters the water to parasitize aquatic Pyralidae larvae (Resh & Jameison, 1988; Jamieson & Resh, 1998).

Much of what we know of cremastine biology stems from studies on very few species, which may or may not be representative of the majority of cremastines. The following account is mainly abstracted from Bradley & Burgess's (1934) study of the non-British *Trathala flavoorbitalis* (Cameron) and Rosenberg's (1934) work on *Pristomerus vulnerator*. Host-searching *T. flavoorbitalis* are attracted to larval frass and webbing and usually ignore exposed larvae; oviposition is into the thorax, just behind the host's head. *Pristomerus vulnerator* attacks small larvae and Rosenberg (1934) reports that a small host larva containing five *P. vulnerator* eggs was killed by this superparasitism. Several authors report a preference in cremastines for oviposition in early instar hosts, at least in *Pristomerus* and *Temelucha* (Force, 1989; Paull & Austin, 2006). In Bradley & Burgess's study, fourth instar *Ostrinia nubilalis* (Hübner) (Lepidoptera: Crambidae) were used as hosts but this may not reflect the parasitoid's preference in the wild. Oviposition is into the haemocoel, where the egg floats freely before eclosion after 3.5 to 20 days, depending on species. Cremastine eggs are unsculptured and curved, more or less kidney-shaped. The first instar larva has a long caudal appendage and long, simple, sickle-shaped mandibles. Bradley & Burgess (1934) report only three larval instars, with the second and third instars lasting only one and two days respectively. The first instar larva grows considerably in size, the expansion aided by cuticular 'pleats' that impart a ruffled appearance to the newly eclosed larva. As the first instar expands, the caudal appendage proportionally shrinks. In subsequent instars this appendage is much reduced and the mandibles are shorter and stouter. In *P. vulnerator* the winter is spent in the diapausing, cocooned host prepupa, with parasitoid development completed in the spring. The tropical *T. flavoorbitalis*, introduced to North America for control of the introduced *Ostrinia nubilalis*, overwinters either in a host larva or, occasionally, in its cocoon (Bradley & Burgess, 1934).

Although *Pristomerus vulnerator* accounts for a significant proportion of mortality of *Cydia pomonella* in Europe (e.g. Russ & Rupf, 1974; Subinprasert, 1987), this is a relatively recent host association as *C. pomonella* is native to Central Asia, where *P. vulnerator* was not found until recently (Kuhlmann & Mills, 1999). Several European studies list *P. vulnerator* as a parasitoid of Sesiidae species but these host associations presumably stem from substrate rearings, where the parasitized tortricid larva had pupated in a crevice in wood.

Superparasitism is frequent in at least some cremastines (Tsankov, 1988; Jamieson & Resh, 1998), which may be a mechanism that allows a superparasitising female to take advantage of a weakened host immune system. Only one parasitoid larva develops beyond the first instar in cases of superparasitism; the first instar larva's elongate mandibles are presumably used in the elimination of supernumeraries. *Temelucha* species have been found to cleptoparasitize a braconid, *Orgilus obscurator* (Nees), when attacking the tortricid *Rhyacionia buoliana* (Denis & Schiffermüller); Arthur *et al.* (1964) and Schröder (1974) report *T. interruptor* (Gravenhorst) and Tsankov (1988) reports *T. confluens* (Gravenhorst) as the *Temelucha* species and it is not clear if more than one species is really involved. An unidentified *Pristomerus* species (probably *P. armatus* (Lucas); Horstmann, 1990c) also acts cleptoparasitically in the *R. buoliana* system (Schröder, 1974) and the possibility of cleptoparasitism was the main reason why *Pristomerus vulnerator* was not among the ichneumonids released to control *Cydia pomonella* populations in North America (Mills, 2005).

Unlike their putative sister-group, the Campopleginae, polydnaviruses or virus-like particles have not been found in cremastines, although few species have been screened.

Identification. Fitton & Gauld (1980) revised the British fauna although they acknowledged that the specimen base was small and that further discoveries would undoubtedly be made. Their key is still useful but now incomplete, as two additional species of *Pristomerus* have since been found and some unidentified *Temelucha* and *Cremastus* in BMNH and NMS probably represent additional species. European *Pristomerus* can be identified using Horstmann (1990c), Šedivý's (1970, 1971) revisions of the European cremastine fauna should prove useful, and Vas (2016) provides an updated key to European *Temelucha*. Klopfstein (2016) separates the two *Dimophora* species.

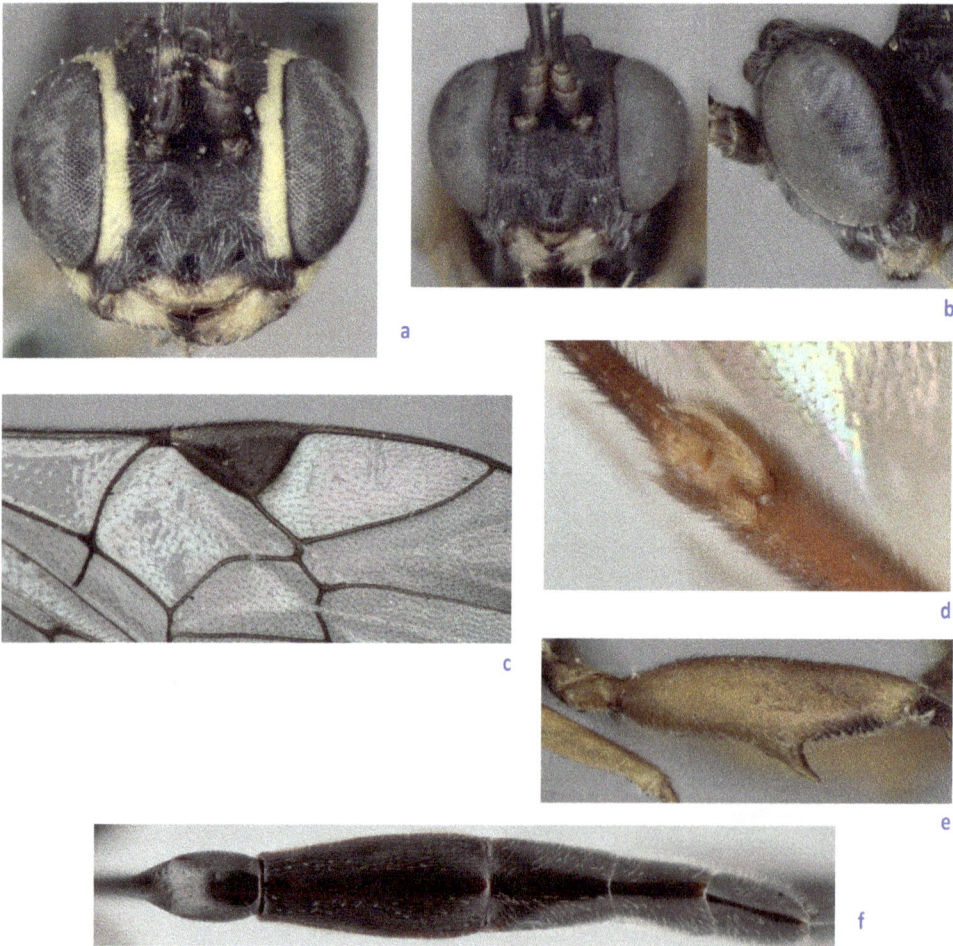

Figure 63. Morphology of Cremastinae: (a) *Cremastus geminus* Gravenhorst face; (b) *Pristomerus vulnerator* (Panzer) head; (c) *P. vulnerator* part of fore wing; (d) *Cremastus spectator* Gravenhorst hind tibial spurs; (e) *P. vulnerator* hind femur; (f) *C. spectator* metasoma.

Subfamily CRYPTINAE

Tribes
Aptesini
Cryptini

The Cryptinae is species-rich globally and across many habitats, but particularly so in tropical regions. About 3,100 species have been described, in 275 genera, with cryptines second only to Ichneumoninae in described species richness. Most cryptines in the tropics remain undescribed and there is a steady trickle of new genera being recognised. The British fauna of 137 species in 40 genera is about one third of the European fauna and far smaller than the closely related Phygadeuontinae, which are more species-rich in temperate regions. In contrast to Phygadeuontinae, only a very small proportion of the European cryptine fauna probably awaits description, and modern keys cover a good percentage of the species.It should be noted that the recent (Santos, 2017) split of the former 'Cryptinae' into three subfamilies, which we follow here (see below), has outdated many published subfamily placements in the recent literature (including in Broad, 2016).

Figure 64. *Cubocephalus distinctor* (Thunberg), tribe Aptesini.

Recognition. Most Cryptinae are readily recognisable by the combination of a long sternaulus, which in the great majority of species extends over at least half of the length of the mesopleuron and ends (or at least the course of the sternaulus can be traced back to) below the posterior ventral corner of the mesopleuron (above the posterior ventral corner in Phygadeuontinae); the petiolate first metasomal segment, never with a glymma, with the spiracles beyond the middle of the tergite; the pentagonal areolet (which may very occasionally be open, i.e. when vein 3*rs-m* is absent); the ovipositor usually extending well beyond the metasomal apex at rest and with distinct teeth and a nodus (species of the ichneumonine genus *Crypteffigies* have long ovipositors but are

Figure 65. *Hoplocryptus confector* (Gravenhorst), tribe Cryptini.

otherwise typical ichneumonines); and the rather narrow (compared to many Ichneumoninae) and convex clypeus. Compared to Phygadeuontinae, in cryptines fore wing vein *2m-cu* always has one bulla (often two in Phygadeuontinae) and the propodeum generally has a reduced number of carinae, with the area superomedia absent in the vast majority of species. Most cryptines are larger than most phygadeuontines, but exceptions occur. Males of some Ichneumoninae can be confused with cryptines but can generally be recognised by the wider, flatter clypeus; the subtly different form of the areolet; the very short sternaulus (but a long sternaulus is present in *Dicaelotus* of the Ichneumoninae: Phaeogenini and *Listrodromus* of the Ichneumoninae: Listrodromini); and the often deeply impressed, often striate gastrocoelus and frequently large, wide thyridium (the latter short and usually ovoid in cryptines; the former not apparent). A few species are anomalous within the subfamily, especially some genera (*Echthrus, Helcostizus, Hemiphanes* and *Sphecophaga*) that have a stouter first metasomal tergite, with the spiracle at about the mid-length. In *Helcostizus, Hemiphanes* and *Sphecophaga*, the sternaulus is weak, although the propodeum is still typically cryptine (with only transverse carinae). *Sphecophaga vesparum* (Curtis) is a fairly common British species that is hard to recognise as a cryptine as the sternaulus is weak, the first metasomal segment not conspicuously petiolate, and fore wing vein *3rs-m* is absent. The hypopygium of *Sphecophaga* is unusally large for a cryptine (inviting confusion with Banchinae) but the short ovipositor has a nodus on the dorsal valve, the clypeus is of the typical Cryptinae shape (rather evenly convex) and the propodeum has a single transverse carina, as in many Cryptini. *Helcostizus* has a swollen fore tibia, as do several other cryptines attacking deeply concealed hosts. Both of these genera are keyed out separately in the subfamily key.

Systematics. The Cryptinae (*s.l.*) was one of the 'traditional five' subfamilies of Ichneumonidae and there has generally been little doubt that this is a natural grouping. However, recent phylogenetic work (Laurenne *et al.*, 2006; Santos, 2017), has suggested that the Cryptinae *s.l.* is probably monophyletic only if the Adelognathinae (which are very different biologically) and Ichneumoninae are included within the subfamily. In order to resolve this, Santos (2017) elevated the Phygadeuontini and Ateleutina to subfamily status and restricted Cryptinae to the tribes Aptesini and Cryptini. The genus *Hemigaster* was transferred to Phygadeuontinae, with the former tribe Hemigastrini taking the next available name, Aptesini. Additionally, Klopfstein *et al.* (in press) have transferred *Hemiphanes* from Orthocentrinae to the Cryptinae, tribe Aptesini. These studies, and that of Quicke *et al.* (2009), found that the Adelognathinae, Agriotypinae, Alomyinae, Ateleutinae, Cryptinae, Ichneumoninae, Phygadeuontinae and possibly Microleptinae formed a monophyletic assemblage, which corresponds to much of the ichneumoniformes grouping of Wahl (1993a). This is in accordance with their overall similarity in appearance (arguably with the exception of Adelognathinae), particularly the petiolate form of the first metasomal segment in the great majority of species. Striking differences in larval morphology between the Cryptinae *s.l.* and Ichneumoninae led Townes (1969) and Short (1978) to believe that they were only distantly related; however, these differences can be ascribed to markedly different parasitoid behaviour, with cryptines and phygadeuontines being predominantly ectoparasitoids and Ichneumoninae all endoparasitoids. Gokhman (1992) argued that the subfamilies Cryptinae (*s.l.*) and Ichneumoninae, and their constituent tribes, were not well defined by apomorphic characters and proposed an evolutionary pathway from the cryptine tribe Phygadeuontini (i.e. Phygadeuontinae as recognised by Santos, 2017) to Ichneumoninae. Gokhman's hand-drawn cladogram is similar in many respects to the results of Laurenne *et al.* (2006) and Santos (2017).

As found by both Laurenne *et al.* (2006) and by Santos (2017), the Phygadeuontinae seem to be paraphyletic with respect to at least Adelognathinae, Ateleutinae, Cryptinae and Ichneumoninae (Alomyinae were not included in Santos, 2017), but the phylogeny and classification of phygadeuontine genera needs a lot more work before stable monophyletic groups can be circumscribed. Note that Gupta (1970) had earlier, rather arbitrarily, elevated Townes's (1970a) tribes to subfamily level (as Gelinae, Hemigasterinae and Mesosteninae). To add to the potential confusion, there has been little consensus over the names that the tribes or the subfamily should take, mainly because of different opinions regarding priority of names. The following names have all been in use over the last few decades (current names in bold):

Cryptinae *s. l.*	**Phygadeuontinae.**	**Aptesini**	**Cryptinae *s. s.***	**Ateleutinae**
Phygadeuontinae	Gelini	Echthrini	Mesostenini	Ateleutina
Hemitelinae	Hemitelini	Hemigastrini		
Gelinae				

In addition to the tribes, and in an attempt to further partition an unwieldy mass of genera in a useful way, Townes (1970a) divided the Phygadeuontinae (as Gelini) and the Cryptinae (as Mesostenini) into a number of subtribes, with keys to subtribes and then to genera. This has arguably made it more difficult to identify many cryptine genera, given the rather subtle nature of many of these subtribes and the fact that many of them are probably not monophyletic anyway (Laurenne *et al.*, 2006; Santos, 2017). These subtribes are not recognised here. The tribe Aptesini can be separated from Cryptini by the small triangular projections of the metanotum, opposite the anterior ends of the lateromedian longitudinal carinae of the propodeum (Fig.69h), absent in Cryptini (Fig. 69i,j), and by the usually more complete propodeal carination.

Biology.
Host associations and life histories.
As with Phygadeuontinae, the biology of Cryptinae is fairly uniform in that almost all species, where known, are idiobiont ectoparasitoids of more or less concealed hosts, though extralimital records suggesting idiobiont endoparasitism of particularly robust hosts such as some Lepidoptera pupae warrant further investigation for the British fauna. The range of host groups is narrower than for Phygadeuontinae, but with areas of overlap as well as distinct differences in host foci

across the subfamilies. As with various other ichneumonid subfamilies, little is known of the hosts of many cryptines and host ranges are particularly poorly understood for those species attacking weakly cocooned Lepidoptera (pre)pupae in the soil (e.g. *Cryptus*) and sawfly (pre)pupae (most of the Aptesini). As a very broad generalisation, the subfamily has a profound relationship with silk-producing hosts, and Cryptini are centred on Lepidoptera cocoons, but with radiations onto spider egg sacs, nests of aculeate Hymenoptera, sawfly (Hymenoptera) pupae and prepupae, and Coleoptera larvae/pupae; Aptesini are predominantly parasitoids of sawfly cocoons but with several attacking Coleoptera and Lepidoptera. Šedivý & Ševčik (2003) report that *Hemiphanes gravator* Förster was reared from a fungus, but with no evidence as to its host.

Figure 66. *Agrothereutes abbreviatus* (Fabricius) (J. Voogd).

Figure 67 (right). *Listrognathus obnoxius* (Gravenhorst) ovipositing into a *Zygaena filipendulae* (L.) (Lepidoptera: Zygaenidae) cocoon (M.R. Shaw).

Figure 68 (above). *Sphecophaga vesparum* (Curtis) larva feeding on a pharate adult of *Vespula vulgaris* (L.) (Hymenoptera: Vespidae) (R. Brown).

The majority of British Cryptini are, or are presumed to be, parasitoids of Lepidoptera pupae and prepupae. Some are well known as parasitoids of conspicuous cocoons of toxic zygaenid moths, e.g. *Listrognathus obnoxius* (Gravenhorst) (Fig. 67) and *Gambrus ornatus* (Gravenhorst). Although most Cryptini are solitary, several species of *Agrothereutes* attacking large, cocooned hosts produce large broods, such as *A. leucorhaeus* (Donovan) on *Lasiocampa quercus* (L.) (Lasiocampidae) and *A. saturniae* (Boie) on *Saturnia pavonia* (L.) (Saturniidae). *Agrothereutes mandator* (L.) is unusual in being a gregarious parasitoid of cimbicid sawflies (Schwarz & Shaw, 1998), but it is noteworthy that these have tough cocoons very like those of the above saturniid and lasiocampid. Physically smaller taxa, such as species of *Ischnus*, develop in 'microlepidopteran' (such as Tortricidae and Yponomeutidae) cocoons (Cole, 1979; Schwarz & Shaw, 1998). Some genera, particularly *Cryptus*, concentrate their host searching at or near ground level (Schwarz & Shaw, 1998), probably usually seeking Lepidoptera that pupate in the soil layer, although they are rarely reared despite often being very common. One European (but not British) species, *Cryptus genalis* Tschek, is a gregarious parasitoid of subterranean cetoniine beetle (Coleoptera: Scarabaeidae) pupae (Schwarz *et al.*, 2013). *Apsilops* species are parasitoids of aquatic Pyralidae (Sawoniewicz, 2008). Parasitism of aculeate Hymenoptera nests seems to have arisen independently in several lineages of Cryptini (Santos, 2017), including *Hoplocryptus*, *Nematopodius* and allied genera, and a small group of genera (*Sphecophaga* group of genera) represented in Britain by *Sphecophaga vesparum*, a well-known parasitoid of pupae in the nests of *Dolichovespula* and *Vespula* social wasps (Vespidae: Vespinae) (Schwarz & Shaw, 1998). Several Cryptini have been shown to act as cleptoparasites with distinctly predatory tendencies: for example, *Buathra tarsoleuca* (Schrank) was observed to consume the provisioned prey as well as, presumably, the larva of its host Sphecidae (Hymenoptera) (Casiraghi *et al.*, 2001); and *Nematopodius debilis* (Ratzeburg) kills the host aculeate larva and goes on to consume the prey items stored in its cell (C. Vardy, pers. comm.). The larva of at least one species of *Hoplocryptus* is known to consume the contents of more than one cell of its aculeate host during its development (Daly, 1983). Various genera are known to lay their eggs near (rather than on) the host (e.g. Cole, 1979; Barthélémy & Broad, 2012) and in some cases it seems that the host pupa is not permanently paralysed by the ovipositing female (Cole, 1979), though this does not necessarily mean that such hosts are not stung and perhaps developmentally arrested. At least three genera (*Hidryta*, *Idiolispa* and *Trychosis*) are essentially predators in spider egg sacs and the non-British *Thrybius togashii* Kusigemati (the biology of the British *T. brevispina* (Thomson) and *T. praedator* (Rossi) is poorly known) acts as a predator of gregarious broods of an endophytic phytophagous *Tetramesa* species (Hymenoptera: Eurytomidae) (Matsumoto & Saigusa, 2001). In contrast, some Neotropical Cryptini have been shown to be idiobiont endoparasitoids of lepidopteran prepupae and pupae (e.g. *Polycyrtus* Spinola, Zúñiga Ramírez, 2004) and some are possibly koinobionts (I.D. Gauld, pers. comm.); the slender ovipositors of various extralimital taxa would be evidence in favour of this. However, whether oviposition is external or internal for *Polycyrtus* has not been reported and it is not clear whether this form of endoparasitism is essentially different from, for example, *Ischnus*, in which the larva is often 'inside' the ruptured host pupa but the egg is placed externally. Some *Listrognathus* species can have a particularly strong effect on the population of their zygaenid hosts by preferentially selecting the larger host cocoons, which tend to be female (Shaw, 1975, as *Mesostenidea*).

Aptesini have been little studied, except for a few species that are parasitoids of sawflies regarded as forestry pests. They are most diverse in the Holarctic although a few taxa are found in the Old and New World tropics. Sawoniewicz (2008) recently summarised the recorded host associations of aptesines. It is difficult to know how reliable the data are but the host records reported do at least seem to indicate consistent trends within genera. The most frequently recorded hosts are sawfly cocoons of the superfamilies Tenthredinoidea and Pamphilioidea, particularly the families Tenthredinidae, Diprionidae and Pamphiliidae (non-cocooned pupae in the latter case), sometimes as facultative secondary parasitoids via primary parasitoids that already occupy the host cocoon. Some small genera, e.g. *Aconias*, *Megaplectes*, *Schenkia*, and a few species of *Aptesis* and *Cubocephalus*, have been recorded predominantly as parasitoids of Lepidoptera pupae whilst several other species of *Cubocephalus* are recorded as parasitoids of concealed Coleoptera pupae. However, some of the records probably arose from poorly controlled rearings from large woody substrates in which the true host could easily be mistaken. The rather aberrant *Demopheles corruptor* (Taschenberg) is a regular parasitoid of stem-nesting solitary aculeate wasps (Sawoniewicz, 2008; C. Vardy, pers. comm.). The genus *Rhembobius* is exceptional in the tribe for attacking

Syrphidae (Diptera) puparia; whilst Townes (1970a) placed this genus in the Phygadeuontinae (as the tribe Gelini), Laurenne *et al.* (2006) found that *Rhembobius* was a member of the Aptesini in phylogenetic analyses, with Sawoniewicz (2008) also treating the genus as belonging in this tribe, as did several earlier authors. Whilst Townes (1970a) and Sawoniewicz (2008) have regarded *Polytribax* as belonging to the Aptesini, presumably mainly on the basis of propodeal structure, Laurenne *et al.* (2006) found that it belonged in the Cryptini, in keeping with the known host range, of cocooned Lepidoptera. However, the American *Polytribax contiguus* (Cresson) was included in the phylogenetic study by Santos (2017) and recovered within Aptesini (*Rhembobius* was not included). In general, there are various intriguing published host records and many apparently self-consistent rearings but, unfortunately, most published observations lack any detail and too little is known to draw any conclusions about trends in host utilisation across the Aptesini.

The *Gabunia* genus-group (represented in Britain by *Echthrus* and *Helcostizus*) seem, on the basis of antennal and fore leg modifications, to have adapted to attacking deeply concealed Coleoptera larvae in wood (and stem-nesting aculeate Hymenoptera), located by vibrational sounding, a form of echo-location via a solid medium (Broad & Quicke, 2000). Despite diverse morphology, which led to Townes (1970a) classifying two of the more aberrant genera in Phygadeuontinae and Aptesini respectively, these shared morphological traits associated with host location correlate with the molecular phylogenetic results of Laurenne *et al.* (2006) and Santos (2017), which placed them in Cryptini. Otherwise, little is known about host detection mechanisms employed when the host is deeply concealed in the substrate.

Developmental Biology.
Notwithstanding the taxa that consume successive items such as spider eggs in the sacs, the vast majority of cryptine species are likely to be idiobiont ectoparasitoids, although some species of *Goryphus* and *Cryptus* may be endophagous in lepidopterous pupae. It is unclear in these cases whether they are strictly endophagous or feeding from a position in the ecdysal space, inside the pupal or puparial case but outside the body of the host itself. Cole (1979) noted that the larva of *Ischnus migrator* (Fabricius) (as *I. inquisitorius* (Müller)) at first feeds on its tortricid host pupa externally, and can leave the host for moulting, before finally cleaning it out from an internal position. Like other idiobiont groups, Cryptinae (and also Phygadeuontinae) are probably mainly synovigenic, as has been confirmed for a few Cryptini (e.g. Cole, 1979).

Several species of Cryptinae have been studied in some detail because of their potential in biocontrol programmes, although few data are available on their efficacy. *Sphecophaga vesparum* is a semi-gregarious, Holarctic parasitoid of the pupae (usually) of Vespinae (Fig. 67), studied because of its potential for controlling populations of the invasive *Vespula germanica* (Fabricius) and *V. vulgaris* (L.) (Hymenoptera: Vespidae) in New Zealand, although its life-cycle is complex and only relatively recently elucidated (Donovan, 1991). Mating is brief, lasting from 20-40 seconds. Larval feeding is rapid, taking 3-4 days for larvae that would become brachypterous females, 4-5 days when they would become macropterous adults. Several generations of *Sphecophaga* can develop in a vespine nest in one season and larvae can be solitary or gregarious. overwintering cocoons have thicker ridges and more layers of silk than those from which adults hatch quickly. A very unusual feature is the production of three different cocoon types from larvae that can develop from oviposition events by the same female *Sphecophaga*: white cocoons rapidly give rise to brachypterous females, 10-13 days following oviposition; thin-walled yellow cocoons produce macropterous females and males 12-15 days after oviposition; and thick-walled yellow cocoons overwinter and may produce macropterous females and males in any of the subsequent four years. Brachypterous females are thelytokous (macropterous females can reproduce thelytokously but can also mate with males that emerge from yellow cocoons); the offspring of brachypterous females can develop into any of the three modes. This complicated life-style has presumably evolved in response to the concentration of hosts in one place, the relatively short duration of this host resource and the very patchy distribution of vespine wasp nests. However, the strategy of emergence from the cocoon after often more than one year contributes towards the species' poor success rate as a biocontrol agent in New Zealand, where models predict an ultimate outcome of only about 10% nest suppression (Barlow et al., 1996).

Identification. Townes's (1970a) treatment of world genera covers both tribes. For Cryptini, van Rossem's (1969a) key to European genera is easier to use for the more restricted fauna. Horstmann (1993a) provides keys to the few known European brachypterous species, alongside the more numerous taxa now placed in the subfamily Phygadeuontinae. The available works which cover the British fauna are listed below, alphabetically by tribe and genus.

Aptesini: other than Jonaitis (in Kasparyan, 1981) there is very little available for identification of aptesines; *Hemiphanes* (van Rossem 1981, 1987, 1988); *Oresbius*, brachypterous females (Horstmann, 1993a); *Pleolophus*, brachypterous females (Horstmann, 1993a), parasitoids of Diprioninae (Oehlke, 1966); and *Rhembobius* (Horstmann, 2000a).

Cryptini: *Acroricnus*, single species included in Schwarz (2005); *Apsilops* (Yoshida et al., 2011); *Aritranis*, partial key (Schwarz, 2005); *Buathra* (van Rossem, 1971; Schwarz, 1990); *Caenocryptus* (Schwarz, 1991a); *Cryptus* (Schwarz, 2015; van Rossem, 1969a is still useful); *Enclisis* (Schwarz, 1989; Bordera et al., 2007); *Gambrus*, separation of *incubitor* and *ornatus* (Schwarz, 2005); *Helcostizus*, single British species included in Townes (1983); *Hidryta* (Schwarz, 2005); *Hoplocryptus* (Schwarz, 2007); *Idiolispa* (Schwarz, 1988); *Listrognathus* (Horstmann, 1990a); *Meringopus*, partial key (van Rossem, 1969b; Schwarz, 2005); *Nematopodius*, single British species included in Horstmann (1990a); *Trychosis*, van Rossem (1966) will be of some use; *Xylophrurus*, British species provisionally separated by Schwarz & Shaw (1998). A handbook to the species of British Cryptini is in preparation by M. Schwarz et al. (in prep.).

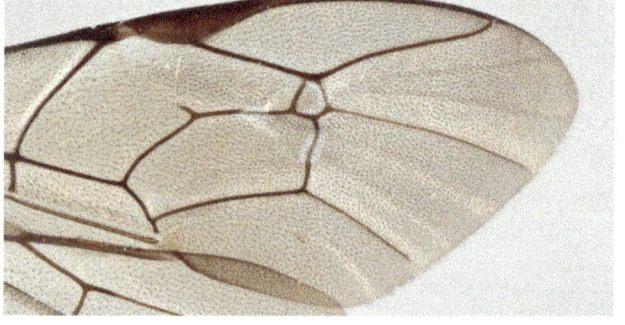

Figure 69. Morphology of Cryptinae: (a) *Agrothereutes abbreviatus* (Fabricius) (Cryptini) brachypterous female; (b) *Cryptus armator* Fabricius (Cryptini) head; (c) *Echthrus reluctator* (L.)(Cryptini) head; (d) *Helcostizus restaurator* (Fabricius) (Cryptini) head; (e) *Cryptus dianae* Gravenhorst part of fore wing; (f) *C. dianae* mesopleuron; (g) *C. dianae* sternaulus highlighted; (continued overleaf).

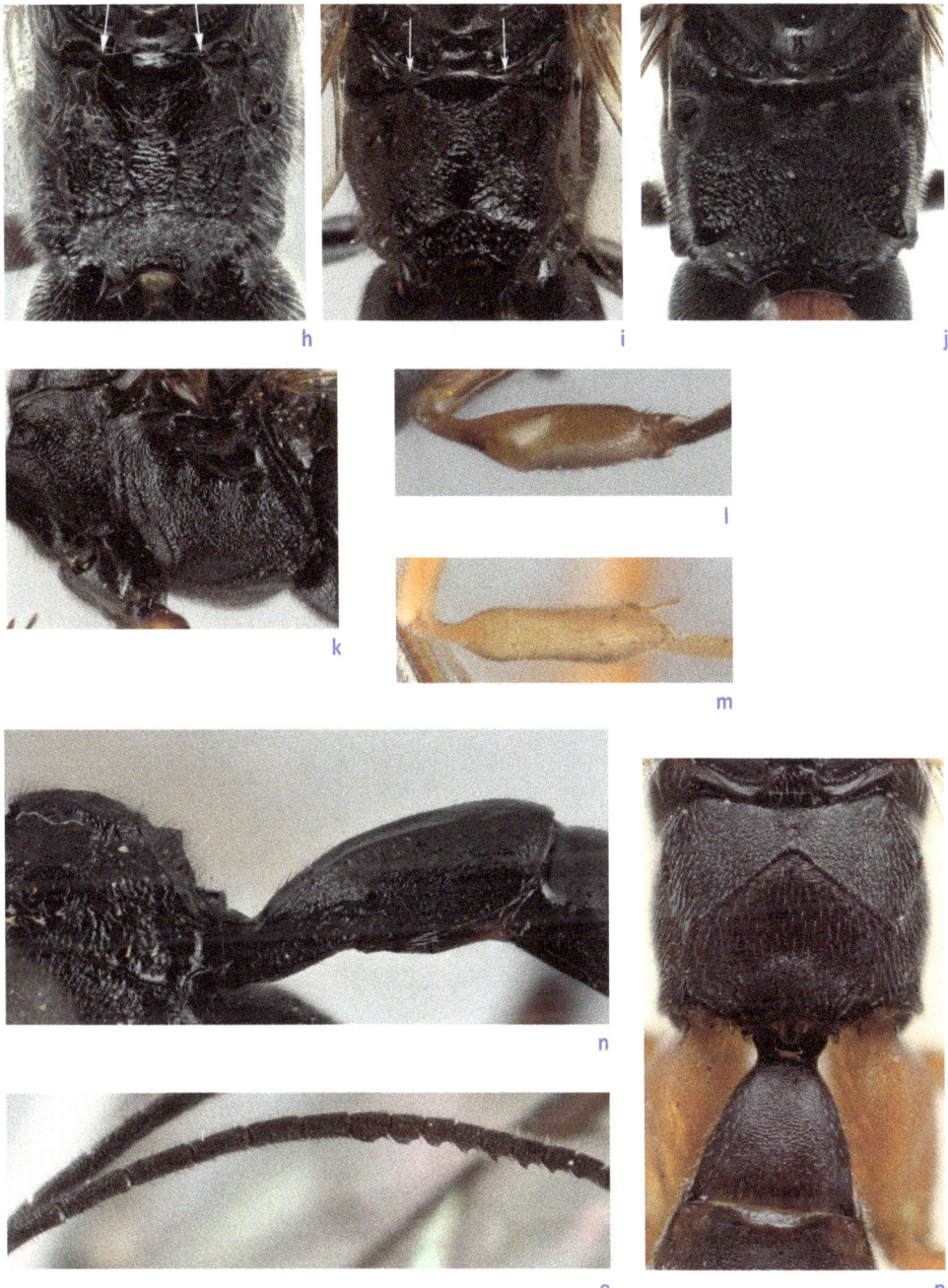

Figure 69 continued: Morphology of Cryptinae: (h) *Cubocephalus anatorius* (Gravenhorst) (Aptesini) propodeum and metanotum, with arrows pointing to triangular projections of the metanotum; (i) *E. reluctator* propodeum and metanotum with arrows pointing to rounded projections of the metanotum; (j) *C. dianae* propodeum and metanotum, with no projections of the metanotum; (k) *E. reluctator* mesopleuron with sternaulus ventral; (l) *E. reluctator* fore tibia; (m) *H. restaurator* fore tibia; (n) *E. reluctator* propodeum and first metasomal segment (lateral); (o) *H. restaurator* male part of antenna; (p) *H. restaurator* propodeum and first metasomal tergite (dorsal); (continued overleaf).

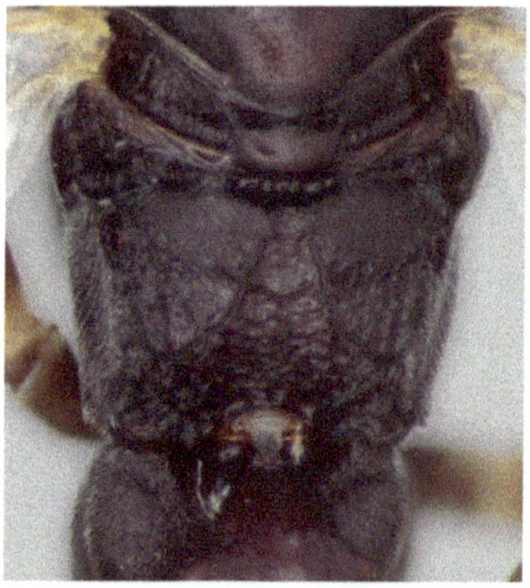

q

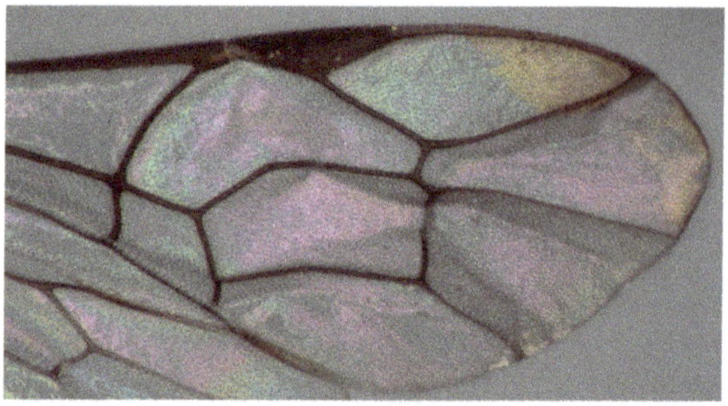

r

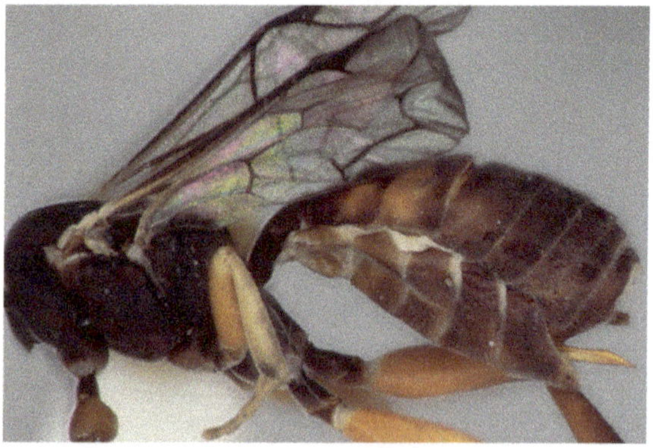

t

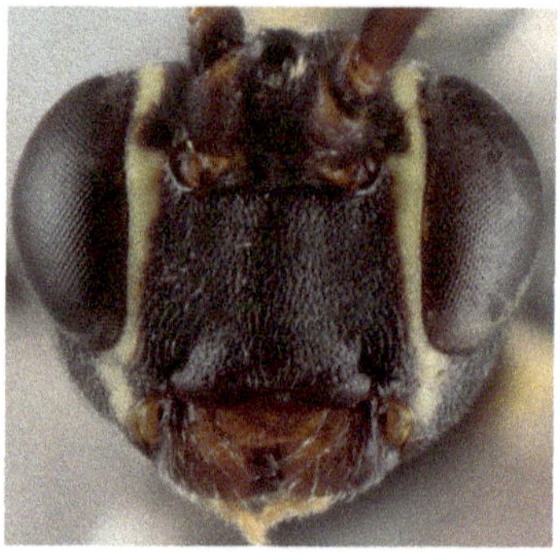

s

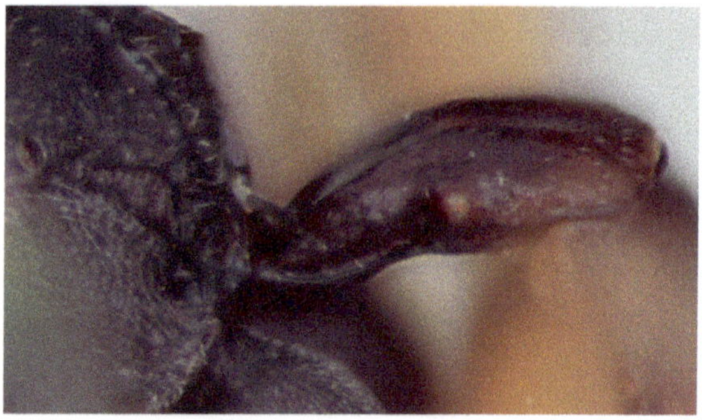

u

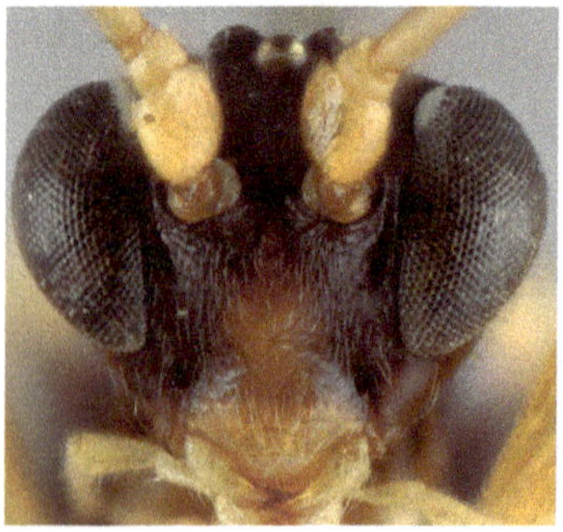

v

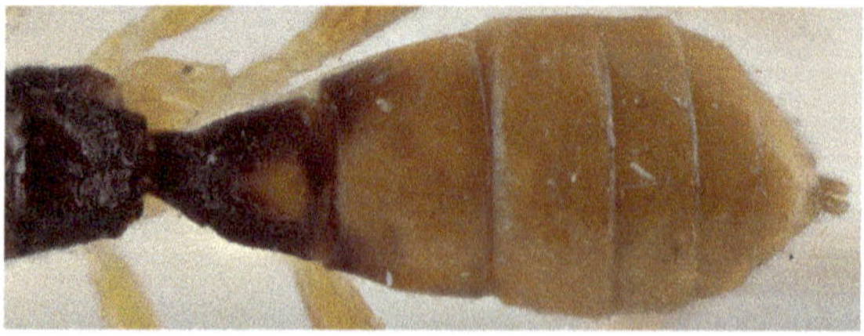

w

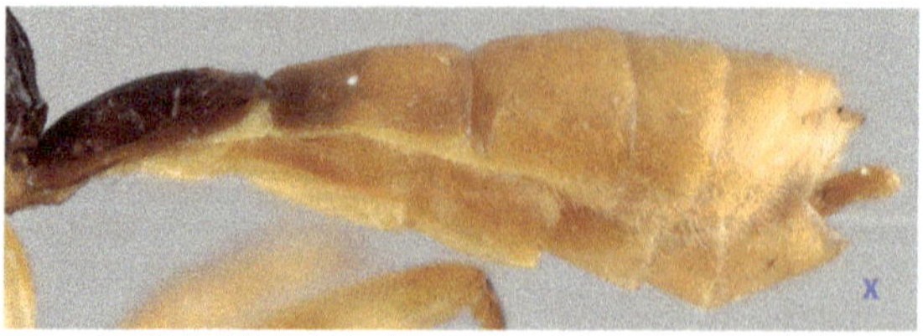

x

Figure 69 continued: Morphology of Cryptinae: (q) *Sphecophaga vesparum* (Curtis) (Cryptini) propodeum; (r) *S. vesparum* part of fore wing; (s) *S. vesparum* head; (t) *S. vesparum* brachypterous female; (u) *S. vesparum* first metasomal tergite (lateral); (v) *Hemiphanes erratum* Humala, head; (w) *H. erratum,* propodeum and metasoma, dorsal; (x) *H. erratum*, metasoma, lateral.

Subfamily CTENOPELMATINAE

Tribes
Chrionotini (=Olethrodotini)
Ctenopelmatini
Euryproctini
Mesoleiini
Perilissini
Pionini
Scolobatini
*Seleucini
*Westwoodiini

The Ctenopelmatinae comprises one of the two major radiations of parasitoids of sawflies, the other being Tryphoninae. Although best represented in the north temperate regions, ctenopelmatines are globally distributed and morphologically diverse. The approximately 1,350 species are currently classified in 105 genera and nine tribes (note that Yu *et al.*, 2012, erroneously include *Ischyrocnemis* and *Scolomus* in Ctenopelmatinae, and Watanabe *et al.*, 2015, described a new genus, *Tanzawana* Watanabe & Kasparyan). The British fauna of 295 species in 61 genera and seven tribes is fairly small compared to the Scandinavian fauna (e.g. Koponen *et al.*, 2000), which is a reflection of the relative sizes of the sawfly faunas, but still means that the Ctenopelmatinae is one of the larger subfamilies of Ichneumonidae in Britain. Taxonomically, the Ctenopelmatinae is one of the more neglected subfamilies, although several useful species-level revisions have been published.

Figure 70. *Olethrodotis modesta* (Gravenhorst), tribe Chrionotini.

Recognition. Ctenopelmatines are recognised by the tooth-like process on the outer, apical face of the fore tibia in combination with the notched or needle-like ovipositor, and the lack of characters defining other subfamilies(!). Note, though, that the fore tibial tooth varies in size (in some species it is less conspicuous than in others), it can be difficult to see in small ctenopelmatines or those with the fore legs folded against the body, and it is present in some other ichneumonids (e.g., some Metopiinae, Mesochorinae, Sisyrostolinae, Campopleginae and Tryphoninae). The most frequent confusion is with male Tryphoninae (particularly Tryphonini). Most Tryphoninae (but not *Eclytus* or *Sphinctus*) have a fringe of regularly spaced setae on the clypeus (many Ctenopelmatinae have a strongly setose clypeus but never a well-defined, discrete apical fringe) and most Tryphoninae have two bullae in fore wing vein *2m-cu*, whereas Ctenopelmatinae often have a single bulla. A useful 'first glance' difference is that ctenopelmatines usually have a more or less matt surface sculpture whereas tryphonines are often quite shiny. In addition, ctenopelmatines always have two hind tibial spurs (spurs absent in the *Exenterus* group of Tryphonini) and ctenopelmatines often have pale colour on the face or frons (quite uncommon in the Holarctic Tryphonini that have two hind tibial spurs, i.e., all genera except the *Exenterus* group). Most ctenopelmatine tribes have some distinctive features: Chrionotini (with a single British species, *Olethrodotis modesta* (Gravenhorst)) have the ovipositor, which is usually short in Ctenopelmatinae, about as long as the metasoma and the eye surface with distinct setae; Ctenopelmatini females have the eighth tergite produced backwards between the base of the ovipositor sheath and the cercus, and the tribe has also been characterised by a longitudinal ridge linking the spiracle on tergite two to the base of the tergite, although a less distinct ridge is discernible in some other ctenopelmatines; Euryproctini lack glymmae on the first metasomal tergite and this segment is often conspicuously elongate; Mesoleiini lack distinctive characters (the ovipositor is usually notched, the first tergite has glymmae that do not meet) but several genera have the edge of the clypeus medially thick and laterally thin (as, though, do some members of other tribes); Perilissini have an oval patch of closely aggregated sensilla on the outer edge of the first flagellar segment (as do Scolobatini) and very deep glymmae on the first tergite, that meet medially at a translucent 'window' (as do Mesochorinae) or meet at a basal, dorsal pit; Pionini females have sharp, needle-like ovipositors with no preapical notch; Scolobatini lack the dorsal section of the occipital carina, the propodeum lacks carinae, and some (including the single British species, *Scolobates auriculatus* (Fabricius)) have a median tooth on the clypeus.

Systematics. In reality, Ctenopelmatinae is almost certainly paraphyletic with respect to some of the smaller subfamilies of ophioniformes (Bennett *et al.*, in prep.) and is a decidedly polythetic group. The genera are grouped together mainly on account of their biology and the fore tibial tooth, which is also found in various members of the Mesochorinae and Metopiinae, and occasionally in more distantly related subfamilies (e.g. *Rhimphoctona* of the Campopleginae and some Tryphonini). Several features are common to many ctenopelmatines, such as the lower mandibular tooth being longer than the upper tooth (very rare in Tryphoninae), the deep glymmae on metasomal tergite one and the apically thick clypeal edge, but these only pertain to subsets of the subfamily. It is clear that the Ctenopelmatinae belongs to the mainly endoparasitoid ophioniformes clade. Pampel (1914) recognised that these groups shared a characteristic ovary form and Gauld (1985) placed the Ctenopelmatinae as a basal group of the ophioniformes (as named subsequently by Wahl, 1991). Quicke *et al.* (2009) found that the (ectoparasitoid) Tryphoninae formed the basal-most clade (or grade, since they were not monophyletic) of the ophioniformes, with the next basal-most clades being the Banchinae, Ctenopelmatinae and various other, smaller, subfamilies; however, Ctenopelmatinae was not monophyletic. The topology of these phylogenies suggest that the ctenopelmatines are defined mainly by a plesiomorphic biological feature, namely parasitism of sawfly larvae, as is also practised by the Tryphoninae (Kasparyan, 1996). The 'higher' ophioniformes (and Banchinae) have radiated extensively as parasitoids of Lepidoptera larvae, which appears to be an apomorphic biological feature. The Ctenopelmatinae is probably paraphyletic with respect to Mesochorinae, Metopiinae, Oxytorinae and Tatogastrinae (Quicke *et al.*, 2009); Gauld & Wahl (2006) also suggested that Ctenopelmatinae is paraphyletic with respect to the Metopiinae, which are parasitoids of Lepidoptera. It is likely that the Ctenopelmatinae will need to be split into smaller, monophyletic, subfamilies, so as to retain biologically distinctive subfamilies such as Mesochorinae and Metopiinae, but at the moment it is far from clear how the group should be divided. The paraphyly of the Ctenopelmatinae needs to be borne in mind when interpreting any data below – what is the case for some genera may not pertain for others.

Figure 71. *Xenoschesis resplendens* (Holmgren), tribe Ctenopelmatini.

Although Townes (1970b) made a very useful start on a meaningful reclassification of the many ctenopelmatine genera, there is little evidence for the monophyly or otherwise of any of the ctenopelmatine tribes, except that Zhaurova & Wharton (2009) cladistically defined the small tribes Scolobatini and Westwoodiini (the latter is entirely Australasian); the Chrionotini (often referred to as Olethrodotini) contains only two genera that share various unusual features. The Seleucini contains only the genus *Seleucus* (Vikberg & Koponen, 2000), with two included species (Shimizu, 2018), an aberrant genus rarely collected in several European countries (but not Britain) that had been classified for many years in the Phrudinae – but Quicke *et al.*'s (2009) analyses supported the placement of this genus in the Ctenopelmatinae, as first proposed by Kolarov (1987), possibly as the sister group to Euryproctini. Otherwise, each tribe (with the possible exception of the relatively distinctive Ctenopelmatini) seems to consist of a core of 'typical' genera surrounded by a cloud of less distinctive genera that blur the boundaries between tribes. Males can be particularly difficult to place to tribe. Several interesting and rather phylogenetically isolated genera are found in the southern temperate areas of South America.

Biology[1]. Ctenopelmatines are basically koinobiont endoparasitoids of sawfly larvae (Hymenoptera: Pamphilioidea, Tenthredinoidea). The only known exceptions are within the perilissine genus *Lathrolestes*, where several species are parasitoids of leaf-mining tenthredinids and some have evidently budded off from that original host range to specialize on leaf-mining members of other insect orders, namely Eriocraniidae (Lepidoptera) in the case of *L. clypeatus* (Zetterstedt) (Heath, 1961)

---

[1] In this subfamily of many tribes we have not separated *Host associations and life histories* from *Developmental biology* overall, in the way done for other large subfamilies, because the often large differences between tribes make it simpler to deal with the tribes separately.

and the American *L. mnemonicae* (Rohwer) (Rohwer, 1914; Carlson, 1979), and *Zeugophora* (Coleoptera: Megalopodidae) in the case of the American *L. zeugophorae* Barron (Barron, 1994). All ctenopelmatines studied so far are solitary and often very host specific. They are also synovigenic (Cummins *et al.*, 2011). Although oviposition is in the larva (or egg), parasitoid development is largely in the cocooned prepupa.

Host ranges of tribes and genera are poorly understood and casual rearings are much less frequent than is the case with parasitoids of the more popular Lepidoptera, exacerbated by the greater difficulties often encountered in rearing sawfly larvae. Most of the detailed biological data come from studies of Ctenopelmatinae as parasitoids of forestry and other crop pests. Broad host associations are described below for each tribe in turn. Chrionotini have never been reared; the long ovipositor of *Olethrodotis modestus* implies deeply concealed hosts. The small tribe Ctenopelmatini use Pamphilioidea as hosts, usually Pamphiliidae but one unidentified *Ctenopelma* sp. was reared from *Megalodontes cephalotes* (Fabricius) (Megalodontesidae) in Austria (Pschorn-Walcher, 1990). Females of some species-groups of *Ctenopelma* have slender ovipositors which lack a subapical notch and it is known that at least one of these species oviposits in the host egg (Barron, 1981; Kasparyan, 2004a); by morphological analogy, this is likely to be the case in the British *C. ruficorne* Holmgren. Other species of *Ctenopelma* oviposit into early instar host larvae. Kasparyan (2002) remarks on the absence of other ctenopelmatines from the guild of pamphiliid parasitoids and surmises that oviposition into the embryo or very young larva, and precise egg placement, are adaptations to avoid an effective immune response. The egg of at least one *Homaspis* species is laid in a ganglion or salivary gland and is surrounded by a trophamnion (Eichhorn, 1988).

Euryproctini have most often been reared from Tenthredinidae but *Zemiophora scutulata* (Hartig) is a parasitoid of economically important sawflies of the family Diprionidae, including *Neodiprion sertifer* (Geoffroy) (Pschorn-Walcher, 1967). *Hadrodactylus* and *Anisotacrus* species are particularly slender and are parasitoids of grass and horsetail-feeding Dolerini (Tenthredinidae) (Hinz, 1961; Idar, 1975). Other genera have been reared from a variety of tenthredinid hosts, including *Tenthredo* (*Mesoleptidea prosoleuca* (Gravenhorst)) and *Caliroa* (*Euryproctus ratzeburgi* (Gorski)) (Heitland & Pschorn-Walcher, 1992). *Hypamblys albopictus* (Gravenhorst) is unusual within the subfamily in apparently attacking several species of Nematinae but Heitland & Pschorn-Walcher (2005) suggest that this wide host range may reflect a complex of species, with different 'populations' of *H. albopictus* preferentially attacking a range of different tenthredinid hosts but restricting their hosts to those on particular food plants. Where known, euryproctine larvae develop initially surrounded by a trophamnion (Zinnert, 1969a; Heitland & Pschorn-Walcher, 1992).

The host associations of Mesoleiini are rather poorly known. The large numbers of species classified in *Mesoleius* and related genera (e.g. *Campodorus* and *Saotis*), seem most frequently to be parasitoids of Nematinae (Tenthredinidae). *Saotis* is particularly associated with gall-forming Nematinae (Kasparyan & Kopelke, 2010) such as the former *Phyllocolpa* and *Pontania* (now classified in *Euura*; Prous *et al.*, 2014). One small species-group of *Mesoleius* comprises parasitoids of leaf-mining *Heterarthrus* (Tenthredinidae: Blennocampinae) larvae (Shaw & Kasparyan, 2003) and Teunissen (1945) records two *Mesoleius* species as parasitoids of Diprionidae. Some physically large mesoleiines, of the genera *Himerta*, *Perispuda* and *Protarchus*, are parasitoids of Cimbicidae larvae (Bauer, 1961; Leblanc, 1999; Aubert, 2000) and some *Lamachus* species have been fairly well studied as parasitoids of diprionid forestry pests (e.g. Pschorn-Walcher, 1973). *Otlophorus* species are known to be parasitoids of tenthredinid genera (*Apethymus* and *Caliroa*) of the subfamily Allantinae (Horstmann, 1999). Zinnert (1969a) found *Mesoleius* species to be generally host specific, with a minority of the species studied attacking two or more host species. The European *Mesoleius tenthredinis* Morley was intensively studied prior to releases in Canada in an attempt to control the nematine sawfly *Pristiphora erichsonii* (Hartig), which devastated large parts of the eastern larch,

Larix laricina, timber industry in eastern North America (Graham, 1953). The parasitoid spread and established quickly after release (Criddle, 1928) but later declined, apparently the victim of competition from *Olesicampe melanogaster* (Thomson) (Ichneumonidae: Campopleginae) (Ives & Muldrew, 1984). When parasitoids were released in the vicinity of host larvae, host location and oviposition by *Mesoleius tenthredinis* was rapid; females oviposit into the posterior, ventral surface of the early instar host larva, with the egg sitting freely in the haemocoel (Graham, 1953). Parasitoid larvae remained quiescent in the host larva in either their first or second instar, with the latter cohort emerging earlier; this results in two peaks in emergence. Location of host populations is much more efficient than in other parasitoids of *P. erichsonii*, which, together with high fecundity, is thought to compensate for the high levels of superparasitism and locally high rates of encapsulation by the host (Pschorn-Walcher & Zinnert, 1971). *Protarchus* species have been recorded as releasing a strong odour when handled, which is unusual in ichneumonids; Gauld (1991), Townes (1971) and Townes & Townes (1978) record that *Pimpla* (Pimplinae), *Exochus* (Metopiinae) and *Banchus* (Banchinae) species, respectively, release pungent odours when handled. It can be surmised that this is a defensive reaction but there is no information on whether this is off-putting to potential predators.

Hosts of Perilissini are poorly known. Most of the reliable host records come from a variety of Tenthredinidae. The very common *Perilissus variator* (Müller) is a parasitoid of grass-feeding *Dolerus* species (Hinz, 1961) whereas other *Perilissus* species have been recorded from a variety of tree/shrub-feeding tenthredinids. Possibly reliable host records indicate that the closely similar, nocturnal *Absyrtus vernalis* Bauer and *A. vicinator* (Thunberg) attack different genera of Tenthredininae (*Macrophya* and *Tenthredo*, respectively) (Hinz, 1961). A few perilissines have diverged from using Tenthredinidae as hosts; the large, impressive and nocturnal *Opheltes glaucopterus* (L.) is a parasitoid of *Cimbex* larvae (Cimbicidae) (Aubert, 2000) and *Lophyroplectus oblongopunctatus* (Hartig) is a parasitoid of diprionid larvae. The genus *Lathrolestes* includes several parasitoids of sawfly leaf miners, particularly of the tribe Fenusini (Tenthredinidae: Blennocampinae) but also parasitoids of Lepidoptera and Coleoptera, as detailed above. A female of *Lathrolestes nigricollis* (Thomson) will search the surface of birch leaves, apparently using her antennae to find mines of the host *Fenusa pusilla* (Lepeletier) and only showing signs of locating a host when the

Figure 72. *Mesoleptidea cingulata* (Gravenhorst), tribe Euryproctini.

mine is directly contacted. Once the host is located she uses her ovipositor to stab randomly into the mine and Quednau & Guevremont (1975) presumed that the host is recognized by its convulsive movements. Oviposition is accomplished in a few seconds. Host feeding occurs when, as a consequence of oviposition, a droplet of host haemolymph appears on the surface of the mine. *Lathrolestes ensator* (Brauns), a parasitoid of the apple pest *Hoplocampa testudinaria* (Klug) (Tenthredinidae), was studied in the Netherlands by Zijp & Blommers (1993), who reported that the black, comma-shaped egg is clearly visible in the fully grown host larva but does not hatch until the host has become a cocooned prepupa in the soil. The parasitoid then develops to the prepupal stage during the summer, and the onward progression towards the adult stage continues slowly through the winter, apparently without an evident diapause. They also found that eggs were distributed over the host population at random, with no avoidance of superparasitism, but the host stage at oviposition was not stated. Whereas euryproctines and at least some Ctenopelmatini develop as eggs and young larvae in a trophamnion, no trophamnion is present in Perilissini (Heitland & Pschorn-Walcher, 1992). The two largest perilissine genera, *Lathrolestes* and *Perilissus*, are very vaguely defined and any analysis of host utilisation patterns must await the delineation of monophyletic groups and, of course, resolve the status of the tribes as presently deployed. In a study of parasitoids of leaf-mining Fenusini (Tenthredinidae) sawfly larvae, Pschorn-Walcher & Altenhofer (1989) found a very high degree of host specificity amongst the *Lathrolestes* species that specialise on Fenusini. Interestingly, the parasitoid fauna of heterarthrines, currently classified in the same tribe (Blennocampinae: Fenusini) as fenusines, showed virtually no similarity in parasitoid fauna and Pschorn-Walcher & Altenhofer (1989) used this difference to argue against the close relationship of these leaf-mining genera.

Figure 73. *Himerta sepulchralis* (Holmgren), tribe Mesoleiini.

Figure 74. *Oetophorus naevius* (Gmelin), tribe Perilissini.

Members of the tribe Pionini have very slender, un-notched ovipositors and deposit their egg in the egg of the host sawfly or (in some *Rhorus*) in early instars. All reliable host records are from Tenthredinidae, of several subfamilies. There have been few observations on European Pionini but the North American species *Glyptorhaestus tomostethae* (Cushman) has been investigated by McConnell (1938). The adults of the ichneumonid and its host *Tomostethus* (Tenthredinidae: Blennocampinae) are active in spring. The egg is placed in the yolk of the host egg and the parasitoid egg or newly hatched larva thus becomes enveloped in the mesenteron of the host. At about the time of host hatching, the parasitoid larva migrates to the thoracic segments and takes up a position below the gut but above the fat body. It maintains this position during the two week feeding period of the host. Although the *Glyptorhaestus* larva does not moult during this time it does increase greatly in size. The first instar larva has an anal vesicle, which becomes invaginated after the first moult. The parasitoid larva makes this moult to the second larval instar at about the time the host enters the ground to form its earthen pupation chamber, and the remainder of parasitoid development (second to fifth larval instars) occupies about the same time as the first instar. It is full-grown and spins its cocoon in early summer, then remains quiescent until about mid September when it pupates. The adult 'emerges' in autumn but remains in the cocoon until the following spring. He makes the intriguing observation that adults spend the winter 'with the tip of their abdomens imbedded in the liquid meconial mass', which might perhaps be interpreted as antifreeze, as in the braconid *Acampsis alternipes* (Nees) (Shaw & Quicke, 2000). Eggs of *Rhorus* species are very small (Cummins *et al.*, 2011) and oviposition, at least in *R. lapponicus* (Roman), is very precise, through the larval ocellus (Pschorn-Walcher & Zinnert, 1971).

Figure 75. *Pion nigripes* Schiødte, tribe Pionini.

With respect to Scolobatini, little is known. *Scolobates auriculatus* is a parasitoid of larvae of several species of *Arge* (Hymenoptera: Argidae) (Pschorn-Walcher & Kriegl, 1965); the only host record for another (exotic) genus of Scolobatini is also from an argid (Gauld, 1997). The Australasian tribe, Westwoodiini, parasitize Pergidae.

Finally, the Palaearctic *Seleucus cuneiformis* Holmgren (Seleucini), has recently been reared from *Blasticotoma filiceti* Klug (Blasticotomidae), so far the only ctenopelmatine known to be a parasitoid of this fern-feeding family (van Achterberg & Altenhofer, 2013). Throughout its range, *S. cuneiformis* is rarely collected and it is possible that it occurs in Britain, although its host is rare here.

Details of the life histories of Ctenopelmatinae are known for only a few species and the following summarizes some of the known facts and differences. Females of the perilissine *Lathrolestes nigricollis* (Thomson) did not mate until about 45 hours following emergence (Quednau & Guevremont, 1975) whereas in *Mesoleius tenthredinis* females mated quickly following emergence, but males required one or two days maturation first (Graham, 1953). Females of *Lathrolestes nigricollis* would not mate if they were still virgins after a few days, by which time their first eggs had matured (Quednau & Guevremont, 1975). Whereas most Ctenopelmatinae dissected by Cummins *et al.* (2011) had rather large numbers of eggs that were fairly small, *Euryproctus* had fewer, larger eggs and were the only ctenopelmatines that these authors considered may be host-limited rather than time-limited. Superparasitism is common in euryproctines and Heitland & Pschorn-Walcher (1992) document some field examples where superparasitism was much more frequent than would be expected by chance. Given these observations, and the potentially vulnerable, trophamnion-encased larvae (or large, sclerotized larvae), Heitland & Pschorn-Walcher (1992) surmise that superparasitism might be a response to competition within the hosts. The Euryproctini are developmentally distinct in that the embryo and young larva are enclosed in a trophamnion for a relatively long time. An exception to this is *Phobetes atomator* (Müller), the first instar larva of which is heavily sclerotized (Heitland & Pschorn-Walcher, 1992). Interestingly, Cummins *et al.* (2011) found that females of *Phobetes* differ from other euryproctines in their relatively small eggs. Eggs of Ctenopelmatinae can be entirely encapsulated by the host's haemolymph cells but Adam (1966) showed that *Alexeter niger* (Gravenhorst) eggs were still viable when encapsulated, as long as they were not melanized as well.

The parasitoid assemblage of the diprionid sawfly *Neodiprion sertifer*, a pest of conifer forestry, has been the focus of much study. Pschorn-Walcher (1988) points out that this species, with its nearest relatives in the Nearctic, is unique in the European diprionid fauna in that it overwinters in the egg stage and aestivates in the cocoon. Nevertheless, it has a rich complex of parasitoids including two ctenopelmatines apparently specific to this host, *Lamachus eques* (Hartig) (Mesoleiini) and *Lophyroplectus oblongopunctatus* (Perilissini). The nearest relatives of *L. eques* are parasitoids of other diprionids and Pschorn-Walcher (1988) surmises that *Microdiprion pallipes* (Fallén) is the source of the parasitoid assemblage that adapted to *Neodiprion sertifer* following its arrival in Europe. The sawfly's populations fluctuate and fairly regularly produce outbreaks. One or two years after peak host abundance, parasitoid-induced mortality can reach 90% (Pschorn-Walcher, 1987). Under outbreak conditions, populations of *L. oblongopuntatus* build up considerably but almost invariably these endoparasitoids are outcompeted by the ectoparasitoid tryphonine ichneumonid *Exenterus abruptorius* (Thunberg). *Lamachus eques* is the weakest competitor, accounting for less than 5% of mortality in conditions of high host density. However, when host density is low, *L. eques* can account for up to 65% mortality as it can find hosts far more efficiently than either of its two competitors (Pschorn-Walcher, 1987).

There are indications that some Ctenopelmatinae may harbour polydnaviruses, although this has not been demonstrated. Cummins *et al.* (2011) found that the *Barytarbes* and *Rhorus* species they examined had an enlarged calyx region (at the junction of the ovarioles and lateral oviduct), which is consistent with polydnavirus production in Banchinae and Campopleginae (but other Ctenopelmatinae examined lacked this). Stoltz (1981) found baculoviruses in the ovaries of *Mesoleius tenthredinis*, pre-dating the description of polydnaviruses. This is clearly an area worth following up.

Identification. Identification of many ctenopelmatines remains difficult or impossible without a very good reference collection as there are few revisions available. Townes's (1970b) key to genera is the only such key available but is difficult to use, with the initial separation of the genera into mainly poorly defined tribes as the first hurdle. Aubert (2000) catalogued the western Palaearctic Ctenopelmatinae and is thus a very useful guide to the literature (although this should be used along with Yu *et al.*, 2012), with some notes on separation of a few species. Shaw & Kasparyan (2003) detailed the occurrence of some British Mesoleiini and Shaw *et al.* (2003) revised the British list of Ctenopelmatini. The only British representative of the Chrionotini, *Olethrodotis modestus*, was discussed by Shaw & Kasparyan (2002). The Ctenopelmatini consists of small genera (in Britain), three of which have been recently revised by Kasparyan (2004a: *Ctenopelma* and *Homaspis*, 2002: *Notopygus*). There has been very little revisionary taxonomic work on the Euryproctini: *Hadrodactylus* (Idar 1979, 1981; Kasparyan & Shaw, 2009; Kasparyan, 2011); *Phobetes*, partial key (Kasparyan, 2004b); *Synomelix* (Idar, 1983). The tribe Mesoleiini is large and the genera difficult to recognise, but revisions exist for the following genera: *Campodorus* (Kasparyan, 2003, 2005, 2006); *Himerta* (Horstmann, 2002a); *Lamachus*, parasitoids of Diprioninae (Oehlke, 1966); *Hyperbatus* (Kasparyan, 1998); *Mesoleius* (Kasparyan, 2000, 2001); *Protarchus* (Viitasaari, 1979); *Rhinotorus* (Reshchikov, 2016); *Saotis* (Kasparyan & Shaw, 2003; Kasparyan & Kopelke, 2010); Gauld & Mitchell (1977a) separated two nocturnal species of *Alexeter*. The tribe Perilissini is again poorly served by taxonomic revisions. Broad & Wharton (in prep.) provide a key to the genera of the tribe; only the small genera *Absyrtus* (Broad, 2012, online), *Trematopygodes* (Hinz & Horstmann, 1998) and *Zaplethocornia* (Aubert, 1985) have been revised, although Reshchikov (2013) provides a key to the European *Lathrolestes* species. The Pionini are better served by revisions and keys exist for species of the following genera: *Glyptorhaestus* (Hinz, 1975); *Lethades* (Hinz, 1996); the single British *Phaestus* is included in Kasparyan (1998); *Rhorus*, revision of Western Palaearctic species by Kasparyan (2012, 2014, 2015, 2017), partial key (Aubert, 1988); *Sympherta* (Hinz, 1991); *Trematopygus* (Hinz, 1986). Van Achterberg & Altenhofer (2013) provide good illustrations of the non-British *Seleucus cuneiformis*. The single British representative of the Scolobatini (*Scolobates auriculatus* (Fabricius)) can be recognised through Townes's (1970b) keys to tribes and genera and through Zhaurova & Wharton's (2009) diagnosis of the tribe and genus. Nocturnal (i.e. predominantly testaceous) ctenopelmatine genera can be recognised using Broad (2012, online) (note that the key by Huddleston & Gauld (1988) contains an error; couplet 23 has only one half of the couplet, with the characters diagnosing *Priopoda* but labelled as *Perilissus*; *Priopoda* isn't mentioned).

Figure 76. *Scolobates auriculatus* (Fabricius), tribe Scolobatini.

Figure 77. Morphology of Ctenopelmatinae: (a) *Olethrodotis modesta* (Gravehorst) (Chrionotini) head; (b) *Homaspis analis* (Holmgren) (Ctenopelmatini) head; (c) *Hadrodactylus indefessus* (Gravenhorst) (Euryproctini) head; (d) *Alexeter multicolor* (Gravenhorst) (Mesoleiini) head; (e) *Pion fortipes* (Gravenhorst) (Pionini) head; (f) *Rhorus chrysopus* (Gmelin) (Pionini) head; (g) *Scolobates auriculatus* (Fabricius) (Scolobatini) head; (h) *Perilissus variator* (Müller) (Perilissini), scanning electron micrograph, first flagellar segment, arrow pointing at aggregation of sensilla; (i) *Campodorus holmgreni* (Schmiedeknecht) (Mesoleiini) fore wing; (j) *A. multicolor* fore tibia, arrow pointing at tooth; (k) *H. indefessus* fore tibia with arrow pointing at tooth; (continued overleaf).

Figure 77 continued: Morphology of Ctenopelmatinae continued: (l) *H. indefessus* metasoma (lateral); (m) *H. indefessus* metasoma (dorsal); (n) *Notopygus emarginatus* Holmgren (Ctenopelmatini) first metasomal segment (lateral); (o) *A. multicolor* first metasomal segment (lateral); (p) *P. fortipes* metasoma; (q) *P. fortipes* propodeum and metasoma (dorsal); (r) *Protarchus testatorius* (Thunberg) (Mesoleiini) propodeum and first metasomal segment (lateral); (s) *Perispuda facialis* (Gravenhorst) (Mesoleiini) head; (t) *Rhorus longicornis* (Holmgren) (Pionini) metasoma.

Subfamily **CYLLOCERIINAE**

This is a very small subfamily which includes only 38 described species in four genera, *Allomacrus*, *Cylloceria*, *Hyperacmus* and *Rossemia*. It is Holarctic, Oriental and Neotropical in distribution. Seven species, representing *Allomacrus*, *Cylloceria* and *Hyperacmus*, are known from Britain (G. Broad, unpublished).

Figure 78. *Cylloceria caligata* (Gravenhorst).

Recognition. Cylloceriines are rather difficult to characterise but are best recognised by a combination of the flattened clypeus; rather elongate propodeum with median longitudinal carinae but no transverse carinae; the granulate first metasomal tergite; and the well-defined notauli. Females of *Allomacrus* and *Cylloceria* have long, notched ovipositors whereas *Hyperacmus* have very short ovipositors and are generally very compact insects, somewhat resembling Metopiinae. Males of *Cylloceria* and *Hyperacmus* have one or two sub-proximal flagellar segments with very characteristic tyloids in the form of semi-circular excavations.

Systematics. There has been little recent consensus over the composition of the subfamily and its systematic position. The genera have been regarded as belonging to the Orthocentrinae *s.l.* (Plectiscinae of Perkins, 1959, Microleptinae of Townes, 1971), but Wahl (1990) and Wahl & Gauld (1998) recognised that *Allomacrus*, *Cylloceria* and *Rossemia* (which they described under the synonymous name *Sweaterella*) formed a monophyletic group and separated the Cylloceriinae from Orthocentrinae. Humala (2003, 2007) disagreed with this position and regarded Cylloceriini (also including *Apoclima* and *Entypoma*) as a tribe within Orthocentrinae. *Hyperacmus* was regarded as belonging to Orthocentrinae (e.g. Wahl & Gauld, 1998) or Microleptinae in the narrowly defined sense (Humala, 2003, 2007a). Quicke *et al.* (2009) established that *Hyperacmus* shared some significant apomorphic characters with *Cylloceria* (a unique divided venom gland, the structure and sculpture of the propodeum and first tergite, and the form of the male flagellar tyloids) and transferred this genus to Cylloceriinae. Settling whether or not this subfamily also includes *Apoclima* and *Entypoma*, and whether the Cylloceriinae should indeed be separated from the Orthocentrinae, requires more detailed phylogenetic analyses than have yet been undertaken.

Biology. Little is known of the biology of this group although there are host records for a few species of *Cylloceria* and one of *Hyperacmus*. Species of *Cylloceria* have been reared from larvae of *Tipula* (Diptera: Tipulidae), and Wahl (1986) surmised that they must be acting as endoparasitoids, based on the larval head capsule morphology (also supported by the notched ovipositor). In NMS are three series of *Cylloceria melancholica* (Gravenhorst) (one doubtfully identified) reared from tipulids collected as larvae in decaying wood. Six certainly identified specimens were reared from *Tipula irrorata* Macquart larvae collected in June. The accompanying cocoons are rather substantial, subcylindrical, dense and coarsely woven, and buff in colour. The adults in all cases emerged in July of the year. *Hyperacmus crassicornis* (Gravenhorst) has recently been recorded as a parasitoid of the fungus gnat, *Sciophila varia* (Winnertz) (Diptera: Mycetophilidae) (Šedivý & Ševčik, 2003), but without details of its development, and the supposed host seems too small. Records of *Cylloceria* as parasitoids of Lepidoptera are probably due to confusion with *Lissonota* species, or possibly the result of uncontrolled substrate rearings in which the true host was overlooked.

Identification. Wahl & Gauld (1998) define the Cylloceriinae and give a key to genera, but not including *Hyperacmus*; keys to the wider orthocentrine genera, including cylloceriines, are given by Townes (1971) and van Rossem (1990) (but the latter includes some unrelated genera described as orthocentrines); Western Palaearctic *Allomacrus* and *Cylloceria* are keyed by Humala (2002), but not all specimens of *Cylloceria* key easily and the genus is in need of revision. Only single described species of *Allomacrus* and *Hyperacmus* have been found in Britain but there are additional *Allomacrus* species in Europe that might occur here. G. Broad (in prep.) is working on keys to the British species.

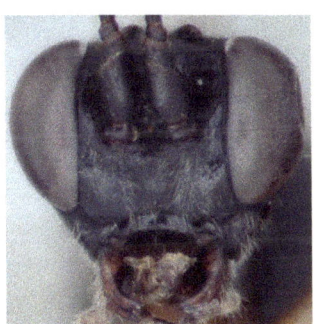

Figure 79. Morphology of Cylloceriinae: (a) *Allomacrus arcticus* (Holmgren) head; (b) *Cylloceria caligata* (Gravenhorst) head; (c) *Hyperacmus crassicornis* (Gravenhorst) head, lateral; (continued overleaf).

Figure 79 continued: Morphology of Cylloceriinae: (d) parts of antennae of *C. caligata* female (left), *C. caligata* male (middle) and *H. crassicornis* male (right); (e) *H. crassicornis* mesoscutum; (f) *C. caligata* propodeum and first metasomal segment (dorsal); (g) *H. crassicornis* metasoma; (h) *A. arcticus* ovipositor, detail of tip inset; (i) *C. caligata* ovipositor, detail of tip inset.

Subfamily DIACRITINAE

Recognition and systematics. This very small group comprises only seven known species in three genera, found across the Palaearctic and Nearctic. It was formerly treated as a tribe of Pimplinae (for example, Townes, 1969; Fitton *et al.*, 1988) but Eggleton (1989) elevated the Diacritinae to subfamily rank, which was followed by Gauld (1991). The single European species, *Diacritus aciculatus* (Vollenhoven), is readily recognised by a combination of the flattened clypeus with a basal ridge, cylindrical first metasomal segment (tergite and sternite fused) with spiracles at the mid-length, the propodeum with strong, parallel longitudinal carinae and posterior transverse carina (anterior transverse carina absent) and long ovipositor lacking a dorsal, subapical notch, nodus or conspicuous ventral teeth. The metasomal tergites are conspicuously cream-coloured apically.

Biology. Nothing is known of the life history of any diacritine; the observation behind the suggestion in Fitton *et al.* (1988) that an exotic species of diacritine had been reared from a chrysomelid beetle is now known to refer to a species of *Pachymelos* (Pimplinae) (Gauld *et al.*, 2002b). Although Diacritinae clearly belongs to the pimpliformes group of subfamilies (Wahl & Gauld, 1998; Quicke *et al.*, 2009), its nearest relatives are uncertain and may lie in either the predominantly ectoparasitoid or endoparasitoid groups. *Diacritus aciculatus* can be common in damp woodland.

Identification. The single British species is covered by Fitton *et al.* (1988) and can be recognised using the key to subfamilies.

Figure 80. *Diacritus aciculatus* (Vollenhoven).

Figure 81. Morphology of Diacritinae: (a) *Diacritus aciculatus* (Vollenhoven) head; (b) part of fore wing (to show areolet); (c) propodeum and first metasomal segment, dorsal; (d) first metasomal segment, lateral; (e) metasoma apex and ovipositor, detail of tip inset.

Subfamily **DIPLAZONTINAE**

One of the more readily recognisable subfamilies of Ichneumonidae, Diplazontinae can be found throughout the world, although species-richness seems to be highest in the north temperate regions, where aphids are at their most species-rich (almost all known hosts of diplazontines are aphidophagous Syrphidae). Of a global fauna of about 355 described species in 22 genera, 60 species in 15 genera have been found in Britain.

Figure 82. *Homotropus pallipes* (Gravenhorst).

Recognition. Diplazontines are distinctive in that the mandible is tridentate (the upper tooth is subdivided); the head in facial view is broad; the clypeus is usually flattened, at least apically; and the first metasomal tergite is invariably sub-rectangular in dorsal view (parallel-sided, narrowed only very basally). Despite significant variation in metasoma shape across the subfamily, diplazontines are generally instantly recognisable. A very few other ichneumonids have vaguely tridentate mandibles (with the upper tooth broad and weakly divided), in Britain including *Banchus* and *Rhynchobanchus* of the Banchinae, and some *Lamachus* of the Ctenopelmatinae. In other respects, there is little similarity between these genera and diplazontines; the shape of the first tergite differs, the nervellus of Banchini is intercepted very high up, most *Banchus* have a spine on the hind edge of the scutellum, *Lamachus* have a distinct antero-distal tooth on the fore tibia. Most diplazontines are rather small insects, with a wing length of 3-8 mm (Klopfstein, 2014), usually around 4-5.5 mm. The only diplazontine that lacks tridentate mandibles is *Episemura ensata* (Bauer), which also has an unusually long ovipositor (Kasparyan & Manukyan, 1987), although this species can still be recognised as a diplazontine on the basis of its general morphology. It has not yet been found in Britain but *E. ensata* is a very rarely collected species (Klopfstein, 2014) and should be searched for in larch (*Larix*) foliage.

Systematics. Their biology, together with morphological features including the tridentate mandibles, wide face and parallel-sided first metasomal tergite, strongly suggest that the Diplazontinae is monophyletic and Wahl (1990) lists several larval features that also support the monophyly of the subfamily. On the basis of a shared character of the larval morphology, the hypostomal-stipital plate, Wahl (1990) demonstrated that the Diplazontinae are both a distinctive, monophyletic group and form part of a clade, together with Cylloceriinae and Orthocentrinae, of koinobiont endoparasitoids of Diptera within the pimpliformes group of subfamilies. Although the 28S D2-D3 rDNA region is widely used in ichneumonoid phylogenetics, this region has not provided much phylogenetic resolution within the pimpliformes (Belshaw *et al.*, 1998; Quicke *et al.*, 2009); nevertheless, monophyly of the Diplazontinae has been corroborated with independent morphological, molecular and combined analyses (Wahl & Gauld, 1998; Quicke *et al.*, 2009; Klopfstein *et al.*, 2010a). The Cylloceriinae, Diplazontinae and Orthocentrinae together potentially comprise by far the most species-rich clade of ichneumonid endoparasitoids of Diptera.

The internal classification of Diplazontinae is better resolved than for any other medium-sized ichneumonid subfamily; Klopfstein *et al.* (2010a, 2011), on the evidence provided by four genes and a morphological dataset, proposed three major genus-groups, the *Diplazon*, *Sussaba* and *Syrphoctonus* genus-groups. These studies did not include several small, mainly extralimital genera, and it is likely that *Homotropus*, and possibly *Phthorima* and *Tymmophorus*, are not monophyletic. This phylogenetic framework has proved useful for examining rates of change in sexually selected characters (Klopfstein *et al.*, 2010b) and would offer a valuable opportunity to examine shifts in host use, if our knowledge of diplazontine hosts were not so limited. Although most of the genera deployed in previous identification keys (e.g. Dasch, 1964; Townes, 1971; Fitton & Rotheray, 1982) are thought to be monophyletic, the formerly polythetic *Syrphoctonus* has been split, with the majority of species placed (formally for the Western Palaearctic, only implicitly for much of the world) in *Homotropus* which, on a global basis, is doubtfully monophyletic (Klopfstein *et al*, 2011). There have been convergences in the morphology of diplazontines, notably concerning the lateral compression of the metasoma, connected to oviposition behaviour, and some supposed genera with widely disjunct ranges may not comprise closely related species (Gauld & Hanson in Gauld, 1997). Additionally, various southern hemisphere species have been referred to genera known primarily from the northern hemisphere and their generic placements are untested (Klopfstein, pers. comm.). On the whole, though, the British diplazontines can be readily accommodated in mostly well-defined genera.

Biology. All reliably reared diplazontines are solitary, koinobiont endoparasitoids of Syrphidae (Diptera), ovipositing in the egg or the larva and emerging from the host's puparium. Given current uncertainties in the subfamily classification of Syrphidae, it is difficult to say whether diplazontine hosts are closely related but they certainly share, for the most part, a common biology. In all but one genus, *Bioblapsis*, the known syrphid hosts are aphidophagous. It may be more accurate to say that these hosts of diplazontines are all syrphids that feed on Aphidoidea, given that some feed on the family Adelgidae; however, there are, as yet, no reported associations of diplazontines parasitizing adelgid-feeding syrphids on conifers. Diplazontine host range is thus remarkably constrained for Ichneumonidae, by host taxon and habitat. The Orthocentrinae might be considered to be similarly constrained, in that hosts are generally mycophagous Bibionomorpha, although their known hosts (most orthocentrines have yet to be reared) span a wider range of Diptera families, including predatory Diptera, and would be further expanded if the Cylloceriinae were included within the Orthocentrinae (see the relevant subfamily chapters in this handbook). The known exceptions to parasitism of aphidophagous Syrphidae are two species of the small genus *Bioblapsis*; *B. cultiformis* (Davis) is a parasitoid of a mycophagous syrphid, *Cheilosia longula* (Zetterstedt), whilst *B. polita* (Vollenhoven) has been reared from a *Ferdinandea* species, which feed as larvae in sap runs on trees (Rotheray, 1990, in which *B. cultiformis* was described as *B. mallochi* Rotheray). Given the diversity of syrphid life histories, hosts of some other diplazontines could conceivably be found within the non-aphidophagous Syrphidae. Nevertheless, the association with aphidophagous Syrphidae is so strong that it apparently limits the distributions of diplazontines, which are scarce in lowland tropical forests, where aphids are sparsely represented (e.g. Gauld, 1997). The exception is the near ubiquitous *Diplazon laetatorius* (Fabricius) which, being thelytokous (over most of its range; males have been found in India and North America: Klopfstein, 2014) and having a wide host range of syrphids associated with agricultural pest aphids, has been accidentally transported around the world and can be found in all sorts of climates, principally in anthropogenically altered habitats.

Host ranges of diplazontines are generally not narrowly constrained by host taxon, at least in species that are frequently reared, but instead are defined more by host ecology and the season (Dušek *et al.*, 1979). There are exceptions; for example, *Homotropus pictus* (Gravenhorst) is a parasitoid of *Platycheirus* species only (Rotheray, 1979, cited in Fitton & Rotheray, 1982). Species of several genera have strongly laterally compressed metasomas; presumably these are adaptations towards attacking concealed hosts, although there are few host records for these taxa. Rotheray (1990) found that *Bioblapsis cultiformis* probes with its compressed metasoma in the semi-liquid medium of decomposing fruiting bodies of *Boletus* fungi. Some other diplazontines with laterally compressed metasomas may be attacking syrphid larvae feeding on concealed aphids (e.g. in galls and leaf curls), as is the case for *Phthorima* species reared from species of Pipizini feeding on wax-secreting aphids (Fitton & Boston, 1988). *Sussaba erigator* (Fabricius) is a parasitoid of *Pipizella* species predating root-feeding aphids (Fitton & Rotheray, 1982). A good number of diplazontine species can be found very frequently around aphid colonies in the herb layer, on common garden plants or in agricultural areas. However, numerous species are rarely collected and there is still much to be discovered about host niches of these species. Some very common and widespread species have host ranges that encompass several genera (host ranges, as far as known, were summarised by Fitton & Rotheray, 1982); *Diplazon laetatorius* is a regular parasitoid of genera such as *Episyrphus*, *Metasyrphus* and *Sphaerophoria*; *Enizemum ornatum* (Gravenhorst) attacks *Metasyrphus luniger* (Meigen) and *Scaeva pyrastri* (L.); *Syrphoctonus tarsatorius* (Panzer) parasitises the abundant *Episyrphus balteatus* (DeGeer) and *Syrphus ribesii* (L.).

Host-searching diplazontines (probably the vast majority of species, at least) initially orientate towards aphid odours, which can be emanating from various species of aphids. The syrphid host is then located by contact with the antennae when searching the aphid colony, with the host being recognised through chemicals associated with its cuticle (Rotheray, 1981a). In the few well-studied diplazontines, the host stages attacked either encompass eggs with mature embryos and first or second instar larvae, or more mature, second or third instar larvae; species that would oviposit in third instar larvae ignored eggs of hosts whilst those that would oviposit in eggs ignore third instar larvae (Rotheray, 1981a). Rotheray's study suggests differences between genera in host stage preference, with the *Diplazon* and *Syrphophilus* species studied ovipositing in eggs or young larvae and the *Enizemum* and *Homotropus* species attacking older larvae. Although few species have been studied, if these host stage preferences are true of most species in these genera then this difference accords with the generic groupings found by Klopfstein et al. (2011).

When oviposition is into a larval host, the diplazontine egg is laid in the haemocoel, where larval development takes place (Fitton & Rotheray, 1982). Diplazontine larvae are able to cope with the immune responses of different host species and genera and, at least in *Syrphoctonus fissorius* (Gravenhorst), this is achieved by an unknown mechanism whereby the ichneumonid egg, or something introduced at oviposition, hinders the transformation of haemocytes into capsule tissue (Schneider, 1951). Presumably this explains their wide potential host ranges. Host feeding is frequent and eggs or larvae of potential hosts can be destructively consumed by the diplazontine female, often following ovipositor insertion (Schneider, 1950; Rotheray, 1981a). Schneider (1950) estimated total potential fecundity of a female *Syrphoctonus fissorius* as 600-700 eggs and a female *Diplazon laetatorius* in the laboratory can live for around three weeks (Scott, 1939).

In a comparative study of several species of Diplazontinae, only *Enizemum ornatum* (Gravenhorst) was found to exhibit any preference for oviposition site, always ovipositing just behind the mouthparts, as a response to the host rearing up (Rotheray, 1981a). Various biological observations have been ascribed to *E. ornatum* (e.g. Rotheray, 1981a, 1984); however, at least some of the specimens have turned out to be the closely similar *E. scutellare* (Lange) (Klopfstein, 2014), although *E. ornatum* seems to be by far the most frequent *Enizemum* species in Britain.

One of the few detailed life history studies of a diplazontine species is that of Schneider (1950, 1951) on *Syrphoctonus fissorius* and much of the following account of diplazontine development is taken from those sources. As in many ichneumonids, the first instar larva's mandibles are relatively large but become proportionally smaller as the larva grows. Presumably the large mandibles are mainly for combat as the young larvae do not feed via their mouths but instead absorb nutrients through their cuticle (Schneider, 1950) and indeed the middle three instars lack mandibles altogether. There are four ecdyses in rapid succession, between 24 and 36 hours at 20°C. Feeding commences once the host has pupariated, with the diplazontine remaining inactive if the host enters diapause (Schneider, 1951). There are five larval instars and development from egg to adult in diplazontines generally takes about three weeks when there is no diapause (Fitton & Rotheray, 1982). The winter is passed in the host larva or puparium, depending on the host generation (Scott, 1939). Early pupariation of the host is triggered by secretions from the diplazontine larva, at least in *Diplazon pectoratorius* (Thunberg) (Schneider, 1950), and parasitized puparia can readily be recognised as they are smaller than non-parasitized puparia (Dušek et al., 1979) and the anterior spiracles do not protrude, as they do in non-parasitized syrphids (Rotheray, 1990). If a diplazontine larva dies at an early stage, the adult syrphid still emerges but with severe developmental abnormalities (Schneider, 1951). Diplazontines emerge from the host puparium by cutting small circular and then progressively larger semicircular strips from the puparium cuticle (Rotheray, 1981b), so it may be that the distinctive mandible allows the broad upper tooth to act as a blade, with the upper point of this subdivided tooth initially piercing the cuticle. Emergence from Diptera puparia has also given rise to the distinctive (although very different) mandibles of the braconid subfamily Alysiinae, again featuring extra mandibular teeth, although in this case the exodont alysiine mandibles seem to function by levering apart the puparium along the fissures that are used by the fly (Griffiths, 1964).

Females of *Diplazon pectoratorius*, at least, mate only once (Rotheray, 1981c). Males of this species, and of *Diplazon tetragonus* (Thunberg) and *Syrphophilus tricinctorius* (Thunberg), aggregate in numbers, although this seems not to be the case for many diplazontine species (Rotheray, 1981c). Whether these aggregations are leks or gatherings in response to female-produced pheromones is unknown. Males of many diplazontines coil their antennae around the female's antennae during courtship, with the area of coiling corresponding to the tyloid-bearing segments (Steiner *et al.*, 2010). Taking advantage of the fact that this coiling can be reproduced in preserved insect specimens, when the antennae are softened and subjected to increased hydrostatic pressure, Klopfstein *et al.* (2010b) demonstrated that diplazontine male courtship ancestrally consisted of an antennal coil, with associated tyloids, and that a clade consisting of *Homotropus* and relatives all exhibit double coiling (i.e. the male antenna twists through two complete coils). Males of a further clade, of *Diplazon* and relatives, lack antennal tyloids and do not coil their antennae. Whilst most ichneumonid tyloids bear pores and presumably secrete substances involved in courtship, tyloids of *Enizemum* lack pores and Klopfstein *et al.* (2010b) surmised that diplazontine tyloids are involved only in structural deformation of the antenna during courtship.

Large populations of syrphids, such as arise following immigration of *Episyrphus balteatus* and *Scaeva pyrastri*, can lead to the rapid build-up of large populations of diplazontines such as *Diplazon laetatorius* and *Enizemum ornatum*, although there have been no studies on the population dynamics of diplazontines or whether these species exert any control on the populations of their host hoverflies. Notionally, diplazontines could be regarded as potentially injurious to the natural biological control exerted by syrphids on their economically important aphid hosts, although this is not likely to be of major importance, given that syrphid populations are not artificially augmented in biological control programmes; Scott (1939) reported only a 3-6% parasitism rate of syrphids collected from field-grown cabbages in England. Several diplazontines are frequent enough in light traps to suggest that they are at least partly nocturnal or crepuscular (e.g. *Promethes sulcator* (Gravenhorst), *Sussaba flavipes* (Lucas) and *Syrphoctonus tarsatorius*), although these species display none of the morphological adaptations associated with nocturnal habits (Gauld & Huddleston, 1976) as seen in, for example, Ophioninae and *Netelia* (Tryphoninae).

Identification. Klopfstein (2014) revises the Western Palaearctic fauna of Diplazontinae, with complete keys to genera and species. Other keys are listed here although they are probably redundant for identification purposes. Fitton & Rotheray (1982), as well as providing information on the biology of diplazontines and an overview of the British fauna, include a key to the genera which is simpler than Klopfstein's but obviously pre-dates Klopfstein *et al.* (2011) who divided *Syrphoctonus s.l.* into three genera (*Fossatyloides*, *Homotropus* and *Syrphoctonus*) and does not include one rare genus collected subsequently in Britain (*Eurytyloides*). Beirne's (1941b) keys are now redundant, with their over-reliance on simple colour characters and a simplified taxonomy compared with current knowledge. Fitton & Boston (1988) provide a key to British *Phthorima* species.

Figure 83. Morphology of Diplazontinae: (a) *Enizemum ornatum* (Gravenhorst) head; (b) *Sussaba cognata* (Holmgren) head; (c) *Syrphophilus tricinctorius* (Thunberg) clypeus and mandibles; (d) *S. cognata* first and second metasomal tergites; (e) *E. ornatum* propodeum, first and second metasomal tergites; (f) *Phthorima compressa* (Desvignes) metasoma, lateral; (g) *Diplazon laetatorius* (Fabricius) metasoma and hind tibia.

Subfamily EUCEROTINAE

Eucerotinae is a small subfamily of only two genera, *Barronia*, a monotypic Chilean genus (Gauld & Wahl, 2002), and *Euceros*, world-wide in distribution (but absent from the neotropics and south-east Asia) with 48 described species and a few additional known but undescribed species. Three species of *Euceros* occur in Britain.

Figure 84. *Euceros serricornis* Haliday male.

Recognition. Males of *Euceros* (at least, those found in Europe) are easily recognised by their antennae, with the central flagellar segments strongly flattened. A similar shape is found in a European (but not British) pimpline, *Dolichomitus sericeus* (Hartig), but in other respects that is an unremarkable *Dolichomitus*. Both sexes of *Euceros* can be recognised by the bilobate process on the centre of the dorsal margin of the pronotum, but this can be hard to see if the head is tilted back. Other features, in combination, will allow recognition of *Euceros*: the clypeal sulcus is absent with the face and clypeus only weakly differentiated; the hypostomal and occipital carinae meet at the mandible base (in European species at least); females have the antenna with the middle section of the flagellum slightly expanded and flattened; the tarsal claws are pectinate; fore wing vein 3*rs-m* is absent (areolet open); cross vein 2*m-cu* has one bulla; the metasoma is rather broad with the first tergite stout and the spiracles in the anterior half; the ovipositor is very short and weakly sclerotized.

Systematics. The Eucerotinae was often associated with the Tryphoninae or Ctenopelmatinae, and Townes (1969) treated Eucerotini as a tribe of the former, although Perkins (1959) had already recognised the isolated position of *Euceros* and treated it as a separate subfamily. Placements in Tryphoninae or Ctenopelmatinae were based on plesiomorphic features such as the stout first tergite and development in sawfly larvae (which proved to be incorrect) and on the homoplasious feature of the pectinate claws. Although both *Euceros* and the Tryphoninae possess stalked eggs, these are very different in other respects. Barron (1976, 1978) proposed the relationship Tryphoninae+ (Ctenopelmatinae+Eucerotinae) but this was based entirely on plesiomorphic biology and morphology. Recent phylogenetic work (Belshaw *et al.*, 1998; Quicke *et al.*, 2009) has shown that, while Ctenopelmatinae and Tryphoninae are relatively basal (and probably not monophyletic) groups of ophioniformes, under different analytical methods *Euceros* either groups with a small clade of predominantly southern hemisphere groups (Brachycyrtinae, Claseinae, Pedunculinae) which are loosely associated with the ichneumoniformes group of subfamilies, or it is placed in the middle of the Tryphoninae. The latter result seemed to be an artefact of a rather weak molecular signal and the consequent dominance of a few morphological characters in the analysis. The former scenario, with a clade of subfamilies that seem to have had a Gondwanan origin, is more consistent with the findings of Gauld & Wahl (2002), who recovered Eucerotinae as the sister group to Brachycyrtinae and Labeninae, but with limited outgroup taxon sampling. Clearly, however, Eucerotinae remains a rather enigmatic group in terms of its relation to other subfamilies.

Biology. Species of *Euceros* are hyperparasitoids and they have an extraordinary and complex life-history, quite unlike that of any other ichneumonids. Finlayson (1960) was the first to realise that *Euceros* are hyperparasitoids and Tripp (1961) completed a picture of the whole life cycle. He studied *E. frigidus* Cresson attacking the parasitoids of *Neodiprion swainei* Middleton (Hymenoptera: Diprionidae) on *Pinus banksiana* in Canada. Host remains preserved with adults of other species of *Euceros* have allowed their host associations to be worked out and confirmation that they have comparable habits (Finlayson, 1960; Varley, 1965; Barron, 1976, 1978). At least two of the British species, *E. albitarsus* Curtis and *E. pruinosus* (Gravenhorst), regularly have Lepidoptera as the secondary host (i.e. the host of the primary parasitoid that is the actual (primary) host of the *Euceros*), with various aspects of their biology explored by Shaw (2014). All of the detailed biological observations that follow relate to these studies on *E. albitarsus*, *E. frigidus* and *E. pruinosus*.

Females of *Euceros frigidus* lay stalked eggs in groups on the pine foliage, near to the sawfly larvae. The egg stalk, or pedicel, seems not to be chorionic (as is the case in Tryphoninae), although it is unclear whether it is formed from a secretion by the ovipositing female. The pedicel is attached by a cup-like, presumably sticky, structure to the plant. Compared with other ichneumonids, the ovipositor of *Euceros* is extremely feeble. The first larval instar is very heavily sclerotized and adapted to withstand long periods of exposure without dehydrating. It is of the 'planidium' type (cf. Clausen, 1940; Askew, 1971) found, albeit uncommonly, in a wide range of parasitoids; though in this case the static habit of at least some *Euceros* larvae does not meet the etymology of 'diminutive wanderer'. Instead, the larva of *E. albitarsus* and *E. frigidus* uses the egg pedicel (with the collapsed egg) as a platform from which it can transfer to a passing sawfly larva or caterpillar. It is possible that *E. pruinosus* leaves the platform to wander actively, although observations have been incomplete (Shaw, 2014). This secondary host larva serves as a carrier and minor source of nutrition, but no further development of the *Euceros* larva can take place unless it is parasitized by another ichneumonid (or possibly braconid). The two adaptations needed to ensure the success of this hazardous strategy are the production of an adequate number of eggs and the timing and location of oviposition. In *E. frigidus* the female has the exceptionally large number of about 100 ovarioles and this corresponds well to the number of eggs laid in a single cluster. A minimum estimate of total fecundity obtained by counting oocytes was 900. Shaw (2014) found that a female of *E. albitarsus* laid not more than 3000 eggs and Iwata (1960) estimated a total of 5000 in *E. pruinosus*. Tripp found that emergence of adult *E. frigidus* remained synchronized with the life cycle of the sawfly, even in a year when the sawfly emerged about two weeks later than expected. The egg clusters extended over about 15 cm of foliage starting about 11 cm from the colonies of young sawfly larvae.

The sawfly larvae proceed towards the *Euceros* eggs as they feed and their passage through the cluster occurs just after the eggs have hatched. If a planidium does not come into contact with a sawfly or lepidopteran larva it will eventually die, still attached to the platform that was its egg.

The planidium usually makes its way to an integumental fold on the carrier host and attaches itself using its mouthparts. Before attachment has taken place, planidia can transfer from one larva to another (Tripp, 1961). Just prior to each host ecdysis the planidium passes through the old integument and re-attaches itself, maintaining the same position on the host. Within the cocoon spun by the mature sawfly larva the planidium of *E. frigidus* transfers to its true host (if one has materialised), the larva of a primary ichneumonid parasitoid of the sawfly, which might be a cocoon parasitoid. Tripp deduced that *Euceros frigidus* passes through 7 or 8 larval instars and that the first three (after the planidial instar) develop as endoparasitoids and the last three or four as ectoparasitoids. However, this has not been substantiated and a more parsimonious interpretation is that there are five larval instars in *Euceros*, as in most other ichneumonids, with the first instar being the planidial stage, followed by one endoparasitoid instar and three ectoparasitoid instars (Shaw, 2014); however, there has been no careful study of the number of larval instars. The primary parasitoid larva survives to become fully grown and to spin its own cocoon, within which the *Euceros* spins its flimsy cocoon and pupates. In one case, a planidial exuvia of *E. albitarsus* was found in the head capsule of its host (Shaw, 2014), but it is not known if the *Euceros* larva habitually travels to the head capsule. Of course, the carrier host is not always parasitized by a primary parasitoid and in such cases the planidium transfers to the pupa and eventually to the adult insect but fails to develop further. At least when Lepidoptera act as secondary hosts (Shaw, 2014), the *Euceros* larva enters the body of the lepidopteran at the time of its pupation, remaining then under the wing case of unparasitised pupae. The moment of transfer of *Euceros* larvae to endoparasitoid or ectoparasitoid (idiobiont or koinobiont) hosts has not been observed.

The secondary hosts with which *Euceros* species are associated are Lepidoptera and sawflies. The primary ichneumonid parasitoids which act as hosts include ecto- and endoparasitoid species, of the subfamilies Anomaloninae, Banchinae, Campopleginae, Ctenopelmatinae, Ophioninae, Cryptinae and Tryphoninae (Phytodietini) (Yu *et al.*, 2012). Except for Ctenopelmatinae, these are groups that mainly or exclusively parasitize Lepidoptera – perhaps not surprisingly, as Lepidoptera larvae are much more often collected and reared than those of sawflies, but nevertheless it is rather remarkable that this list does not include the tryphonine parasitoids of sawflies. There are also a few records from braconid primary parasitoids (Barron, 1976). All *Euceros* species seem to be associated with secondary hosts on trees and shrubs (both angiosperms and gymnosperms) and the species associated with sawflies appear to have more restricted distributions and host ranges than those associated with Lepidoptera (Barron, 1976). The available data suggest that the north temperate species, at least, are all univoltine.

In Britain, *E. pruinosus* is fairly common in woodlands and has been reared from Campopleginae, Ctenopelmatinae and Ophioninae via a range of sawfly and Lepidoptera larvae. Both *E. albitarsus* and *E. serricornis* (Haliday) seem to be rather scarce; the former has been reared from *Dusona* and *Ophion*, via arboreal Geometridae, the latter has been reared extralimitally from an argid sawfly (Hinz, 1961) and from a lymantriine (Lepidoptera: Erebidae) (Constantineanu & Constantineanu, 1994) but with no indications as to the true host or the reliability of these secondary host records.

Identification. The British species are treated by Fitton (1984). The world species have been revised by Barron (1976, 1978), with the latter paper covering two additional species known in Europe but omitting two species either described subsequently (*E. unispina* Kasparyan) (Kasparyan, 1984) or raised from synonymy (*E. superbus* Kriechbaumer) (Horstmann, 2006c).

Figure 85. Morphology of Eucerotinae: (a) *Euceros albitarsus* Curtis male antennae; (b) *Euceros serricornis* Haliday pronotum, dorsal; (c) *Euceros pruinosus* (Gravenhorst) female metasoma, lateral; (d) *E. pruinosus* ovipositor and sheaths.

Subfamily **HYBRIZONTINAE**

This Holarctic subfamily includes 13 extant, described species classified in three genera (van Achterberg, 1999), with seven belonging to *Hybrizon*. There are two British species: *Hybrizon buccatus* (de Brébisson) is fairly common in the south and most frequently encountered by Malaise trapping, while *Ghilaromma fuliginosi* (Donisthorpe & Wilkinson) has very rarely been found. Males of *H. buccatus* are very seldom collected, and it seems never by Malaise traps. *Hybrizon buccatus* is often common on light, sandy soils.

Figure 86. *Hybrizon buccatus* (de Brébisson).

Recognition. The two British species are easily recognised as hybrizontines by the unique wing venation (fore wing with areolet obliterated by fusion of *RS* and *M*, cross vein *2m-cu* absent, pterostigma long and narrow). In addition, hybrizontines have reduced mouthparts (vestigial mandibles and maxillary and labial palps both three-segmented) and short antennae, with just 13 segments. Their general habitus (Fig. 86) is reminiscent of some nematocerous Diptera, and without familiarity the main difficulty in identification may be to recognise hybrizontines as ichneumonids (rather than braconids).

Systematics. Hybrizontinae, formerly called Paxylommatinae, was traditionally placed as a subfamily of the Braconidae, based on superficial similarity in wing venation. Recently more attention has been paid to a wider suite of characters and it has been treated as a separate family (Tobias, 1968, 1988a, 1988b; Marsh, 1989; Mason, 1981; Kasparyan, 1988) or as a subfamily of Ichneumonidae (Rasnitsyn, 1980; Gauld, 1984). The published gene sequences of *Hybrizon* are as anomalous as their morphology (Gillespie *et al.*, 2005) but there is some evidence that Hybrizontinae is best classified within the ophioniformes group of ichneumonid subfamilies, possibly closely related to Anomaloninae (Quicke *et al.*, 2009). Fossil species, known from Baltic amber (Kasparyan, 1988), show greater structural variation than extant species and are classified in two additional tribes, Ghilarovitini and Tobiasitini. They are of particular interest because one has cross vein *2m-cu* present in the fore wing (as in most ichneumonids) and another has 23-segmented antennae (a much higher count than in extant species). Two spellings of the subfamily name, 'Hybrizoninae' and 'Hybrizontinae', are found in the literature. As *hybrizon* is the present participle of the verb *hybrizo*, the plural would be *hybrizontes* (A.C. Galsworthy, pers. comm.), therefore we use Hybrizontinae.

Biology. Species of Hybrizontinae are now known to be endoparasitoids of ant (Formicidae) larvae but with most details of their biology unknown. No hybrizontines have ever been reared from isolated hosts although oviposition into ant larvae has been observed in three species representing all three genera (*Ghilaromma*: Hisasue *et al.* (2018)). Females of *Ogkosoma* (=*Eurypterna*) *cremieri* (Romand) in Japan (the species also occurs in Europe) were observed to hover over *Lasius nipponensis* Forel larvae being transported outside the nest and to oviposit in the larvae during rapid, diving attacks (Komatsu & Konishi, 2010). Komatsu & Konishi found the parasitoid egg within the body of the host, in the abdominal part of the haemocoel. Intriguingly, the host ant larvae do not reach a sufficient size to account for the large size of the adult ichneumonid, which may suggest either that the parasitoid larva moves on to predate further individuals, or that it is nourished in some other way by the ants. In Europe, Cobelli (1906) reported what he presumed to be oviposition by *O. cremieri* into larvae of *Lasius fuliginosus* (Latreille) as they were being carried by workers in the open, also recently observed by Holy *et al.* (2017). Both of these *Lasius* species belong to the subgenus *Dendrolasius*, which nest in trees and transport their larvae down the trunk in the autumn, presumably accounting for the narrow phenology of adult *O. cremieri* (e.g. Ratzeburg, 1852; Komatsu & Konishi, 2010). However, as species such as *H. buccatus* in Britain, and *Ghilaromma orientalis* Tobias in Japan (Komatsu & Konishi, 2010), are on the wing for much of the summer, it would be surprising for this to be the only means of access, although disturbed ant colonies often seem to attract hybrizontines. Gómez Durán & van Achterberg (2011) found that *Hybrizon buccatus* oviposits rapidly (the oviposition sequence takes 0.4 seconds) in final instar larvae of *Lasius grandis* Forel being transported outside of the nest by worker ants. There are several records of adult Hybrizontinae hovering, sometimes in numbers, over ants' nests, including common species such as *Lasius niger* (L.) nesting under stones (e.g. Donisthorpe, 1912). The ants have mainly been species of *Lasius* and *Formica*; there are published observations of hybrizontines being associated with Myrmicinae too, although these should be regarded as unsubstantiated. Gómez Durán & van Achterberg (2011) report that *H. buccatus* was attracted to ant trails even in the absence of ants, suggesting an olfactory cue for host location. Donisthorpe (in Donisthorpe & Wilkinson, 1930) reared specimens of *H. buccatus* from an observation nest of *Lasius alienus* (Förster) in June 1913, almost a year after its collection from the wild. The pupae of *Hybrizon* were found naked among the ant cocoons. The much reduced mandibles of adult hybrizontines may well be correlated with this lack of a cocoon.

Identification. Van Achterberg (1999) keys the European species. The two British species (*Ghilaromma fuliginosi* and *Hybrizon buccatus*) can also be separated using Watanabe's (1984) key to the species found in Japan.

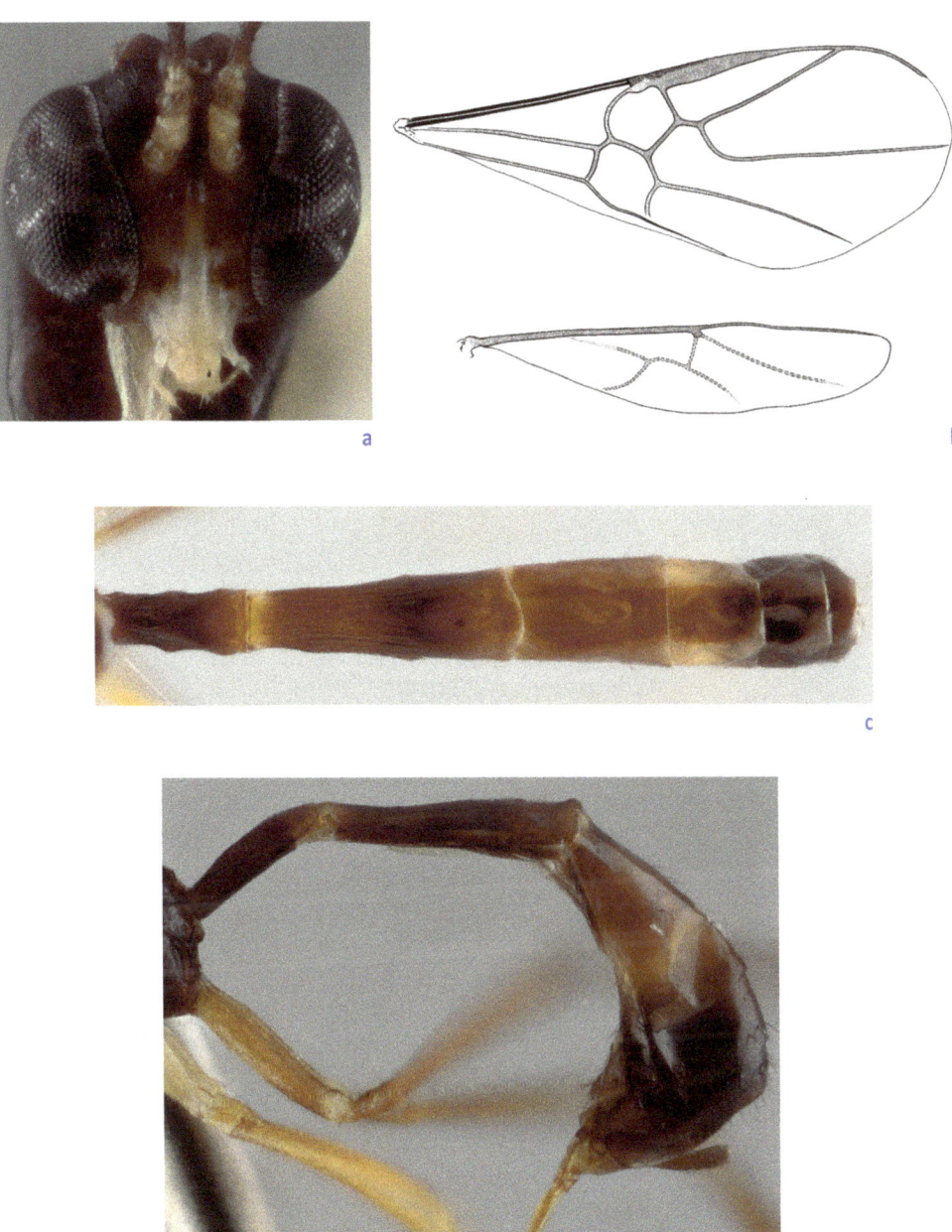

Figure 87. Morphology of Hybrizontinae: (a) *Hybrizon buccatus* (de Brébisson) head; (b) wings; (c) metasoma, dorsal; (d) hind coxa and metasoma, lateral.

Gavin R. Broad, Mark R. Shaw and Michael G. Fitton

Subfamily ICHNEUMONINAE

Tribes
*Ceratojoppini
*Clypeodromini
*Compsophorini
*Ctenocalini
Eurylabini
Goedartiini
Heresiarchini
Ichneumonini
*Ischnojoppini
*Joppocryptini
Listrodromini
Oedicephalini
Phaeogenini
Platylabini
Zimmeriini

Figure 88. *Eurylabus torvus* Wesmael, tribe Eurylabini.

The Ichneumoninae comprises about 15% of the British ichneumonid fauna, being the second most species-rich subfamily in Britain, after Phygadeuontinae, and globally the largest subfamily in terms of described species. As it includes a relatively high proportion of large, colourful species it has always attracted attention, and in various areas of the world is better known than any other large subfamily. The British fauna of 383 species are arranged in 95 genera and nine tribes, with six mainly tropical tribes not represented here. Globally, around 4,100 species are classified in a little over 440 genera with the generic limits more narrowly defined than in many other ichneumonid subfamilies. This is the one major subfamily omitted by Henry Townes in his generic revisions of the Ichneumonidae (he also omitted the Hybrizontinae, which he regarded as braconids), but Heinrich (1961-1978) published a series of regional monographs that established a working tribal and generic classification, although Townes and co-workers adopted broader limits to some of the genera in their regional catalogues (e.g. Townes *et al.*, 1961; Townes & Townes, 1973). Although ichneumonines are better known than most ichneumonid subfamilies in Britain, additions to the British list are still frequent (e.g. Diller & Shaw, 2014). For some larger, more conspicuous species this may represent recent colonisation or introduction (Jones, 2001; Broad & Davis, 2015) but for others it is a sign of the still poor state of our knowledge of the British fauna.

Figure 89. *Goedartia alboguttata* (Gravenhorst) male, tribe Goedartiini.

Figure 90. *Protichneumon pisorius* (L.) female, tribe Heresiarchini.

Recognition. Ichneumonines are most liable to be confused with Cryptinae and Phygadeuontinae, particularly in the case of some Phaeogenini males. Ichneumoninae almost invariably have short ovipositors, usually barely extending beyond the metasomal apex at rest (longer in most Cryptinae and Phygadeuontinae), and barely a trace of a sternaulus at the anterior edge of the mesopleuron (usually extending to at least half the length of mesopleuron in Cryptinae and Phygadeuontinae). Most ichneumonines have diverged relatively little from a common *gestalt* and most can be recognised by the well-developed gastrocoeli and thyridia on the second metasomal tergite and the wide, rather flat clypeus. However, two genera particularly apt to cause confusion are *Crypteffigies* (Ichneumonini) and *Dicaelotus* (Phaeogenini). *Crypteffigies* species have relatively long ovipositors, conspicuously exserted. As Perkins (1959) notes, the form of the ovipositor sheath is not like that of cryptines or phygadeuontines, being rather stiff-looking and smooth, as is normal for ichneumonines, and in other characters the genus is typical of Ichneumoninae. *Dicaelotus* have a pronounced, sternaulus-like (but probably not homologous to the sternaulus) groove across the mesopleuron and lack obvious gastrocoeli and thyridia. While females have a typical ichneumonine ovipositor and sheaths, males are less conspicuously ichneumonine but have the upper tooth of the mandible obviously longer than the lower tooth and a heavily punctate mesoscutum on a shiny background, not shown by any British Phygadeuontinae with which they might be confused. Alomyinae (represented by two species of *Alomya* in Britain) also resemble Ichneumoninae and have often been classified there, either as a separate tribe or as belonging to the Phaeogenini (which then takes the name Alomyini). As discussed in the Alomyinae section, we take the view, based on recent phylogenetic work (Laurenne *et al.*, 2006; Quicke *et al.*, 2009), that Alomyinae is best regarded as a separate subfamily within the ichneumoniformes clade. Separation from ichneumonines is fairly straightforward, based on the structure of the first metasomal segment, the mandibles (although rather similar to *Colpognathus* of the Phaeogenini) and the structure of the female occiput (although rather similar to the non British *Pseudalomya* Telenga of the Phaeogenini).

Systematics. The tribal classification of Ichneumoninae has been unstable and is far from resolved. Perkins (1959) noted some instances of taxa that could almost equally easily be accommodated in one tribe or another, and this remains the case. For example, *Ichneumon deliratorius* L. has been placed either in the Heresiarchini, in the genus *Coelichneumon*, or in the Ichneumonini, in *Ichneumon* (as at present: Tschopp *et al.*, 2013). Various specialised lineages have been recognised as tribes (e.g. the Trogini of Heinrich) but have been shown (e.g. Sime & Wahl, 2002) or are suspected to be nested within larger groupings. Traditionally, the subfamily was divided into the Ichneumoninae Cyclopneusticae, equivalent to the tribe Phaeogenini, and the Ichneumoninae Stenopneusticae, encompassing all other tribes. This was based on the rather different appearance of the phaeogenines, with their small size and rounded propodeal spiracles (elongate in most other ichneumonines), which are parasitoids of 'microlepidoptera', as opposed to (predominantly) 'macrolepidoptera' in the other tribes. Wahl (in Sime & Wahl, 2002) states that both groupings are monophyletic, which is supported by the phylogenetic analyses of Quicke *et al.* (2009) (but is not in agreement with the topology of Laurenne *et al.*, 2006). However, sequencing of ichneumonines to date has been sparse, given the size of the subfamily, and almost entirely restricted to one gene (28S rRNA), so nothing definitive can be said regarding the phylogeny of the subfamily. Heinrich (1934) and Hilpert (1992) (the latter based on the former) produced intuitive phylogenies of the Ichneumoninae; however, the single thorough cladistic analysis of part of the Ichneumoninae is that of Sime & Wahl (2002), and their study demonstrated the paraphyly of the older concept of Heresiarchini (paraphyletic with respect to 'Trogini') and of various previously postulated subtribes.

Figure 91. *Hepiopelmus variegatorius* (Panzer) male, tribe Ichneumonini.

Figure 92. *Listrodromus nycthemerus* (Gravenhorst), tribe Listrodromini.

Figure 93. *Notosemus bohemani* (Wesmael), tribe Oedicephalini.

Biology.

Host associations

Ichneumonines, especially the males, can often be seen flying, usually low down in the vegetation, where the bright yellow or white markings of many species render them conspicuous. A few genera are frequent visitors to umbellifer flower-tables such as those of *Angelica* and *Heracleum*. Sexual dimorphism in colour pattern and structure is often pronounced and this gives rise to difficulties in associating the sexes of some species, particularly within *Ichneumon* where males are rather uniform in appearance. All Ichneumoninae are endoparasitoids of Lepidoptera and this is the most taxonomically restricted host association of any of the larger ichneumonid subfamilies. Emergence is almost always (there is one known exception) from the host pupa, with oviposition either into the pupa or the larva. The latter habit, undisputedly a form of koinobiosis, is termed larva-pupal parasitism, and is generally a consistent feature of the taxa that practise it. In the other case, of species that oviposit into pupae, there is usually a rather short period during which this takes place, in fact flexibly either side of the host's actual moult to the pupa. In the case of oviposition into the prepupa it could be argued that, as the host's pupal moult follows later, there is also koinobiosis. However, because such hosts are already in apolysis, with pupal cuticle already forming, it is unhelpfully pedantic to regard this as anything other than idiobiosis. As histolysis proceeds, the pupal stage in holometabolous insects is physiologically less well-defended than the larval stage, and the essential condition of idiobiont parasitoidism, of developing on a physiologically inert host, is met.

Within the smaller tribes, there is often pronounced specialisation on certain host taxa (though this is not true of the larger tribes). Not surprisingly, reliable host records are mainly available for larva-pupal parasitoids, which is a feature of most of the following small tribes. Two of the three British Eurylabini are regularly reared from Noctuidae (*Eurylabus tristis* (Gravenhorst) from *Hadena bicruris* (Hufnagel) and *E. torvus* Wesmael from *Eremobia ochroleuca* (Denis & Schiffermüller) (Elzinga et al., 2007; collections of NMS)), while our much rarer third species, *E. larvatus* (Christ), has several times been recorded from Notodontidae. The single British representative of the small tribe Goedartiini, *Goedartia alboguttata* (Gravenhorst), has been reared in Britain from *Calliteara pudibunda* (L.) (Erebidae: Lymantriinae) (only one other goedartiine has been reared, also from Lymantriinae: He et al., 1996). Listrodromini are all parasitoids of Lycaenidae; in Britain, *Anisobas cingulatellus*

Horstmann is a parasitoid of several low-feeding Polyommatini, *A. platystylus* Thomson of *Callophrys rubi* (L.), *Listrodromus nycthemerus* (Gravenhorst) of *Celastrina argiolus* (L.) and responsible for the notable population cycles of the host butterfly (Revels, 1994), and *Neotypus nobilitator* (Gravenhorst) of *Cupido minimus* (Fuessly) (Shaw *et al*., 2009). Zimmeriini comprises only two species in the aberrant genus *Cotiheresiarches*; *C. dirus* (Wesmael) is a parasitoid of *Eriogaster lanestris* (L.) (Lasiocampidae) and is very rarely collected and possibly now extinct in Britain. Of the tribes represented in Britain, hosts are entirely unknown only for Oedicephalini, represented by one European species, *Notosemus bohemani* (Wesmael), which is rather atypical within the tribe. This is a small species, included by Perkins (1959) within the Phaeogenini. Platylabini is considerably more species-rich than the preceding tribes but is specialised on Geometridae, with the exception of *Platylabus histrio* Wesmael and *P. tenuicornis* (Gravenhorst), parasitoids of the closely related Drepanidae (Perkins, 1959; Riedel, 2008). The above tribes generally comprise koinobionts, so the potential host range of species is usually small. Note, though, that the few reared *Goedartia* specimens we have seen are all from hosts collected as cocoons and it is not known whether oviposition is into the larva or pupa, and also there is no reason to believe that the unknown hosts of Oedicephalini are attacked as larvae.

Within the larger tribes, Heresiarchini, Ichneumonini and Phaeogenini, there has been much shifting between sometimes distantly related host groups. Some of the largest British ichneumonids are found in the Heresiarchini, including parasitoids of Sphingidae (*Amblyjoppa*, *Callajoppa* and *Protichneumon*), of which the commonest are *Amblyjoppa proteus* (Christ) and *A. fuscipennis* (Wesmael), regular parasitoids of *Deilephila elpenor* (L.) and *D. porcellus* (L.) respectively. These and some other heresiarchine genera are larva-pupal parasitoids and a few can attack the host long before pupation. In the case of *Psilomastax pyramidalis* Tischbein, a parasitoid of *Apatura iris* (L.) (Nymphalidae), oviposition is into very small host larvae before their hibernation (Dell & Burckhardt, 2004). This conspicuous species, incidentally, is known as British from maybe only two (reared) specimens: a reflection of how little we may know. The related *Trogus lapidator* (Fabricius) can similarly oviposit into its *Papilio* hosts in their early instars (Dupuis *et al*., 2016), though in this case the parasitoid overwinters in the host pupa. Within the *Callajoppa* genus-group koinobiont taxa have shifted hosts from ancestral parasitism of Sphingidae to attacking Papilionidae and Nymphalidae; Sime & Wahl (2002) hypothesise that the ability to metabolise alkaloids toxic to most

Figure 94. *Nematomicrus tenellus* Wesmael, tribe Phaeogenini.

Figure 95. *Linycus exhortator* (Fabricius), tribe Platylabini.

insects has allowed this group to exploit hosts feeding on certain families of plants. Note, however, that although there have been several host shifts within the *Callajoppa* group, individual genera and, usually, species have very narrow host ranges. Some other heresiarchine genera (e.g. *Coelichneumon* and *Syspasis*), are idiobionts and consequently much less is known of their exact hosts. While at least most *Coelichneumon* parasitise noctuids, *Syspasis lineator* (Fabricius) is a well known parasitoid of the geometrid *Abraxas grossulariata* (L.) and *S. scutellator* (Gravenhorst) was found by Carl (1968) to be a parasitoid of the hesperiid *Thymelicus lineola* (Ochsenheimer).

Figure 96. *Cotiheresiarches dirus* (Wesmael), tribe Zimmeriini.

The Ichneumonini, a much larger tribe in Britain, similarly comprises a mixture of idiobiont and koinobiont taxa, usually consistently so within a genus but the large genus *Ichneumon* is an exception. Its species parasitise a very wide range of hosts overall (Hinz & Horstmann, 2007), including butterflies (Shaw *et al.*, 2009), various noctuid and erebid groups, Lasiocampidae and even Hepialidae. Indeed, in the tribe as a whole the span of hosts used by the large and medium-sized genera tend to be wide. Host ranges in idiobiont taxa are of course the least well known and, as idiobiosis undoubtedly means that physiological constraints are not as important as they are to larva-pupal parasitoids, it is probable that at a species level host ranges of the idiobionts will tend to be the broadest. There are very few reliably known regular host associations of species within some commonly collected idiobiont genera such as *Aoplus*, *Cratichneumon* and *Barichneumon*, although *Cratichneumon culex* (Müller) was deduced by Varley and co-workers to be a major pupal parasitoid of *Operophtera brumata* (L.) in their classic life-table studies describing mortality of that geometrid (summarised in Varley *et al*., 1973), and *Barichneumon heracliana* (Bridgman) is one of the commonest species to be found parasitising *Depressaria radiella* (Goeze) (=*D. pastinacella*; Depressariidae) pupae in Apiaceae stems (*Vulgichneumon suavis* (Gravenhorst) is another). *Virgichneumon callicerus* (Gravenhorst) and *V. tergenus* (Gravenhorst) parasitise a range of Lycaenidae (Shaw *et al*., 2009). Among the koinobionts, most of which parasitise the host only late in its larval life, the genus *Hoplismenus* parasitises various nymphalid butterflies; the supposedly three British species all having been reared from *Coenonympha*. Other regular associations of rather distinctive species include *Achaius oratorius* (Fabricius) from *Diarsia* species (Noctuidae), *Amblyteles armatorius* (Forster) from *Xestia* and *Noctua* species (Noctuidae), *Diphyus quadripunctorius* (Müller) from *Noctua fimbriata* (Schreber), *Eutanyacra glaucatoria* (Fabricius) from *Cucullia* species (Noctuidae), *Hepiopelmus melanogaster* (Gmelin) from *Spilosoma* and *Diaphora* species (Erebidae: Arctiinae), *Thyrateles camelinus* (Wesmael) from vanessines, especially *Vanessa cardui* (L.), and *Tricholabus strigatorius* (Gravenhorst) from *Euclidia mi* (Clerck) (Erebidae). Further examples are given under 'Life histories' below.

Phaeogenini are generally parasitoids of 'microlepidopteran' pupae or prepupae although host associations are poorly known; Perkins (1959) notes that females of most species overwinter as adults. At least in the well-studied *Diadromus collaris* (Gravenhorst), a parasitoid of *Plutella xylostella* (L.) (Plutellidae), oviposition into prepupal or fresh pupal hosts is preferred (Lloyd, 1940) although older pupae are accepted up until the development of imaginal organs. Lloyd (1940) notes that *D. collaris* females were unable to detect and avoid superparasitism. Host acceptance was triggered by the presence of the silken cocoon and the oviposition stance is one that enables the female to gain access through the small gaps at the ends of the cocoon, such that females had difficulty ovipositing in pupae that had shifted from the usual position. Whilst at least some host associations of several phaeogenine genera are fairly well known, hosts of the large group of species centred around *Dirophanes*, *Phaeogenes* and *Tycherus* are poorly understood. Several species in the largest genus *Tycherus* are known to be parasitoids of Tortricidae, others of Depressariidae, Argyresthiidae and Choreutidae (Diller & Shaw, 2014). *Hemichneumon* and *Trachyarus* species are parasitoids of Psychidae, *Herpestomus brunnicornis* (Gravenhorst) attacks *Yponomeuta* species (Yponomeutidae), *Oiorhinus pallipalpis* Wesmael various Choreutidae and *Heterischnus* species are parasitoids of Pterophoridae. One unusual exception to pupal or prepupal parasitism is found in *Colpognathus*, which differs from other Phaeogenini in being amblypygous, as noted by Perkins (1959). Shaw & Bennett (2001) found that *C. celerator* (Gravenhorst) attacks its host crambid (Crambidae) as an active larva (stage unknown), then mummifies and emerges from the prepupa. An American *Colpognathus* species is also known to have emerged from a mummified prepupal crambid (Shaw & Bennett, 2001). This strategy of host mummification is otherwise unknown in Ichneumoninae but is known in *Alomya* (Alomyinae). Some morphological similarities between adults of *Alomya* and *Colpognathus* have been used to posit a close relationship between these taxa (Wahl & Mason, 1995) but, if the molecular phylogenetic results that place these taxa apart are to be trusted, they could be due to convergence in two taxa that mummify their host lepidopteran larvae. One British phaeogenine, *Epitomus proximus* Perkins, is known to be a koinobiont larva-pupal parasitoid of its elachistid hosts (Diller & Shaw, 2014). A few ichneumonine species have been considered for their potential in biological control of agricultural pests; Jenner *et al.* (2010) demonstrated the efficacy of augmented releases of *Diadromus pulchellus* Wesmael within their native European range in assessing the likely parasitism rates when released in Canada, against the introduced *Acrolepiopsis assectella* (Zeller) (Glyphipterigidae). Cole (1970) studied mate finding by male *Dirophanes invisor* (Thunberg); males have an olfactory response to females up to a few metres away, responding to chemicals released by unmated females as soon as they have eclosed, and empty host pupae that housed females are also attractive to males.

Life histories

It was thought that species whose female metasomal apex is 'amblypygous' invariably oviposit into larval hosts and those which are 'oxypygous' into the pupa (Heinrich, 1978a; Hinz, 1983), but Sime & Wahl (2002) demonstrate that this functional correlation remains suggestive, at least within the *Callajoppa* group. However, it does seem to be a useful clue to the biology of taxa within the Ichneumonini and some other tribes. Amblypygous describes a relatively shorter, less robust ovipositor nearly concealed by the hypopygium whereas oxypygous species have a relatively longer, more robust ovipositor exposed by a short hypopygium. Within Ichneumonini, many of the genera that Perkins (1960) treated as *Amblyteles* (e.g. *Achaius*, *Amblyteles* s.s., *Diphyus*, *Eutanyacra*), oviposit into larvae, as do several other, small genera, such as *Exephanes* (see below).

Figure 97. Ecdysal space in opened *Celastrina argiolus* (L.) (Lepidoptera: Lycaenidae) pupa, where the partly grown larva of *Listrodromus nycthemerus* (Gravenhorst) (Ichneumoninae) passes the winter, interrupting its feeding in the histolysed host pupa (M.R. Shaw).

Figure 99 (above). *Trogus lapidator* (Fabricius) recently emerged from a pupa of *Papilio machaon* (L.) (Lepidoptera: Papilionidae) (G.N. Nobes).

Figure 98 (left). *Listrodromus nycthemerus* freshly emerged from a pupa of *Celastrina argiolus* (R. Revels).

Ichneumon and related genera mostly oviposit into pupae but some *Ichneumon* species oviposit into larvae, including *Ichneumon caloscelis* Wesmael which is an uncommon parasitoid of satyrines (Shaw, 1977). The non-British *Ichneumon eumerus* Wesmael, which is a parasitoid of lycaenid butterflies of the genus *Maculinea*, is a particularly interesting species. *Maculinea* larvae are predatory on ant (Formicidae) larvae and the female *Ichneumon eumerus* gains access to the host larva by entering the ant nest and, by chemical means, inducing the worker ants to attack each other, leaving the parasitoid unmolested (Thomas *et al.*, 2002). The genus *Ichneumon* is very species-rich and possibly paraphyletic with respect to some other genera (Sime & Wahl, 2002). Hilpert (1992) divided the European species of the genus into several groups and Hinz & Horstmann (2007) listed the host associations of numerous species based, however, largely on presentation of potential hosts to parasitoids and therefore probably not an entirely accurate representation of natural host ranges. Nevertheless, given the difficulties in ascertaining the identities of hosts of pupal parasitoids, this is a valuable summary of the biology of a diverse and commonly collected genus.

The females of *Limerodops elongatus* (Brischke) (Ichneumonini) use their specialized elongate metasoma to reach the early larval instars of *Apamea* species hidden in the inflorescences of certain grasses (Hinz & Horstmann, 1999), and the parasitoid larva overwinters within that of the host. *Exephanes* (Ichneumonini) species exhibit merely apically elongated metasomas, which they use to reach their fully endophytic grass-feeding noctuid hosts (Hinz & Horstmann, 2000). However, females of many groups search for hosts low down in the undergrowth or in the litter layer and do this essentially on foot. Some have fossorial fore limbs and may tunnel into loose soil; at least some of those which oviposit only into prepupae or fresh pupae apparently locate them by following the ephemeral scent trail left by the host larva as it burrowed into the ground (Hinz, 1983), these allomones perhaps arising from the hormonal changes associated with the host's apolysis.

Most European species of Ichneumoninae are univoltine. Fertilized females of some groups hibernate in tree stumps, grass tussocks, caves and similar situations during the winter (Rasnitsyn, 1964; Hinz, 1968; Vas & Kutasi, 2016). Other species overwinter as a first instar larva in the fatty tissue of the host pupa (Klomp & Teerink, 1978), although *Listrodromus nycthemerus* larvae were found to be half-grown in their pupal hosts in the winter (M.R. Shaw, unpublished obs.: Fig. 97). Those that overwinter as larvae are sometimes plurivoltine and some may parasitize different host taxa in the spring and autumnal generations, as in the platylabine *Poecilostictus cothurnatus* (Gravenhorst) (van Veen, 1981). Species that overwinter as adult females generally, unsurprisingly, attack hosts that are active in the spring but Hinz (1983) records that *Chasmias paludator* (Desvignes) overwinters as an adult then waits until August before parasitizing noctuid pupae in *Typha* and *Phragmites* stems. Female ichneumonines feed from the host's exuded haemolymph following oviposition (Lloyd, 1940; Hinz, 1983). There are thought to be three or four larval instars and larvae grow rapidly. The emergence of *Trogus* and close relatives from the host pupa is unusual, not only for Ichneumoninae but also for other subfamilies of Ichneumonidae that hatch as adults from Lepidoptera pupae, both in taking place through a wing case and in involving a secretion that dissolves the host cuticle (Shaw et al., 2015). In other cases the mandibles are used, and often a cap is cut from the host's capital extremity, though as Shaw et al. (2015) illustrate, there is considerable variation.

Developmental biology
Very little seems to be known about egg placement in ichneumonines, other than several reports that the egg may be deposited free in the host haemocoel, but the report by van Veen (1981) that *Poecilostictus cothurnatus* attaches its egg to the inner lining of the host's hindgut suggests that specializations in the subfamily probably exist. It might also be significant that the egg in this koinobiont species is protected from the haemolymph of the host, perhaps permitting its unusually wide host range. *Diadromus pulchellus*, at least, transmits ascoviruses to its host at oviposition. Unlike polydnaviruses, ascoviruses are not assimilated into the wasp's genome but are found throughout the wasp's tissues (Bigot et al., 1997). There seems to be a mutualism between the parasitoid and virus which alters the host lepidopteran's physiology to the advantage of *D. pulchellus*. The virus precludes development of a pimpline ichneumonid, *Itoplectis tunetana* (Schmiedeknecht), within the host (see review by Whitfield & Asgari, 2003). *Diadromus pulchellus* also introduces teratocyte-like cell bodies into the haemolymph of the host pupa, derived from the serosal membrane surrounding the embryo (Rouleux-Bonnin et al., 1999). Given (1944) found that host-feeding was essential for egg development in *Diadromus collaris*, but Jackson (1937) reports that *Sypasis lineator* (Fabricius) does not host-feed. Given's (1944) synthesis of the internal and external anatomy of adult *D. collaris* is unusual in its thoroughness but, unfortunately, equivalent attention has not been devoted to the developmental biology of ichneumonines.

Identification. The two handbooks to the British species (Perkins, 1959, 1960) cover most of the fauna and are still very reliable for most of our ichneumonines. They are, though, difficult to use, especially for the beginner, and need to be used alongside the current checklist (Broad, 2016) because several of Perkins's larger genera have been split up. Perkins (1960) did not include a key to males of the large genus *Ichneumon*. Hilpert's (1992) revision of *Ichneumon* should be consulted instead, and includes the additional British species not known to Perkins. Since Perkins's (1959) handbook many additional phaeogenines, particularly in the genus *Tycherus*, have been found (see Diller & Shaw, 2014) and the genus *Phaeogenes* has been split. There are no comprehensive keys to these difficult genera but Selfa & Diller's (1994) key to phaeogenine genera is an alternative to Perkins's and Diller (2006) differentiates *Baeosemus*, *Phaeogenes* and *Tycherus*; Ranin (1983) revised the species of the *Tycherus elongatus* group. Other useful works include Townes *et al*.'s (1965) key to Eastern Palaearctic genera, which includes some of the generic splits that Perkins (1960) anticipated but did not enact; Diller's (1981) revisions of *Dilleritomus* and *Epitomus*; Horstmann's (2000b) revision of *Probolus*; Hinz & Horstmann's (2000) revision of *Exephanes*; Horstmann's (2002b) revision of the *Coelichneumon orbitator* group; Sime & Wahl's (2002) key to the genera of the *Callajoppa* group (Heresiarchini); Diller & Schönitzer's (2003) revision of *Colpognathus*; Riedel's (2008, 2012) revisions of *Platylabus* and *Coelichneumon*; Broad & Davis's (2015) characterisation of *Lymantrichneumon disparis* (Poda) as a genus new to Britain; and Tereshkin's (2009, 2011) keys to, respectively, ichneumonine tribes and genera of the 'Amblytelina' (Ichneumonini). Diller & Shaw (2014) summarize the distribution, hosts and taxonomy of some British Phaeogenini (and the single British representative of Oedicephalini), and include descriptions of several new species.

Figure 100. *Amblyteles armatorius* (Forster) male nectaring (A.M. Broome).

Figure 101. Morphology of Ichneumoninae. (a) *Eurylabus torvus* Wesmael, head; (b) *Diphyus quadripunctorius* Müller, head; (c) *Listrodromus nycthemerus* (Gravenhorst), head; (d) *Crypteffigies lanius* (Gravenhorst), head; (e) *Dicaelotus pumilus* (Gravenhorst), head; (f) *Platylabus heteromallus* (Berthoumieu), head; (continued overleaf).

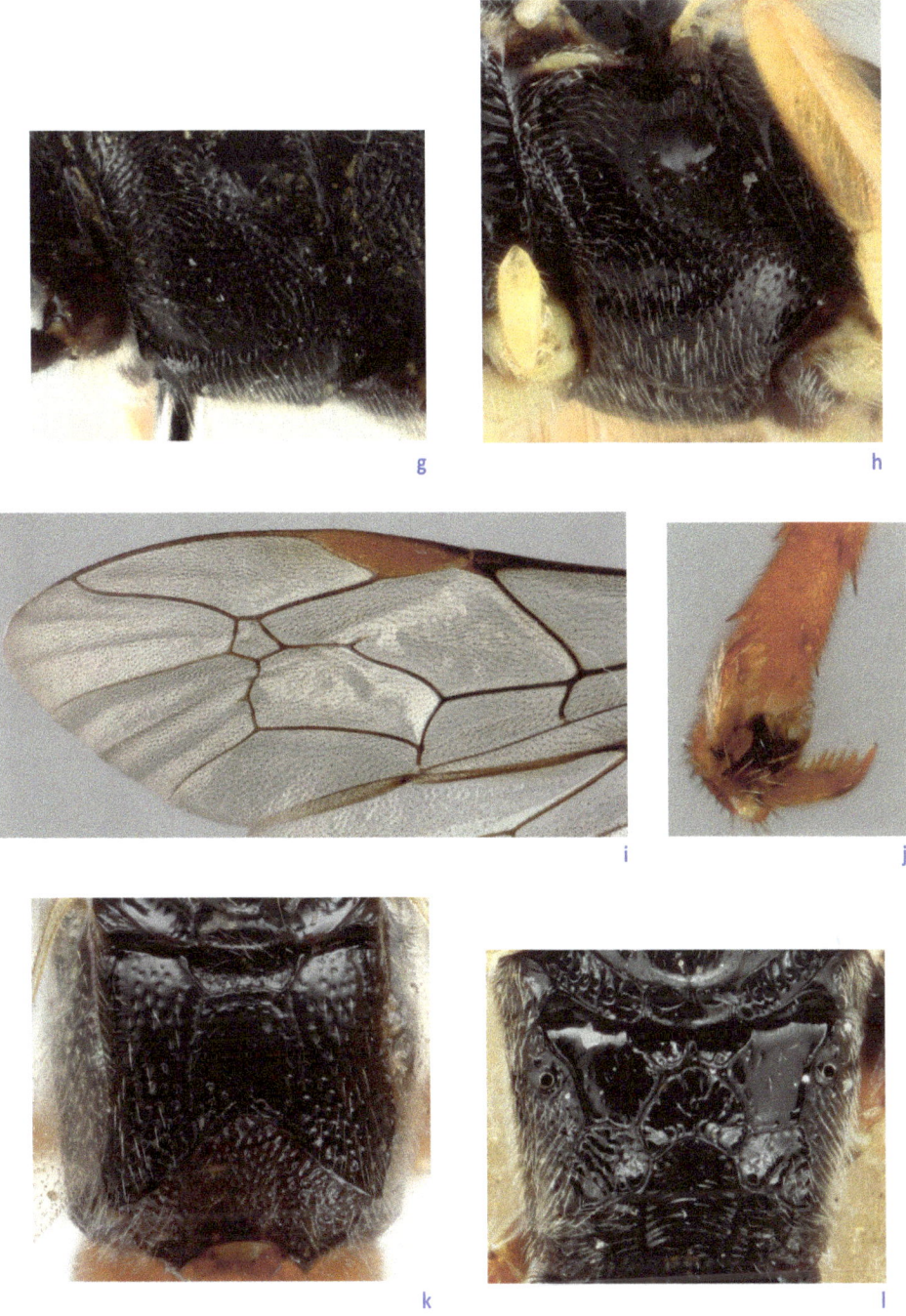

Figure 101 continued: Morphology of Ichneumoninae. (g) *Listrodromus nycthemerus*, mesopleuron with sternaulus; (h) *Dicaelotus pumilus*, mesopleuron with sternaulus; (i) *Diphyus quadripunctorius*, part of fore wing, (j) *Listrodromus nycthemerus*, tarsal claw; (k) *Ichneumon oblongus* Schrank, propodeum; (l) *Dicaelotus pumilus*, propodeum; (continued overleaf).

Figure 101 continued: Morphology of Ichneumoninae. (m) *Ichneumon oblongus*, first and second metasomal tergites; (n) *Dicaelotus pumilus*, first and second metasomal tergites; (o) *Platylabus heteromallus*, propodeum, first and second metasomal tergites; (p) *Crypteffigies lanius*, ovipositor sheaths; (q) *Ichneumon oblongus*, apex of female metasoma; (r) *Cotiheresiarches dirus* (Wesmael), ovipositor and sheaths.

Subfamily LYCORININAE

This is a small subfamily of only 34 described species in the single genus *Lycorina*, world-wide in distribution (Gauld, 1984; Rousse & van Noort, 2014a; Sheng & Sun, 2014).

Figure 102. *Lycorina triangulifera* Holmgren.

Recognition. There is one species in Britain and Europe, *Lycorina triangulifera* Holmgren, which is readily recognisable by two apomorphies: the pattern of grooves on the metasomal tergites, delimiting central, triangular areas bordered by sublateral swollen regions, and the hook-shaped projection of the metapostnotum. Other useful recognition features include the strongly sagittate ovipositor tip, the pectinate claws and the lack of an areolet in the fore wing.

Systematics. Perkins (1959, 1960) treated *Lycorina* as a distinct subfamily but Townes *et al.* (1965) and Townes & Townes (1966) lumped the genus within the banchine tribe Glyptini, although Townes (1970b) later regarded Lycorininae as a subfamily too, and this status has subsequently been generally accepted. Although superficially resembling Glyptini or Pimplinae (with respect to the grooves and swellings of the metasomal tergites), the affinities of Lycorininae to other ichneumonid subfamilies remain obscure. Quicke *et al.* (2009) were unable to say much about the phylogenetic placement of the subfamily, its relationships being masked in analyses of molecular data by long branch attraction to other taxa with anomalous sequences. Some derived features of the ovipositor indicate that *Lycorina* belongs in the ophioniformes group of subfamilies (Quicke *et al.*, 2009).

Biology. Despite a number of rearings, it has still not been certainly established whether *Lycorina* are ecto- or endoparasitoids, although some recent observations (Coronado-Rivera *et al.*, 2004; Shaw, 2004) strongly suggest that the larva resides in the host's hind gut and is thus essentially ectoparasitoid but concealed. Development is certainly as a koinobiont (Shaw, 2004). The recorded hosts are 'microlepidoptera' of several families, with a Costa Rican species reared from Crambidae and Ethmiidae (Coronado-Rivera et al., 2004). All of the feasible European host records relate to species with larvae, particularly of Tortricoidea, that construct leaf rolls or similar shelters but Shaw (2004) found no published detail to firmly support any of the reported host records. A female of *L. triangulifera* readily accepted various tortricid larvae for oviposition, and was attracted to faecal pellets of other lepidopterans, although development was completed in only one species of Tortricidae, *Acleris schalleriana* (L.) (Shaw, 2004). Although the ovarian egg has been described as being of an unusual, 'leech-like', shape (Iwata, 1958) this appears to be incorrect, as both Coronado-Rivera *et al.* (2004) and Shaw (2004) found that the egg is unremarkably elongate ovoid but possesses a distinct, stalked anchor. The high number of ovarioles were considered more typical of an endoparasitoid koinobiont life history by Coronado-Rivera *et al.* (2004), but the anchored egg, structure of the ovipositor, form of the head sclerites of the final larval instar, with denticulate mandibles and cuticular clothing of dense setae (Finlayson, 1976), suggest that the larvae are ectophagous. Shaw (2004) observed oviposition in the host's anus but was unable to establish whether the larva resided in the hind gut (which is anatomically formed by an invagination of the cuticle and is shed at moulting) or entered the body of the host. A Costa Rican species certainly completes its feeding externally, soon after the host spins its cocoon (Coronado-Rivera *et al.*, 2004). The cocoon of *Lycorina* (Finlayson, 1976; Shaw, 2004) is cylindrical with rounded ends, transparent, and very thin and soft, with the exit hole at the end. If *Lycorina* is essentially ectoparasitoid then this would support a phylogenetically basal position for the subfamily within the ophioniformes, either close to the Tryphoninae or retaining some plesiomorphic characters (particularly the stalked egg) with the Tryphoninae (which may not be monophyletic anyway: Quicke *et al.*, 2009).

Identification. The single British species can be identified using the key to subfamilies and a description of the female is given by Perkins (1960).

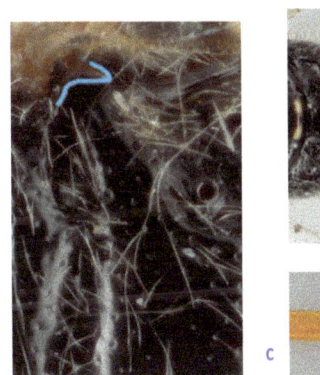

Figure 103. Morphology of Lycorininae. *Lycorina triangulifera* (a) part of fore wing; (b) part of metasoma, lateral; (c) anterior of propodeum and metapostnom, with projection outlined; (d) propodeum and metasoma, dorsal; (e) ovipositor tip.

Gavin R. Broad, Mark R. Shaw and Michael G. Fitton

Subfamily MESOCHORINAE

This is a moderately large subfamily with nine genera recognized by Wahl (1993b). Almost 90% of the approximately 880 described species are placed in the cosmopolitan genus *Mesochorus*, and six of the other eight genera are monotypic. Of the 88 species known from Britain, 65 belong to *Mesochorus* with 18, three and one species of, respectively, *Astiphromma*, *Cidaphus* and *Dolichochorus*.

Figure 104. *Mesochorus fulgurans* Curtis.

Recognition. Mesochorines comprise one of the more distinctive subfamilies of Ichneumonidae. The ovipositor (relatively straight, needle-like) and male parameres (spine-like) are both unusual features, although not unique within the family. Most mesochorines have a very distinct fore wing areolet, which is large and regularly rhombic, i.e. all sides are roughly equal in length (but not so in *Cidaphus*). The clypeus and face are only vaguely differentiated in mesochorines and have a distinctive appearance; the tarsal claws are fully pectinate; and the first metasomal segment has deep glymmae which are separated medially only by a transparent 'window' of cuticle. Many Perilissini, of the Ctenopelmatinae, have a similar first metasomal segment and claw pectination but differ in other features (e.g. ovipositor, parameres, face). Final instar larvae of *Astiphromma* and *Mesochorus*, but not *Cidaphus*, are unique within Ichneumonidae in that the mandible is basally desclerotized (Short, 1978; Wahl, 1993b); the early stages of other mesochorine genera are unknown.

Systematics. The Mesochorinae form part of the ophioniformes clade of, predominantly, koinobiont endoparasitoids (Wahl, 1991; Quicke *et al.*, 2009). They seem most likely to be closely related to the Ctenopelmatinae, quite possibly nested within the current concept of Ctenopelmatinae (Quicke *et al.*, 2009). Mesochorines fairly closely resemble some Pionini and Perilissini in features of the first metasomal tergite, ovipositor and wing venation, but also display striking apomorphies that have precluded definite placement of the group. Within the subfamily, there has been much disagreement over the limits of the genera. Some authors, such as Townes (1971) and Schwenke (1999) have adopted a larger number of genera than Wahl (1993b), whose cladistic analysis of the relationships between a number of mesochorine groups concluded that several taxa – *Stictopisthus* is the most relevant in a British context – are merely specialized offshoots from within *Mesochorus*, recognition of which would render the latter paraphyletic. This has not prevented other authors (e.g. Schwenke, 1999) from recognizing some of these distinctive species-groups as genera. This multiplication of genera, leaving a paraphyletic mess of *Mesochorus*, can add little to our understanding of mesochorine relationships and host associations. Wahl (1993b) recognized nine genera on the basis of his morphological cladistic analysis; we additionally recognize *Dolichochorus* as a valid genus, rather than as a junior synonym of *Astiphromma* (Broad & Watanabe, in prep.). On morphological grounds, *Cidaphus* occupies a basal position within the Mesochorinae (Wahl, 1993b). Most other British mesochorines, in the genera *Astiphromma* and *Mesochorus*, occupy an apparently highly derived position within Wahl's proposed phylogeny.

Biology. Species of Mesochorinae are probably invariably obligate hyperparasitoids, developing as koinobiont endoparasitoids of primary parasitoid larvae, most frequently Braconidae and Ichneumonidae (much less often Tachinidae (Diptera)). The primary parasitoid is killed only after making its cocoon (or puparium), from which the adult mesochorine emerges in due course. The small subfamily Eucerotinae aside, Mesochorinae are unique within Ichneumonoidea in constituting a significant radiation of obligate, true hyperparasitoids, i.e. the host of the mesochorine is always still acting as a parasitoid when it serves as host to the mesochorine. Although Dasch (1971), Townes (1971) and other authors have stated that there is evidence of some mesochorines being primary parasitoids, we are not aware of any detailed studies of mesochorine development that would support this contention, and it seems best to reject it until there is clear evidence. The hosts of mesochorines are almost always endoparasitoids, but there are a few records of *Mesochorus* having been reared from ectoparasitoids, such as *Phytodietus* (Ichneumonidae: Tryphoninae) on Lepidoptera larvae (Shaw, 1993); however in these cases the *Mesochorus* involved is believed not to be specialized to attacking only ectoparasitoids, and indeed *Mesochorus atriventris* Cresson has been recorded from both endoparasitoid (Braconidae: Microgastrinae) and ectoparasitoid (*Phytodietus polyzonias* (Forster)) primary hosts (Horstmann, 2006a). In contrast with many groups of koinobiont endoparasitoid Ichneumonoidea, Tachinidae are only occasionally hosts of mesochorines (at least in Britain) and, in this case too, such associations are most often probably merely incidental rather than a result of specialisation; Horstmann's (2006a) host list for *Mesochorus fulgurans* Curtis, for example, includes both ichneumonids and a tachinid (*Lypha ruficauda* (Zetterstedt)) through Geometridae larvae; additionally, in NMS, there are just two specimens of *Mesochorus temporalis* Thomson reared from the very frequent tachinid parasitoids of *Zygaena* (Lepidoptera: Zygaenidae) among many specimens reared from the various koinobiont ichneumonoid species parasitising *Zygaena*. However, in the *politus*-group (which includes the British *M. politus* Gravenhorst), Tachinidae appear to be the usual hosts (Dasch, 1971). Both the

Mesochorus politus-group and the *Astiphromma dorsale*-group, which seems also to specialise on Tachinidae, share a posterior protrusion on the scutellum, which Riedel (2015) states is used for emergence from the host puparium, although details are lacking. Primary hosts of mesochorines can often be identified to some level by the (unaltered) cocoon of the primary parasitoid from which the mesochorine adult emerged or, with a bit more effort, from the head capsule of the final instar primary host larva to be found alongside that of the mesochorine within that cocoon. The adult mesochorine invariably emerges from its host's cocoon by chewing a subapical hole, so when the host belongs to one of the several groups of Braconidae which cut neat apical caps from their cocoons to emerge (cf. Shaw & Huddleston, 1991), it is easy to see which cocoon(s) have produced the hyperparasitoid. This applies only to (many but not all) braconid primary hosts as most ichneumonids make similar emergence holes to those of Mesochorinae, but it is sometimes useful when the braconid host is gregarious and only some of the primary parasitoids have been attacked (though pseudohyperparasitoids, which oviposit into the cocoon stage, make similar emergence holes to those of Mesochorinae).

Many species of Mesochorinae exhibit a strong relation with a particular more or less narrow range of secondary hosts (i.e. the hosts of their primary parasitoid hosts), often with a clearly taxonomic definition. This often means that many of the koinobiont primary parasitoids of a particular secondary host species (or genus, or wider group) will be used by the same single or few species of Mesochorinae (e.g. Shaw, 2017b), and the impression is that the initial cues used for host-searching by Mesochorinae are based on finding the secondary host, and thus may not be dissimilar from the cues used by primary parasitoids. Overall, the secondary hosts of mesochorines are usually holometabolous insect larvae, mainly Lepidoptera and sawflies (Hymenoptera: Tenthredinoidea and Pamphilioidea), and to a lesser extent, Coleoptera. Mostly these secondary hosts feed in exposed positions, though a few leaf-mining Gracillariidae (Lepidoptera) regularly support a very small species of *Mesochorus* (formerly *Stictopisthus*) through *Pholetesor* (Braconidae: Microgastrinae). However, secondary hosts include Heteroptera too; for example, in North America the Holarctic species *Mesochorus curvulus* Thomson has been reared from *Leiophron* species (Braconidae: Euphorinae) parasitizing a number of species of Miridae (Dasch, 1971; note that Dasch's records of *M. curvulus* from Coleoptera and Lepidoptera refer to other species – Horstmann, 2006a). In Britain, Waloff (1967) found *Mesochorus* parasitizing *Leiophron* in two species of *Orthotylus* (Miridae) on broom and reared *Mesochorus* from primary parasitoids of four other species of Miridae. There are also records of *Mesochorus* species reared from Psocoptera (Carlson, 1979:703, citing a pers. comm. from C. C. Loan), presumably via their euphorine parasitoids. Apart from the Pimplinae, Cryptinae and Phygadeuontinae attacking spiders and their egg sacs (also a few other egg masses, including the egg nests of a pseudoscorpion) and *Phygadeuon* (Cryptinae) parasitising tachinid parasitoids of Dermaptera, Mesochorinae are the only ichneumonids associated, albeit indirectly, with hosts other than immature stages of holometabolous insects. No mesochorines have been reared from the rather large guild of tryphonine ectoparasitoids of sawfly larvae, possibly because in this group the development of the parasitoid larva is minimal until the host constructs its cocoon or pupation chamber, thereby becoming inaccessible. A similar explanation might be offered for the apparent absence of parasitism by Mesochorinae of the large radiations of cyclostome braconid (Alysiinae and Opiinae) endoparasitoids of Diptera, many of their hosts being only thinly concealed as leaf-miners, as in both these groups the parasitoid does not progress beyond its first instar until the dipteran host starts to form its puparium (cf. Shaw & Huddleston, 1991). The feeding biology of the Diptera concerned seems unlikely to be the root of this failure, as some braconid (Microgastrinae) parasitoids of leaf-mining Lepidoptera such as Gracillariidae are susceptible to mesochorines. However, as noted below, some authors have stated that *Mesochorus* species readily, or even preferentially, oviposit into first instar larvae of primary parasitoids, so factors other than the absence of a suitable host instar might be involved. A similar failure to use Aphidiinae (Braconidae) parasitoids of Aphididae (Homoptera) might be rooted in the small size of the hosts. Either of the above explanations might apply to the absence of Mesochorinae associated with the various braconid groups that parasitise Nepticulidae (Lepidoptera).

The fine ovipositor of a female mesochorine is used to probe for the primary parasitoid larva within the secondary host. Oviposition sites can be precise, with the American *Mesochorus agilis* Cresson preferring to oviposit in the fat body of abdominal segments six to nine of its host, *Bathyplectes curculionis* Thomson (Ichneumonidae: Campopleginae) (Ellsbury & Simpson, 1978). *Mesochorus discitergus* (Say) will often oviposit into first instar larvae of its microgastrine braconid hosts but can also attack other instars successfully (Yeargan & Braman, 1989). Zinnert (1969b) notes that *Mesochorus* species attacking nematine (Tenthredinidae) sawfly parasitoids prefer to oviposit in the first instar ichneumonid larva. Cross & Simpson (1972) found that *Mesochorus agilis* in culture will attempt to oviposit in exposed *Bathyplectes curculionis* larvae that have emerged from their weevil host but that the mesochorine ovipositor did not penetrate the *Bathyplectes* cuticle. When mesochorines attack gregarious endoparasitoids, such as *Cotesia* (Braconidae: Microgastrinae) in Lepidoptera larvae, usually several individual larvae within a host will be parasitized. In view of their high trophic level, and also the low impetus to study in detail species whose potential effect in biological control is deleterious, it is not surprising that there have been few studies of mesochorine development and life history; the few detailed studies are mainly concerned with *Mesochorus* species attacking primary parasitoids of the invasive alfalfa weevil (*Hypera postica* (Gyllenhal)) in North America (e.g., Cross & Simpson, 1972; Ellsbury & Simpson, 1978) and there have been no published studies of life histories of mesochorine genera other than *Mesochorus*. Given that *Mesochorus* seems to be a rather derived genus within the subfamily (Wahl, 1993b), it should be borne in mind that the life history parameters of other genera could differ substantially.

Figure 105. *Mesochorus pallipes* Brischke drinking (M.R. Shaw).

Whereas it has been reported that *M. agilis* can often discriminate between parasitized and unparasitized weevil larvae prior to probing with the ovipositor (Ellsbury & Simpson, 1978), Coseglia *et al*. (1977) and Yeargan & Braman (1989) found that *M. arenarius* Haliday and *M. discitergus* discriminate only after probing the secondary host larva with the ovipositor. When a primary host is present in the caterpillar or weevil larva, the mesochorine will often probe in several locations to find the best spot for oviposition. When *M. agilis* probe early instar *Hypera* larvae, the weevil, and so also its parasitoid, is frequently killed. *Mesochorus discitergus* employs two methods of contacting Lepidoptera larvae that have dropped on silken threads from their foodplants: either walking down the silk to the caterpillar (frequent for older instars) or reeling in the caterpillar by its thread (most frequently first and second instar secondary hosts) (Yeargan & Braman, 1989). In the laboratory, *M. arenarius* oviposited in 2.5% of unparasitized *Hypera* larvae when offered as pairs of parasitized and unparasitized weevil larvae, with first instar *M. arenarius* larvae sometimes consequently being found in the *Hypera* larva (Coseglia *et al.*, 1977), although the mesochorine cannot develop in the weevil larva. There seem to be no documented cases of primary parasitoids encapsulating mesochorine eggs or larvae, although there has been little relevant published work on the early stages of mesochorines. The alecithal egg is stalked, with the stalk diminishing as the egg expands prior to hatching (Coseglia *et al.*, 1977). Although some studies report three larval instars, detailed studies (e.g. Coseglia *et al.*, 1977) have found four instars, which are difficult to distinguish beyond the first instar with its well sclerotized head capsule, long mandibles and caudate terminal abdominal segment. *Mesochorus discitergus* retards the growth of its studied host, *Cotesia marginiventris* (Cresson) (Braconidae: Microgastrinae), delaying the onset of cocoon formation (Yeargan & Braman, 1989). *Mesochorus arenarius* overwinters as a first instar larva within diapausing *Bathyplectes* larvae. If oviposition is into non-diapausing hosts then the development of the *Bathyplectes* larva is arrested at the final larval instar and the *Mesochorus* larva fails to develop beyond the first instar (Coseglia *et al.*, 1977). Note that the European species *Mesochorus arenarius* was inadvertently introduced to North America along with its host *Bathyplectes* and has been the subject of several studies (as *Mesochorus nigripes* Ratzeburg). A thelytokous form was introduced (although this has not been recorded in Europe) which differs behaviourally from the arrhenotokous population, from which it is reproductively isolated (Day & Hedlund, 1988). Hung *et al.* (1988) could find no morphological or isozyme differences between the 'forms' but suggest that these are sibling species, or cryptic species as they would now be termed.

While *Mesochorus* species often have high degrees of association with particular secondary host taxa (reared specimens in NMS; Horstmann, 2006a; Shaw, 2017b), we do not know if the same applies to *Astiphromma* and *Cidaphus*, for which there are too few reliable host records. It is often not the case, though, that all of the primary koinobiont ichneumonoids will be parasitised to the same degree, and some may escape altogether (Shaw, 2017b suggests some reasons). For example, *Mesochorus lilioceriphilus* Schwenke is a parasitoid of two *Lemophagus* species (Campopleginae) in any of three *Lilioceris* species (Coleoptera: Chrysomelidae) but does not parasitize the tersilochine, *Diaparsis jucunda* (Holmgren), which is a frequent endoparasitoid within these beetle larvae (Haye & Kenis, 2004). An unidentified *Mesochorus* sp. parasitizes about a quarter of the *Hyposoter horticola* (Gravenhorst) population associated with *Melitaea cinxia* (L.) (Lepidoptera: Nymphalidae) but is an infrequent parasitoid of the gregarious microgastrine (*Cotesia melitearum* agg.) involved in the host-parasitoid system in Finland (van Nouhuys, 2005). Similarly, *Mesochorus iniquus* Schwenke parasitizes comparable proportions of the solitary parasitoids *Hyposoter notatus* (Gravenhorst), *Aleiodes bicolor* (Spinola) and *Cotesia saltatoria* (Balevski) which attack polyommatine lycaenid butterfly larvae in grassland habitats in Britain, but is a much less frequent parasitoid of the two gregarious species *Cotesia tenebrosa* (Wesmael) and *Cotesia astrarches* (Marshall) (M. R. Shaw, unpublished data). Heitland & Pschorn-Walcher (2005) report that *Mesochorus* is almost always reared when an *Olesicampe* species (Ichneumonidae: Campopleginae) is reared in any numbers, but unfortunately there are few reliable data on how many *Mesochorus* species are involved in these interactions. There is a little information on host range for *Mesochorus* species attacking *Olesicampe* through sawflies, with differences in host specificity apparent from the published records. Pschorn-Walcher & Zinnert (1971) found that *Mesochorus dimidiatus* Holmgren attacks *Olesicampe* parasitoids of several Nematinae species, although only attacking

Olesicampe melanogaster (Thomson) within the well-studied parasitoid assemblage of *Pristiphora erichsonii* (Hartig) (Hymenoptera: Tenthredinidae). This taxonomic restriction in host range arises from the fact that *M. dimidiatus* only parasitizes hosts within first instar sawfly larvae, which usually means that only *O. melanogaster* is available to it. Horstmann's (2006a) host records indicate that some *Mesochorus* species can be reared through different sawfly genera and also that some *Mesochorus* are parasitoids of a wider range of ichneumonoid taxa. For example *M. dimidiator* Aubert is a parasitoid of *Olesicampe* spp. and *Lethades laricis* Hinz (Ichneumonidae: Ctenopelmatinae) through *Nematus* and *Pristiphora* (Hymenoptera: Tenthredinidae) species. Some *Mesochorus* species (e.g. *M. pallipes* Brischke) seem to have a very wide range of lepidopteran secondary hosts (perhaps based on the nature of the silken webs they inhabit) and consequently very many ichneumonoid primary hosts, and *Mesochorus olerum* Curtis has a host range focused on parasitoids of Pieridae (Lepidoptera) larvae, also attacking some other secondary host taxa that feed on the same plants, such as *Plutella xylostella* (L.) (Lepidoptera: Plutellidae), but with some seemingly anomalous hosts too, e.g. Braconidae through *Zygaena filipendulae* (L.) (Lepidoptera: Zygaenidae) (Horstmann, 2006a). While there is always the possibility that more than one *Mesochorus* species is involved in cases such as these, the general question of how a given mesochorine species is able to use such a disparate range of primary parasitoids is interesting. Askew & Shaw (1986) suggested that it might be either a consequence of poor haemolytic (encapsulating) defenses in the host groups, or an abnormally high ability of mesochorines to withstand this. In this context it would be interesting to know whether the infrequency with which mesochorines parasitise tachinids is a consequence of ovipositional restraint, or of a possibly superior encapsulation response of tachinids.

There are few reliable host records for *Astiphromma* species; hosts of European species are summarised by Riedel (2015), who reports several species reared from Tenthredinidae and Lepidoptera (primarily Noctuidae and Geometridae) but only a few primary host records. The exceptions are the non-British (but European) *A. dorsale* (Holmgren) from Tachinidae and *A. nigriceps* (Brischke) from a chrysomelid beetle, *Colaphellus sophiae* (Schaller) (Riedel, 2015; Schwenke, 1999). *Cidaphus* species are much larger than any other mesochorines in Britain; the few reliable host records are from large endoparasitoid ichneumonids via noctuid Lepidoptera (*Banchus* and *Dusona*: Fitton, 1985b; *Enicospilus*: in NMS). Only *Cidaphus atricillus* (Haliday) is reasonably widespread in Britain, with *C. areolatus* (Boie) particularly restricted in range, so far known only from parts of Scotland and one site in Kent. Even *C. atricillus* is localised within its range, although all three species can be numerous where they occur. Their distributions are being mapped by the nocturnal Ichneumonoidea recording scheme (http://nocturnalichs.myspecies.info/). Several mesochorines are nocturnal and are frequently found at light. This applies particularly to the three British *Cidaphus* species and several *Mesochorus* which are predominantly testaceous. *Astiphromma splenium* (Curtis) and some additional *Mesochorus* are also frequent at light but do not display any chromatic adaptation (cf. Gauld & Huddleston, 1976).

Identification. Wahl (1993b) provides a key to world genera, but omitting *Dolichochorus*, represented in Britain by *D. longiceps* (Strobl). Lawton's (1981) key is useful but the taxonomy does not entirely correspond with ours. Schwenke (1999, with supplements in 2002, 2004) published a revision of the entire European fauna but the work suffers from some major drawbacks, such as an outmoded taxonomy, over-simplified keys, insufficient diagnoses and descriptions, and misinterpretations/misidentifications of several taxa. Schwenke's revision is probably most useful for *Astiphromma* but should be treated with caution, and in conjunction with Horstmann (2006a), when dealing with *Mesochorus*. Riedel (2015) provides a more complete and user-friendly key to European *Astiphromma* species. Riedel (2018a) revises the testaceous species of the *Mesochorus fulvus* species-group. Fitton (1985b) provides a key to the three British *Cidaphus* species (treating *C. areolatus* under its synonym *brischkei* (Szépligeti)).

Figure 106. Morphology of Mesochorinae. (a) *Cidaphus atricillus* (Haliday), head; (b) *Mesochorus punctipleuris* Thomson, head; (c) *C. atricillus*, part of fore wing; (d) *M. punctipleuris*, part of fore wing; (e) *C. atricillus*, first metasomal tergite, lateral; (f) *Mesochorus politus* Gravenhorst, first metasomal tergite, lateral; (continued overleaf).

Figure 106 continued: Morphology of Mesochorinae. (g) *C. atricillus*, ovipositor and sheaths; (h) *M. politus*, ovipositor and sheaths; (i) *C. atricillus*, male parameres.

Subfamily METOPIINAE

With 76 British and Irish species in 13 genera, out of a world fauna of about 840 described species in 27 extant genera, the Metopiinae is a medium-sized subfamily of Ichneumonidae. Yu *et al.* (2012) listed 24 extant genera but omitted *Scolomus* and the enigmatic *Ischyrocnemis* (see below), and Khalaim *et al.* (2012) described *Ojuelos* Khalaim from Mexico. Species-richness may be a little higher in tropical than in temperate climes (e.g. Gauld & Sithole, 2002) but metopiines are encountered rather frequently over much of the world, including Britain.

Figure 107. *Hypsicera curvator* (Fabricius).

Recognition. For the most part, Metopiinae can be readily recognised as such by the uniformly convex face and clypeus, the dorsally-pointing, intra-antennal projection, the short, stout legs and antennae and their generally compact body shape; most genera have the fore trochantellus partly or completely fused with the femur. However, there are some less distinctive genera and confusion is possible with several other ichneumonid taxa that possess a convex face plus clypeus, particularly the *Orthocentrus* genus-group of the subfamily Orthocentrinae, which however can be distinguished by their characteristically long antennal scape. Species of the metopiine genus *Scolomus* look rather like mesochorines or ctenopelmatines (the type species of *Scolomus* was described in the ctenopelmatine tribe Pionini by Townes, 1970b) but can be recognised as metopiine by the shortened fore tarsal segments (in the British species), lack of a sulcus between the clypeus and face and its biology, as a primary parasitoid of Lepidoptera. The genus *Metopius* is immediately recognisable by the unique face, where ridges define a shield-shaped area covering much of the face. The genus *Ischyrocnemis* is much harder to recognise as belonging to any particular subfamily and, indeed, most probably does not belong in the Metopiinae (see below). The lack of a division between face and clypeus and lack of evidence for a placement in any other ichneumonid subfamily are the reasons for its classification in Metopiinae (Townes, 1971).

Systematics. With their notched ovipositor, high ovariole number and koinobiont, endoparasitoid larvae, Metopiinae are undoubtedly part of the large ophioniformes assemblage of subfamilies, but the exact systematic position of the subfamily is unclear. Townes (1971) placed Anomaloninae and Metopiinae next to each other in his generic monograph, based on similarities in larval head capsule morphology, but stated that this could well be a convergence based on emergence from the host pupa, which is otherwise unusual within the ophioniformes group of subfamilies. Gauld (1976a) and Short (1978) both concluded that reductions in the larval hypostomal spur and labial sclerite in anomalonines and metopiines are indeed convergent, associated with the spinning of only a feeble cocoon within the protection of the host pupa and the reduction in associated larval head musculature. Gauld (1985), mainly on the basis of Pampel's (1914) work on internal female reproductive anatomy, established that Metopiinae belong to the ophioniformes clade but not to the clade of 'higher' ophioniformes, comprising (now) the Anomaloninae, Campopleginae, Cremastinae, Nesomesochorinae and Ophioninae. This finding has been borne out by the combined morphological and molecular phylogenetic results of Quicke *et al.* (2009), where Anomaloninae are recovered amongst a clade of subfamilies that all share a petiolate first metasomal tergite and (usually) laterally compressed metasoma, to the exclusion of Metopiinae. Quicke *et al.* (2009) recovered the Metopiinae as a clade of Lepidoptera parasitoids that rendered the Ctenopelmatine plus Mesochorinae (hence parasitoids of Hymenoptera) paraphyletic, a relationship that had been posited by Gauld & Wahl (2006), although these results are far from robust.

Whilst there has been little change to the generic classification of the 'core' (i.e. distinctively metopiine) genera since Townes (1971), some peripheral genera have been variously excluded or included. Otherwise, Townes's (1971) generic classification of the Metopiinae has remained mostly unchanged. Gauld & Sithole (2002) and Khalaim *et al.* (2012) each described a new genus, from Costa Rica and Mexico, and the only other genus-level changes have involved synonymisation of the ctenopelmatine genus *Scolomus* with *Apolophus* (Gauld & Wahl, 2006) (the former having priority) and the synonymy of *Metopius* (*Tylopius*) under *Metopius* (*Peltastes*) (Horstmann, 2001b). However, the limits of some genera are vague, as acknowledged by Townes (1971). British genera that seem not to be satisfactorily separated are *Chorinaeus* and *Trieces*, and *Hypsicera* and *Stethoncus*. There is no published phylogeny of metopiine genera, which makes it impossible to reconstruct any of the interesting host shifts that can be seen within the subfamily. Most genera are relatively easy to diagnose for Ichneumonids. Although there has been little doubt over the monophyly of most of the subfamily, Townes (1971) placed three aberrant Palaearctic genera, *Bremiella* Dalla Torre, *Ischyrocnemis* and *Lapton* Nees, in the Metopiinae, seemingly for want of anywhere else to place them. Unfortunately, none of these three genera have been reared as knowledge of their biology would doubtless help in placing them. Aubert (2000) transferred *Bremiella* to the subfamily Ctenopelmatinae (which could well be correct), tribe Perilissini (almost certainly incorrect). Quicke *et al.* (2009) were able to offer little in the way of suggested placements for the non-British *Bremiella* and *Lapton* within the Ichneumonidae, other than that they are probably associated with some elements of Ctenopelmatinae, which is, though, another poorly supported subfamily. *Ischyrocnemis* completely defied classification and remains *incertae sedis* within Ichneumonidae, being most often recovered by Quicke *et al.* (2009), surprisingly, within the pimpliformes group of subfamilies, and is included here in Metopiinae in the absence of any persuasive alternative suggestions. Aubert (2000) transferred *Ischyrocnemis* to the ctenopelmatine tribe Pionini, but without explanation, although this was probably based on Perkins (1962), who believed that *Ischyrocnemis* may be close to *Rhorus*. *Ischyrocnemis* have a prominent tooth on the outer distal edge of the fore tibia, as in Ctenopelmatinae but also several other ichneumonids.

Biology. Metopiinae are all koinobiont, larva-pupal endoparasitoids of Lepidoptera. In most genera, eclosion from the egg usually happens only once the host has pupated; in a few cases (e.g. *Chorinaeus*; Aeschlimann, 1974a) eclosion occurs shortly before the host pupates but development beyond the first instar only takes place once the host has pupated. Hosts of the smaller metopiines (the majority of the subfamily) are centred on leaf-rolling and web-inhabiting 'microlepidoptera', including Choreutidae, Crambidae, Depressariidae, Epermeniidae, Ethmiidae, Gelechiidae, Glyphipterigidae, Oecophoridae, Pyralidae, Tineidae, Tortricidae, Yponomeutidae and Ypsolophidae. This is true of the more species-rich genera such as *Chorinaeus*, *Exochus*, *Hypsicera* and *Triclistus*. The distinctive morphology of most metopiines seems to represent a suite of adaptations to allow metopiines to push their way through silk webbing and in to leaf-rolls, to reach their lepidopteran hosts. There are some notable exceptions to this biological sketch: the larger and distinctive *Metopius* species are parasitoids of more free-ranging Arctiinae (Erebidae), Lasiocampidae and Notodontidae (e.g. Clément, 1930), and *Colpotrochia cincta* (Scopoli), another relatively large species, is recorded from noctuid hosts (e.g. Haeselbarth, 1989). There are published host records of *Synosis* and *Chorinaeus* from Geometridae but these are unsubstantiated. A few species of *Triclistus* have been recorded from Geometridae (Aeschlimann, 1973a) and *Triclistus areolatus* Thomson is a parasitoid of *Earias chlorana* (L.) (Nolidae) in Britain (Aeschlimann, 1973a). The morphologically rather isolated *Scolomus borealis* (Townes) is a parasitoid of the distinctive, and isolated in the British fauna, *Schreckensteinia festaliella* (Hübner) (Lepidoptera: Schreckensteiniidae), the larvae of which graze the upper surface of leaves (typically of *Rubus*) (Broad & Shaw, 2005). Several genera, including the British *Ischyrocnemis* and *Periope*, each of which is morphologically distinctive and presumably specialised in the manner in which they access their hosts, have yet to be reared. These two small genera each contain only one very rarely collected British species; in the case of *Periope auscultator* Haliday occurring in northern upland areas. Although the single European *Carria* species, *C. paradoxa* Schmiedeknecht, a rather distinctive elongate species that flies early in the year, has yet to be reared, extralimital *Carria* species are known to be parasitoids of Tortricidae (e.g. Bradley, 1974; Munro & Henderson, 2002). Some *Hypsicera* and *Stethoncus monopicida* Broad & Shaw are known to be parasitoids of detritivorous Tineidae (Aeschlimann, 1989; Broad & Shaw, 2005). Other *Hypsicera* have been reared from Oecophoridae, and one is recorded from Tortricidae (Aeschlimann, 1989).

Aeschlimann (1973a, 1975a) records several species of *Triclistus* and *Chorinaeus* as being parasitoids of more than one family of Lepidoptera. Although there is little in the way of supporting information to infer reliability, the wide taxonomic spread of host records for some metopiine species is at least partly real, as evidenced by occasional rearings of *Triclistus epermeniae* Shaw & Aeschlimann from *Acrolepia autumnitella* Curtis, of the family Glyphipterigidae, when it is much more frequently reared from *Epermenia chaerophyllella* (Goeze) (Epermeniidae) (Shaw & Aeschlimann, 1994). Some flexibility in host taxon may be afforded through avoiding the host's encapsulation responses. Oviposition in metopiines is often precise and into tissues that do not have an immune response, such as the epithelium of the hind gut by *Chorinaeus funebris* (Gravenhorst) (Aeschlimann, 1974a) or into the sub- or supra-oesophageal ganglion by *Triclistus* species (e.g. Gerig, 1960; Aeschlimann, 1975b; Dijkerman, 1988). Shaw (2017b) illustrates the egg of *Triclistus anthophilae* Aeschlimann in the supra-oesophageal ganglion of its choreutid host, *Anthophila fabriciana* (L.). The adult parasitoid grasps the host firmly with its stout legs, aligning itself head to tail with the host larva, during this process of precision oviposition. At least in *Trieces tricarinatus* (Gravenhorst), eggs deposited in the host's haemocoel are invariably encapsulated (Dijkerman, 1988), which is possibly the case for metopiines more generally. However, this can be avoided if the host already harbours a koinobiont endoparasitoid of a different species that has to a large extent subdued the host's encapsulation capacity. *Exochus tibialis* (Holmgren), a parasitoid of *Epinotia tedella* (Clerck) (Tortricidae), is only found in hosts already parasitized by *Lissonota dubia* (Holmgren) and Münster-Swendsen (1979) surmised that the species is acting as a cleptoparasitoid, taking advantage of the tortricid's pre-weakened immune response following initial parasitism by the banchine. It is not known whether this form of cleptoparasitism is obligatory for other *Exochus* species; despite their being species-rich and commonly collected, very little is known of the developmental biology of *Exochus*.

The biology of a few metopiines has been studied in detail because of their potential in controlling populations of defoliating Tortricidae or, in the case of *Trieces tricarinatus* and *Triclistus yponomeutae* Aeschlimann, as part of a wider study looking at factors promoting host plant switching and speciation in their host *Yponomeuta* species (Yponomeutidae) (Dijkerman, 1988). *Chorinaeus funebris* has been studied as a parasitoid of *Zeiraphera diniana* (Guénée) (Aeschlimann, 1974a), a significant pest of conifers, although it has been reared from other tortricid hosts too. A summary for *C. funebris* will serve as a general description of metopiine biology, although differences in other taxa are highlighted here. Handling of the final instar host and oviposition takes only 20 seconds. In other metopiines, oviposition stance and reaction to host movement differ between species; however, Dijkerman (1988) reports that, in both *Trieces tricarinatus* and *Triclistus yponomeutae*, acceptance of host larvae depends on the host's movements, inactive larvae being ignored. Oviposition by *C. funebris* is via the anus of the host, into the hind gut (this is not the case in other studied Metopiinae), and the ovipositing female usually concurrently host-feeds, i.e. feeds on the individual parasitized. At 18°C, eclosion from the egg takes place after 96 hours and timing of oviposition is critical as the *C. funebris* egg does not develop if the host pupates within this four day period; unlike the majority of metopiines, it presumably needs to hatch from the egg before the host's pupation in order to avoid being sloughed at ecdysis. This may be general for *Chorinaeus*, though evidence is lacking. As in other metopiines, larvae are initially caudate, until the fourth instar in *C. funebris*, the appendage being lost in the fifth, final instar. Feeding is initially in the pupal host abdomen, usually dorsally, but after eight or nine days the larva adopts a more central position. The fifth instar is reached after 20-22 days of larval development and the host ceases movement by this stage. *Chorinaeus funebris* overwinters as a fully-grown larva, presumably prepupa, within the host pupa. The cocoon is flimsy, white and spun within the host's pupa, as is the case in other metopiines. Superparasitism is frequent in *Z. diniana*, and *C. funebris* is generally the inferior competitor except when the other parasitoid is a *Triclistus* species. Superparasitism between two metopiine species was also recorded by Dijkerman & Koenders (1988), who found that the outcome depended on the length of time between oviposition and pupation of the host. In this case, the larva of *Trieces tricarinatus* can seek out the larva of *Triclistus yponomeutae* and kill it whilst it is still within the host ganglion; usually, though, *T. yponomeutae* is the superior competitor. Oviposition by *Trieces tricarinatus* is through the larval abdominal cuticle, the egg being placed either into the hind gut epithelium or directly below the cuticle, presumably between layers of epidermis (Dijkerman, 1988). Although *Exochus tibialis* (and possibly other *Exochus*) may be an exception, with its apparent requirement for an already-parasitised host, it appears that in general, metopiines may not be able to detect the presence of pre-existing parasitoids – as indeed might be expected if oviposition is into a site other than the haemocoel.

The parasitoid complex attacking *Z. diniana* includes three metopiines, two closely related: *Chorinaeus funebris*, *Triclistus podagricus* (Gravenhorst) and *Triclistus pygmaeus* Cresson. Each is essentially univoltine (although a variable proportion of individuals emerge in the year of oviposition) but overwintering is achieved in different ways; *C. funebris* and *T. pygmaeus* usually pass the winter as prepupae and *T. podagricus* as an unemerged adult within the host pupa (Aeschlimann, 1974b). With its high fecundity, parthenogenesis and potential (in the laboratory, at least) for bivoltinism, *T. pygmaeus* was considered to have the greatest potential for controlling populations of *Z. diniana* (Aeschlimann, 1974b). In experimental releases of *T. pygmaeus*, the parasitoid built up large population densities but did not provide control against damage to larch

(*Larix decidua*), although Aeschlimann (1975b) posited that population augmentation by releases three to four years before the population of *Z. diniana* built up would be more successful.

Rather little has been published on the biology of other metopiines, beyond host records. Townes (1939a) notes that North American *Colpotrochia*, which are rather large and brightly patterned, produce a distinct odour, presumably for defensive purposes, as do the smaller *Exochus* species (Townes, 1971). *Metopius* species are aposematically coloured and make a buzzing noise when captured (Townes & Townes, 1959), presumably to add to the vespid wasp resemblance. *Metopius dentatus* (Fabricius) is a particularly large species and a fairly regularly reared parasitoid of *Lasiocampa quercus* (L.) (Lasiocampidae). However, of the six *Metopius* species on the British list, three do not seem to have been collected for many years, despite their distinctive appearance.

Identification. Townes's (1971) key to metopiine genera is still adequate. Available species-level identification keys cover most of the British metopiine fauna, although many *Exochus* specimens will remain difficult to identify. Aeschlimann provides keys to *Chorinaeus* (1975a, 1981), *Hypsicera* (1989), *Triclistus* (1973a) and *Trieces* (1973b), with Aeschlimann (1983) and Shaw & Aeschlimann (1994) describing additional *Triclistus* species. Tolkanitz (2011) provides an alternative key to Palaearctic *Hypsicera*. Perkins's (1936) key to *Metopius* still covers our limited fauna and Tolkanitz (2015) provides keys to the Palaearctic species. Tolkanitz's (2007) revision of Palaearctic *Exochus* is not comprehensive but is useful. Broad & Shaw (2005) cover the identification and biology of species of four small genera, *Ischyrocnemis*, *Scolomus* (as *Apolophus*), *Stethoncus* and *Synosis*. The genera *Carria*, *Colpotrochia* and *Periope* are (like *Ischyrocnemis*, *Scolomus* and *Stethoncus*) each known from only one British species and can thus be recognised from Townes's (1971) generic key.

Figure 108. *Metopius dentatus* (Fabricius).

Figure 109. Morphology of Metopiinae. (a) *Ischyrocnemis* sp., head; (b) *Metopius citratus* (Geoffroy), head; (c) *Scolomus borealis* (Townes), head; (d) *Stethoncus monopicida* Broad & Shaw, head; (e) *Triclistus globulipes* (Desvignes), head, dorsal; (f) *Periope auscultator* Haliday antennae; (continued overleaf).

Figure 109 continued: Morphology of Metopiinae. (g) *Triclistus globulipes*, fore leg; (h) *Periope auscultator*, hind leg; (i) *Ischyrocnemis*, male metasoma; (j) *Scolomus borealis*, first metasomal segment (continued overleaf).

Figure 109 continued: Morphology of Metopiinae. (k) *Scolomus borealis*, hypopygium and ovipositor; (l) *Triclistus globulipes*, first and second metasomal tergites; (m) *Trieces tricarinatus* (Holmgren), propodeum and first to third metasomal tergites.

Gavin R. Broad, Mark R. Shaw and Michael G. Fitton

Subfamily MICROLEPTINAE

This subfamily comprises only the Holarctic and Oriental genus *Microleptes*, with 14 described species (Yu *et al.*, 2012; Sheng & Sun, 2014), three of which occur in Britain.

Figure 110. *Microleptes rectangulus* (Thomson).

Recognition. *Microleptes* species are fairly easily recognised by the following combination of characters: antennal sockets very prominent (protruding anteriorly); head ventrally with angular corners; mandible with the teeth blunt, the upper tooth larger than the lower; mesoscutum with the lateral carina posteriorly not continuous with that enclosing the scuto-scutellar groove; tergite 1 with the spiracles at or just in front of the midpoint, the sternite extending beyond the spiracle, and this segment rather cylindrical; ovipositor short, not or only just projecting beyond the apex of the metasoma; hypopygium large; inner side of apex of hind tibia with a sloping comb of dense long setae.

Systematics. *Microleptes* is of uncertain phylogenetic position (Quicke *et al.*, 2009). Wahl (1986) demonstrated that *Microleptes* does not share larval synapomorphies with the other parasitoids of Diptera, of the subfamilies Cylloceriinae, Diplazontinae and Orthocentrinae, in the pimpliformes. Santos (2017) found that *Microleptes* belongs with the ichneumoniformes group of subfamilies, possibly nested within Phygadeuontinae. The genus *Hyperacmus* (with the synonymous name *Cushmania* sometimes treated separately) shares some superficial similarities with *Microleptes* and has been regarded as belonging to Microleptinae by Humala (2003, 2007a), but Quicke *et al.* (2009) demonstrated that *Hyperacmus* is closely related to *Cylloceria*, of the Cylloceriinae.

Biology. Few details of the biology of *Microleptes* are known. Wahl (1986) described the structure of the final larval instar of an unidentified North American species of *Microleptes* reared from the stratiomyid *Allognosta fuscitarsis* (Say) (Diptera). No cocoon was spun, pupation being within the larval skin of the host. The European (and British) species *M. splendidulus* Gravenhorst has also been reared from a stratiomyid (Wahl, 1986) and Schwarz (1991b) records the European (non-British) *M. obenbergeri* Gregor as a parasitoid of *Chloromyia formosa* (Scopoli) (Stratiomyidae). The available information is insufficient to distinguish between the possibilities that *Microleptes* species are larva-pupal koinobionts or idiobionts attacking the host pupa. Because the pupa in Stratiomyidae is enclosed within the larval skin (Oldroyd, 1969) it is not even clear whether *Microleptes* is properly endoparasitoid. *Microleptes* species are most commonly collected in pitfall or water traps in damp habitats in Britain, much less commonly in Malaise traps.

Identification. The three British species are keyed, under three separate generic names, by Stelfox (1961). Another key that includes these species, with brief descriptions, is given by van Rossem (1981). Schwarz (1991b) gives a key to the five species now known in Europe.

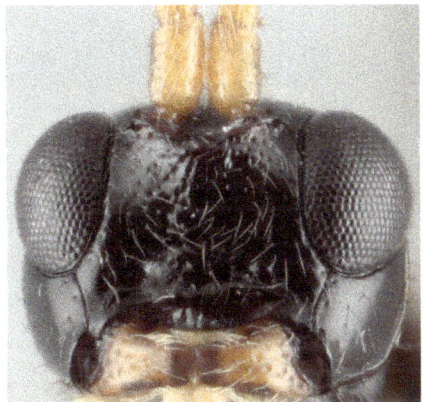

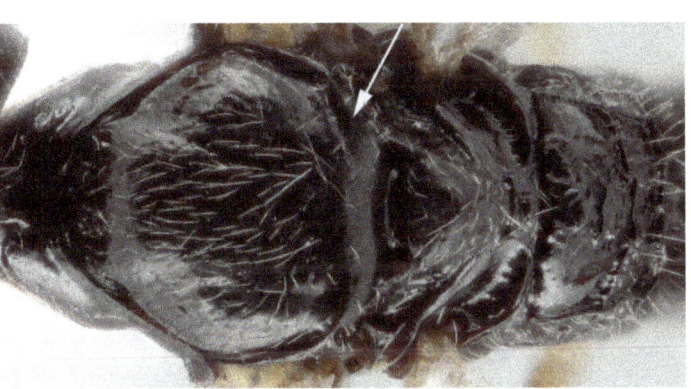

Figure 111. Morphology of Microleptinae. (a) *Microleptes rectangulus* (Thomson), face and mandibles; (b) *M. rectangulus*, head; (c) *M. rectangulus* mesoscutum, arrow pointing to gap in lateral carina of the mesoscutum; (continued overleaf).

Figure 111 continued: Morphology of Microleptinae. (d) *M. aquisgranensis* (Förster), hind tibia, inner face (upper) and posterior face (lower); (e) *M. rectangulus*, metasoma; (f) *M. rectangulus*, first metasomal segment.

Subfamily **NEORHACODINAE**

The Neorhacodinae is a very small subfamily comprising only six described species in three genera found in the Palaearctic, Nearctic and Neotropical regions. There is only one species known from Britain, *Neorhacodes enslini* (Ruschka); the other European species, *Eremura perepetshaenkoi* Kasparyan, occurs in arid areas of Spain (Notton & Shaw, 1998).

Figure 112. *Neorhacodes enslini* (Ruschka).

Recognition. *Neorhacodes* is a distinctive genus of small ichneumonids and unlikely to be confused with any other ichneumonid: the wing venation is reduced, with the *rs-m* cross veins completely absent, such that the areolet is obliterated by the fusion of *RS* and *M*, and cross vein *2m-cu* only faintly indicated and never tubular; the antenna has only 13 segments; and the basal metasomal tergites have distinctive transverse and sculptured grooves.

Systematics. At one time the neorhacodines had been placed as a tribe of Banchinae (Perkins, 1959; Townes, 1970b); however, the discovery of the final instar larva (Horstmann, 1968; Danks, 1971), which is quite unlike that of any Banchinae, led Townes (1971) to modify his view and regard the

Neorhacodinae as a separate subfamily. Rather surprisingly, Quicke *et al.* (2009) found *Neorhacodes* (no specimens of *Eremura* Kasparyan or the third genus *Romaniella* Cushman were available for sequencing) to nest within the Tersilochinae. Whilst dissimilar in most respects from the tersilochines (including the *Phrudus* genus-group now included in the Tersilochinae), Quicke *et al.* (2009) identified two potential morphological synapomorphies supporting the inclusion of the neorhacodines within an expanded Tersilochinae, namely the large, rather equilaterally triangular fore wing pterostigma and the largely unpigmented hind wing vein *M+CU*. However, we retain Neorhacodinae as a distinct subfamily given the significant difference in biology between the Tersilochinae (parasitoids of beetle larvae, applying also, where known, to the *Phrudus* group) and *Neorhacodes* (the biology of *Eremura* and *Romaniella* is unknown), the lack of substantial phylogenetic support for the monophyletic Tersilochinae including *Neorhacodes* (which was based on only one gene plus a morphological matrix) and the forthcoming phylogenetic results of Bennett *et al.* (in prep.), in which the placement of *Neorhacodes* is very variable but never groups with the Tersilochinae. The monophyly of the Neorhacodinae has not been tested.

Biology. Little is known of the biology of *Neorhacodes*. Danks (1971) studied *Spilomena* species (Hymenoptera: Crabronidae) in southern Britain and in most areas in which they were found he detected parasitism by *N. enslini*. Host species identified with some certainty by Danks (1971) were *S. enslini* Blüthgen and *S. troglodytes* (Vander Linden) and it has also been reared from *S. differens* Blüthgen (Notton & Shaw, 1998). *Neorhacodes* appears to attack the host larva; a fairly thick elongate-ovoid cocoon is made, and adult emergence takes place about the same time as that of the hosts. Short (1978) inferred that the larva will be found to be endophagous and the presence of a subapical notch on the dorsal valve of the ovipositor lends support to this suggestion. Nothing is known of the biology of *Eremura* or *Romaniella*.

Identification. *Neorhacodes enslini* can be recognised using the key to subfamilies and is dealt with by Perkins (1959) and by Fitton (1984). The European fauna of neorhacodines is treated by Notton & Shaw (1998).

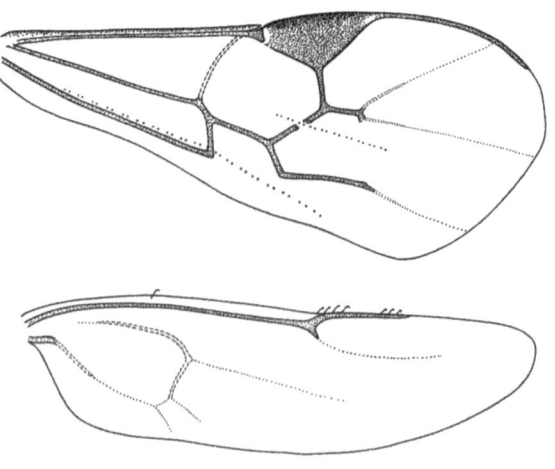

Figure 113. Morphology of Neorhacodinae. *Neorhacodes enslini*: (a) wings; (continued overleaf).

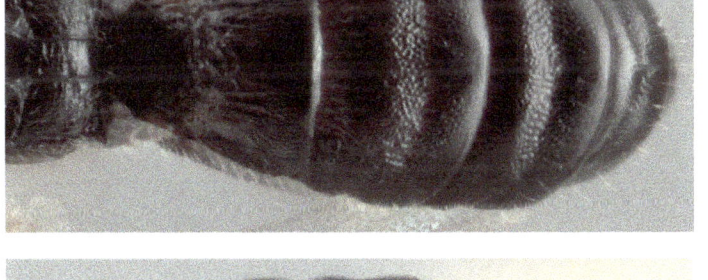

Figure 113 continued: Morphology of Neorhacodinae. (b) head, frontal; (c) head, lateral; (d) propodeum and metasoma, dorsal; (e) metasoma, lateral.

Subfamily **OPHIONINAE**

The Ophioninae is, worldwide, a large subfamily with over 1,000 described species in 31 genera. In many parts of the world, the vast majority of specimens will be found to belong to *Enicospilus*, although with *Ophion* predominating in the northern temperate zones. As the adults of most species are big and attracted to lights they are relatively easy to sample and the group is well represented in collections. On a global front it is one of the taxonomically best known subfamilies, mainly thanks to the monographing efforts of Ian Gauld (especially Gauld & Mitchell, 1978, 1981; Gauld, 1988a). There are many more species in tropical than temperate zones, mainly owing to the massive diversification of *Enicospilus* in the tropics. In Britain there are currently 27 species representing 4 genera out of a European fauna of 67 species in 7 genera (although several species are of dubious status). Many species of *Ophion* are morphologically highly conserved, making identification difficult and probably masking many additional species.

Figure 114. *Ophion longigena* Thomson.

Recognition. Ophioninae are readily recognised as such by their overall testaceous colour (in most species) and other nocturnal adaptations (such as long antennae, large ocelli) in combination with their characteristic wing venation, with the fore wing discosubmarginal cell extending beyond vein 2*m-cu* (vein 2*rs-m* absent and 3*rs-m* joining *M* distal to 2*m-cu* by more than 0.5 the length of 3*rs-m*) and the presence of a spurious vein extending from the posterodistal corner of the 1st subdiscal cell, subparallel to the wing margin. In addition, the tarsal claws are pectinate, the metasoma is strongly laterally compressed, the first tergite has the spiracles well behind the midpoint with the tergite and sternite fused, and the ovipositor is very short, with a subapical notch. Various other genera of nocturnal ichneumonids (especially *Netelia* (Tryphoninae)) have a superficially similar appearance but all can be distinguished by their very different wing venation and structure of the first metasomal tergite. The aberrant genera *Hellwigia* and *Skiapus* have been transferred to Ophioninae (Quicke *et al.*, 2005) but may not immediately be recognised as ophionines. There are two species of *Hellwigia* in Europe, either of which could conceivably turn up in Britain. Superficially they resemble Campopleginae and are not testaceous, but they have some of the wing venational features of Ophioninae (in particular, the discoscubmarginal cell extends beyond vein 2*m-cu*).

Systematics. Ophioninae is a well-defined, monophyletic group (Gauld, 1985; Wahl, 1991; Quicke *et al.*, 2009) that has arisen within the 'higher ophioniformes' group of subfamilies, comprising koinobiont endoparasitoids mainly of Lepidoptera larvae. The classification of the Ophioninae has been the subject of detailed study for some considerable time, the major works being by Cushman (1947), Townes (1971), Gauld & Mitchell (1978; 1981) and Gauld (1977; 1979; 1985). Gauld (1985) proposed a generic classification and phylogeny supported by explicit accounts of both the characters and methods used in their construction. However, the results of combined analyses of morphological and molecular data (Rousse *et al.*, 2016) have supported some of Gauld's clades (the *Enicospilus* group, for example) but not other relationships. Rousse *et al.* (2016) included more genera and analysed more genes than were included in Quicke *et al.* (2009) but more than half of ophionine genera remained unplaced in their proposed tribal reclassification of Ophioninae, owing to limited taxon sampling; we therefore do not employ tribes here (Rousse *et al.*, 2016, recognise Enicospilini, Ophionini and Thyreodonini). Schwarzfeld *et al.* (2016) have proposed several monophyletic species-groups within *Ophion* that only partly correspond to those of Gauld (1985).

Biology. Where known, practically all Ophioninae are koinobiont endoparasitoids of Lepidoptera larvae. A surprising exception to this pattern of host associations occurs in North America, where *Ophion nigrovarius* Provancher reputedly parasitizes the soil dwelling larvae of the scarabaeid beetle *Phyllophaga fusca* (Froelich) (Townes & Townes, 1951, as *O. bifoveolatus*). Many ophionine species attack caterpillars in their middle or late instars but some oviposit into very young larvae, with a species of the *Enicospilus americanus* complex, of the latter group, having twice as many (24) ovarioles per ovary as the former (6 to 15) (Price, 1975). With few, but notable, exceptions adult ophionines are nocturnal or crepuscular and are pale in colour, have long slender bodies and appendages, enlarged ocelli and large wings. These adaptations are found in several other, unrelated, groups of nocturnal ichneumonoids (Gauld & Huddleston, 1976). A small number of ophionine genera comprise species that do not exhibit some of these features, as in the apparently very basal (Quicke *et al.*, 2005) non-British (but European) genus *Hellwigia*, which is diurnal. Some of the nocturnal species may also be found flying during the day, particularly in dull weather and particularly males – for example males of the large, dark and rather rare *Eremotylus marginatus* (Jurine) can be particularly conspicuous as they fly around bushes by day.

Despite the economic importance of some tropical species, and the relatively large numbers of rearings from known hosts, very few life histories have been investigated; two of the more complete accounts are of *Enicospilus sesamiae* Delobel parasitizing *Sesamia vuteria* Stoll (Noctuidae) in Mauritius (Moutia & Courtois, 1952) and *Ophion bilineatus* Say parasitizing *Spodoptera frugiperda* (Smith) (Noctuidae) in Texas (Vickery, 1929). However, several features of these two life histories will not apply to the majority of species in temperate areas such as the British Isles. In *E. sesamiae* copulation usually lasts for about two and a half hours, but sometimes as long as eight hours. Moutia & Courtois (1952) do not record any courtship activity preceding copulation but, even so, such a long copulation time is most unusual in the Ichneumonidae. Females search for nocturnally active caterpillars of various families of Lepidoptera feeding in exposed positions. In *O. bilineatus* the female rises up on her legs, brings the metasoma forward beneath her body and stabs the caterpillar quickly and without hesitation, withdrawing the ovipositor immediately (Vickery, 1929). Along with *Netelia*, Ophioninae are often very quick to sting when handled, twisting their mobile metasomas to effect purchase on the skin, and this seems to be a well differentiated defensive manouvre, also exhibited by males which, of course, cannot sting. *Enicospilus sesamiae* lays a single, elongate cylindrical egg in the haemocoel of, mainly, the fourth and fifth larval instars of its host (Moutia & Courtois, 1952). In this species the parasitoid larva hatched in two to three days and larval development, which comprised three instars, took 25 to 35 days. In *O. bilineatus* the time from oviposition to eruption of the fully grown larva was about 10 days, but was 3 to 5 days longer if oviposition had been into a second instar caterpillar, rather than the preferred third or fourth instars.

In most ophionines the fully grown larva erupts from its host's prepupa and makes its cocoon in the host's pupation site, typically in the soil or litter layer or, if the host is one that makes a discrete tough cocoon, then within that (e.g. *Enicospilus inflexus* (Ratzeburg) parasitizing the lasiocampid *Lasiocampa quercus* (L.)). *Ophion parvulus* Kriechbaumer is exceptional in forming its cocoon within the ruptured pupal remains of its *Orthosia* (Noctuidae) host (Brock, 1982). Ophionine cocoons are elongate-ovoid, have a rather coarse external surface overlying remarkably finely-spun inner layers, and are generally dark brown in colour with a slightly paler equatorial band. The multi-layered nature of the cocoon might guard against invasion by microorganisms during the often protracted period spent in the soil.

In Europe most species of the subfamily are univoltine. One conspicuous apparent exception is the abundant and widespread *Ophion obscuratus* Fabricius, which on the face of it has three generations, with some rather vague morphological differences between them. Brock (1982) thought that these generations probably represent good biological species but could not find any entirely discriminatory characters. However, rearings indicate that individuals of each 'generation' seem to behave as a univoltine species would, and work by Johansson & Cederberg (in prep.) in Sweden will result in taxonomic changes. Similarly, the first 'generation' of *O. parvulus* on *Orthosia* spp. (Noctuidae) will overwinter and emerge in time for the next year's 'first generation' (Brock, 1982), and there is significant morphological variation between different 'generations' of *O. parvulus*. There is circumstantial evidence that some species overwinter as fully grown larvae (or prepupae) within their cocoons, while others, possibly only those with a flight period very early in the year, are known to pass the winter as adults within their cocoons (Morley, 1915).

Figure 115. *Ophion obscuratus* Fabricius (G.R. Broad).

Host ranges of the British species are fairly well understood only for a few species whilst other, often common, species have very seldom, if ever, been reared. For example, *Ophion luteus* (L.), an ubiquitous species in late summer and early autumn, has only very recently been shown to be a parasitoid of the very common and widespread *Agrotis exclamationis* (L.) (Noctuidae) (Broad et al., 2015). While most British *Ophion* species probably parasitize noctuids, three rather aberrant species are parasitoids of other families of Lepidoptera, with *O. ocellaris* Ulbricht (often treated as belonging to a separate genus *Platophion*) having been reared from Drepanidae (Thyatirinae) and both *O. minutus* Kriechbaumer and *O. ventricosus* Gravenhorst being parasitoids of arboreal Geometridae (*Agriopis* and *Apocheima* respectively). The phylogeny of *Ophion* in Schwarzfeld et al. (2016) has these geometrid parasitoids basally on the tree whereas the *O. ocellaris* group is nested within the noctuid-parasitizing clade. Among the species parasitizing Noctuidae, some appear to have narrow host ranges (e.g. *O. longigena* Thomson on *Cucullia*) but others have a wider scatter of known hosts (e.g. *O. pteridis* Kriechbaumer which is regularly reared from several genera of Hadeninae). The two common early spring species, *O. scutellaris* Thomson and *O. obscuratus*, parasitize practically fully-grown grassland noctuid larvae (respectively *Xestia* and *Mythimna* species) that have been feeding up through the winter. However, for many of our *Ophion* species, rearings have been at best too few for host ranges to be deduced – and for several species no host is known. A similar situation exists for *Enicospilus*: of the smaller species only *E. ramidulus* (L.) is commonly reared, from a wide range of summer-feeding noctuids (perhaps especially Hadeninae) but sometimes recorded also from more disparate hosts. Examples of host specificity seem to arise from ophionines attacking hosts in specialised biotopes where there is a limited range of potentially suitable host taxa, such as *Enicospilus merdarius* (Gravenhorst) (usually known by the synonymous name *E. tournieri* (Vollenhoven): Broad & Shaw, 2016) having been reared only from *Agrotis ripae* (Hübner) (Noctuidae) on sand dunes. Our largest *Enicospilus* species, *E. inflexus* and *E. undulatus* (Gravenhorst), are parasitoids of large Lasiocampidae. *Stauropoctonus bombycivorus* (Gravenhorst) is a parasitoid of *Stauropus fagi* (L.) (Notodontidae) and *Eremotylus curvinervis* (Kriechbaumer) has been reared from *Dryobotodes eremita* (Fabricius) (Noctuidae), whilst the host of the large and conspicuous *Eremotylus marginatus* is unclear, although there is an old record from the noctuid *Eupsilia transversa* (Hufnagel) in France (de Gaulle, 1908). Some extralimital genera of Ophioninae specialize on other host groups, for example the New World genus *Thyreodon* parasitizes Sphingidae (Gauld & Janzen, 2004) and the African genus *Euryophion* uses Bombycoidea (Saturniidae and Eupterotidae) (Gauld & Mitchell, 1978). Worldwide, however, there are numerous genera for which the hosts remain completely unknown.

Several species of British Ophioninae seem to be dependent on ancient, or at least mature, deciduous woodland, such as *Ophion brevicornis* Morley and *Eremotylus marginatus*; others have less demanding woodland requirements, such as *O. minutus* and *O. costatus*; some are strictly coastal, such as *O. forticornis* Morley and *Enicospilus merdarius*; at least one species, *Enicospilus repentinus* (Holmgren), seems to be restricted to chalk downland; one species is a heath and moorland specialist (*Enicospilus inflexus*) – in at least some of these cases probably just reflecting specialization on a narrow range of hosts with strong biotope associations. Another set of species are essentially ubiquitous, including *Enicospilus ramidulus*, *O. luteus* and *O. obscuratus*. Being easily sampled through light traps, we have more of an idea of distribution patterns for British Ophioninae than for any other group of parasitoids (other than *Netelia*) and these data are being gathered on a continuing basis by G. Broad as part of a nocturnal Ichneumonoidea recording scheme (http://nocturnalichs.myspecies.info/). It is clear that some species, such as *S. bombycivorus*, have distributions much more restricted than those of their hosts and understanding the ecology behind these distribution patterns is a challenge. Further, despite the abundance of some species and the ease with which they can be found, there is very little known about most aspects of their biology, and ophionines would be a particularly rewarding group for experimental investigation.

Identification. Keys to British genera are given by Gauld (1973) and draft keys to all British species by Broad (2012, online). *Platophion* is now regarded as a synonym of *Ophion* (Gauld, 1979; 1985). Gauld's keys to British species of Ophioninae unfortunately contain too many misidentifications and omissions to be reliable. Brock's (1982) key to *Ophion* is recommended but is difficult to use.

Broad & Shaw (2016) revise the British *Enicospilus* species. Molecular phylogenetic work by M. Schwarzfeld *et al.* is revealing a number of cryptic species within *Ophion* (e.g. Schwarzfeld & Sperling, 2015).

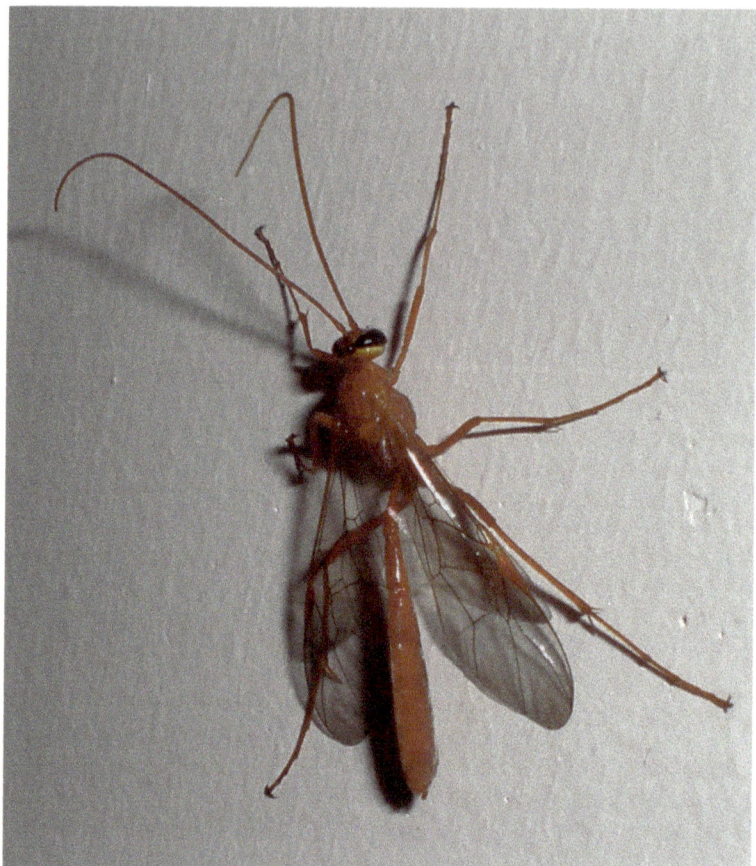

Figure 116. *Ophion luteus* (L.) in typical resting pose, with the metasoma raised above the wings held flat (G.R. Broad).

Figure 117. Morphology of Ophioninae. (a) *Ophion pteridis* Kriechbaumer, head; (b) *Enicospilus ramidulus* (L.), head; (c) *Ophion mocsaryi* Brauns, part of fore wing; (d) *E. ramidulus*, part of fore wing to show discosubmarginal cell; (e) *E. ramidulus*, first metasomal segment; (f) *E. ramidulus*, ovipositor.

Subfamily **ORTHOCENTRINAE**

This large but poorly studied subfamily comprises mainly small ichneumonids which are often only weakly sclerotized. Many dry specimens in collections have a distorted, shrivelled appearance because of this. The subfamily is world-wide in distribution and most species occur in damp habitats. Orthocentrines are diverse in both temperate and tropical areas but the tropical faunas are almost completely undescribed (Veijalainen *et al.*, 2012). The approximately 470 described species (of 30 genera; Yu *et al.*, 2012; Villemant *et al.*, 2016b) are overwhelmingly from the Palaearctic and North America. Rather unusually, a high proportion of the genera are cosmopolitan in distribution. Although little serious work has been done on the British fauna, so far 109 species representing 22 genera have been found.

Figure 118. *Megastylus cruentator* Schiødte.

Recognition. There is wide morphological variation between genera but most are readily recognizable as orthocentrines. Orthocentrines are generally small and relatively weakly sclerotized; the mandibles are narrow and tapered, with the lower tooth sometimes absent; there is usually a distinct subocular sulcus; the clypeus (when defined) is strongly convex and often small; the inner side of the apex of the hind tibia has a strong fringe of closely spaced setae; and the ovipositor lacks ventral teeth, with the dorsal valve being either simple or notched subapically. There is much variation in the form of the first tergite and sternite (whether fused or not, and in the position of the spiracles), and the presence or absence of the areolet, propodeal carination and other characters that are often useful for recognising ichneumonid subfamilies. The *Orthocentrus* group of genera is particularly distinctive as the clypeal sulcus is absent, with the face and clypeus forming a single convex surface, the scape is elongate and subcylindrical and the mandibles are strongly twisted or vestigial. Superficially, members of the *Orthocentrus* group resemble small metopiines or *Hyperacmus* (Cylloceriinae), but can be recognized easily by their characteristic elongate scape. Some other orthocentrines can be confused with small phygadeuontines, particularly the genus *Aclastus*, which has narrow tapered mandibles, a simple ovipositor, a rather weakly sclerotized metasoma, and a strong fringe of setae on the hind tibia. However, *Aclastus* invariably have at least the anterior section of the sternaulus and a characteristically pentagonal areolet, discernible even if the outer vein 3*rs-m* is missing.

Figure 119. *Batakomacrus caudatus* (Holmgren).

Systematics. The genera that are currently or have been included in Orthocentrinae have had a confused taxonomic history. Townes (1971) arranged the genera of Orthocentrinae, as treated here, in two subfamilies, Orthocentrinae for the *Orthocentrus* group of genera and Microleptinae for the remaining groups; however, he acknowledged that the latter was partly a 'wastebasket', and it includes taxa that have since been transferred to other subfamilies. To confuse matters, Townes's Microleptinae assemblage has often been referred to as Oxytorinae (the name has priority over Microleptinae: Fitton & Gauld, 1976), but in any case both *Microleptes* and *Oxytorus* are now placed in their own separate subfamilies) or Plectiscinae (although *Plectiscus* belongs to the *Orthocentrus* group of genera, in the older literature the name was applied to what is now *Plectiscidea*). Wahl (1986, 1990), mainly on the basis of larval morphology, did much valuable work in establishing monophyletic subfamilies and reassigning misplaced genera; the helictine group of 'oxytorines' and Orthocentrinae *sensu* Townes were merged in the subfamily Orthocentrinae, which he showed to belong within the pimpliformes group of subfamilies. There are still differences of opinion regarding the composition of the Cylloceriinae (i.e. whether or not including *Apoclima*, *Entypoma* and *Hyperacmus*) and whether it is better placed as a tribe within Orthocentrinae (Humala, 2003, 2007a) or as a separate subfamily (Wahl, 1990; Wahl & Gauld, 1998), but the morphological evidence either way is weak. Wahl (1990) and Wahl & Gauld (1998) employed a cladistic approach and we prefer to follow their conclusions. Humala's (2007a) argument for treating them all as one subfamily seems to rest mainly on the shared trait of Diptera parasitism, but this is also the case in Diplazontinae, which could therefore also be merged into a larger group (possibly monophyletic: see Wahl & Gauld, 1988). However, this would not be definable on adult morphological characters, whereas the three subfamilies recognised at present are definable and diagnosable on the basis of external morphology. Humala (2007a) also treated Diacritinae and Microleptinae as tribes within Orthocentrinae, but this is also not supported by the phylogenetic results of Wahl & Gauld (1998) and Quicke *et al.* (2009). Here we refer to the orthocentrines which do not belong to the *Orthocentrus* group as the *Helictes* group, but this is not a presumption of monophyly. Klopfstein *et al.* (in press) have found that *Hemiphanes* was misplaced in Orthocentrinae and have transferred this small genus to Cryptinae.

Biology. Very few species of Orthocentrinae have been reared from known hosts, but most workers, on the basis of the few published records, confidently describe them as parasitoids of 'fungus gnats', i.e. Mycetophilidae (Diptera) (Carlson, 1979; Short, 1978; Townes, 1971). Although many probably are parasitoids of Mycetophilidae there are records of some species being reared from Sciaridae (e.g. Roman, 1939) as well as other families of bibionomorph Diptera that have been split off from Mycetophilidae *sensu lato*. Orthocentrines are almost certainly solitary koinobiont endoparasitoids (Short, 1978; Shaw & Askew, 2010; Wahl, 1986) but this has not been proven for more than a very few species. In *Orthocentrus* at least, adult emergence is from the host pupa, showing that the larva is endophagous, and some *Megastylus* have been shown to be koinobiont endoparasitoids (Wahl, 1996; Baker *et al.*, 2008).

Rearing records are particularly scant for the *Orthocentrus* group, despite their abundance in many habitats. *Neurateles papyraceus* Ratzeburg has a long, telescopic metasoma which bears very conspicuous erect bristles. It has been reared from larvae of *Xylosciara lignicola* (Winnertz) (Sciaridae) found under bark of pine and spruce (Waterston, 1929b). The few available rearings of *Stenomacrus* and *Plectiscus* indicate that they are also parasitoids of Sciaridae (Roman, 1939; Vilkamaa & Komonen, 2001), including two specimens of a *Plectiscus* species in NMS that were reared from the pupae of *Bradysia giraudii* (Egger) in stems of *Filipendula ulmaria*. A few species of *Orthocentrus* have been recorded as parasitoids of *Sciophila* (Mycetophilidae) in fungi (Kolarov & Bechev, 1995; Šedivý & Ševčik, 2003), but these were not rearings from isolated hosts. Examination of host pupal remains (Fitton & Shaw, unpublished) has revealed no traces of a cocoon made by *Orthocentrus* and the feeble mandibles of the genus probably reflect this lack of a cocoon. In contrast, species of the *Helictes* group pupate within tightly woven cocoons (Wahl, 1986). Whilst it has been suggested (Shaw & Askew, 1979) that some species are highly specialised (for example, *Proclitus edwardsi* Roman is supposed to attack only *Brachypeza radiata* Jenkinson, a mycetophilid developing only in the fungus *Pleurotus cornucopiae*, which is confined to *Ulmus*

(Roman, 1923)), there is no real evidence as to host specificity. Various species have been reared from fungal fruiting bodies but these are almost invariably bulk substrate rearings and the host larvae have not been isolated. There are several unpublished rearings from reliably identified mycetophilid hosts in the collections of NMS, including *Aperileptus vanus* Förster from *Mycetophila alea* Laffoon, *Aperileptus* sp. from *Rondaniella dimidiata* (Meigen), and *Plectiscidea collaris* (Gravenhorst) from *Allodiopsis rustica* (Edwards). While eight orthocentrine genera are listed as being reared from mycetophilids in fungi in one study (Šedivý & Ševčik, 2003), some other genera have been reared from other substrates and it is likely that a variety of Sciaroidea groups are utilised as hosts. All reliable rearings for *Megastylus* are from the predatory Keroplatidae (Dasch, 1992; Humala et al., 2016; Mansbridge, 1933; Wahl, 1996); these include the British *M. cruentator* Schiødte and *M. orbitator* Schiødte and an Australian species which has been reared from a bioluminescent keroplatid inhabiting caves (Baker et al., 2008). Given the wide range in ovipositor lengths and metasomal morphology within the subfamily it is likely that their hosts will be found among a diverse range of Mycetophilidae and related families living in a variety of substrates.

Identification. Townes (1971) gives keys to the genera that he regarded as comprising two subfamilies (Microleptinae and Orthocentrinae). A revised key to European genera of the *Helictes* group (i.e. excluding the *Orthocentrus* group) was provided by van Rossem (1990) (but note that he included the genera *Epitropus*, *Fetialis*, *Pantomima* and *Phosphoriana* which have been shown to be synonyms of genera in other subfamilies, and some other genera subsequently transferred to Cylloceriinae, and *Microleptes* (Microleptinae)); a revised key to the genera of the *Orthocentrus* group was provided by Broad (2010). For the *Orthocentrus* group, Aubert (1978a, 1981) produced preliminary treatments of the European species, which deal with a good proportion of *Neurateles*, *Orthocentrus* and *Stenomacrus* species and all of the known *Plectiscus*. There are probably numerous undescribed species, however. Horstmann (1994b) revised the European *Picrostigeus* and Broad (2010) and Humala (2010) revised the European *Batakomacrus*. In a series of papers, van Rossem revised the European fauna of *Helictes* group genera: *Aniseres* (1981), *Aperileptus* (1985), *Apoclima* (1981, 1987), *Catastenus* (1981), *Dialipsis* (1981), *Entypoma* (1981, 1988), *Eusterinx* (1982, 1987, 1988, 1990, 1991), *Gnathochorisis* (1981, 1987), *Helictes* (1987), *Megastylus* (1983b, 1987, 1988), *Pantisarthrus* (1981, 1987), *Plectiscidea* (1987, 1988, 1991), *Proclitus* (1983a, 1987, 1988), *Proeliator* (1982, 1987) and *Symplecis* (1981, 1988). However, it will often be impossible to reach a satisfactory identification when using van Rossem's keys for the larger genera (particularly *Aperileptus*, *Plectiscidea* and *Proclitus*) and in places they can be positively misleading. Useful additional papers are those of Dasch (1992) (for the *Helictes* group in the Nearctic), who found numerous shared species between the Nearctic and the Palaearctic and uses some more reliable taxonomic characters than van Rossem, Humala (2003, 2007b) for various groups of Russian Orthocentrinae, and Jussila (1994), who described an additional species of *Aniseres*.

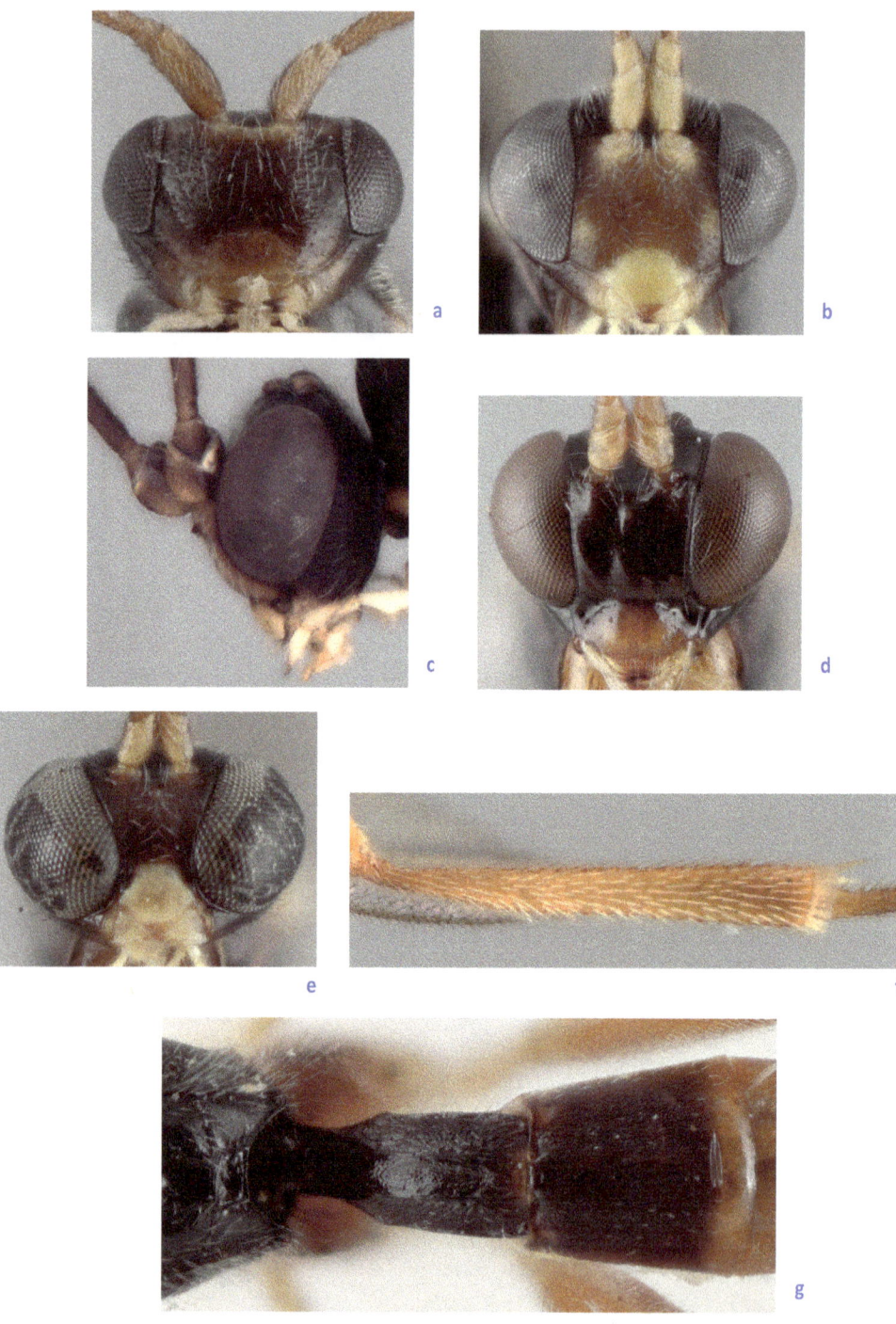

Figure 120. Morphology of Orthocentrinae: (a) *Batakomacrus caudatus* (Holmgren), head; (b) *Gnathochorisis dentifer* (Thomson), head; (c) *Megastylus cruentator* Schiødte, head, lateral; (d) *Proclitus praetor* (Haliday), head; (e) *Symplecis breviuscula* Roman, head; (f) *S. breviuscula*, hind tibia, inner aspect; (g) *Gnathochorisis crassula* (Thomson), propodeum, first and second metasomal tergites; (continued overleaf).

Figure 120 continued: Morphology of Orthocentrinae: (h) *Pantisarthrus lubricus* (Förster), metasoma; (i) *P. praetor*, first and second metasomal segments, lateral; (j) *P. praetor*, ovipositor and sheaths, tip detail inset; (k) *P. praetor*, propodeum and hind wing subbasal cell.

Subfamily ORTHOPELMATINAE

Orthopelmatinae comprises a single, Holarctic genus, *Orthopelma*. Eight described species are known (Yu *et al.*, 2012), of which two occur in Europe and both of them in Britain.

Figure 121. *Orthopelma mediator* (Thunberg).

Recognition. Orthopelmatines are small ichneumonids, readily recognised by the following combination of characters: the edge of the clypeus is concave with the labrum conspicuously exposed; the apex of the mandible is twisted so that the smaller lower tooth is hidden behind the upper tooth; the antenna has only 16-22 segments; the mesopleuron has a longitudinal furrow (above the position of a sternaulus) extending in an arc from the upper end of the epicnemial carina; the tarsal claws have a basal tooth; the first tergite is narrow, with the spiracles well in front of the midpoint and the sclerotized part of the sternite extending far behind the spiracles; and the ovipositor is slender, lacking a notch or teeth.

Systematics. Similarities between the final instar larvae of orthopelmatines and ichneumonines have led some authors to postulate a close relationship. Barron (1977) boldly affirmed two sets of relationships in different parts of his paper: Cryptinae *s.l.* + Orthopelmatinae and Ichneumoninae + Orthopelmatinae. However, there was very little evidence for either relationship, other than some convergent larval morphology (both ichneumonines and orthopelmatines are endoparasitoids) and an overall vague similarity between *Orthopelma* and some Phygadeuontinae. Subsequent phylogenetic analyses (Quicke *et al.*, 2000; 2009) have found the Orthopelmatinae to be an isolated subfamily, prompting Quicke *et al.* (2000) to establish an informal name, the orthopelmatiformes, for this subfamily, in recognition of the fact that it does not group with the other major groups of subfamilies, namely the ophioniformes, pimpliformes or ichneumoniformes.

Biology. *Orthopelma* species are parasitoids of gall-forming Cynipidae (Hymenoptera) – *Diplolepis* on *Rosa* and *Diastrophus* on *Rubus*. The rarer of the two British species, *O. brevicornis* Morley, attacks pea galls, and of the three species of *Diplolepis* which make these, *D. nervosa* (Curtis) may be its most frequent host (Blair, 1945). Only some outline details of the development of *Orthopelma* are known. The common species, *O. mediator* (Thunberg), is univoltine and adults emerge from the bedeguar galls of its regular host, *Diplolepis rosae* (L.) on *Rosa*, at about the same time as the host itself, from about mid May to mid July (Callan, 1943), although some studies have found a small second emergence in September (e.g. Rizzo & Massa, 2006). The host is attacked as a larva and *Orthopelma* is a solitary koinobiont endoparasitoid (Blair, 1945). There is no external sign of the parasitoid until about the end of September, when it has consumed the entire body contents of its host, to which it imparts its own shape. The host skin is cast off after a few days and in a few weeks, about the end of October, the fully grown *Orthopelma* larva begins to assume a prepupal form. These developments are more pronounced about April, before pupation takes place. Apparently no cocoon is spun.

Within the gall *O. mediator* itself may be subject to attack by the chalcidoid *Pteromalus bedeguaris* (Thomson) (Hymenoptera: Pteromalidae), which develops externally on the prepupa or pupa (Blair, 1945), and some torymid parasitoids have been recorded extralimitally (Nieves Aldrey, 1984). The overall inquiline and parasitoid community of *D. rosae* is extensive and has been much studied (Schröder, 1967). *Orthopelma mediator* may be amongst the first ichneumonids to have been depicted in art, in a 1600 painting by Redi of *D. rosae* galls (Pagliano *et al.*, 1997).

Identification. A key to the British species is given by Gauld & Mitchell (1977b). Callan (1943) presents some morphometric data on reared material of the two British species. There are no other described European species but there are British specimens in NMS that seem to represent an additional species.

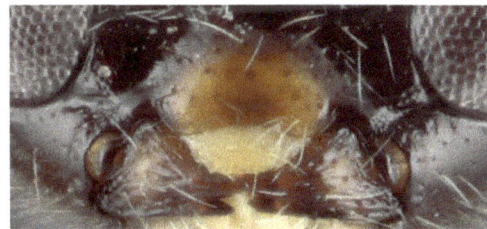

Figure 122. Morphology of Orthopelmatinae: *Orthopelma mediator* (Thunberg): (a) head; (b) mandible; (continued overleaf).

Figure 122 continued: Morphology of Orthopelmatinae: (c) mesoscutum to second metasomal tergite; (d) part of fore wing; (e) first metasomal segment, lateral; (f) ovipositor.

Subfamily **OXYTORINAE**

This subfamily has been restricted by Wahl (1990) to include only *Oxytorus*, a small genus of 23 described (and several known but undescribed) species spread across the Palaearctic and Americas (Alvarado *et al.*, 2011; Yu *et al.*, 2012; Kasparyan *et al.*, 2014; Sheng & Sun, 2014), two species of which occur in Europe, and both of them in Britain.

Figure 123. *Oxytorus luridator* (Gravenhorst) female.

Recognition. *Oxytorus* females are readily recognised by their very short, leaf-like ovipositor sheaths and elongate metasoma, in combination with their white-banded antennae, flattened clypeus, and long maxillary palps extending back far beyond the fore coxae. Males are more difficult to recognise, being easily confused with Cryptinae, Phygadeuontinae or Ctenopelmatinae (Euryproctini). They can best be recognised by the long maxillary palps and flat clypeus, combined with the absence of both a sternaulus and an apical tooth on the fore tibia.

Systematics. Other genera previously referred to the 'Oxytorinae' (or 'Microleptinae' in Townes, 1971) mainly comprise, where known, koinobiont endoparasitoids (often larva-pupal, but see Cylloceriinae) of Diptera and have been reassigned to the Microleptinae, Cylloceriinae, Orthocentrinae, Tryphoninae (for the misplaced genus *Acaenitellus* Morley) and Tatogastrinae. Wahl (1990) recognised that *Oxytorus* was not closely related to the subfamilies which he included in his pimpliformes. Quicke *et al.* (2009) found that *Oxytorus* species were associated with a polyphyletic Ctenopelmatinae, possibly belonging to the tribe Euryproctini, but *Oxytorus* has not been formally transferred to the Ctenopelmatinae as the molecular evidence is weak.

Biology. Nothing is known of the biology of *Oxytorus*. The presence of a dorsal subapical notch on the ovipositor suggests that it is an endoparasitoid (Townes in Wahl, 1990), an inference also supported by the apparent phylogenetic position of *Oxytorus* amongst a clade of koinobiont endoparasitoids. *Oxytorus* species can be abundant in woodland localities, and males of both British species are fairly commonly found in Malaise traps and light traps. Females are much less frequently trapped and seem to keep close to the ground. It is of interest that the ovipositor sheaths of the two British species are rather different, perhaps suggesting a substantial difference in host relations.

Identification. The two British *Oxytorus* (*O. armatus* Thomson and *O. luridator* (Gravenhorst)) are satisfactorily keyed by Kerrich (1939) and van Rossem (1981, 1987).

Figure 124. *Oxytorus luridator* (Gravenhorst) male.

Figure 125. Morphology of Oxytorinae, *Oxytorus luridator* (Gravenhorst): (a) head; (b) mesosternum and maxillary palps; (c) metasoma apex and ovipositor sheaths.

Gavin R. Broad, Mark R. Shaw and Michael G. Fitton

Subfamily PHYGADEUONTINAE

The British fauna of 392 species in 59 genera is a little over 40% of the European total and, even after splitting off the Cryptinae and Ateleutinae, this makes it the largest ichneumonid subfamily in Britain, just ahead of Ichneumoninae, and it is likely to contain more currently unrecognised species. Parts of the current generic classification that have focused on the European fauna are likely to need revision as and when a wider world view is taken. The generic classification of phygadeuontines of other biogeographic regions is in particular need of revision and a far smaller proportion of species have been described than is the case for Cryptinae.

Figure 126. *Mesoleptus laevigatus* (Gravenhorst).

Recognition. Most Phygadeuontinae are readily recognisable by the combination of a long sternaulus, which in the majority of species reaches the posterior edge of the mesopleuron distinctly above the postero-ventral corner (reduced in a few species); the petiolate first metasomal segment, never with a glymma, with the spiracles usually beyond the middle of the tergite; the pentagonal areolet (which may sometimes be open, i.e. when vein 3*rs-m* is absent); the ovipositor usually extending well beyond the metasomal apex at rest and with distinct teeth and a nodus; and the rather narrow (compared to many Ichneumoninae) and convex clypeus. In contrast to Cryptinae, the propodeum usually has more complete carination, including a well-defined area superomedia, and fore wing vein 2*m-cu* often has two bullae (always one bulla in Cryptinae). However, all of these characters are prone to exception and, with such a variety of genera, there are various taxa with the potential to confuse. Other than Cryptinae, males of some Ichneumoninae are the main source of potential difficulty but can generally be separated from phygadeuontines by the wider, flatter clypeus; the subtly different form of the areolet (also, in only a very few Ichneumoninae: Phaeogenini is vein 3*rs-m* absent, whereas it is fairly frequently absent in comparably sized Phygadeuontinae); the very short sternaulus (but a long sternaulus is present in *Dicaelotus* of the Ichneumoninae: Phaeogenini and *Listrodromus* of the Ichneumoninae: Listrodromini); and the often deeply impressed, often striate gastrocoelus and frequently large, wide thyridium (the latter short and usually ovoid in phygadeuontines; the former not apparent). Two genera have males that can be particularly difficult to recognise as phygadeuontines, namely *Aclastus* and *Gelis*. In both, males frequently have the sternaulus vestigial and the fore wing areolet open, with vein 2rs-m fairly steeply sloping. *Aclastus* have narrowed mandibles and a conspicuous apical fringe of setae on the inner surface of the hind tibia, inviting confusion with orthocentrines. Some *Gelis* males (macropterous males associated with apterous females) are morphologically particularly indistinct but can best be recognised as phygadeuontines by the form of the clypeus and of the first metasomal segment, and the rather large pterostigma that is proximally white (a feature of many *Gelis* males in particular).

Systematics. Our concept of Phygadeuontinae follows Santos (2017), who split Cryptinae, as used by almost all recent authors, into three subfamilies (Ateleutinae, Cryptinae and Phygadeuontinae). The classification of the ichneumoniformes assemblage is dealt with in more detail in the 'Cryptinae' chapter. Even after this reclassification, it seems that Phygadeuontinae probably represents a paraphyletic group, ancestral to Adelognathinae, Cryptinae and Ichneumoninae, and possibly Microleptinae (Laurenne *et al.*, 2006; Santos, 2017). A better understanding of the phylogeny of Phygadeuontinae would illuminate some of the factors behind radical shifts in developmental biology within this clade. Gokhman (1992) had originally argued that the subfamilies Cryptinae *sensu lato* and Ichneumoninae, and their constituent tribes, were not well defined by apomorphic characters and proposed an evolutionary pathway from the Phygadeuontinae (as a tribe of Cryptinae) to Ichneumoninae. Gokhman's hand-drawn cladogram is similar in many respects to the results of Laurenne *et al.* (2006) and Santos (2017).

As discussed further in the 'Cryptinae' chapter, three family-level names have been regularly applied to phygadeuontines: Gelinae, Hemitelinae and Phygadeuontinae. Townes (1970a) divided the Phygadeuontinae (as Gelini) into a number of subtribes, with keys to subtribes and then to genera. As with Cryptinae, this has not aided identification of phygadeuontines and the majority of subtribes are probably not monophyletic anyway (Laurenne *et al.*, 2006; Santos, 2017). We do not recognise tribes or subtribes here; nevertheless, this was a valiant attempt to impose a natural order on the group and it recognised distinct biological trends within the subfamily, such as parasitism of Diptera larvae or sawfly cocoons, as indicative of evolutionary history.

Biology.
Host associations and life histories
As with Cryptinae (to which we make frequent comparison as the subfamilies are closely related; indeed Cryptinae may be nested within Phygadeuontinae), phygadeuontines could be considered to be biologically fairly uniform in that the vast majority of species, where known, are idiobiont ectoparasitoids of more or less concealed hosts. However, the host associations of the subfamily cover a greater taxonomic range than any other subfamily and exhibit some striking evolutionary pathways in terms of niche specificity and, frequently, taxon specificity. Whilst often being involved in essentially unspecialised physiological relationships with their hosts, even to the point of predation of successive spider eggs within a sac, there seem to be very definite constraints acting on host selection, even when acceptance can encompass both primary hosts and cocooned parasitoids of those hosts. It should be emphasised, though, that very little is known about the biology of most phygadeuontines beyond a few host records that give an indication of host ranges (and most species have yet to be reared). As a very broad generalisation, Phygadeuontinae attack a very wide range of holometabolous insect pupae and prepupae (including cocooned Ichneumonoidea), spider egg sacs and Diptera larvae and puparia. However, there have been various biological off-shoots from these trends.

Whilst the Cryptinae predominate at lower latitudes, at higher latitudes the Phygadeuontinae are much more species-rich. In Britain, Phygadeuontinae are nearly three times as species-rich as Cryptinae. This diverse array of Phygadeuontinae species, most of which are relatively small, reflects use of a wide range of small cocoons and cocoon-like structures, in a manner that parallels the radiation of Pimplinae. However, narrow niche specialisation seems to have been much more important in the more extensive speciation of phygadeuontines. Some genera have probably radiated as specialised parasitoids of certain host taxa, such as *Charitopes* and *Dichrogaster* as parasitoids of cocooned Neuroptera, and *Ethelurgus* attacking Syrphidae puparia. Many species of *Endasys* are parasitoids of sawfly cocoons. In contrast, some of the larger phygadeuontine genera, such as *Gelis* and *Bathythrix*, exhibit mostly very narrow host ranges at the species level but remarkably broad host range across the genus (Schwarz & Shaw, 1999, 2010). For example, the closely related (and only fairly recently separated) species *Bathythrix formosa* (Desvignes) and *B. fragilis* (Gravenhorst), have widely divergent hosts, the former attacking spider egg sacs of the genus *Agroeca* and the latter developing on Eumeninae wasp larvae in stem nests (Horstmann, 1998a); the use of soil particles to make these aerial structures by both host groups providing a possible link. Other *Bathythrix* species are known to be parasitoids of (separately) cocooned ichneumonoids, sawflies, Lepidoptera, Coleoptera, and puparia of Diptera. Most, possibly all, *Bathythrix* species are solitary: Schwarz & Shaw (2010) did not consider the possibility that *B. alter* (Kerrich) may be a solitary pseudohyperparasitoid via gregarious pimpline ichneumonid cocoons within *Euthrix potatoria* (L.) (Lepidoptera: Lasiocampidae) cocoons, rather than being a gregarious primary parasitoid as they stated.

Figure 127. *Gelis melanocephalus* (Schrank).

Many phygadeuontines attack ichneumonoid cocoons, and can then be interpreted as pseudohyperparasitoids of the original host, usually a lepidopteran (this categorization as a form of hyperparasitoid is helpful chiefly from the viewpoint of the host's population dynamics). Some of these associations are highly specialized: as examples, different species of *Acrolyta* parasitize solitary and gregarious cocoons of microgastrine Braconidae, and seem never to be reared from the cocoons of other Ichneumonoidea (Schwarz & Shaw, 2000); *Mastrulus marshalli* (Bridgman & Fitch) may be a specialist parasitoid of the subspherical cocoons of *Scirtetes robustus* (Woldstedt) (Ichneumonidae: Campopleginae) (Schwarz & Shaw, 2010); *Phygadeuon vexator* (Thunberg) is, rather regularly, a pseudohyperparasitoid of earwigs (*Forficula auricularia* L.) via puparia of the parasitoid fly *Triarthria spinipennis* (Meigen) (Diptera: Tachinidae) (Schwarz & Shaw, 2011).

Another strategy that has evolved on many occasions is the use of spider egg sacs, which is a characteristic of many of the *Hemiteles* group, some of the *Acrolyta* group and of various *Gelis* species (as well as being found in Cryptini, and indeed in Pimplinae: Fitton *et al.*, 1987). Spider egg sacs can suffer high mortality rates from phygadeuontine parasitoids, up to 60% in one species in one study (Finch, 2005). Within the *Hemiteles* group, *Obisiphaga stenoptera* (Marshall), uniquely within Ichneumonoidea, attacks the egg sacs of a pseudoscorpion (Morley, 1907). The extremely wide host spectrum across the Phygadeuontinae as a whole also encompasses Raphidioptera under tree bark (*Tropistes*), Trichoptera cases exposed at pond edges (*Sulcarius*: Wisseman & Anderson, 1984) and gyrinid Coleoptera cocoons exposed at pond edges (*Medophron*, *Leptocryptoides* and a species of *Bathythrix*). A few extralimital species are known to parasitize the egg masses of other insects, with the African *Sozites kerichoensis* Kerrich attacking the gelatinous egg masses of the weevil *Entypotrachelus meyeri* Kolbe (Coleoptera: Curculionidae) (Benjamin & Demba, 1969), the Indian *Nipponaetes haeussleri* (Uchida) having been reared from the egg mass of *Scirpophaga excerptalis* (Walker) (Lepidoptera: Crambidae) (Sivaraman & Manickavasagam, 2011), and a Bornean *Palpostilpnus* species has been reared from eggs of Phasmida (Darling & Broad, in prep.), but whether these are acting as primary or secondary parasitoids is often uncertain. There are probably numerous other surprising host associations to be discovered; however, Diptera, Hymenoptera and Lepidoptera predominate as hosts. Whilst there seems to have been much recruitment of new host groups in some groups of Phygadeuontinae (although the relevant phylogenetic hypotheses are not available to test this), there have been large radiations in the taxa that utilise Diptera, particularly the genus *Phygadeuon*, and also *Stilpnus* and related genera. Also notably rich is the radiation of *Gelis* species, many of which have apterous females and may narrowly focus their host searches (Schwarz & Shaw, 1999). Certainly, the apterous species of *Gelis* far outnumber the macropterous species. It could be surmised that searching for hosts low down has promoted the repeated evolution of flightlessness in both Cryptinae and Phygadeuontinae, where it is much more prevalent than in other subfamilies.

Many *Gelis* species behave as pseudohyperparasitoids by parasitizing the cocoons of microgastrine braconids and other ichneumonoids that are primary parasitoids of Lepidoptera larvae (Schwarz & Shaw, 1999). Nouhuys & Hanski (2000) have demonstrated the potential for apparent competition between two primary parasitoids (a solitary *Hyposoter* (Ichneumonidae: Campopleginae) and a gregarious *Cotesia* (Braconidae: Microgastrinae)) via a shared species of *Gelis* that parasitizes both.

Developmental Biology
Although most phygadeuontine species seem to be idiobiont ectoparasitoids, there have been changes in a few lineages. The 'stilpnine' genera (*Atractodes*, *Mesoleptus* and *Stilpnus*) are presumed to all be koinobiont endoparasitoids of Diptera, ovipositing into the larva, as is also the case in at least a few (though not all) species of *Phygadeuon* (Rotheray, 1988; Schwarz & Shaw, 2011) also attacking Diptera. Apparent idiobiont endoparasitism in Lepidoptera pupae is known in several genera, such as *Isadelphus*, *Mastrus* and *Zoophthorus*, that were classified by Townes (1970a) in his Mastrina (Schwarz & Shaw, 2010). Larvae of *Gelis vicinus* (Gravenhorst) (formerly *Blapsidotes vicinus*: Schwarz, 2016) develop more or less endophagously as gregarious broods in the naked pupae of butterflies (Aubert, 1955; Blunck & Janssen, 1957).

Biological information on the development of koinobiont phygadeuontines is sketchy. The most reliable and complete account relates to an unidentified species of *Mesoleptus* which is a solitary parasitoid of *Sarcophaga nigriventris* Meigen (Diptera: Sarcophagidae) feeding in dead snails (*Cepaea nemoralis* (L.)) (Beaver, 1972). The female *Mesoleptus* searches for and detects the larvae of the fly by probing the decaying body of the snail with her ovipositor. The antennae are not used at this stage and neither are the tarsi important, as females were frequently observed to stand on a host apparently without detecting it. The larva may attempt to escape attack by wriggling movements when touched by the ovipositor, and its thick cuticle is not easily pierced. Fairly mature fly larvae are selected for oviposition and they continue to develop normally and to form a puparium (which is of a similar size to those formed by unparasitized larvae). The parasitoid larva pupates within the host puparium and the adult *Mesoleptus* escapes by biting its way out at the anterior end. There are possibly several generations during the summer and the winter is passed within a host puparium (perhaps as a fully grown larva). Another interesting fact noted by Beaver was that this particular *Mesoleptus* species did not attack two other species of *Sarcophaga* found in the dead *C. nemoralis*, even when they were present in the same snail as *S. nigriventris*. Other apparently reliable host records for *Mesoleptus* are from Sarcophagidae and Sciomyzidae (Jussila *et al.*, 2010). Some *Stilpnus* are larva-pupal endoparasitoids of muscid flies and have been studied for the possible biological control of synanthropic pest Diptera (Legner, 1995).

Figure 128. *Gelis agilis* Fabricius (J. Voogd).

Figure 129. *Acrolyta*(?) sp. ovipositing in *Microplitis ocellatae* (Bouché) (Braconidae: Microgastrinae) cocoons ex *Smerinthus ocellatus* (L.) (M. Boddington).

Like other idiobiont groups, Phygadeuontinae are probably mainly synovigenic, as confirmed for a few Phygadeuontinae (e.g. Harvey *et al.*, 2009). Host location has been little-studied. The importance of host allomones in locating hosts of the correct stage (prepupae in this case) was shown for *Mastrus ridibundus* (Gravenhorst), a gregarious parasitoid of the Codling Moth, *Cydia pomonella* (L.) (Lepidopera: Tortricidae). Jumean *et al.* (2005) demonstrated that effectiveness of host location depended on a complex combination of 11 chemicals, the deletion of more than three of which drastically reduced the ability of the parasitoid to locate hosts. One of the more detailed studies is that of van Baarlen *et al.* (1996) who looked at the host searching behaviour of *Gelis festinans* (Fabricius), a parasitoid of egg sacs of the linyphiid spider *Erigone atra* (Blackwall). They found that *G. festinans* females responded positively to long-range habitat cues (vegetation odours) and short-range host cues (spider webbing). Despite being essentially predatory on spider eggs within the egg sac, both wild-caught and laboratory-reared *G. festinans* responded only to the webbing of *E. atra*, ignoring several other linyphiid species present. Contact chemicals, or physical properties of the silk, seemed to be involved in the parasitoid's detection and acceptance of egg sacs, not any long range cues; egg sacs placed away from webbing were parasitized only if encountered by chance. Females were also able to avoid searching on pre-searched areas and areas searched by other females, and thus avoided superparasitism. By contrast, some other *Gelis* species, such as *G. agilis* (Fabricius) and *G. areator* (Panzer), parasitize an extraordinarily wide range of insects (Schwarz & Shaw, 1999), with the apterous *G. agilis* and the macropterous *G. areator* exploiting ranges of hosts in different niches. However, a particular *Gelis* species either attacks spider egg sacs or cocooned insect hosts, not both. In a study on two commonly reared pseudohyperparasitoids of *Cotesia glomerata* (L.) (Braconidae: Microgastrinae) cocoons, Harvey & Witjes (2005) found that success of the niche generalist parasitoid *Gelis agilis* was much lower than that of the more specialised *Lysibia nanus* (Gravenhorst), the result of lower emergence rates and lowered fecundity; further evidence of constraints acting upon parasitoids in ways that are not obvious from a simple perspective of, for example, koinobiosis versus idiobiosis. In the case of *Lysibia* parasitizing *Cotesia*, Harvey *et al.* (2004) demonstrated that developmental constraints arose from the identity of the *Pieris* butterfly primary host, presumably as a result of differences in the glycoside toxins present in the caterpillars, via their food plants. Two specialist parasitoids of *Cotesia* cocoons, *L. nanus* and *Acrolyta nens* (Hartig), proved to be similar in various life history traits and seemed to share an abundant resource (Harvey *et al.*, 2009; see also Schwarz & Shaw, 2000), often emerging from the same cocoon clusters.

Identification. For genus-level identification, the only reasonably comprehensive treatments are Townes's (1970a) key to world genera and the provisional treatment of the Russian fauna by Jonaitis (in Kasparyan, 1981), although unfortunately the former is difficult to use and the latter is in Russian, and neither has kept pace with generic splits. Easier to use, but incomplete for the British fauna, is the generic key in Townes (1983). References are provided below for the recognition of genera omitted by Townes (1970a, 1983). Horstmann (1978b) gives a revised key to the 'Mastrina' genera, which supersedes that of Townes (1970a). Horstmann (1993a) provides keys to all the known European brachypterous species and Schwarz (1995) keys the genera with apterous species. Revisions of genera and species-level keys are scattered and often in German. Although an increasing number of species are included in these keys there are still some significant gaps, particularly in large and commonly collected genera such as *Mastrus* and *Phygadeuon*. The available works which cover the British fauna are listed below, alphabetically by genus; in some cases there have been subsequent changes in nomenclature, traceable through Broad (2016).

Aclastus (Horstmann, 1980b); *Acrolyta* (Schwarz & Shaw, 2000); *Agasthenes* (Horstmann, 1998a); *Amphibulus*, single British species included in Sawoniewicz (1990) and Luhman (1991); *Arotrephes* (Horstmann, 1995); *Atractodes* (Jussila, 1979, 2001); *Bathythrix* (Sawoniewicz, 1980); *Cephalobaris*, notes on the single species provided by Horstmann (2012a); *Ceratophygadeuon* (Townes, 1983; Horstmann, 1993b); *Charitopes* (Townes, 1983; Horstmann, 1998a); *Chirotica*, single British species included in Horstmann (1983); *Clypeoteles*, monotypic genus keyed by Horstmann (1978); *Cremnodes* (Horstmann, 1992); *Dichrogaster* (Townes, 1983; Horstmann, 1992); *Encrateola* (Horstmann, 1998a); *Endasys* (Sawoniewicz & Luhman, 1992); *Ethelurgus* (Horstmann, 2000a); *Eudelus*, some species separated by Schwarz & Shaw (2000); *Fianoniella* (Horstmann, 1998a); *Gelis*, apterous females (Schwarz, 2002b), brachypterous females (Horstmann, 1993a; Schwarz, 1994); macropterous species, including *G. vicinus* (Gravenhorst), formerly in *Blapsidotes* (Schwarz, 2016; see also Horstmann, 1986); *Glyphicnemis* (Sawoniewicz, 1985); *Gnotus* (Horstmann, 1993c); *Grasseiteles* (as *Diaglyptellana*) (Schwarz & Shaw, 2000); *Hemiteles* (Schwarz & Shaw, 2000); *Holcomastrus*, single species redescribed in Horstmann's (2012a) description of the genus; *Isadelphus* (Horstmann, 2009b); *Lochetica*, single British species included in Townes (1983); *Lysibia* (Townes, 1983); *Mastrulus*, monotypic genus keyed by Horstmann (1978); *Mastrus*, *rufobasalis* and *pictipes* species-groups (Horstmann, 1990b); *Medophron*, taxonomic notes by Horstmann (1998a); *Megacara* (Townes, 1983); *Mesoleptus* (Jussila *et al.*, 2010); *Micromonodon*, included in Horstmann's (1978) generic key; *Neopimpla*, single British species included in Schwarz & Shaw (2000); *Odontoneura*, included in Horstmann's (1978) generic key; *Orthizema*, macropterous females (Schwarz & Shaw, 2011), brachypterous females (Horstmann, 1993a); *Phygadeuon*, brachypterous females (Horstmann, 1993a), '*Iselix*' species-group (Horstmann, 2001a), species easily confused with *Theroscopus* (Schwarz & Shaw, 2011); *Platyrhabdus* (Horstmann, 1998a); *Pleurogyrus* (Horstmann, 1995); *Pygocryptus*, single British species included in Townes (1983); *Stibeutes* (Horstmann, 2010a; additional species described by Schwarz & Shaw, 2011); *Stilpnus* (Jussila,

1987; 1999); *Sulcarius* (Townes, 1983, additional species described by Horstmann, 1992); *Thaumatogelis* (Schwarz, 2001); *Theroscopus*, brachypterous females (Horstmann, 1993a), macropterous females (Schwarz & Shaw, 2011); *Tropistes* (Schwarz & Shaw, 2011); *Uchidella* (Horstmann, 1993c); *Xenolytus* (Townes, 1983); *Xiphulcus*, single British species included in Townes (1983); *Zoophthorus*, species groups (Horstmann, 1978).

Schwarz & Shaw (1999, 2000, 2010, 2011) summarise the distribution, hosts and taxonomy of British Phygadeuontinae (as Phygadeuontini), and include some keys to species as well as descriptions of several new species. However, the genera *Endasys*, *Atractodes*, *Mesoleptus* and *Stilpnus* (the latter three comprising the distinctive *Stilpnus* genus-group) were not included.

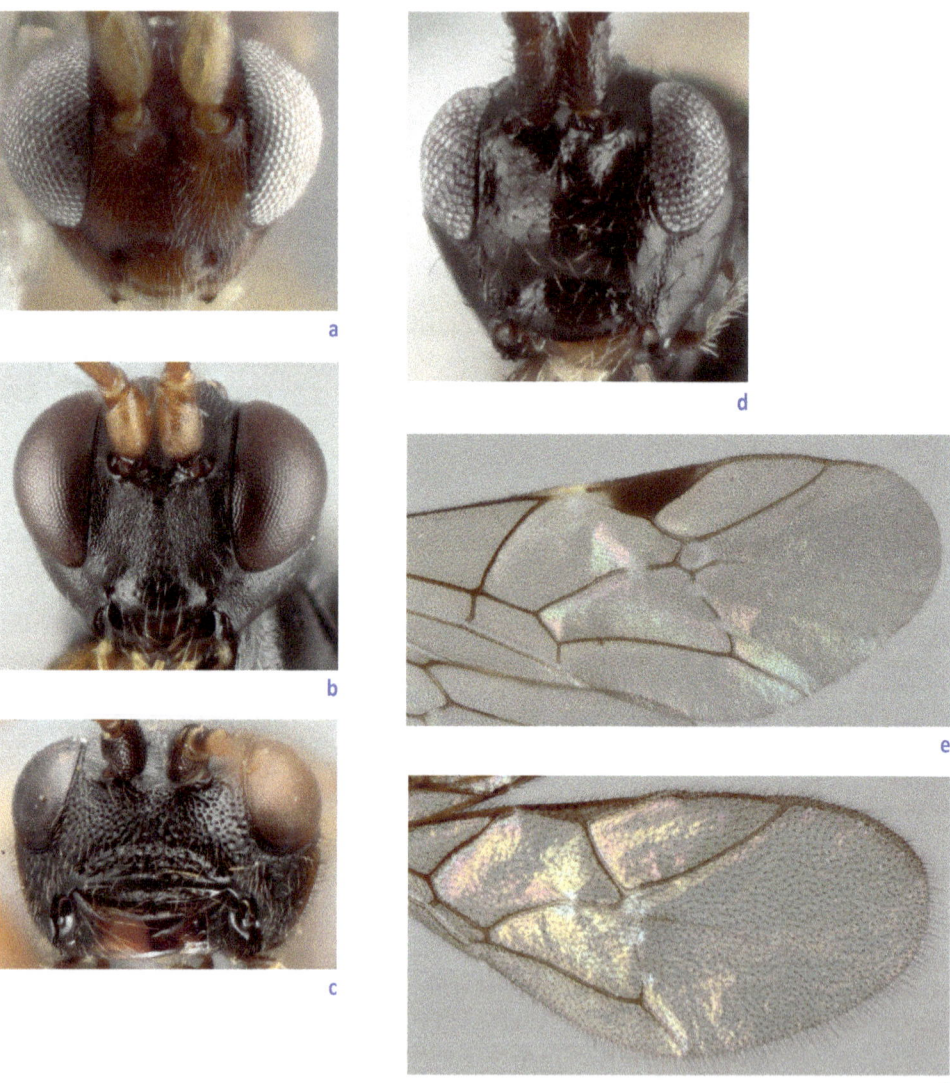

Figure 130. Morphology of Phygadeuontinae: (a) *Aclastus solutus* (Thomson), head; (b) *Gelis* sp., head; (c) *Phygadeuon flavimanus* Gravenhorst, head; (d) *Polyaulon paradoxus* (Zetterstedt), head; (e) *Gelis* sp., male, part of fore wing; (f) *P. paradoxus*, male, part of fore wing; (continued overleaf).

Figure 130 continued: Morphology of Phygadeuontinae: (g) *Gelis rufogaster* Thunberg, female mesosoma; (h) *Gelis* sp., male mesosoma; (i) *Dichrogaster liostylus* (Thomson), mesosoma; (j) *P. flavimanus*, mesosoma to first metasomal segment, lateral; (k) *P. flavimanus*, mesopleuron, sternaulus highlighted; (l) *Phygadeuon cephalotes* Gravenhorst, propodeum; (m) *Aclastus eugracilis* Horstmann, ovipositor sheaths; (n) *A. eugracilis*, ovipositor tip; (o) *D. liostylus*, ovipositor tip.

Gavin R. Broad, Mark R. Shaw and Michael G. Fitton

Subfamily **PIMPLINAE**

Tribes
Delomeristini
Ephialtini
Pimplini
Theroniini

Pimplinae is a moderately large subfamily, well represented in both temperate and tropical zones, currently comprising about 1,700 described species in 79 genera. Because of the relatively large size of many species, and their often conspicuous habits, they tend to be well-represented in insect collections. All four tribes occur or occurred in Britain, and there are currently 111 species representing 32 genera on the British list, of a European fauna of about 205 species in 37 genera. The highest known species-richness per unit area is in the neotropics (Gaston & Gauld, 1993; Sääksjärvi et al., 2004; Gómez et al., 2017).

Figure 131. *Perithous septemcinctorius* (Thunberg), tribe Delomeristini.

Recognition. Although several large pimplines, such as *Pimpla rufipes* (Miller) and species of the genus *Dolichomitus*, are relatively familiar insects to entomologists and sometimes feature in field guides, there is no simple combination of characters that diagnoses the Pimplinae. Generally, pimplines are either robust, with the ovipositor exserted by less than the length of the metasoma, or elongate, with long or very long ovipositors; some of the small species parasitising spiders (the *Polysphincta* group), however, do not fall into either category. One of the most useful recognition features, and an apomorphy of the subfamily (Gauld et al., 2002b), is that the tergites have impunctate, posterior bands, in contrast to the usually heavily sculptured remainder of the tergites. Most pimplines have a flattened clypeus, which is distinctively medially apically emarginate in the tribes Delomeristini and Ephialtini. Pimplines frequently have oblique grooves on at the second and third tergites. Males have a characteristically elongate last metasomal sternite. The metasoma of almost all pimplines is approximately parallel-sided to the base, the ovipositor always lacks a notch, and in the majority of species there are conspicuous teeth on the ventral valves and often a raised nodus on the upper valve. In the vast majority of pimplines, the first metasomal segment is stout and sessile, i.e. not petiolate. The female tarsal claws frequently have a basal, sharply angulate lobe and the fore wing areolet, usually, is obliquely rhombic. *Pseudorhyssa* can be distinguished from all other British ichneumonids by a combination of the transversely rugose mesoscutum in conjunction with the complete occipital carina (which separates it from Rhyssinae). Note that a few non-ichneumonoid Hymenoptera (*Ibalia* Latreille (Ibaliidae) and *Aulacus* Jurine (Aulacidae)) that similarly parasitize wood-boring hosts share the first feature.

The *Polysphincta* group are decidedly aberrant compared to other pimplines, with simple, needle-like ovipositor tips, frequent lack of fore wing vein 3*rs-m* and with tergites often not heavily sculptured. Females can be recognised by the swollen fifth segment of their tarsi and lobed claws. Males, and females of *Acrodactyla* and *Megaetaira*, are frequently confused with Orthocentrinae and Phygadeuontinae (especially the genus *Aclastus*), but can be separated from the former by the lack of the apical fringe of setae on the inner side of the hind tibia and the broader mandibles, and from the latter by the wing venation (vein 3*rs-m* missing, 2*rs-m* short) and lack of a sternaulus. Unlike phygadeuontines, the propodeum of the *Polysphincta* group lacks the anterior transverse carina and is elongate anteriorly, so the median longitudinal carinae (when present) are long. A useful recognition feature for *Acrodactyla* is the dorsally projecting crest at the anterior end of the notaulus.

Figure 132. *Pseudorhyssa alpestris* (Holmgren), tribe Delomeristini.

Systematics. The "Pimplinae" was one of the traditional five subfamilies into which the Ichneumonidae was divided, but in that concept it included a welter of extraneous groups. Even after the classification of the Ichneumonidae was overhauled by Townes (1969-1971), the Pimplinae as defined by Townes (1969) was still a diverse group defined by plesiomorphic characters. Eggleton (1989), in an unpublished thesis, refined the classification of the pimplines and related groups, proposing monophyletic subfamilies, namely the Diacritinae, Pimplinae, Poemeniinae and Rhyssinae. Eggleton's (1989) work formed the basis of Gauld's (1991) formal reclassification of the 'lower pimpliformes'. Wahl & Gauld's (1998) cladistic analysis of the pimpliformes clade resulted in hypotheses of relationships between the pimpliformes groups, including the definition of a monophyletic Pimplinae. Wahl & Gauld (1998) proposed that the Pimplinae be classified in four tribes, later reduced to three by Gauld et al. (2002b) with the merging of Delomeristini and Perithoini. Gauld et al.'s (2002b) system recognises three putatively monophyletic tribes, Delomeristini, Ephialtini and Pimplini, and a series of informal genus-groups within the Ephialtini and Pimplini. This system differs in numerous details from Townes's (1969) classification as it recognises that the *Polysphincta* group comprises a specialised group of Ephialtini genera, rather than a separate tribe (Polysphinctini), and the heterogeneous Delomeristini was split, with the *Theronia* group transferred to the Pimplini. The phylogeny and classification of the genera of Pimplinae have been addressed by Gauld *et al.* (2002b) and, dealing with the *Polysphincta* group, Gauld & Dubois (2006) and Matsumoto (2016). The resulting hypotheses regarding the classification of pimplines are thus unusually thorough for an ichneumonid subfamily, and based on cladistic principles. However, many clades within these morphology-based phylogenies are supported by rather few, and often homoplastic, morphological characters. Molecular phylogenetic studies (Quicke *et al.*, 2000, 2009) have so far relied on 28S rRNA, which has very little phylogenetic signal in pimpliformes. Klopfstein *et al.* (in press), using many more genes, have also found low rates of sequence divergence in basal branches of the pimpliformes, although there are some stongly supported results that resulted in some changes to the tribal classification of Pimplinae: the *Theronia* group is recognized as a separate tribe, Theroniini, and *Pseudorhyssa* is transferred from Poemeniinae to the pimpline tribe Delomeristini. Together with the rather weak differentiation of most genera of Pimplinae, this is all evidence that the pimpliformes, and Pimplinae, underwent a rapid radiation. Much of the initial diversification of pimpline lineages probably took place in the northern hemisphere (Wahl & Gauld, 1998: Gauld *et al.*, 2002b); generic richness (but not necessarily species richness) of the pimpliformes, and especially of the subfamilies Pimplinae, Poemeniinae, Rhyssinae and Acaenitinae, is considerably higher in the Palaearctic and Oriental regions than in the Nearctic and Neotropics.

Figure 133. *Endromopoda detrita* (Holmgren), tribe Ephialtini.

Because of Henry Townes's differing views on the type species of certain ichneumonid genera (Townes, 1969), particularly that of *Ephialtes*, in defiance of the imperfect ICZN Opinion 159 (1945), Townes and various other workers applied some well-known names in Pimplinae, namely *Apechthis*, *Ephialtes* and *Pimpla*, differently from us and most other authors. For an explanation see Wahl & Mason (1995). However, usage of genus names as applied here is now stable.

Figure 134. *Schizopyga frigida* Cresson, tribe Ephialtini.

Figure 135. *Itoplectis maculator* (Fabricius), tribe Pimplini.

Biology.
Host associations and life histories

The Pimplinae, even as now restricted, is probably biologically more diverse than any other subfamily, although rivalled by the much larger Phygadeuontinae. A very wide range of hosts are attacked and the subfamily includes ecto- and endoparasitoids, idiobionts and koinobionts. Species of Delomeristini and Ephialtini are all ectoparasitoids whilst most Pimplini are endoparasitoids (the exception being, supposedly, a few extralimital species of the *Theronia* genus-group). All but the *Polysphincta* group of the Ephialtini are idiobionts; the *Polysphincta* group are koinobiont ectoparasitoids, a trait shared in the Ichneumonidae only by the Tryphoninae and at least some species of Adelognathinae (functionally, the Lycorininae are ectoparasitic koinobionts too, with oviposition and at least initial larval development taking place in the anus; Shaw, 2004). The account of pimpline host relations that follows is summarised largely from Fitton et al. (1988) and Shaw (2006b). Gauld et al. (2002b) also provide an extensive discussion on the likely evolutionary history of life history transitions in the Pimplinae, including non-British species-groups.

There are few reliable host records for Delomeristini but what we do know suggests a host range centred on Hymenoptera; *Perithous* species are parasitoids of stem-nesting aculeates, especially of Crabronidae, whilst *Delomerista* have been reared from sawfly and ichneumonoid cocoons. Best known biologically are *Pseudorhyssa* species, which are cleptoparasitoids of rhyssine Ichneumonids. In Britain, *Pseudorhyssa alpestris* (Holmgren) has been reared as a cleptoparasitoid from the alder woodwasp *Xiphydria camelus* (L.) (Hymenoptera: Xiphydriidae), which it attacks by means of the drill holes made by *Rhyssella approximator* (Fabricius). The *Pseudorhyssa* female inserts her ovipositor down the hole created by the ovipositing *Rhyssella* and lays an egg near the *Rhyssella* egg. The first instar *Pseudorhyssa* larva has much larger mandibles than *Rhyssella* and almost invariably wins the contests between them and so kills the *Rhyssella* before consuming the *Xiphydria* larva (Skinner & Thompson, 1960). The non-British *P. nigricornis* (Ratzeburg) cleptoparasitises *Rhyssa persuasoria* (L.) and was studied by Spradbery (1968) (as *P. sternata*); the details related here are derived from that study but the interactions between *P. alpestris* and *Rhyssella approximator* appear to be similar (Skinner & Thompson, 1960). The *Pseudorhyssa* female locates drill holes by detecting secretions from the *Rhyssa* female; it could paralyse the host siricid larva under experimental circumstances whereby an active siricid larva was placed below a *Rhyssa* boring being used by the *Pseudorhyssa*, but generally it will oviposit on an already paralysed host. The *P. nigricornis* larva moves across the body of the siricid larva using a pair of caudal appendages as a prop. Eggs of *Rhyssa* are not attacked but when the smaller *Rhyssa* larva is encountered it is killed and some feeding on the mangled *Rhyssa* larva takes place before consumption of the siricid larva. Mature *Rhyssa* larvae can be attacked and development can then take place on the rhyssine rather than the (mostly consumed) siricid larva. There are five larval instars and development takes on average 12.5 days from first to fifth instar. Feeding continues for three to five days in the final instar. The cocoon lines the host chamber and *P. nigricornis* overwinters as a larva before pupating in the spring and then emerging by chewing through the wood. The non-British *Atractogaster semisculptus* Kriechbaumer has been recorded from xylophagous Coleoptera (Kasparyan, 1987) but possibly as a substrate rearing without certainty that the true host was not a hymenopteran which had parasitised the beetle.

Figure 136. *Itoplectis aterrima* Jussila host-feeding from a pupa of *Hyposoter* sp. (Ichneumonidae: Campopleginae) (M.R. Shaw).

The hosts of Ephialtini span a wide range of taxa although individual species usually have narrowly focused host ranges, which can encompass widely divergent host taxa but within a particular searching niche. Considering the two British genus groups of Gauld *et al.* (2002b) in turn (three other small genus groups are extralimital), the hosts of the *Ephialtes* group primarily comprise insect larvae concealed in plant tissue, the genera with conspicuously long ovipositors apparently forming a monophyletic group attacking particularly deeply concealed hosts. *Dolichomitus* and *Flavopimpla* are parasitoids of wood-boring or gall-forming Coleoptera and Sesiidae (Lepidoptera). *Ephialtes* and *Townesia* are closely related genera (Gauld *et al.*, 2002b) which specialize on solitary aculeate nests. *Exeristes* and *Liotryphon* are specialised on Lepidoptera under bark, in pine cones and resinous structures, fruit, wood, etc. *Paraperithous gnathaulax* (Thomson) has several times been reared from honey bee hives where it was believed to be parasitising galleriine pyralids, and also from bracket fungi harbouring the large tineid *Morophaga choragella* (Denis & Schiffermüller); a disparate host range which merits further attention, especially as the sometimes large size of the adult makes it unlikely that it has used a single individual of such hosts. The genera of the *Ephialtes* group with shorter ovipositors comprise *Acropimpla*, *Endromopoda*, *Fredegunda* and *Scambus*, with *Gregopimpla* and *Iseropus* very similar in appearance but apparently being basal members of another, related lineage. However, *Acropimpla didyma* (Gravenhorst) is a gregarious parasitoid of flimsily cocooned prepupae or pupae of Lasiocampidae and this habit, together with the male's yellow face, seem to indicate an affinity with members of the *Sericopimpla* group such as *Iseropus* and *Gregopimpla* (see later). *Fredegunda diluta* (Ratzeburg) and *Endromopoda* are associated with hosts in (mainly) Poaceae stems. Both *Endromopoda* and *Fredegunda* may be derived from within *Scambus* whilst *F. diluta* is almost certainly derived from within *Endromopoda* (Gauld *et al.*, 2002b). *Scambus* host associations as a whole are wide, spanning Coleoptera, Diptera, Hymenoptera and Lepidoptera, although species seem to be habitat and/or niche specialists, with different species parasitising similar leaf spinning hosts in the field layer or on trees, and others specialising on hosts in flower and seed heads, or in stems. All *Scambus* species seem able to develop facultatively as either primary or secondary parasitoids, in the latter case most often in ichneumonoid primary parasitoid cocoons. *Scambus* species attacking Coleoptera and Hymenoptera hosts seem to have the narrowest primary host ranges. *Scambus calobatus* (Gravenhorst) has two morphologically distinct generations per year, being reared in spring from *Curculio* (Coleoptera: Curculionidae) and *Cydia* (Lepidoptera: Tortricidae) larvae in fallen acorns (*Quercus*) and then developing on *Acrobasis* (Lepidoptera: Pyralidae) larvae, or their primary parasitoids, in webs on oaks in late summer (Shaw *et al.*, 2011). Ecological separation of closely related species can be seen in the *S. inanis* (Schrank), *S. signatus* (Pfeffer), *S. tenthredinum* (Goeze) complex of very closely related species: *S. tenthredinum* has been reared only from the stem-galling sawfly *Euura amerinae* (L.) (Hymenoptera: Tenthredinidae); *S. inanis* attacks small, arboreal lepidopterous hosts, especially Gracillariidae and Tortricidae, whereas *S. signatus* targets lepidopterous hosts in the field layer (Horstmann, 2005; Shaw, 2006b). The three species are very difficult to separate on morphological characters. *Scambus pomorum* (Ratzeburg) is of interest in parasitising larvae of the weevil *Anthonomus pomorum* (L.) feeding in apple (*Malus*) blossom in spring, the resulting adult females, which emerge in early summer, then feeding voraciously on various small Lepidoptera larvae before overwintering as adults (Zijp & Blommers, 2002; see also Shaw, 2006b).

Figure 137. *Theronia atalantae* (Poda), tribe Theroniini.

The *Sericopimpla* group comprises two clades, one of usually gregarious parasitoids of cocooned Lepidoptera and the second clade attacking spider egg sacs and, in the specialised *Polysphincta* group, active spiders; the *Polysphincta* group is nested within the group associated with spider egg sacs (*Clistopyga*, *Tromatobia* and *Zaglyptus*) (Gauld *et al.*, 2002b). The British representatives of the first clade, *Gregopimpla inquisitor* (Scopoli) and *Iseropus stercorator* (Fabricius), are gregarious parasitoids of prepupal and pupal cocooned Lasiocampidae, Lymantriinae (Erebidae) and probably some Noctuidae. *Sericopimpla*, not found in Europe, are parasitoids of Psychidae. There is thus a clear association with hosts that spin silk, offering a link between parasitoids of Lepidoptera, spider egg sacs and then active spiders, as was suggested by Townes (Townes *et al.*, 1960; Townes, 1969) and subsequently supported by phylogenetic studies (Wahl & Gauld, 1998; Gauld *et al.*, 2002b). *Clistopyga*, *Tromatobia* and *Zaglyptus* are mainly predators, or pseudo-parasitoids, in spider egg sacs and, in the case of at least some *Zaglyptus*, also the adult female spider, which is stung to death within its nest (Nielsen, 1935). However, these genera can more conveniently be reconciled as parasitoids if the host egg-nest is regarded as a single host item. Host ranges are generally rather poorly known but *Tromatobia* species attack exposed egg sacs, especially of Araneidae, whilst *Clistopyga* and *Zaglyptus* species attack concealed egg sacs and nests; in Britain especially of Clubionidae though the commonest *Clistopyga* (*C. incitator* (Fabricius)) is a regular parasitoid of Segestriidae. Frítzen & Sääksjärvi (2016) demonstrated that an unidentified species of *Clistopyga* (the genus is apparently the sister-group to the *Polysphincta* group: Gauld *et al.*, 2002b, but see Matsumoto, 2016), is an ectoparasitoid of adult salticid spiders within their nests, paralyzing the host. The remaining clade of the *Sericopimpla* group is the highly specialised and distinctive *Polysphincta* group, more simply called the polysphinctines. Where the life history is known, all polysphinctines are parasitoids of active spiders, with a correspondingly pointed, needle-like ovipositor tip and distinctively enlarged fifth tarsal segments, which are involved in manouvering around spider webs. For many British species we have a fairly good idea of host ranges following extensive rearing efforts (Shaw, 1994). Most polysphinctine species seem to have rather narrow host ranges, attacking a few congeneric spider species, or species within closely related genera. A clade of polysphinctine genera representing putatively ancestral host associations (Matsumoto, 2016) place the egg on the cephalothorax whereas the majority of genera place the egg at the front of the abdomen or further back. The mode of oviposition, and use of the ovipositor, also differs substantially between these two groups (Takasuka *et al.*, 2018). The most frequent host groups are Araneidae, Clubionidae, Linyphiidae, Tetragnathidae and Theridiidae. The rarely collected genus *Piogaster* comprises parasitoids of Salticidae (Takasuka *et al.*, 2018).

Figure 138. Pupa of female *Dolichomitus* sp. (Pimplinae: Ephialtini) found under bark of dead *Quercus* (M.R. Shaw).

Unlike Delomeristini and Ephialtini, Pimplini are endoparasitoids of Lepidoptera pupae or of the pupae or prepupae of ichneumonoid primary parasitoids. Host ranges are generally broad, often comprising any host pupa of a suitable size within a particular searching niche. *Apechthis* and *Itoplectis* species (the former may represent a specialised lineage of the latter: Gauld *et al.*, 2002b), attack a variety of exposed or weakly concealed 'microlepidopteran' pupae (most commonly Tortricidae). Some *Apechthis* are also frequent parasitoids of butterfly pupae (Aubert, 1969; Shaw *et al.*, 2009) whilst *Itoplectis* are frequently pseudohyperparasitoids, in some species perhaps obligatorily so, in which case development is as an endoparasitoid of the cocooned primary parasitoid (Shaw, 2009). The down-turned ovipositor tip of *Apechthis* is used rather like a crowbar, inserted between the overlapping plates of often wriggly pupae (Cole, 1967). Most *Pimpla* species seem to attack more concealed hosts than *Apechthis* or *Itoplectis*, sometimes in soil or stems or crevices in tree bark, though the largest British species, *P. rufipes* (Miller), is often reared from exposed butterfly pupae. There are occasional reliable host records from Coleoptera and Diptera; indeed experimental cultures of some *Pimpla* species have been maintained on mealworms.

Figure 139. *Acrodactyla* sp. larva feeding on *Tetragnatha* sp. (Araneae: Tetragnathidae) (M.R. Shaw).

The single British (although probably extinct here) representative of the Theroniini, *Theronia* group, *Theronia atalantae* (Poda), is a parasitoid of medium-sized to large Lepidoptera pupae; although sometimes regarded as a primary parasitoid its pseudohyperparasitoid role is well-known, and it may function in that capacity exclusively. Some related non-British species, such as *Theronia maculosa* Krieger and *Neotheronia tacubaya* (Cresson), are believed to be always pseudohyperparasitoids, and apparently ectoparasitoids, of ichneumonid primary parasitoids within Lepidoptera cocoons (Gauld et al., 2002b); however, we have seen no published evidence that these species are indeed ectoparasitoids and it is possible that this assumption has been made on the basis of their being pseudohyperparasitoids, which in some Pimplini does not correspond with ectoparasitism (e.g. Shaw, 2009), although it should be noted that Theroniini may be more closely related to the Delomeristini ectoparasitoids than the Pimplini endoparasitoids (Klopfstein et al., in press).

In summary, many pimplines are niche-specialists, with several examples of niche separation of closely related species. Idiobiont development, in which the host is essentially immobilised to a state of fresh meat, means that pimpline development can very frequently take place on either the primary host or a parasitoid of that host or, indeed, on insects of other orders that are encountered. The basic requirements as larvae of most species render them relatively easy to culture in the laboratory, sometimes on unnatural surrogate hosts (*Pimpla* on *Tenebrio* pupae, Sandlan, 1980) or even on completely artificial diets (e.g. *Itoplectis*, House, 1978). One Costa Rican *Calliephialtes* species is exceptional in being facultatively phytophagous, sometimes completing its development, presumably when its host gall-forming lepidopteran larva is too small, by feeding on the gall tissue (Nishida in Gauld *et al.*, 2002b), which could also potentially be the case in at least one species of the closely related genus *Scambus*, in which the host sawfly often seems too small to have supported development of *Scambus vesicarius* (Ratzeburg) (Shaw in Gauld *et al.*, 2002b).

Developmental biology
Like most Hymenoptera, adults of many (probably all) species depend for their activity on an intake of carbohydrate, and feed on honeydew, nectar (Leius, 1960; Cole, 1967) and sometimes other plant secretions (Juillet, 1959). As is usual for synovigenic parasitoids, which produce a succession of fully yolked eggs but normally emerge as adults with relatively undeveloped ovaries, they also feed on haemolymph of larval insects. Host-feeding can be on non-host as well as potential host larvae (e.g. *Scambus pomorum*, see above), and can include prepupae and pupae of primary parasitoids (Shaw, 2009). Larvae that are fed on by the adult pimpline are not subsequently used for oviposition. Destructive host-feeding can be responsible for a more significant portion of host mortality than parasitism alone, and in one North American species of *Itoplectis*, regular mutilation and host feeding attacks on pupae of the introduced moth *Lymantria dispar* (L.) (Erebidae: Lymantriinae) have been found to cause up to 200 times the mortality that results from actual parasitism (Campbell, 1963). When host-feeding, the ovipositor is inserted deep in the pupa and moved around in a churning motion, to break up part of the prey's body; the hole may be widened by the mandibles (Jackson, 1937; Shaw, 2009). There have been several laboratory studies on Pimplini and some Ephialtini demonstrating that both host-feeding and intake of carbohydrates is essential (Leius, 1961a,b; Sandlan, 1979a; Osman, 1978). There is nothing known regarding the nature and frequency of food intake in those Ephialtini which attack hosts deeply concealed in wood. Although destructive host-feeding is essential for females of at least some species, nutritional needs of males can be met by sugars alone (Fox, 1927). The non-British, but European, *Exeristes roborator* (Fabricius) has been seen to use its ovipositor to spear otherwise inaccessible larvae to bring them up to the mouthparts for feeding on in rearing cages (Jackson, 1937). A *Clistopyga* species has also been observed to use its ovipositor, which is upcurved and furnished with small teeth, to limit the movements of an evenomed host spider and then physically move the paralysed spider (Fritzén & Sääksjärvi, 2016). The ovipositor tip in this species also functions as the equivalent of a felting needle, in picking silk threads up on the upstroke, thus closing up the hole made for oviposition in the spider's egg sac.

Although essentially idiobionts, Pimplini do not paralyse the host at oviposition (e.g. Shaw, 2009) but immobilisation occurs 20-24 hours post-oviposition, as is especially striking with the jumping cocoons of *Scirtetes robustus* (Kriechbaumer) (Ichneumonidae: Campopleginae) which continue to be capable of jumping after oviposition by *Itoplectis clavicornis* (Thomson) (Stelfox, 1929). Although Jackson (1937) reports that host immobilisation occurs before the parasitoid larva emerges from the egg, Führer & Kilincer (1972) found that paralysis is caused by the newly hatched parasitoid larva, which quickly migrates to the host's brain. Selective egg placement (Carton, 1978) and injected secretions from the female's accessory glands (Führer, 1975; Osman, 1978) are also important in helping to overcome the host's defences. *Pimpla* venoms comprise a large and complex array of chemicals, including several components thought to cause paralysis, i. e. neurotoxins, proteases, etc. (Parkinson *et al.*, 2001, 2002; Uçkan *et al.*, 2004). *Pimpla rufipes* venom prevents the aggregation

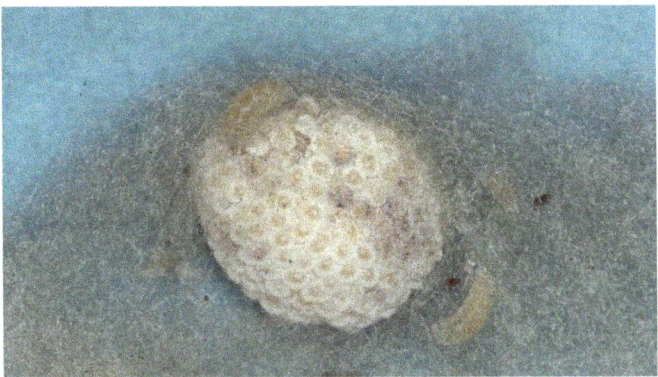

Figure 140. Larvae of *Tromatobia lineatoria* (Villers) feeding slowly through the winter on eggs of *Araneus diadematus* Clerck (Araneae: Araneidae) (M.R. Shaw).

of host haemocytes (Richards & Parkinson, 2000; Parkinson *et al.*, 2002). Pimplini generally prefer younger pupae (or even prepupae, but Sandlan (1982) found that in such cases development was less successful) and larval survival is lower in older pupal hosts (Ueno, 2004). Polysphinctines are known to use a venom which causes the temporary paralysis of the host spider, enabling the egg to be affixed (Nielsen, 1923; Cushman, 1926). In the polysphinctines the egg does not issue direct from the genital opening, as had been supposed by several authors (see Takasuka *et al.*, 2009), but is either expelled ribbon-like from the middle part of the ovipositor and guided onto the host by sliding down the parted lower valves in the supposedly basal taxa that oviposit onto the cephalothorax of spiders or, in the supposedly more advanced clade ovipositing onto the host's abdomen, it issues from near the base of the ovipositor where in most species there is a clear sub-basal ventral node (Takasuka *et al*, 2018). The egg is glued to the spider's cuticle by a secretion applied by the ovipositor (Eberhard, 2000; Takasuka *et al.*, 2009).

Adult pimplines find their hosts by a variety of successive environmental cues. Odours from the general substrate supporting the host (Thorpe & Caudle, 1938) as well as odours and contact chemicals stemming from the host itself (Sandlan, 1980) and vision (Sandlan, 1980) may all play a part. Orientation towards, and recognition of, the searching niche – 'host habitat finding' – by parasitoid Hymenoptera is usually recognised as mechanistically distinct from the discovery and recognition of an actual host individual (Doutt, 1964; Vinson, 1976, 1981; see also Weseloh, 1981 and Arthur, 1981). The location of hosts in the immediate searching environment is achieved by vibrational sounding in at least some (probably many) Pimplini (Henaut & Guerdoux, 1982; Wäckers *et al.*, 1998; Broad & Quicke, 2000). This host detection process involves tapping on a substrate with the antennal tips (which are furnished with columnar structures: Gauld *et al*, 2002b) and detecting the 'echoes' of a host in a space below. This has been best demonstrated in the laboratory by *Pimpla turionellae* (L.), which will attempt to oviposit in cigarette filter tips beneath paper, and explains observations such as Halstead's (1987) of *Pimpla rufipes* attempting to oviposit in bean seeds in a packet. Some pimplines are known to be capable of simple associative learning (Jackson, 1937; Arthur, 1966; Wardle & Borden, 1985), responding to both visual (Fischer *et al.*, 2004) and odour cues (Sato & Takusa, 2000), enabling them to concentrate their efforts on the most productive sections of the environment, and this suggests a mechanism whereby the realised host range of a particular species might differ radically from place to place. As is usual for idiobionts, many pimplines have adult flight periods that correspond to peaks of potential host availability rather than being attuned to a particular host species. Bivoltine pimplines often therefore have different hosts at different times of the year (Cole, 1967), although some of the more niche-specialised species may make double use of a single host generation (Danks, 1971).

In common with most Hymenoptera, mated females of several pimpline genera are known to be able to select the sex of their progeny by controlling the access of stored sperm to the egg as it passes down the oviduct, a process which is detectable by direct observation in at least one species (Cole, 1981). This enables them to use the larger host individuals preferentially for female progeny (Jackson, 1937; Aubert, 1959; Arthur & Wylie, 1959; Sandlan, 1979b), even in some cases when the size of the concealed host cannot be directly assessed visually (e.g. Askew & Shaw, 1986). As with *Rhyssa persuasoria* (L.) (Ichneumonidae: Rhyssinae), the difference in size between small males of a species such as *Ephialtes manifestator* (L.) and large females can be striking, although, unlike in *Rhyssa*, this is probably largely due to the use of different taxa as hosts, rather than the same species at different stages of development. When a range of suitable hosts is freely available it often happens in the field that the two sexes develop on substantially different groups of hosts (Kishi, 1970; Askew & Shaw, 1986).

There are five larval instars in *Pimpla* (Rojas-Rousse & Benoit, 1977), the middle three being very similar and hard to distinguish, and this seems likely to be the case for all Pimplinae. Much remains to be discovered about the larval adaptations of pimplines. However, various interesting scraps of information on the larval behaviour of particular species are reported in the literature, including consuming the host's skin after all the haemolymph and internal organs have been sucked out by *Perithous* (Danks, 1971), and Nielsen's (1923 *et seq.*) various observations on polysphinctines and the Ephialtini most closely related to them. The larva of *Pimpla* produces a biochemically complex anal secretion that has antibiotic properties and presumably helps to keep the wounded host aseptic while it is being consumed (Führer & Willers, 1986).

The majority of species of the *Polysphincta* group overwinter as immature larvae on their fully functioning immature spider hosts, but overwintering as an early instar larva is far less accessible to the idiobiont pimplines. However, *Tromatobia* females do not kill the spider eggs in the sacs they attack, and some *Tromatobia* species can therefore overwinter as small larvae feeding on fresh diapausing spider eggs. Otherwise the most usual method of overwintering is as a prepupa, although females of a few species overwinter as adults, carrying with them the male's contribution of stored sperm (e.g. *Itoplectis maculator*, Cole, 1967; possibly *Pimpla turionellae* in part, Jackson, 1937; *Scambus pomorum* and *Exeristes ruficollis*, Hancock, 1925). Adults of the univoltine *I. maculator* are also known to aestivate (Cole, 1967). The influences of the host species and of environmental conditions on the onset of winter diapause by the prepupae of *Pimpla rufipes* have been investigated by Claret & Carton (1975) and Claret (1973; 1978) respectively. Many of the pimplines which are known to overwinter as adults have a pronounced reddish coloration, which may be more cryptic than black in the winter months. Another, even clearer, ecological correlate of reddish body coloration is an association with reed beds and fens: this is easily seen in pimplines and is found through most groups of parasitoid Hymenoptera, but it has not been satisfactorily explained.

Polysphinctines are the only Pimplinae known to allow the host some further development after being attacked, and have extremely specialized larvae for their unusual niche. These are able to remain in position even through the host's moult (though dead larvae are invariably sloughed when the host moults), fixing themselves to a caked saddle of their progressively accumulated exuviae on the host integument by means of ventral protuberances (Nielsen, 1923). In addition, polysphinctine larvae in their final instars develop paired wart-like dorsal protuberances on several consecutive body segments, each retractile and furnished with outwards-projecting hooks. These structures differ according to genus (Nielsen, 1923, 1928, 1935, 1937), and are used by the larva to grip the silk spun by the host as the parasitoid larva releases its hold on the host itself to finish its feeding, and then to move rather easily among strands of the host's silk as it constructs its cocoon. In the tribe Ephialtini the final instar larvae of the genera *Zaglyptus* (Nielsen, 1923, 1935) and *Clistopyga* (Nielsen, 1929) are furnished with broadly similar hook-bearing warts to those of polysphinctines. *Tromatobia*, which similarly feeds in spider egg cocoons but is probably more basal (Gauld *et al.*, 2002b), has less elaborate structures bearing only unhooked spines (Nielsen, 1923). Fields of cuticular hooks forming holdfasts also occur in the larvae of *Sericopimpla*, a mainly tropical genus which attacks Psychidae (Smithers, 1956). In all cases these structures seem to aid movement and orientation via the host's silk during cocoon formation, albeit in slightly different ways. Although a number of bivoltine species seem to oviposit onto mature spiders for one of their generations (e.g. *Schizopyga*), polysphinctines are particularly associated with immature spiders. In general, the egg appears to hatch quickly but development of the parasitoid larva is at first very slight so that the immature host continues to feed and usually moults. Spiders with polysphinctine larvae on them are most often seen in autumn and spring because so many species overwinter as minute larvae on active hosts.

Polysphinctines display a range of behavioural adaptations in attacking their spider hosts, and can manipulate their hosts' behaviours, in ways that are only beginning to be uncovered. *Zatypota albicoxa* (Walker), a widespread Palaearctic species, has been studied by Takasuka and colleagues (e.g. Takasuka *et al.*, 2009). There is a pronounced plasticity in the behaviours exhibited by adult females when tackling potential hosts; females can pounce directly onto spiders on their webs and sting them or they can lie on their dorsum at the base of the web and lure the spider towards them by plucking a thread of silk, as an entrapped insect would do (Takasuka & Matsumoto, 2011a). Polysphinctines cannot detect, prior to stinging, whether or not a spider is already host to anther polysphinctine egg or larva. However, being ectoparasitoids of exposed hosts, polysphinctine females can detect superparasitism relatively easily when handling the parasitised host in at least a few species (Takasuka & Matsumoto, 2011b; Eberhard, 2000; Gonzaga & Sobczak, 2007). Superparasitism is avoided through the elimination of the earlier incumbent before oviposition, by the female scraping her ovipositor along the spider and levering off any eggs or larvae she encounters (Takasuka & Matsumoto, 2011b). When a pre-existing larva is large, it takes the polypshinctine female a considerable time to remove it and Takasuka & Matsumoto (2011b) relate this effort to the high potential costs involved in locating and parasitising a spider. There are some well-documented examples of behavioural modification of spiders by the polysphinctine larva. Eberhard (2000, 2001) studied the Neotropical *Hymenoepimecis argyraphaga* Gauld in Costa Rica, which induces its host tetragnathid spider to spin a markedly atypical web towards the end of the parasitoid's feeding phase. This web is essentially an augmented version of the usual orb web's support struts, with most of the prey-catching orb dispensed with. Excess web production, beyond what is needed for the *H. argyraphaga* larva to safely pupate in, represents resources diverted from potential nutrition for the rapidly feeding final larval phase of the polysphinctine. Subsequent discoveries suggest that behavioural manipulation of hosts is widespread across the *Polysphincta* genus-group (Gonzaga *et al.*, 2010; Eberhard, 2013). Takasuka *et al.* (2015) have demonstrated that in one species, at least, the web produced by parasitised spiders, soon before they are killed, results from hormonal induction by the parasitoid larva of a form of web spinning usually exhibited only by moulting spiders, when prey-catching orb webs are redundant, with certain behavioural steps repeated over and over to produce a novel web form. The manipulation of host behaviour by polysphinctines has become a popular topic in recent years, and Takasuka *et al.* (2018) reference many studies.

Specialized cocoon morphology in the Pimplinae is found only in the polysphinctines. Some polysphinctines construct their cocoons within the host spider's web, in which case the cocoon is usually densely spun, spindle-shaped and sometimes angularly ribbed, no doubt resembling a well-wrapped prey item of little interest to its potential enemies. Other polysphinctines finally kill their hosts in retreats spun or selected by the spider and the cocoon is then usually much frailer, expanded and loosely spun. Most polysphinctine cocoons have a clearly visible caudal opening through which the meconium (the accumulated larval faecal material) is voided. In *Schizopyga* and *Iania* (until recently included in *Dreisbachia*; Matsumoto, 2016), and in the most closely related ephialtine genera, the opening is less clear and faecal material often adheres partly within the cocoon. Elsewhere in the Pimplinae, however, cocoon formation appears to be rather generalised and is well developed only when the host or its substrate fails to provide a really secure or concealed pupation site, particularly if the pimpline passes the winter as a prepupa. Thus the most substantial cocoons are seen in pimpline genera which attack and overwinter in frail structures, like the exposed cocoons of Lepidoptera that are attacked by *Iseropus*. On the other hand, genera attacking hosts living deeply concealed in woody tissues or the hard pupae of Lepidoptera seem often to spin only enough silk to restrict the involuntary movement of the pupa, or sometimes no detectable silk at all. Unlike in the ophioniformes, cocoons of pimplines are relatively simple in that the larva does not spin any guiding marks to aid orientation (Salt, 1977). Another function of a cocoon is to isolate the occupant from micro-organisms, and some species inhabiting plant tissues that are decaying, or are likely to decay, appear to make relatively strong cocoons. The cocoons of most genera remain unstudied, however. In common with many other Hymenoptera, eclosion from the pupa precedes emergence from the cocoon by two or three days, during which the initially teneral adult parasitoid hardens so that it is fully protected when it finally bites its way out.

Many pimplines, particularly in the tropics, possess a vesicle on the hind tarsal claw which Gauld *et al.* (2002b) refer to as the 'poison fang', capable of producing a stinging sensation when the claw breaks off in the skin (Gauld, 1991), and Gauld (1987) describes the irritant, detachable hairs on the tarsi of some African *Pimpla*. Clearly, some pimplines are able to defend themselves and some are appropriately aposematically coloured. Some neotropical pimplines, e.g. some *Neotheronia* and *Pimpla* (Gauld, 1991), mimic paper wasps (Vespidae: Polistinae) and are probably involved in Müllerian mimicry rings. Some pimplines produce a distinctive odour when handled (Townes *et al.*, 1960), as do a few other ichneumonids (see Banchinae, Ctenopelmatinae and Metopiinae chapters). Moderate numbers of pimpline species can be found in many habitats and this, together with the rather stable state of pimpline taxonomy, has resulted in the Pimplinae being used to document parasitoid species richness across different areas of the globe (Gómez *et al.*, 2017) and in studies of habitat heterogeneity and its effect on parasitoid faunas, in the tropics (e.g. Sääksjärvi *et al.*, 2004) and in England (Fraser *et al.*, 2007, 2008a,b). It is also possible, in a reasonably well-studied fauna such as in Britain, to document probable extinctions and colonisations. Several large and conspicuous pimplines have not been collected in Britain for many years and are certainly or probably extinct in this country, e.g. *Dolichomitus diversicostae* (Perkins), *Pimpla aethiops* Curtis, *Pimpla arctica* (Zetterstedt) and *Theronia atalantae* (Poda) (Shaw, 2006b). Conversely, *Perithous albicinctus* (Gravenhorst) is a relatively recent colonist (Brock & Shaw, 1997; Shaw, 2006b).

Identification. An RES handbook to the British species of Pimplinae and related subfamilies has been published (Fitton *et al.*, 1988) and an update is in progress (Shaw *et al.*, in prep.), expanding the treatment to include Acaenitinae and Collyriinae, i.e. the pimpliformes with the exception of the koinobiont endoparasitoids of Diptera larvae (Cylloceriinae, Diplazontinae and Orthocentrinae). In the meantime, Shaw (2006b) should be consulted for various additions and updates to the British fauna. More recently, Horstmann (2011b) has described a new species of *Acrodactyla*, previously confused under *A. degener* (Gravenhorst), Shaw *et al.* (2011) established the synonymy of what had been considered to be three species of *Scambus*, and Fritzén & Shaw (2014) separated two similar species of *Polysphincta*. A small number of undescribed species await description (Shaw *et al.*, in prep.). There are two synonymic and bibliographic catalogues, both of which cover the whole European fauna (Oehlke, 1967; Aubert, 1969); however, these have been superseded by the catalogue of Yu *et al.* (2012). Some papers useful for identification of European pimplines are those of Kasparyan (1973a, 1974) (*Apechthis, Itoplectis, Pimpla*), Horstmann (2008b) (*Ephialtes*), and Zwakhals (2006, 2010) (*Sinarachna, Zatypota, Dolichomitus*). Some pimpline genera, including the large genus *Scambus*, remain unrevised at the regional level.

Figure 141. Morphology of Pimplinae: (a) *Delomerista borealis* Walkley (Delomeristini), head; (b) *Perithous scurra* (Panzer) (Delomeristini), head; (c) *Pimpla turionellae* (L.) (Pimplini), head; (d) *Pimpla rufipes* (Miller) (Pimplini), head and part of antenna, lateral; (e) *Schizopyga frigida* Cresson (Ephialtini), head; (f) *P. turionellae*, part of fore wing; (g) *Ephialtes manifestator* (L.) (Ephialtini), fore tarsal claw; (h) *Polysphincta longa* Kasparyan (Ephialtini), tarsi; (i) *D. borealis*, propodeum and metasoma; (continued overleaf).

Figure 141 continued: Morphology of Pimplinae: (j) *Pseudorhyssa alpestris* (Holmgren) (Delomeristini), mesoscutum; (k) *P. alpestris*, metasomal apex; (l) *Scambus brevicornis* (Gravenhorst) (Ephialtini), propodeum and metasoma; (m) *P. rufipes*, mesosoma and anterior part of metasoma, lateral; (n) *P. rufipes*, first to third metasomal tergites; (o) *Clistopyga incitator* (Fabricius) (Ephialtini), metasoma and ovipositor, lateral; (continued overleaf).

Figure 141 continued: Morphology of Pimplinae: (p) *Zatypota albicoxa* (Walker) (Ephialtini), second to fifth metasomal tergites; (q) *Apechthis rufata* (Gmelin) (Pimplini), ovipositor; (r) *P. rufipes*, ovipositor; (s) *S. brevicornis*, ovipositor; (t) *Z. albicoxa*, ovipositor; (u) *P. rufipes*, male genitalia and apex of metasoma.

Subfamily POEMENIINAE

Tribes
*Rodrigamini
Poemeniini

The Poemeniinae is a small subfamily of approximately 80 species classified in ten genera and three tribes. Represented in both temperate and tropical zones (but not known from Australia), the subfamily is thought to have originated in the Palaearctic, where the basal-most genera are found today (Wahl & Gauld, 1998; Matsumoto & Broad, 2011). The British fauna comprises six species in four genera and one tribe (Poemeniini). The extralimital second tribe Rodrigamini is poorly known but apparently rather widespread (Matsumoto & Broad, 2011; Sheng & Sun, 2014).

Figure 142. *Deuteroxorides elevator* (Panzer), tribe Poemeniini.

Recognition. Poemeniines resemble other ichneumonid groups associated with wood-boring insects and are often confused with Pimplinae, Rhyssinae or Xoridinae. The six British species are classified in the Poemeniini and can be recognized by their lack of the epicnemial carina, elongate propodeum and elongate hind coxa. *Poemenia* species can be recognised by their bidentate mandibles and large, convex clypeus. *Deuteroxorides, Neoxorides* and *Podoschistus* have unidentate mandibles, small, flat clypeus and the eyes ventrally convergent. *Neoxorides* and *Podoschistus* can be recognised by the conspicuous, scale-like denticulate sculpture of the temples, which is reduced to weak rugosity in *Deuteroxorides*. Unlike pimplines, the epomia of Poemeniinae continues ventrally on a raised ridge parallel with the anterior margin of the pronotum and the first tergite lacks glymmae.

Figure 143. *Podoschistus scutellaris* (Desvignes) searching for hosts on dead hazel (*Corylus avellana*) (G.R. Broad); (video available at https://www.youtube.com/watch?v=G8cpPenkz-w).

Systematics. Poemeniines were traditionally included in the Pimplinae (e.g. Perkins, 1959; Townes, 1969; Fitton *et al.*, 1988) but Eggleton (1989), in an unpublished thesis, recognised the polyphyletic nature of this assemblage and separated Diacritinae, Poemeniinae and Rhyssinae from Pimplinae. Gauld (1991) formally enacted the subfamily-level changes and Wahl & Gauld (1998) defined the Poemeniinae on the basis of several characters, including the lateral expansion of the foramen magnum (at the back of the head), the epomia continuing ventrally on a raised ridge parallel with the anterior margin of the pronotum, and the dorsal surface of the hind tibia with stout spines. Even so, the subfamily is morphologically diverse, prompting Wahl & Gauld (1998) to recognise three tribes within this small group of genera. Both Wahl & Gauld (1998) and Quicke *et al.* (2009) recovered the same relationships amongst the 'lower' pimpliformes, i.e. Pimplinae+(Poemeniinae+Rhyssinae). Recent work by Klopfstein *et al.* (in press), however, has shown a very strong molecular signal for *Pseudorhyssa* being better classified in Pimplinae, in the tribe Delomeristini, where Townes (1969) had placed the genus. The biology of *Pseudorhyssa* is rather well known (see Pimplinae chapter), compared to the limited data on Poemeniini.

Biology. The hosts of poemeniines all seem to be associated with wood and it is assumed that they are idiobiont ectoparasitoids of larvae, but this has not been demonstrated for any species other than *Neoxorides nitens* (Gravenhorst) which has been reared (but not in Britain) from young larvae found feeding solitarily on paralyzed larvae of the beetles *Phymatodes testaceus* (L.) (Cerambycidae) and *Ovalisia mirifica* (Mulsant) (Buprestidae) (M. R. Shaw, pers. obs.), though whether or not as cleptoparasitoids was not ascertained. There are many unreliable host associations in the literature that stem from mixed substrate rearings of wood containing beetles and hole-nesting aculeates. Species of *Poemenia* are parasitoids of solitary wasps (Crabronidae) nesting in old beetle burrows, with several reliable records of *Passaloecus* species as hosts (Fitton *et al.*, 1988; Shaw, 2006b). *Deuteroxorides elevator* (Panzer) has been recorded as a parasitoid of cerambycid beetles (e.g. *Molorchus minor* (L.), *Aromia moschata* (L.), *Rhagium inquisitor* (L.)) (Fitton *et al.*, 1988; Campadelli & Scaramozzino, 1994). There are no reliable host records for *Podoschistus scutellaris* (Desvignes) but it has been filmed inserting its ovipositor in pre-existing cracks in wood infested with *Clytus arietis* (L.) (Coleoptera: Cerambycidae), which had also attracted other ichneumonids (https://www.youtube.com/watch?v=G8cpPenkz-w). There is one host record for the extralimital Rodrigamini, with *Rodrigama longissima* (Sheng & Sun) having been reported to be a parasitoid of a cerambycid beetle (Sheng & Sun, 2014).

Identification. A handbook covering the British species has been published (Fitton *et al.*, 1988) and there is a revised edition in preparation (Shaw *et al.*, in prep.). Horstmann (1998b) provides some useful characters for separating males of *Poemenia collaris* Haupt and *P. hectica* (Gravenhorst). J. Boparai has recently found *Neoxorides nitens* in England and this will be formally recorded as British by Boparai *et al.* (in prep.). Wahl & Gauld (1998) provide a key to world genera.

Figure 144. Morphology of Poemeniinae: (a) *Poemenia hectica* (Gravenhorst) (Poemeniini), head; (b) *Podoschistus scutellaris* (Desvignes) (Poemeniini), head, lateral; (c) *P. scutellaris*, hind tarsal claws; (d) *P. hectica*, fore tibia; (e) *P. scutellaris*, ovipositor tip.

Gavin R. Broad, Mark R. Shaw and Michael G. Fitton

Subfamily RHYSSINAE

The Rhyssinae comprises about 250 species in eight genera, with the highest generic and species diversity in the Oriental region (Kamath & Gupta, 1972), although the numbers of species being described from the neotropics and the Afrotropical region are rising rather rapidly (e.g. Porter, 1978; Gauld, 1991; Rousse & van Noort, 2014b; Gómez *et al.*, 2015). Only two species in two genera are known in Britain, from a European fauna of nine valid species in three genera (Horstmann, 1998c, 2002a).

Figure 145. *Rhyssella approximator* (Fabricius).
(NB: in some dry-preserved specimens, such as this one, the three valves of the ovipositor become separated.)

Recognition. The subfamily includes some of the ichneumonids most familiar through general insect books. *Rhyssa persuasoria* (L.) and the non-British *Megarhyssa* species are frequently illustrated as 'typical' ichneumonids because of their large size, spectacularly long ovipositors, and oviposition habits; *Rhyssa persuasoria* females can have a combined body and ovipositor length of up to about 85 mm, and the North American *Megarhyssa atrata* (Fabricius) can be up to about 170 mm long. Rhyssines can be distinguished from all other British ichneumonids by a combination of the mesoscutum having conspicuous, transverse rugae and the medially incomplete occipital carina (complete in the pimpline *Pseudorhyssa*). Similar rugae with the same presumed function of aiding emergence from tunnels in wood occur in some non-ichneumonid taxa: in Britain *Ibalia* species (Ibaliidae) and *Aulacus striatus* (Jurine) (Aulacidae). The last metasomal tergite is produced as a truncate 'horn' in rhyssines, but it is simple (although elongate) in *Pseudorhyssa*, and there are also differences between the two in the position of fore wing vein $1cu$-a relative to RS&M. The ovipositor is very long in all rhyssines and these are conspicuous insects when ovipositing into or inspecting dead wood. The rather widespread *Rhyssa persuasoria* is large and conspicuously patterned white or pale yellow on a black background.

Systematics. Along with diacritines and poemeniines, rhyssines were generally included in the Pimplinae until Eggleton (1989) and then Gauld (1991) elevated the group to a separate subfamily. Rhyssines are distinctive, with numerous apomorphies identified by Wahl & Gauld (1998), including the chisel-shaped upper mandibular tooth, the mesoscutum covered in transverse rugae, the elongate propodeal spiracle, the elongate eighth tergite of the female ending in a truncate horn or boss, and the presence of ovipositor guides on the metasomal sternites. In the phylogenetic results of Wahl & Gauld (1998) and Quicke *et al.* (2009), the Rhyssinae, Poemeniinae and Pimplinae comprise a clade within the pimpliformes group of subfamilies, of predominantly idiobiont ectoparasitoids (with some conspicuous exceptions in the Pimplinae) that is the sister-group to a large assemblage of pimpliformes that are mostly koinobiont endoparasitoids. These studies have, however, been based on relatively small datasets and the evolutionary picture could change substantially with future studies.

Biology. Rhyssinae are well specialized for parasitising hosts deep within wood, with metasomal adaptations for supporting the long ovipositor for drilling and with other characteristics (chisel-like mandibles and transversely rugose mesoscutum) which assist the adult in emerging from wood. Quicke *et al.* (1998) found high concentrations of manganese (1-5% dry weight) in the ovipositor and zinc (5-10%) in the mandibles of *R. persuasoria*, undoubtedly adapations towards strengthening cuticle liable to heavy abrasion when penetrating wood. The two British species are relatively well known biologically and all rhyssines that have been studied are idiobiont ectoparasitoids. *Rhyssa persuasoria* has been extensively studied because of its potential in the biocontrol of introduced *Sirex* woodwasps in Australia and New Zealand (Spradbery & Kirk, 1978), where *R. persuasoria* has become established (Miller & Clark, 1935, 1937; Taylor, 1977). *Rhyssa persuasoria* is a specialised parasitoid of Siricidae (Hymenoptera) larvae in coniferous trees; this host specificity being the result of the ichneumonid's reliance on detecting a fungal associate of the woodwasp rather than any physiological specialisation to the host's physiology. The woodwasps introduce a fungal symbiont, *Amylostereum*, into the wood on oviposition, which the woodwasp larva feeds on along with the wood. This fungus is attractive to females of *R. persuasoria* and it is at its most potent when it has been growing in the wood for a period approximately equivalent to the length of time needed for the siricid larva to reach sufficient size for development of the *Rhyssa* larva (Spradbery, 1970).

Successful location of hosts by *R. persuasoria* occurs in only a minority of attempts (Chrystal & Skinner, 1931; Spradbery, 1970) and drilling can take well over an hour (Eggleton, 1989), with the drilling females thus vulnerable to predation from birds – ovipositors with no body attached are sometimes found sticking out of trees (Fig. 146). Eggs of *Rhyssa* (and other rhyssines) are elongate, with a long pedicel. The larva develops through five instars and has strong mandibles from the first instar. Pupation is in a loosely spun cocoon that fills the host chamber and the female's ovipositor is curved over the dorsum of the pupa's body. Despite its very specialised host location behaviour, *R. persuasoria* is able to develop on woodwasp larvae of all sizes, woodwasp pupae, and on the larvae and pupae of the cynipoid primary parasitoid *Ibalia* (Hymenoptera: Ibaliidae) that have completed development on a siricid (Hanson, 1939). It has even been reported to develop successfully on woodwasp adults still in their pupal cells (Hanson, 1939). The great range in potential sizes of hosts, and the fact that the ovipositing female cannot directly gauge the size of the host before commencing drilling, is reflected in the great size variation in *R. persuasoria* specimens, with fore wing length varying from 5.2 mm (in the smallest males) to 24.4 mm (in the largest females) in a study by Spradbery & Ratkowsky (1974). The size variation in males is much greater than in females, however, suggesting that the ovipositing female can assess and respond to the size of the host once it is contacted with the ovipositor. In Britain, the adults of *R. persuasoria* fly in the second half of the summer and it chiefly parasitises our most abundant woodwasp, *Urocerus gigas* (L.), but its introduction to Australia and New Zealand was for biocontrol of introduced pest *Sirex* species.

Rhyssella approximator (Fabricius) is a parasitoid of *Xiphydria* species (Hymenoptera: Xiphydriidae), in Britain having been reared from *X. camelus* (L.) in *Alnus* and *Betula* and from *X. prolongata* (Geoffroy) in *Salix* (Fitton *et al.*, 1988). The biology of *R. approximator* was studied by Chrystal & Skinner (1932) and filmed by Skinner & Thompson (1960). Development of the larva is similar to that of *R. persuasoria*. The first instar larva is active, with a well sclerotized head capsule, moving around the body of the paralysed host until finding a suitable site for feeding. Successive instars lack the prominent head capsule and are essentially sedentary. As in *Rhyssa*, the larva constructs a rough cocoon which fills the width of the host chamber. It overwinters as a larva, pupating in the spring, with adults on the wing from mid-May to early July (Chrystal & Skinner, 1932).

Figure 146. *Rhyssa persuasoria* (L.) (S. Sexton).

The non-British *Megarhyssa* are also, where known, parasitoids of siricid woodwasps (e.g. Spradbery & Kirk, 1978; Schiff *et al.*, 2012). *Megarhyssa* species have much longer ovipositors in relation to the body size than in *Rhyssa* or *Rhyssella* and have striking metasomal adaptations to parasitise hosts up to 14 cm deep in the wood, and in order to deploy a short enough section of the ovipositor that can drill without excessive buckling. Le Lannic & Nénon (1999) describe how the posterior metasomal segments rotate by 270°, pushing the ovipositor up between metasomal segments 7 and 8, into a hugely expandable membrane. In this way the ovipositor is effectively stored whilst a shorter section is guided along the ventral surface of the metasoma and drilling commences. Drilling through the wood seems to be aided by the secretion of substances via the ovipositor that lyse lignin (Le Lannic & Nénon, 1999), although the nature of the chemicals involved has not been investigated. Other, perhaps all, Rhyssinae also rotate the posterior metasomal segments so as to achieve a vertical drilling position, with the ovipositor lying along the ventral side of the metasoma, but it is apparently only in *Megarhyssa* species that the ovipositor is contained within inter-segmental membranes (there are several videos on YouTube documenting the drilling sequence).

In contrast with the British fauna, for most of the subfamily virtually nothing is known about hosts of rhyssines. In view of the limited siricid fauna, it is assumed that many tropical species are parasitoids of wood-boring beetle larvae but there are just a handful of host records for the entire south-east Asian rhyssine fauna. For example, there are host records for *Triancyra galloisi* (Uchida) from two species of cerambycid beetle (Kusigemati, 1981; Sheng & Sun, 2010b) and different species of *Epirhyssa* have been reported as reared from a siricid woodwasp, cerambycid beetle and a hyblaeid moth (e.g. Sonan, 1933; Sheng & Sun, 2010b). However, none of the original sources provide detail on the rearings and whether other potential hosts had been ruled out. There are no published host records for three of the eight rhyssine genera.

Rhyssella approximator and various extralimital species of Rhyssinae exhibit scramble competition for mating, with males aggregating around borings containing females about to emerge, although males of *Rhyssa* species only mate with females when they have emerged from the wood (Eggleton, 1991). Large males of species exhibiting scramble competition have elongate slender metasomas with which they mate with the female whilst she is still in the tunnel. Small males have relatively shorter, stouter metasomas and rely on 'sneaky' mating tactics when the females have emerged (Eggleton, 1991). In the South-east Asian *Lytarmes maculipennis* (Smith), males defend a 'territory' containing a female about to emerge (Eggleton, 1990). There seems to be no courtship in any rhyssine. As with pimpline parasitoids of deeply concealed hosts, there is little information on how or whether rhyssine females obtain sugars and proteins as host-feeding is impossible and they are very rarely observed on flowers, although Eggleton (1991) records a pers. comm. from G.H. Thompson, that he kept females of *R. approximator* alive for an average of 41 days on sugar solution.

Identification. The British species are covered by Fitton *et al.* (1988) and in a revision of that handbook in preparation (Shaw *et al.*, in prep.). There are useful keys to the European species of *Megarhyssa* (Horstmann, 1998c) and *Rhyssa* (Horstmann, 2002a). While additional species of rhyssines are not very likely to turn up as natives, they may do so as rare casuals in imported timber.

Figure 147. Ovipositor of *Rhyssa persuasoria* penetrating wood.

Figure 148. Morphology of Rhyssinae: (a) *Rhyssa persuasoria* (L.), mesoscutum; (b) *R. persuasoria*, metasomal apex; (c) *R. persuasoria*, ovipositor apex.

Subfamily STILBOPINAE

A small subfamily, the Stilbopinae comprises only 30 described species in three genera. Most of the 27 species of *Stilbops* are found in the temperate Palaearctic, with highest species-richness in the Eastern Palaearctic (Kasparyan, 1999; Watanabe & Maeto, 2012) but a few occur in North America, Mexico and Vietnam. *Panteles* contains two Palaearctic species whilst the third genus, *Notostilbops*, is known only from one Chilean species. Five of the eight European species of Stilbopinae occur in Britain, four *Stilbops* and one *Panteles*.

Figure 149. *Stilbops vetula* (Gravenhorst).

Recognition. The Stilbopinae is not easily characterised, despite the small number of species. The mandibles are short and strongly tapered (twisted in *Panteles*); the malar space is usually wide (*Stilbops*); the apical flagellar segment is usually conspicuously longer than the preceding one (*Stilbops*); the epomia is strong and angulate (*Stilbops*); the areolet is obliquely quadrate or narrowly pentagonal (difficult to describe but a fairly distinctive shape; Fig. 150g); fore wing vein 1*cu-a* is postfurcal and strongly sloping; the hypopygium is large and the ovipositor is either simply tapered, lacking a notch or teeth (*Stilbops*), or bears some conspicous dorsal teeth, which is unusual in Ichneumonidae (*Panteles*). Only one species is common in Britain, *Stilbops vetula* (Gravenhorst), which can be recognised by the particularly dense, silvery setae covering the face and, in the female, by the apically and laterally reddish metasoma.

Systematics. The relationships of the stilbopine genera have long been regarded as uncertain and, on the basis of larval morphology, Short (1957; 1978) advocated that *Stilbops* be placed in a subfamily of its own. Townes (1970b) included *Panteles* and *Notostilbops* with *Stilbops* in a tribe, Stilbopini, in the subfamily Banchinae, although he later (Townes & Townes, 1978) recognized the Stilbopinae as a distinct subfamily, but transferred *Panteles* to the tribe Atrophini in the Banchinae. Wahl (1988), on the basis of a cladistic examination of morphological characters, again reunited *Panteles* with *Stilbops* and *Notostilbops* and considered the subfamily to be distinct from Banchinae. Recent phylogenetic analyses of a wide range of ichneumonids has shown *Panteles* and *Stilbops* to form either a clade or part of a grade group basal to part of a polyphyletic Tryphoninae and other, 'higher ophioniformes', subfamilies. This placement has some morphological support in the presence of an aulaciform rod in the ovipositor (shared with several ophioniform groups; Quicke et al., 1994) and the presence of a stalk on the egg of *Panteles* (Quicke, 2005). The final instar larva of *Stilbops* is rather plesiomorphic, with papilliform antennae, sclerotized pleurostoma and other features that are not obviously homologous with the more derived form of the *Panteles* larva (e.g. simple disc-shaped antennae; pleurostoma narrow and weakly sclerotized) (Quicke, 2005). Conversely, Bennett et al. (in prep.) found *Stilbops vetula* to be the sister to *Notostilbops* sp.+ Banchinae, this grouping the sister group to Metopiinae which in turn was the sister group to the higher Ophioniformes.

Biology. As far as is known Stilbopinae are solitary koinobiont endoparasitoids of Adelidae and Prodoxidae (Lepidoptera: Incurvarioidea), with host records for four of the five British species. The two British genera differ quite substantially in their biology. The egg of *Stilbops* is laid in the Adelidae host egg and the adult emerges from the host cocoon (Hinz, 1981). Female *Stilbops ruficornis* (Gravenhorst) can be conspicuous, probing with their ovipositor, which is much longer than in other species of the genus, in the flower heads of *Knautia arvensis* and ovipositing into the eggs of *Nemophora metallica* (Poda) concealed there (Fig. 149). The host overwinters as a part grown larva and continues feeding in the following spring. Parasitoid and host are found as adults in July and August and both are univoltine. *Stilbops vetula* attacks *Adela reaumerella* (L.) and, although its life cycle is known in outline, most of the details need to be elucidated. Adults of *S. vetula* are often abundant in deciduous woodland in late April and May, swarming around fresh foliage. The eggs of its host are laid beneath the epidermis of various plants and, after locating such an oviposition site, the female *Stilbops* spends some time examining it before bringing her metasoma forward between her legs and laying an egg in that of the host (Hinz, 1981). The third British *Stilbops* with a recorded host, *S. limneriaeformis* (Schmiedeknecht), has been reared from *Nematopogon schwarziellus* (Zeller) (Shaw, 1989). Given the available host records, and the nature of the parasitoid-host interaction, Shaw (1989) suggests that *Stilbops* species will all have extremely narrow host ranges.

The sole British (and European) species of *Panteles*, *P. schuetzeanus* (Roman), is a parasitoid of *Lampronia fuscatella* (Tengström) (Prodoxidae), an incuvarioid moth that forms galls on twigs of birches (*Betula*) (Roman, 1925; Bland, 1989). Various aspects of the life history of *P. schuetzeanus* have been elucidated by Quicke (2005). The egg has a hooked 'tail' which is internally embedded in various parts of the host's tissues but usually towards either the anterior or posterior end of the body. Superparasitism was found to be common (55% of cases) but only a single egg hatched. The *Panteles* cocoon is a thin layer of silk formed within the *Lampronia* gall and the pupal stage is probably very short, based on the rapidly acquired pigmentation of the eye, although direct observation is lacking. It seems that larval development takes place entirely within the host larva. Although only larger larvae were dissected, Quicke (2005) thought it unlikely that early instar larvae were attacked as the *Panteles* eggs are comparatively large, and often occur in aggregations (whether from one or more ovipositing females is not known). Given the synchronous flight times of the host and parasitoid (April to June), Quicke (2005) supposed that an alternative host must be used, which would mature earlier in the season than *L. fuscatella*; however, a two year larval development for *L. fuscatella* seems a much more likely explanation, with *Panteles* ovipositing through the tough wall of the gall (hence the toothed ovipositor) in the host larva's second year.

Identification. A key to the British species of *Stilbops* is given by Fitton (1984) with the European species reviewed by Hinz (1981) (and included in the key by Kasparyan, 1999). *Panteles* is keyed separately here in the key to subfamilies.

Figure 150. *Stilbops ruficornis* (Gravenhorst) females and males on *Knautia arvensis* (a, c), oviposition stance (d), and alongside its host, *Nemophora metallica* (Poda) (Lepidoptera: Adelidae) (b) (S. Taylor).

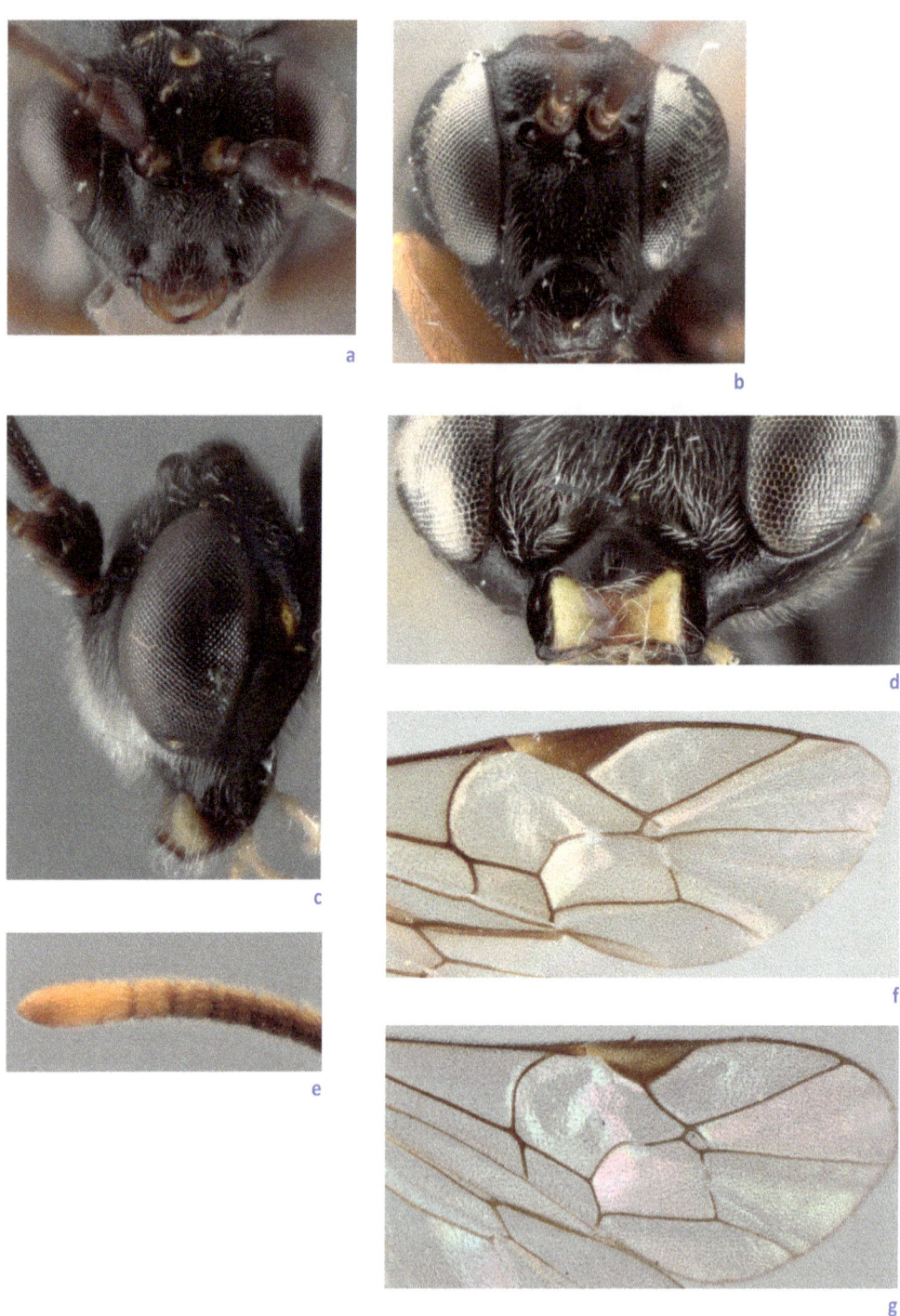

Figure 151. Morphology of Stilbopinae: (a) *Panteles schuetzeanus* (Roman), head; (b) *Stilbops ruficornis* (Gravenhorst), head; (c) *Stilbops vetula* (Gravenhorst), head, lateral; (d) *S. vetula*, mandibles, anteroventral view; (e) *S. vetula*, antenna tip; (f) *P. schuetzeanus*, part of fore wing; (g) *S. vetula* part of fore wing; (continued overleaf).

Figure 151 continued: Morphology of Stilbopinae: (h) *P. schuetzeanus*, mesosoma and metasoma, lateral; (i) *P. schuetzeanus*, metasoma; (j) *S. ruficornis*, metasoma; (k) *S. vetula*, male metasoma; (l) *P. schuetzeanus*, ovipositor tip; (m) *S. vetula*, ovipositor.

Subfamily TERSILOCHINAE

The Tersilochinae is a medium-sized subfamily with a little over 450 described species. So far, 81 species in 16 genera have been found in Britain, which represents a substantial increase in knowledge since the 1978 check list, and many further additions are likely. Globally, Tersilochinae is becoming one of the more completely described, medium-sized ichneumonid subfamilies, mainly thanks to the efforts of A. Khalaim and the late K. Horstmann (e.g. Horstmann, 2010b, 2012b, 2013a,b; Khalaim & Broad, 2012, 2013; Khalaim, 2009, 2011; Khalaim & Sheng, 2009; Khalaim *et al.*, 2013). We include the *Phrudus* group of genera, reviewed in Europe by Vikberg & Koponen (2000). Despite their very different appearance from more mainstream tersilochines, they are not currently formally treated as a separate tribe.

Figure 152. *Phrudus defectus* Stelfox.

Recognition. The subfamily as a whole is rather heterogeneous when accounting for the *Phrudus* group of genera but can be recognised by a combination of a strongly curved and largely (or entirely) unpigmented hind wing vein *M+CU*, a large, triangular pterostigma, a comparatively large antennal pedicel and the proboscidial fossa strongly narrowed. The *Tersilochus* group (the traditional Tersilochinae) are easily recognised by their distinctive wing venation (fore wing veins *2rs-m* and the first abscissa of *M* are very short), fringe of strong setae on the apex of the clypeus, reduced number of palp segments (four maxillary and three labial palp segments) and the petiolate first metasomal segment, with a long sclerotized sternite and the spiracle well behind the middle. Khalaim *et al.* (2017) have described a new genus and species with an extremely long glossa and galeae and further reduced palps (three maxillary and two labial palp segments).

Systematics. Until very recently, the species united here under one subfamily were treated as belonging to two subfamilies, namely Phrudinae (in part) and Tersilochinae. However, in their phylogenetic analyses Quicke *et al.* (2009) found that the 'microphrudine' genera of Phrudinae (the *Phrudus* group) and *Neorhacodes enslini* (Ruschka) (the only included neorhacodine) grouped with the tersilochines, whereas the 'macrophrudines' (which do not occur in Britain) formed a separate clade. Consequently, Quicke *et al.* (2009) resurrected Sisyrostolinae (wrongly named Brachyscleromatinae; see Bennett *et al.*, 2013) for the (non-British) 'macrophrudine' genera and united the 'microphrudines' (including the British genera *Astrenis*, *Phrudus* and *Pygmaeolus*) and neorhacodines (only *N. enslini* in Britain) with the Tersilochinae, where they seem to form a rather basal grade. However, *Neorhacodes* does not form a clade with Tersilochinae in Bennett *et al.*'s (in prep.) phylogeny of the Ichneumonidae which, together with their very different hosts, leads us to recognise Neorhacodinae as a separate subfamily again. Townes (1971) had earlier acknowledged that the physically smaller 'phrudines' may not have been correctly associated with the larger species, and noted a similarity with the Tersilochinae. Gauld (1997) attempted to define the Phrudinae as a monophyletic group (i.e. including the 'microphrudines', the Sisyrostolinae of current usage and *Seleucus*) on the basis of the narrowed proboscidial fossa, whilst Vikberg & Koponen (2000) noted that there were good reasons to suppose the 'microphrudines' are closely related to the Tersilochinae on the basis of distinctive, small projections on the antennae. The genus *Seleucus* had already been removed to the Ctenopelmatinae (Kolarov, 1987; Vikberg & Koponon, 2000). The *Phrudus* group of genera contains a fairly small number of mostly north temperate species (although with some species present as far south as Chile and Australia). Gauld (1985) placed the Tersilochinae in a group of subfamilies which also includes Cremastinae, Ophioninae and Campopleginae but they were not found to be close relatives of this group (mainly parasitoids of Lepidoptera) by Quicke *et al.* (2009). In the analysis of Bennett *et al.* (in prep.), exemplars of the *Phrudus* and *Tersilochus* groups as well as Sisyrostolinae formed a clade that was the sister group to all members of the Ophioniformes except Tryphoninae and Neorhacodinae.

Biology. The Tersilochinae comprises mostly small, inconspicuous species, although some species are numerous on flowers in spring. The majority of Tersilochinae species develop as parasitoids of larval Coleoptera. The known exceptions are *Gelanes* species, which have been reared in North America (Carlson, 1979) and Europe (Schedl, 1997) from larvae of sawflies of the genus *Xyela* (Hymenoptera: Xyelidae); *Tersilochus curvator* Horstmann and an undescribed *Tersilochus* reared from a species of *Eriocrania* (Lepidoptera: Eriocraniidae) (Jordan, 1998a), with an unidentified North American tersilochine also reared from an eriocraniid (Lepidoptera) (Carlson, 1979). The Coleoptera reliably recorded as hosts have a variety of habits and include members of the families Byrrhidae, Ciidae, Chrysomelidae, Curculionidae, Melandryidae, Nitidulidae, and Staphylinidae (Horstmann, 1971b; 1981). The larvae of most of the known hosts feed concealed, often within plant tissue. *Barycnemis* species, which have rather robust ovipositors, attack larvae in the soil. Carlson (1979) suggested that the important link between hosts is the fact that they construct a pupal chamber or cocoon in the soil. All species develop as solitary internal koinobiont parasitoids, at least until their final instar (see below), and kill their host as a prepupa.

Most of our knowledge of the biology of the *Tersilochus* group comes from investigations made in connection with their parasitism of pest beetles. Important studies of European species are by Aubert & Jourdheuil (1959), Osborne (1960), Jourdheuil (1960), Dysart *et al.* (1973) and Montgomery & DeWitt

(1975). The crops in Europe in which tersilochine species help to control populations of pests include oilseed rape and its relatives. A number of workers (Fritzsche, 1957; Winfield, 1963; Lehmann, 1969; Šedivý, 1983; Wyrostkiewicz & Blazejewska, 1985) have assessed the role tersilochines play in these situations but little has been done to enhance their effect, although Kovács *et al.* (2013) have shown that parasitism rates by tersilochines (and the braconid *Diospilus capito* (Nees)) can be increased, and pesticide levels decreased, by intermixing wild *Brassica nigra* with oilseed rape.

Mating takes place very soon after adult emergence and egg laying begins a few days later. Females will apparently oviposit into all larval instars of the host but show a preference for younger larvae. In *Tersilochus conotracheli* (Riley), which attacks *Conotrachelus nenuphar* (Herbst) (Coleoptera: Curculionidae) in North America, the female has no choice but to oviposit into the youngest larvae of its host because on hatching the host larva burrows into the plum in which it lives and soon gets beyond the reach of the ichneumonid's ovipositor: the female searches for the oviposition scars caused by the weevil and oviposits by thrusting her ovipositor down the tunnel made by the host larva (Cushman, 1916). The females apparently could not differentiate between the oviposition sites of the host and other scars on the skin of the plum, into which they also probed with their ovipositor. At least some *Diaparsis* species are exceptional in ovipositing into the exposed larvae of chrysomelids (Dysart *et al.*, 1973).

Figure 153. *Barycnemis gravipes* (Gravenhorst).

There are striking differences in egg placement between the species that have been investigated, and this probably correlates with the considerable range of ovipositor insertion times noted for the subfamily in the literature (from a few seconds to a matter of minutes). The egg may be deposited free in the haemocoel of the host, attached to an internal organ, or placed in the body wall. The egg of *Tersilochus heterocerus* (Thomson) is simple in shape and has a smooth, blackish chorion whereas that of *Phradis morionellus* (Holmgren) has a colourless chorion with a dense covering of minute spines (Osborne, 1960). The eggs of *Diaparsis temporalis* Horstmann and *D. carinifer* (Thomson) have a 'knob-like' protrusion on one side, by means of which they are attached to tissue within the thoracic region of the host (Dysart et al., 1973; Montgomery & DeWitt, 1975). Some individuals of *D. temporalis* produced eggs without, or with only a rudimentary, protrusion and no protrusion is mentioned by Jourdheuil (1960) as being present in *Tersilochus tripartitus* (Brischke), the eggs of which are attached usually to the malpighian tubules of its host. In the South American species *Stethantyx parkeri* (Blanchard) the egg has a weak protrusion and is deposited in the body wall of the host between the cuticle and the epidermis (Parker et al., 1950). Incubation usually takes only a few days and, at least in *T. tripartitus* and *P. morionellus*, is accompanied by a large increase in the volume of the egg. In *Diaparsis* and *Tersilochus* hatching is rapid but in *S. parkeri* development of the posterior part of the embryo is slower than that of the anterior and the larva slowly issues from the egg over a period of about 13 days as development proceeds. In *T. heterocerus* the larva hatches only when the host is fully fed and about to enter the soil to pupate.

The first instar larva is of the caudate type and in *Diaparsis* and *S. parkeri* it bears two pairs of pseudopods near its posterior end. The first instar larva feeds but does not moult to the second instar until the host larva is fully grown and has constructed its pupal cell in the ground. Development of the parasitoid is then rapid and it is fully grown within a few days. Accounts vary between recognizing four or five larval instars. Cushman (1916) states that the last instar larva of *T. conotracheli* erupts from the host and completes its feeding externally, and Osborne (1960) reports the same for *P. morionellus*. This behaviour is unusual (or at any rate unusually reported) for Ichneumonidae, although it occurs in several subfamilies of Braconidae (Shaw & Huddleston, 1991), a necessary condition being that the host is killed in a site of concealment such as its cocoon or some other retreat. Although the behaviour has not been recorded for some other tersilochine species (e.g. Jourdheuil, 1960, contrasts the lack of an ectophagous final feeding phase in *Tersilochus tripartitus* (Brischke) compared to Cushman's (1916) findings), it may well be a widespread characteristic, explaining Carlson's (1979) insight that the group seems restricted to beetle larvae that construct a definite cocoon or pupation chamber in the soil. Parker et al. (1950), however, found that *Stethantyx parkeri* lacks an external feeding phase; it may be of significance that *Stethantyx* seems, on the basis of wing venation, to be phylogenetically rather basal within the *Tersilochus* genus-group. An external feeding phase was not reported for *Earobia*, of the *Phrudus* genus-group (see below), and was definitely ruled out for *Phrudus defectus* Stelfox (F.D. Bennett, pers. comm.). The fully grown *Tersilochus* group larva spins a cocoon within the host pupal cell. The cocoons usually have a distinct median band, for example the cocoon of *Diaparsis stramineipes* (Brischke) is papery in texture and light brown, with a thickened median band of a creamy colour (Carleton, 1939). The adult emerges by biting a small round hole in the end.

Most of the European species that have been studied are univoltine. In *Tersilochus* and *Phradis* pupation immediately follows cocoon formation and the pupal period is brief, but the fully formed adult remains quiescent within the cocoon until the following spring (Jourdheuil, 1960). In *Diaparsis* the prepupa overwinters within its cocoon and pupation does not take place until the following spring (Dysart et al., 1973). Jourdheuil (1960) concluded that species of *Aneuclis* are plurivoltine and that *A. melanarius* (Holmgren) overwinters as a first instar larva in larvae of certain hosts whilst utilizing a different host for its summer generation(s). Horstmann (1971b) records that *Probles gilvipes* (Gravenhorst) is dimorphic in ovipositor length, and a few European species are known to be plurivoltine.

Superparasitism appears to be frequent, suggesting little or no effective mechanism to avoid it. It has been reported that supernumeraries seem to be eliminated by physical aggression between first instar larvae (Parker et al., 1950), and that dead competitors become encapsulated (Cushman, 1916). Successful encapsulation of eggs and larvae sometimes occurs in the normal host of *Diaparsis carinifer* (Thomson) (Montgomery & DeWitt, 1975), and Osborne (1960) noted that *T. heterocerus* usually failed to complete its development when other parasitoids were present and that even in the absence of other parasitoids some eggs failed to hatch and first instar larvae died. He also encountered 'numerous' overwintering host adults with one to three black egg-shells embedded in the fat-body, showing unsuccesful parasitism.

Although most hosts recorded for the *Tersilochus* group are phytophagous Coleoptera larvae, one species, *Barycnemis blediator* (Aubert), has been found to be a parasitoid of a rove beetle, *Bledius spectabilis* (Kraatz) (Coleoptera: Staphylinidae), in tidal mud. Wyatt & Foster (1989) found that *B. blediator* females can occur at high density in saltmarshes and are a significant cause of mortality of *B. spectabilis* larvae. The adult beetle exhibits parental care of larval broods which seems to be a defence against parasitoids, as almost invariably host larvae were attacked only when they had dispersed from their mother. The immature wasp finishes its development in the host's water-resistant pupal chamber, but adult wasps seem to have only a short time in which to find hosts before being drowned by high tides.

Very little is known about host associations of the *Phrudus* group of genera, which includes some of the smallest British ichneumonids. The European species, *Earobia paradoxa* (Perkins), is a koinobiont larval endoparasitoid of a derodontid beetle (Franz, 1958) and in Britain *Phrudus defectus* has recently been found to be a koinobiont endoparasitoid of *Epuraea melanocephala* (Marsham) (Coleoptera: Nitidulidae) on Sycamore (*Acer pseudoplatanus* L.) flowers (F.D. Bennett, pers. comm.). In North America, species of *Peucobius* Townes have been swept in large numbers from immature staminate cones of *Pinus* species (Townes, 1971). There is a record, but with little associated information, that *Pygmaeolus nitidus* (Bridgman) is a parasitoid of *Ceutorhynchus pleurostigma* (Marsham) (Coleoptera: Curculionidae) (Günthart, 1949).

Franz (1958) made extensive observations on *E. paradoxa* in central Europe. The ichneumonid and its host, *Laricobius erichsonii* Rosenhauer, are univoltine. Only females of *E. paradoxa* were found, indicating thelytokous parthenogenesis. From mid May to mid June females search on the bark of conifers for third and fourth instar larvae of *Laricobius erichsonii*, which feed on adelgids. The host is usually stabbed with the ovipositor on the lateral or ventral part of the posterior third of the abdomen. The ovipositor remains inserted for about 20 seconds and the host becomes temporarily paralyzed. The egg is deposited in the host's fat body and by the time it is ready to hatch has increased about 20 times in volume. The females frequently host-feed (whether destructively or concurrently was not stated) and they also feed on adelgid honeydew. The ichneumonid larva remains small while the host continues to be active, growing rapidly only once the host enters the litter layer (although it was not stated whether the host spins a cocoon or makes a pupation retreat) before erupting to spin a whitish cocoon studded externally with soil particles. The adult emerges the following spring; how *E. paradoxa* overwinters within its cocoon has not been investigated.

Identification. Although there are no works dealing specifically with the British Tersilochinae, the whole European fauna of the *Tersilochus* group has been revised by Horstmann (1971b; 1981), with more recent keys to *Aneuclis*, *Barycnemis*, *Diaparsis*, *Epistathmus* (one British species), *Gelanes*, *Probles* (*Rugodiaparsis*) (one British species), *Sathropterus* (one British species) and *Spinolochus* (partially) by Khalaim (2002a,b, 2003, 2004a,b, 2005) and *Phradis* by Khalaim et al. (2009). Gauld & Fitton (1980) revised the British and Irish species of the *Phrudus* group but since then more species have been found (Shaw, 1991; Vikberg & Koponen, 2000) and Vikberg & Koponen (2000) provide a more reliable key.

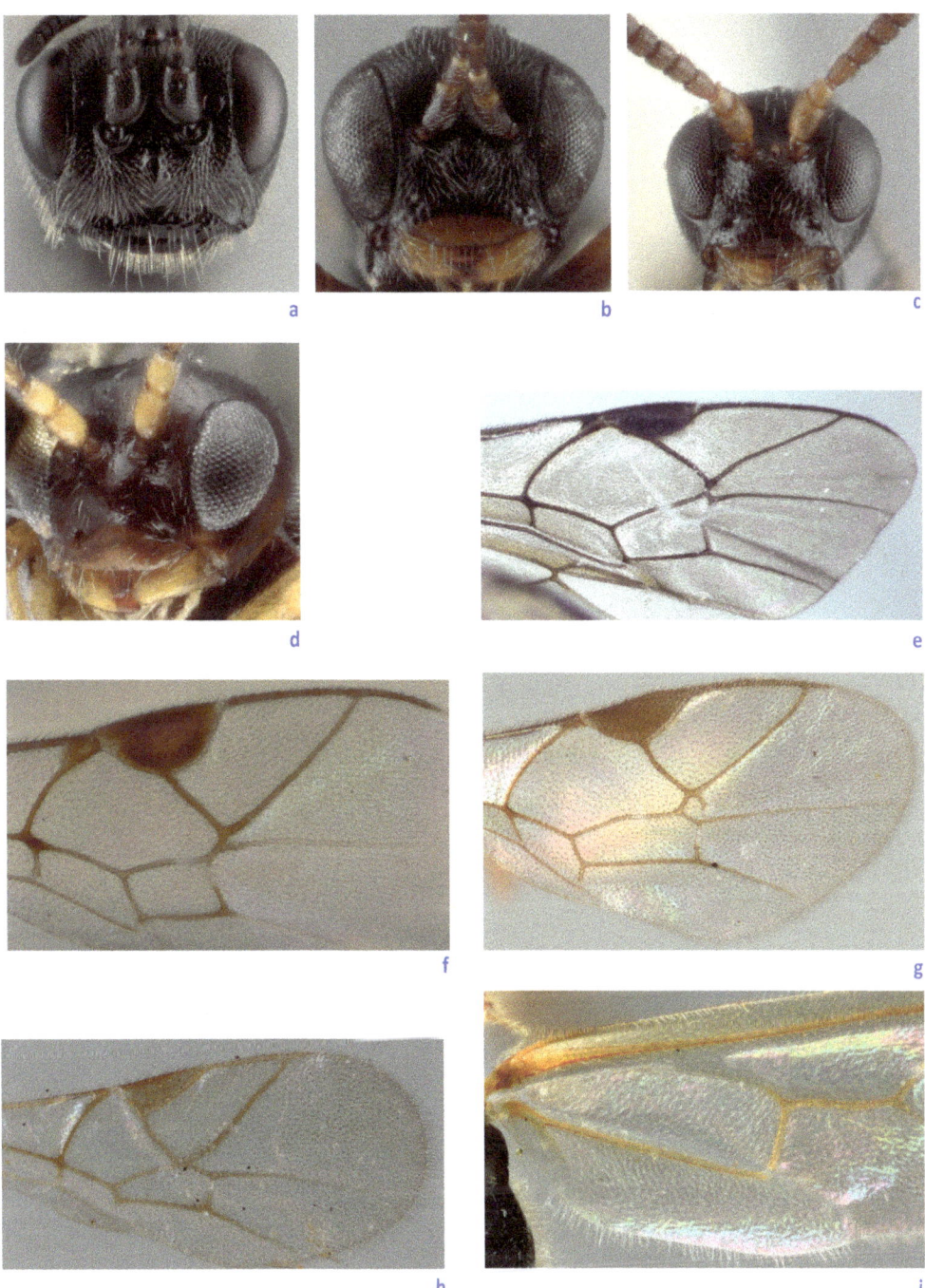

Figure 154. Morphology of Tersilochinae: (a) *Barycnemis gravipes* (Gravenhorst), head; (b) *Heterocola similis* Horstmann, head; (c) *Phrudus monilicornis* Bridgman, head and proximal antennal segments; (d) *Pygmaeolus nitidus* (Bridgman), head; (e) *B. gravipes*, part of fore wing; (f) *H. similis*, part of fore wing; (g) *P. monilicornis*, part of fore wing; (h) *P. nitidus,* part of fore wing; (i) *B. gravipes*, part of hind wing; (continued overleaf).

Figure 154 continued: Morphology of Tersilochinae: (j) *H. similis*, propodeum; (k) *H. similis*, first to third metasomal segments; (l) *Astrenis paradoxus* (Schmiedeknecht), first to fourth metasomal segments, arrows pointing to spiracles; (m) *P. monilicornis*, propodeum and metasoma; (n) *P. monilicornis*, fore tarsal claws; (o) *Sathropterus pumilus* (Holmgren), ovipositor tip.

Subfamily TRYPHONINAE

Tribes
*Ankylophonini
Eclytini
Idiogrammatini
Oedemopsini
Phytodietini
Sphinctini
Tryphonini

Although found world-wide, the Tryphoninae has the highest generic diversity in the north temperate regions and in Britain we have six of the seven tribes represented (only the Australian Ankylophonini is entirely extralimital). Excluding the genus *Netelia*, which is very species-rich in a variety of climes, species-richness is also far higher in the north temperate region, as are the principal hosts of the subfamily, sawflies (Hymenoptera: Tenthredinoidea, Xyeloidea). Of about 1,250 described species in 54 extant genera, 182 species in 30 genera have so far been found in Britain. Although the bulk of the genera are parasitoids of sawfly larvae, three tribes comprise parasitoids of Lepidoptera, although of these only the Phytodietini is particularly species-rich. Adults of some Tryphonini, particularly *Cosmoconus* and *Tryphon*, are frequently found on umbellifer flowers. *Netelia* are most easily recorded by light trapping but other tryphonines are easily collected using the standard collecting techniques, although sawfly larvae are often more difficult to rear to the cocoon stage than Lepidoptera larvae. Some species of *Tryphon* can be abundant in Malaise trap catches in early summer.

Figure 155. *Eclytus exornatus* (Gravenhorst), tribe Eclytini.

Recognition. The Tryphoninae are not readily recognisable as a subfamily owing to their variety of body forms. Many species, particularly of the tribe Tryphonini, are rather stout, with a non-petiolate (sessile) first metasomal segment, the sclerotized part of the first sternite short, and with glymmae on the first tergite; males of these genera are frequently confused with Ctenopelmatinae and Banchinae (Atrophini). Females are more distinctive as tryphonine ovipositors lack a dorsal notch and sometimes have an egg or, in the case of *Polyblastus* species, several eggs hanging from the base of the ovipositor. Individual tribes or genera are usually more distinctive. Almost all tryphonines have the clypeus with a discrete apical row of closely-spaced setae (as do many Tersilochinae, which otherwise look very different), although this row is subapical and medially interrupted in *Eclytus* and absent in *Sphinctus*; many ichneumonid taxa, including Ctenopelmatinae, have many loosely spaced but long setae on the clypeus. The *Exenterus* group of the Tryphonini are distinctive in that they lack the hind tibial spurs. Sphinctini, comprising only *Sphinctus*, are recognisable by the prominent, medial tooth on the apical edge of the clypeus, the clavate metasoma and the hind tibia with one spur (but can be confused with *Ischyrocnemis*, currently classified in the Metopiinae). The Idiogrammatini, comprising the single genus *Idiogramma*, are readily recognisable by the exodont mandibles (they only slightly overlap when closed) which are strongly carinate dorsally and ventrally, and the female's long ovipositor (which can be as long as the body). Most Banchinae that could be confused (mainly males of Atrophini) with Tryphoninae have only the posterior transverse carina of the propodeum present and the submetapleural carina of the metapleuron strongly expanded as a lobe anteriorly, whereas tryphonines usually have more complete propodeal carination and the submetapleural carina not expanded anteriorly. Ctenopelmatinae very frequently have granulate sculpture on the mesosoma, which is shiny in Tryphonini, and have a tooth on the outer edge of the fore tibia (sometimes difficult to see in small specimens or depending on orientation of the leg), which is usually absent in Tryphoninae. Whereas the majority (but not all) Ctenopelmatinae have one bulla in fore wing vein $2m-cu$, Tryphoninae generally have two bullae, or one elongate bulla. Oedemopsini and Eclytini have deep glymmae on the first tergite (at least in European species) and long notauli. All female Tryphoninae lack a dorsal, subapical notch on the ovipositor. Female oedemopsines have a membranous, swollen, sub-apical area on the lower ovipositor valve. Phytodietini can be difficult to recognise as Tryphoninae but the shape of the ovipositor is distinctive (straight and evenly narrowed in *Netelia*, the tip sagittate in *Phytodietus*). The fore wing areolet is triangular (rarely open) and the propodeum lacks carinae, except that there are raised lateral sections of the posterior transverse carina in some species. Unlike many other Tryphoninae, the body microsculpture of Phytodietini is predominantly matt.

Systematics. Typhoninae has long been recognised as one of the 'traditional' subfamilies of Icheumonidae, although formerly with various unrelated groups such as Metopiinae and Ctenopelmatinae included. The modern concept of Tryphoninae assumes monophyly of the subfamily based on two apomorphies: the egg has a chorionic stalk and anchor and passes externally down the ovipositor, and the clypeus has a fringe of closely spaced setae (Gupta, 1988; Bennett, 2015), although the fringe is interrupted in *Eclytus* and absent from *Sphinctus*. The Tryphoninae was long considered to be a 'primitive' subfamily within Ichneumonidae, as evidenced by Townes's (1969) placement of the subfamily in his first volume on the genera of Ichneumonidae, or has been essentially unplaced within Ichneumonidae (Gauld, 1997). The idea that Tryphoninae are basal within the family was based on supposedly plesiomorphic morphology, such as the deep glymmae and anteriorly placed spiracles on the first metasomal tergite, and their biology, as ectoparasitoids (which ignores the fact that they are highly specialised koinobionts). For example, Wahl & Gauld (1998) used *Tryphon* as a rather generalised outgroup in their phylogenetic analysis of pimpliformes relationships. However, Quicke *et al.* (2009) found the Tryphoninae to be a grade at the base of the ophioniformes group of subfamilies, i.e. not basal within the Ichneumonidae, albeit with inconclusive results as to the exact placements of the tribes. Although most of the Tryphoninae formed a monophyletic group (with some currently non-tryphonine taxa included) under many analysis parameters, the Phytodietini were not included in this monophyletic group. Quicke *et al.* (1994) had earlier demonstrated that Tryphoninae shared a characteristic of the ovipositor (the upper valve secondarily divided with the two halves linked by the aulaciform rod) with several other (endoparasitoid) ophioniform taxa.

The tribal composition of the Tryphoninae has changed a little since Townes's (1969) classification of the genera. The Eucerotinae are now universally accepted as a separate subfamily, as had been proposed by Perkins (1959), rather than a tribe of Tryphoninae. Gauld (1988) described a new tribe, Ankylophonini, for a single, distinctive Australian species, *Anykylophon obligatus* Gauld, which has various similarities with the tribe Sphinctini. Gupta separated the tribe Eclytini, with the single genus *Eclytus*, from the Oedemopsini, based on some rather weak morphological characters and their uses of different host groups (sawflies in the case of *Eclytus*, Lepidoptera in Oedemopsini) and included *Acaenitellus* in the Oedemopsini (Townes, 1971, had included *Acaenitellus* in his 'Microleptinae'). Bennett (2002, 2015), on the basis of a morphological phylogenetic study, found that the genera long separated as the tribe Exenterini, easily recognised by their lack of hind tibial spurs, are in fact derived from within the Tryphonini and (Bennett, 2015) formally synonymised the tribes. Other than the descriptions of a few new (extralimital) genera and some changes to the generic ranks and names (see Bennett, 2015), the generic composition of the tribes remains much the same as Townes's (1969) treatment although there is now a phylogenetic framework for the genera.

Biology.
Host associations and life histories
This subfamily constitutes by far the largest group of ichneumonid koinobiont ectoparasitoids (other radiations involve the Adelognathinae, the *Polysphincta* group of the Pimplinae and arguably Lycorininae). Although oviposition is generally (but not invariably) on exposed hosts, development of the tryphonine larva proceeds when the host has prepared a pupation site, thus the ichneumonid secures a protected retreat without having to locate and gain access to a concealed host. Unlike some koinobiont ectoparasitoid cyclostome Braconidae, tryphonines do not cause the host to form a cocoon or pupa precociously; instead, oviposition is onto the final or penultimate instar host larva and eclosion follows when the host is preparing for pupation and the humidity within the retreat is sufficient (Shaw, 2001). The tryphonine egg is characteristic of the subfamily and is adapted to hold fast to the host, potentially through larval ecdyses, via an anchor that is an extension of the egg chorion. The egg is anchored at oviposition, with the anchor passing down the interior of the ovipositor (the notional egg canal) and and the rest of the egg outside of the ovipositor. The anchor fastens the egg, which is otherwise external, to the inside of the host's epidermis. Kasparyan (1973b) published a thorough summary of the biology of the tribe Tryphonini, figuring much variation in egg morphology. There are several detailed studies of exenterine (Tryphonini) species that are parasitoids of economically important defoliators of conifers (e.g. Morris, 1937; Pschorn-Walcher, 1967), and Shaw (2001) described interactions between several species of *Netelia* and their hosts.

The majority of the tryphonine genera are, where known, parasitoids of sawfly (Hymenoptera) larvae, of the families Argidae, Diprionidae, Tenthredinidae (all Tenthredinoidea) and Xyelidae (Xyeloidea), with the majority of host records unsurprisingly from the largest sawfly family, Tenthredinidae. Only one species, the non-British (but European) *Orthomiscus pectoralis* (Hellén) (*Exenterus* group of Tryphonini), has been recorded as a parasitoid of Cimbicidae (Tenthredinoidea) (independently by Jussila, 1975 and Kasparyan, 1976) and it is gregarious: a mixed sex brood of 24 is recorded by Kasparyan (1990) from *Trichiosoma tibiale* Stephens. The British *Orthomiscus unicinctus* (Holmgren) has never been reared. There seem to be no substantiated host records from Pamphilioidea. There are no known tryphonine parasitoids of Pergidae but there are very few host records for tropical tryphonines, where pergids are at their most species-rich. However, there are also significant numbers of tryphonine species attacking Lepidoptera larvae. The host associations of each tribe are outlined here, although it should be borne in mind that there is little evidence for the monophyly of the Tryphonini and the relationships between the tribes are uncertain; Gupta (1988) and Bennett (2015) found evidence for a clade of Lepidoptera parasitoids (Phytodietini, Oedemopsini and Sphinctini, plus the extralimital *Ankylophon* Gauld and *Leptixys* Townes that have not been reared), which would be compatible with various aspects of their host interactions, at least for Oedemopsini and Phytodietini.

Ankylophonini are known from very few specimens and nothing is known of the biology of the single included species.

Eclytus is the sole genus of Eclytini, although there are two described subgenera regarded as valid (*E.* (*Zapedias*) was synonymised under *E.* (*Eclytus*) by Bennett, 2015). Hosts are sawfly larvae, with the distinctive *Eclytus* (*Anoplectes*) *multicolor* (Kriechbaumer) a parasitoid of several *Arge* species (Hymenoptera: Argidae) (Pschorn-Walcher & Kriegl, 1965), and apparently some Tenthredinidae (Nematinae) (Stelfox, 1966), and *E.* (*Eclytus*) having been reared from the tenthredinid subfamily Nematinae, although only three of the 20 described species of the subgenus *Eclytus s.s.* have been reared (Kasparyan, 1977).

Idiogrammatini is a small tribe comprising seven species of the Holarctic (including central Mexico: Khalaim & Ruíz-Cancino, 2017) genus *Idiogramma* and one fossil genus (*Urotryphon* Townes). Three species, including the British *Idiogramma euryops* Förster, have been recorded as parasitoids of *Xyela* Dalman species (Hymenoptera: Xyelidae), the larvae of which feed deep in male staminate conifer cones and are accessed using the unusually long ovipositors of female *Idiogramma*. The Xyelidae is the most basal family of extant Hymenoptera (e.g. Vilhelmsen, 2001), and their parasitoids tend to be rather isolated at the genus or tribe level. Cushman (1937b) studied North

Figure 157 (above). *Oedemopsis scabricula* (Gravenhorst), tribe Oedemopsini.

Figure 156 (left). *Idiogramma euryops* Förster, tribe Idiogrammatini.

American species of *Idiogramma*, parasitoids of *Xyela* in staminate cones of *Pinus virginiana*. There were three species present, apparently attacking some or all of three or four species of *Xyela* known to feed on *Pinus virginiana* (Burdick, 1961); Cushman (1937b) does not say, or did not establish, which species (singular or plural) he is describing in his brief account of the biology of *Idiogramma*, although he lists two species as parasitoids of *Xyela*. The eggs are reported to be white. The distinctive exodont mandibles could be used to spread the bracts of the cone when searching for hosts (Burdick, 1961). The larvae did not eclose until the *Xyela* larvae were in their pupation chambers in the soil. Development was slow and the winter is reportedly passed as a pupa. Although Cushman (1937b) reports that *Idiogramma* pupate in 'thin, shining, transparent cocoons' within the host's cocoon, Burdick (1961) did not find any such cocoons.

Figure 158. *Phytodietus griseanae* Kerrich, tribe Phytodietini.

The Oedemopsini is a tribe of Lepidoptera parasitoids that is virtually cosmopolitan, although not very species-rich. Species are generally small and host records are mainly from 'microlepidopteran' families, particularly Coleophoridae, Tortricidae and Ypsolophidae, that feed in weak concealment. Host ranges, as far as can be inferred from the limited data, are circumscribed by family and habitat; for example, *Hercus fontinalis* (Holmgren) has been reared from a variety of Tortricidae feeding on tree foliage and understorey whilst *Neliopisthus elegans* (Ruthe) attacks a range of arboreal *Coleophora* (Coleophoridae) larvae (Fitton & Ficken, 1990). *Thymaris* species have been reared from unidentified 'microlepidoptera' larvae feeding in rotten twigs. There have been no in-depth studies of the biology of oedemopsines although Gerig (1960) provides detailed descriptions of the egg and the five larval instars of *Hercus fontinalis*.

The tribe Phytodietini is, globally, enormously species-rich (although not especially so in Britain) but comprises only two extant genera, *Netelia* and *Phytodietus*, both of which have been subdivided into several subgenera. These are Lepidoptera parasitoids, *Netelia* being reared usually from 'macrolepidoptera' and *Phytodietus* from 'microlepidoptera' such as pyralids and tortricids. Superficially, the two genera are dissimilar, with *Netelia* being mostly larger, predominantly testaceous and nocturnal, and *Phytodietus* mostly smaller, black, often with yellow markings (some extralimital taxa are entirely yellow) and with a distinctly different ovipositor shape. However, some New Guinean *Netelia* species of the subgenus *Protonetelia* Konishi narrow the morphological gap (Konishi, 1986; undescribed specimens in BMNH). Hosts of British *Phytodietus* are centred on

Depressariidae, Gelechiidae, Pyralidae and Tortricidae that feed in semi-concealed situations, especially in prominent vegetation (Kasparyan & Shaw, 2008). *Netelia* subgenera differ in their host foci; species of the subgenera *Bessobates* Townes, Townes & Gupta, *Paropheltes* Cameron and *Prosthodocis* Enderlein attack predominantly Geometridae, also Notodontidae and Drepanidae (Thyatirinae), although *N. (Bessobates) cristata* (Thomson) has a wide host range that encompasses several families of Noctuoidea; *N. (Netelia)* have been reared, in Britain, from Noctuidae and Notodontidae; the single European *Parabates* Förster, *N. nigricarpa* (Thomson), is the exception in that it has supposedly been reared from a species of Tortricidae (Hedwig, 1939, 1950). As with the nocturnal Ophioninae, many species of *Netelia* can inflict a painful sting, although the effects are very short-lived.

The Sphinctini has just one included genus, *Sphinctus*, and one species on the British list, *Sphinctus serotinus* Gravenhorst. This is a solitary parasitoid of *Apoda limacodes* (Hufnagel) (Lepidoptera: Limacodidae) (Hinz, 1976; Shaw & Voogd, 2016) and it may be extinct in Britain, with the only known specimens collected in the 19th Century despite its host's persistence; although there is a possibility that it has been overlooked, with its black and yellow, vespid-mimicking appearance and autumnal flight period. A Chinese species has also been reared from an unidentified limacodid (He *et al.*, 1996). *Sphinctus* species are poorly represented in most collections, and Kasparyan (1992) noted that these species fly in late summer or early autumn in temperate regions.

Kasparyan (1973b) summarised the host associations of the Tryphonini, the most generically numerous and, at least in temperate areas, species-rich tribe of Tryphoninae. Hosts span a wide range of Tenthredinidae with just a few species in particular genera attacking Diprionidae, Argidae and Cimbicidae. The traditional Tryphonini (i.e. excluding the *Exenterus* group, with respect to which it is paraphyletic) are almost entirely parasitoids of Tenthredinidae, with the exception of the rather aberrant, and non-British, *Boethus* Förster, which are parasitoids of Argidae (e.g. Berland, 1947; Kasparyan, 1973b). Another interesting exception, mirroring the host range expansion of *Lathrolestes* (Ctenopelmatinae), is the adoption of several species of the lepidopteran genus *Eriocrania* (Eriocraniidae) into the host range of *Grypocentrus* (Heath, 1961; Jordan, 1998b), which are otherwise parasitoids of Fenusini sawflies that leaf-mine in similar situations to *Eriocrania*. Although the *Grypocentrus* concerned are supposed to be highly host specific (for both the parasitoids of fenusines and *Eriocrania*) (Altenhofer, 1980; Jordan, 1998b), there has been no thorough taxonomic revision of the genus. Based on oviposition site correlated with host species, many undescribed species are inferred to occur. Other species of Tryphonini seem not to be regularly absolutely host specific. The more species-rich genera of Tryphonini, particularly *Ctenochira* and *Erromenus*, comprise mainly parasitoids of the large subfamily Nematinae, with

Figure 159. *Sphinctus serotinus* Gravenhorst, tribe Sphinctini.

Polyblastus and some *Ctenochira* attacking Blennocampinae. A group of physically larger genera of Tryphonini attack grass and horsetail-feeding Dolerini (some *Tryphon*) and exposed Tenthredininae and Allantinae (*Cosmoconus*, *Dyspetes* and some *Tryphon*) (hosts summarised from Kasparyan, 1973b). The distinctive *Exenterus* group are parasitoids of sawflies of the families Argidae, Cimbicidae, Diprionidae and Tenthredinidae, although the biology of many species is unknown. Whilst numerous species of *Exenterus* have been reared from Diprionidae, almost all other records are from Tenthredinidae, with the exception of one exenterine, *Exyston pratorum* (Woldstedt), reared from an argid, *Arge ustulata* (L.) (Weiffenbach, 1988), and one from a cimbicid (*Orthomiscus unicinctus*, as described above). *Exenterus* parasitoids of Diprionidae have been rather well studied (e.g. Morris, 1937) but little is known of the biology of other exenterines, which are frequently parasitoids of Nematinae (Tenthredinidae) (e.g. Shaw & Kasparyan, 2005).

Figure 160. *Exenterus abruptorius* (Gravenhorst), tribe Tryphonini.

Developmental biology
For the small tribes Idiogrammatini and Oedemopsini the rather sparse knowledge of developmental biology is incorporated in the section above, and the biology of the rather isolated *Sphinctus* is summarized separately below. Most of the following is based on just the few species of Tryphonini and Phytodietini that have been studied in detail.

Temporary paralysis of the host, prior to oviposition, has been reported for very few tryphonines, namely several species of *Netelia* (Shevyrev, 1912; Vance, 1927; Shaw, 2001), *Phytodietus obscurus pulcherrimus* (Cresson) (Simmonds, 1947), *Exenterus abruptorius* (Morris, 1937) and, supposedly, *Idiogramma* (Cushman, 1937b), and Kasparyan (1973b) notes that of the Tryphonini only a few of the *Exenterus* group seem to sting their hosts. *Netelia* venom, which is deployed to temporarily subdue the host larva, has no disruptive effect on the host's development as hosts that have been stung and then had any eggs removed will go on to pupate and produce adults (Shaw, 2001). Two studied species of *Netelia* of the subgenus *Bessobates* do not sting their hosts (and correspondingly have relatively short ovipositors) but grapple with the active host (Shevyrev, 1912; Shaw, 2001).

From the limited observations of other *Netelia* and *Phytodietus*, which all have longer ovipositors, ideal for jabbing caterpillars, it is probable that temporary paralysis of the host is otherwise usual in Phytodietini. Concurrent host-feeding is frequent. If *N.* (*Bessobates*) *cristata* larvae that have been feeding for several days are removed from their hosts, the caterpillar will go on to form pupation chambers but will not pupate, demonstrating that some physiological disruption takes place at this stage of the host-parasitoid interaction (Shaw, 2001). Superparasitism is frequent in Tryphoninae (e.g. Clausen, 1932; Eichhorn & Pschorn-Walcher, 1973), which probably have no means of detecting the presence of another parasitoid larva on a host, but solitary development is the norm. The few known gregarious tryphonines include, in Britain, *Hercus fontinalis* (Holmgren), *Netelia cristata*, *N. fuscicornis* (Holmgren), *N. testacea* (Gravenhorst) and *N. vinulae* (Scopoli), with at least *N. vinulae* invariably so (Fitton & Ficken, 1990; Shaw, 2001; Broad & Shaw, in prep.). Vance (1927) reports that two *Netelia* (*Netelia*) species 'brush' the anterior thoracic segments of the host with their ovipositors, a description reminiscent of the behaviour of some polypshinctine Pimplinae, which remove eggs of conspecifics using the ovipositor as a lever (Takasuka & Matsumoto, 2011b).

Tryphonine eggs have a tough, resistant chorion which is often dark (although colour varies between tribes and genera), and which presumably protects them from mechanical damage and desiccation. The eggs of oedemopsines are generally much smaller than those of other tryphonines (Fitton & Ficken, 1990). The egg stalk in Phytodietini is long and loosely coiled, and does not have a specialised anchor at its tip. When first laid the stalk is elastic but later becomes rigid, holding the egg firmly to the host's cuticle (Vance, 1927). Egg anchors and stalks range in Tryphoninae from relatively simple, coiled and thread-like, to very complex structures, such as are found in *Exenterus abruptorius* (Thunberg), a parasitoid of the important forestry pest, *Neodiprion sertifer* (Geoffroy) (Hymenoptera: Diprionidae). In *E. abruptorius*, the egg is almost entirely embedded in the cuticle of its final instar or prepupal host by means of an enlarged, shield-like anchor that opens up to partially enclose the egg and fix it in the host cuticle at both ends (Pschorn-Walcher, 1967). Female Tryphonini are sometimes found with an egg exposed at the base of the ovipositor and this is particularly obvious in *Polyblastus*, in which a bunch of eggs is found, containing progressively more advanced embryos/larvae towards the posterior (oldest) end of the egg mass. Eclytini and some genera of Oedemopsini (*Hercus*) also often carry an egg at the base of the ovipositor, but in oedemopsines with a centrally swollen ovipositor sheath (such as *Oedemopsis* and *Thymaris*) the egg is not carried externally and it seems likely that it may issue from this area (although direct

Figure 161. *Polyblastus cothurnatus* (Gravenhorst), tribe Tryphonini.

evidence is lacking). As with the anchors, preferred oviposition sites vary between species. In the majority of studied tryphonines, eggs are attached to the host's anterior thoracic segments, where they cannot be reached by its mandibles, although there are numerous exceptions. *Eclytus ornatus* Holmgren attaches its eggs near the base of the mandibles of *Nematus pavidus* (Lepeletier) (Hinz, 1961); *Dyspetes luteomarginatus* Habermehl preferentially oviposits laterally, between the thoracic legs of its *Tenthredo* and *Tenthredopsis* hosts (Hinz, 1961). In *Sphinctus* it is at random on any part of the dorsal carapace of its slug-like caterpillar host, all of which is out of reach of the host's mandibles. Egg placement can vary with the degree of attachment of the egg, as Pschorn-Walcher (1967) described for two exenterines, *Exenterus abruptorius* and *E. amictorius* (Panzer). *Exenterus abruptorius*, with its firmly embedded eggs, shows no preference for oviposition site whereas *E. amictorius*, with more shallowly embedded eggs, oviposits behind the head. Mason (1967) illustrates the different degrees of specialisation in *Exenterus* eggs and hypothesises that the more deeply embedded eggs represent an adaptation towards drier conditions in pine forests, based on Morris et al.'s (1937) idea that the stalks of tryphonine eggs are prone to desiccation and hence being sloughed off at the host's ecdysis. In one extreme case, *Erromenus calcator* (Müller), the female oviposits through the host's anus and attaches the egg to the wall of the hind gut (Zinnert, 1969b). A similar egg placement strategy is deployed by a very few other, unrelated ichneumonids, namely *Chorinaeus funebris* (Gravenhorst), of the subfamily Metopiinae (Aeschlimann, 1974a) and *Lycorina* species (Shaw, 2004), taking advantage of the fact that the hind gut is physiologically external cuticle and eggs are safe from being dislodged by the host. Egg microsculpture and morphology have provided useful taxonomic characters at the species level in genera where the external morphology is rather uniform, as in *Eclytus* (Kasparyan, 1977) and *Netelia* (*Prosthodocis*) (Townes, 1939b; Konishi, 1991, 1992), and Jordan (1998b) inferred from variation in egg morphology the presence of several undescribed species that had been confused under *Grypocentrus basalis* Ruthe.

Figure 162. *Sphinctus serotinus* Gravenhorst searching for host *Apoda limacodes* (Hufnagel) (Lepidoptera: Limacodidae) larva (M.R. Shaw).

Eichhorn & Pschorn-Walcher (1973) studied the biology of *Grypocentrus albipes* Ruthe, which is a host-specific parasitoid of the birch (*Betula*) leaf-miner, *Fenusa pusilla* (Lepeletier). Oviposition is almost always onto the first thoracic segment and larval development (which commences, as is usual, once the host has spun a cocoon) takes 12-13 days. A majority of individuals emerge the same year, after a prepupal and pupal period of 14-15 days, but a minority spend the winter in diapause. This plurivoltinism matches the plurivoltinism of the host. Potential fecundity was about 100 eggs. Fecundity has generally been noted as being rather low in tryphonines but potential egg numbers vary considerably across the subfamily (Kasparyan, 1973b). Larval development generally commences rather quickly, once the host has reached its pupation site, though not in *Sphinctus* (see below). In the univoltine *E. abruptorius*, the first instar larva can enter a period of diapause for up to two and a half months, presumably whilst the host prepupa also diapauses, although this is not stated (Morris, 1937). Unsurprisingly, given that *E. abruptorius* is an ectoparasitoid, it generally out-competes the ctenopelmatine endoparasitoid *Lophyroplectus oblongopunctatus* (Hartig) that can be involved in multiparasitising the same host (Pschorn-Walcher, 1967). Whereas the studied species of Oedemopsini (*Hercus*) and Phytodietini keep the end of the abdomen in the empty egg chorion for most of their larval life, and have specialised caudal setae to enable this (Vance, 1927; Simmonds, 1947; Kasparyan, 1973b), larvae of Tryphonini leave the vicinity of the egg and move around the host's body (Clausen, 1932). First instar larvae of *E. abruptorius*, *Eridolius alacer* (Gravenhorst) and the extralimital *Monoblastus erythrurus* Townes, Townes & Momoi have lateral tufts of long setae and wander actively about the prepupal host in a looping fashion, whereas at least some Tryphonini have more conventional, shorter setae and do not move like this (Clausen, 1932). In *E. abruptorius*, the host is not immobilised until the tryphonine reaches its second instar. In *Polyblastus*, oviposition sometimes occurs when the parasitoid larva is already emerging from the egg and Kerrich (1936) supposed that this propensity for the larva to immediately begin feeding is an adapation to avoid being sloughed off at the host's ecdysis. Egg dumping is frequently observed in *Netelia* when there are insufficient hosts (e.g. Shevyrev, 1912) and such eggs are viable if they are kept in high humidity (Shaw, 2001). Development of the embryo in-utero seems to be frequent across Tryphoninae and Shevyrev (1912) reports that, in the absence of hosts, and presumably if eggs are not dumped fast enough, *Netelia* larvae can hatch in-utero and cause the death of the adult female. Kasparyan (1973b) observed cannibalism within the egg mass of *Polyblastus*. In contrast to *Polyblastus*, embryonic development of *Monoblastus erythrurus* is delayed until the host completes feeding (Clausen, 1932).

Shaw & Voogd (2016) have investigated *Sphinctus serotinus* in the field in the Netherlands and reared it in culture. Its host, *A. limacodes*, has an exceptionally long larval season overall, being present from June to October. By September, when females of *S. serotinus* are active, virtually all remaining *A. limacodes* larvae are in the final instar, and the larval population is by then free of its other parasitoids. In captivity, *S. serotinus* females fed avidly on dilute honey but not on hosts. Eggs were neither dumped nor carried externally on the ovipositor. After only brief antennal investigation the female mounts the host and oviposits, without any pre-oviposition sting, at random and singly on any part of the host larval carapace. The process takes about 10-15 seconds, with the pale brown egg sliding down the exterior of the ovipositor shaft to be anchored in the host epidermis in the usual tryphonine fashion (as had also been witnessed by Hinz, 1976). There is no direct avoidance of superparasitism: in the field host larvae with two, three and occasionally even four eggs were commonly seen, and in culture, supernumerary eggs resulted from repeat visits to a host without detectable inhibition. However, after oviposition the female leaves the scene actively (typically by flight), which presumably reduces at least self-superparasitism. The largest (presumably female) hosts, ca 14 mm long, are more readily accepted than smaller ones, but there is no absolute rejection of hosts as long as they are in the final instar. If eggs are laid in quick succession those after the first are initially pale yellow, darkening to the usual pale brown colour after an hour or so.

The parasitoid failed to develop on one host that had received an egg, and as that host then advanced to the (pharate) adult stage it is clear that there is no physiological venom effect. *Apoda limacodes* overwinters as a cocooned prepupa, and normally (except under captive conditions of high humidity) the parasitoid egg does not hatch until the host has formed its cocoon. Then the parasitoid egg splits, and the setose first instar larva, still partly within the eggshell, starts to feed immediately. In the few cases closely observed by opening host cocoons, feeding continued in this position slowly through the winter but by May the larva had left the eggshell and moulted at least once, but still remained small though moving freely across the still living host and making new feeding lesions. It was not established at which point the supernumeraries were eliminated, or how, but it seems improbable that this could take place while all larvae were still anchored in the eggshell. The parasitoid larva continues to grow steadily through spring and early summer, becoming fully grown in about July and spinning its own diaphanous white cocoon within that of its host. The prepupal and pupal periods appear to be of unexceptional duration, and nor is there a prolonged adult resting phase pre-eclosion. In the Veluwe region of the Netherlands *S. serotinus* has proved to be common and widespread, favouring especially relatively open sunny areas, where the *Quercus*-feeding host larvae that remain by the end of September and early October are usually heavily parasitised.

Figure 163. *Netelia vinulae* (Scopoli) ovipositing on a larva of *Cerura vinula* (Scopoli) (Lepidoptera: Notodontidae) (M.R. Shaw).

The influence of humidity on eclosion means that, in artificial rearing conditions, tryphonine larvae will often hatch prematurely, when the hosts are still active (Clausen, 1932). As in many ichneumonids, the size of the head capsule relative to the body decreases as the larva grows, the setae become less conspicuous (Clausen, 1932) and the number of pairs of spiracles increases after the initial single pair of the first instar (e.g. Simmonds, 1947). Several authors report five larval instars. The cocoons of *Netelia* are closely woven, black, with a median band of rougher silk, and are only spun within a confined space such as is provided by the host's cocoon (Vance, 1927). Those of *Phytodietus* are amber-coloured, with a paler median band (Simmonds, 1947). Most tryphonines pass the winter in the cocoon. Pschorn-Walcher & Zinnert (1971) found that *Polyblastus tener* Habermehl is facultatively univoltine or bivoltine depending on the voltinism of its larch-feeding nematine hosts and altitude (and hence the length of the season). In the intensively studied fauna of parasitoids of *Pristiphora erichsonii* (Hartig), *Polyblastus tener* was a superior competitor compared to all of the endoparasitoids

(Campopleginae and Ctenopelmatinae) but was not responsible for high rates of parasitism, as it has a fairly wide host range of nematine sawflies on larch. Like most ichneumonids, tryphonines are synovigenic and have been reported to take two to three weeks (in *Phytodietus*; Simmonds, 1947) to mature their eggs, with most wild-caught Tryphonini females near the beginning of their flight season containing only immature eggs (Kasparyan, 1973b).

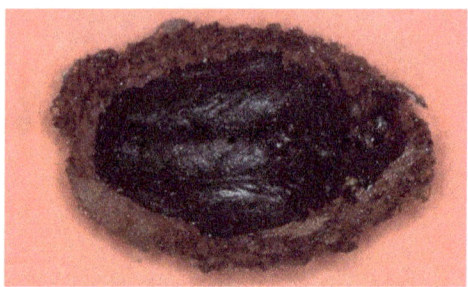

Figure 164. Cocoons of *Netelia vinulae* within the cocoon made by the host *Cerura vinula* (M.R. Shaw).

Identification. Bennett (2015) provides a key to the Palaearctic genera of Tryphoninae, as well as a global key to genera. The tribes are generally each fairly distinctive and can be keyed using Townes (1969) and Fitton & Ficken (1990), but note that these publications pre-date or ignore the separation of *Eclytus* into its own tribe, Eclytini (Gupta, 1988). Most of the British tryphonine fauna is covered by identification keys but the literature is scattered; the major omissions are that there are no complete revisions of European *Eridolius* nor *Phytodietus*. Key works are listed here by tribe. Eclytini (including only *Eclytus*): Kasparyan (1977). Idiogrammatini (only one British species of *Idiogramma*): Perkins (1959). Oedemopsini: Fitton & Ficken (1990) cover the British fauna; *Hercus* (Gupta, 1984); *Thymaris* (Kasparyan, 1993b). Phytodietini: *Netelia* (Delrio, 1975; Broad & Shaw (in prep.), draft keys available online at http://nocturnalichs.myspecies.info/); *Phytodietus* (*Neuchorus*) (Kasparyan & Shaw, 2008), Kasparyan (1994b) provides a partial revision of Palaearctic *Phytodietus*. Sphinctini: one British species in the single genus, *Sphinctus*; Kasparyan & Tolkanitz (1999) provide a key to the Russian and European fauna. Tryphonini: Kasparyan (1973b) covers most of the fauna of the traditional Tryphonini (i.e. excluding the *Exenterus* group); Kerrich (1952) covers much of the exenterine fauna but is of very little use for the large genus *Eridolius*; Kasparyan & Khalaim (2007) cover much of the tryphonine fauna, including the majority of British *Eridolius*; *Acrotomus* (Gupta, 1991); *Cteniscus* (Kerrich, 1953, as *Eudiaborus*); *Dyspetes* (Horstmann, 2006b); *Kristotomus* (Gupta, 1990); *Orthomiscus* (Gupta, 1994b); *Tryphon* (Fitton, 1975); note that *Otoblastus*, with one British species, was separated from *Monoblastus* by Kasparyan (1982). Shaw & Kasparyan (2005) and Kasparyan & Shaw (2008) provide notes on the British tryphonine fauna, including records of several species new to Britain.

Figure 165. Young larva of *Oedemopsis scabricula* (Gavenhorst) on the still-active larva of an unidentified tortricid (Lepidoptera), exposed in its spinning (A.R. Edwards).

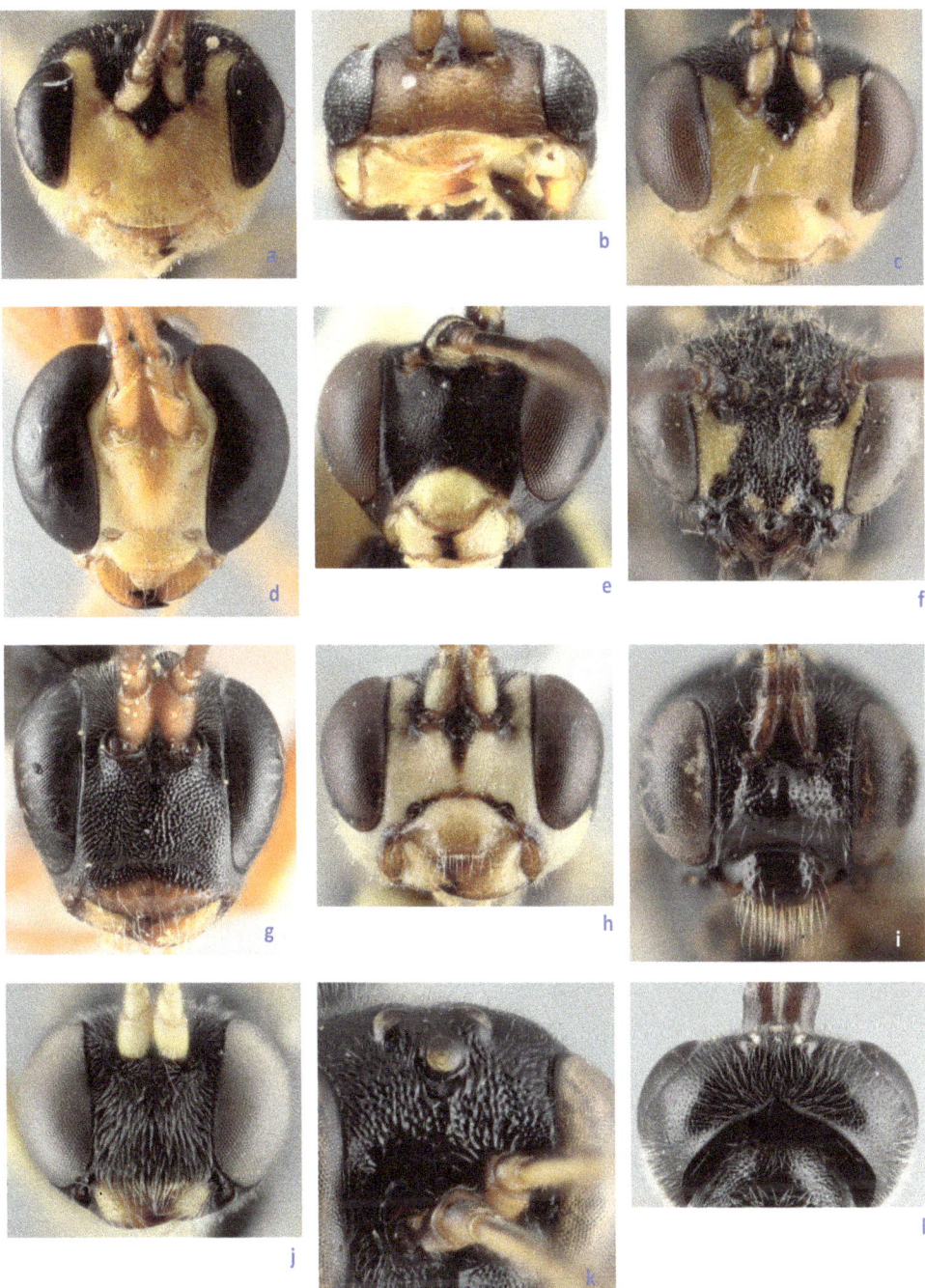

Figure 166. Morphology of Tryphoninae: (a) *Eclytus multicolor* (Kriechbaumer) (Eclytini), head; (b) *Idiogramma euryops* Förster (Idiogrammatini), head, semi-ventral; (c) *Hercus fontinalis* (Holmgren) (Oedemopsini), head; (d) *Netelia infractor* Delrio (Phytodietini), head; (e) *Phytodietus ornatus* Desvignes (Phytodietini), head; (f) *Sphinctus serotinus* Gravenhorst (Sphinctini), head; (g) *Dyspetes luteomarginatus* Habermehl (Tryphonini), head; (h) *Eridolius pachysoma* (Stephens) (Tryphonini), head; (i) *Grypocentrus cinctellus* Ruthe (Tryphonini), head; (j) *Monoblastus caudatus* (Hartig) (Tryphonini), head; (k) *Tryphon latrator* (Fabricius) (Tryphonini), head, anterodorsal view, showing auricles (expanded antennal sockets); (l) *D. luteomarginatus*, back of head; (continued overleaf).

Figure 166 continued: Morphology of Tryphoninae: (m) *D. luteomarginatus*, part of fore wing; (n) *N. infractor*, part of fore wing; (o) *T. latrator*, part of fore wing; (p) *Oedemopsis scabricula* (Gravenhorst) (Oedemopsini), mesoscutum; (q) *Thymaris tener* (Gravenhorst) (Oedemopsini), mesopleuron; (r) *N. infractor*, tarsal claws; (s) *Excavarus apiarius* (Gravenhorst) (Tryphonini), apex of hind tibiae, inner and outer aspects; (t) *S. serotinus*, apex of hind tibia; (continued overleaf).

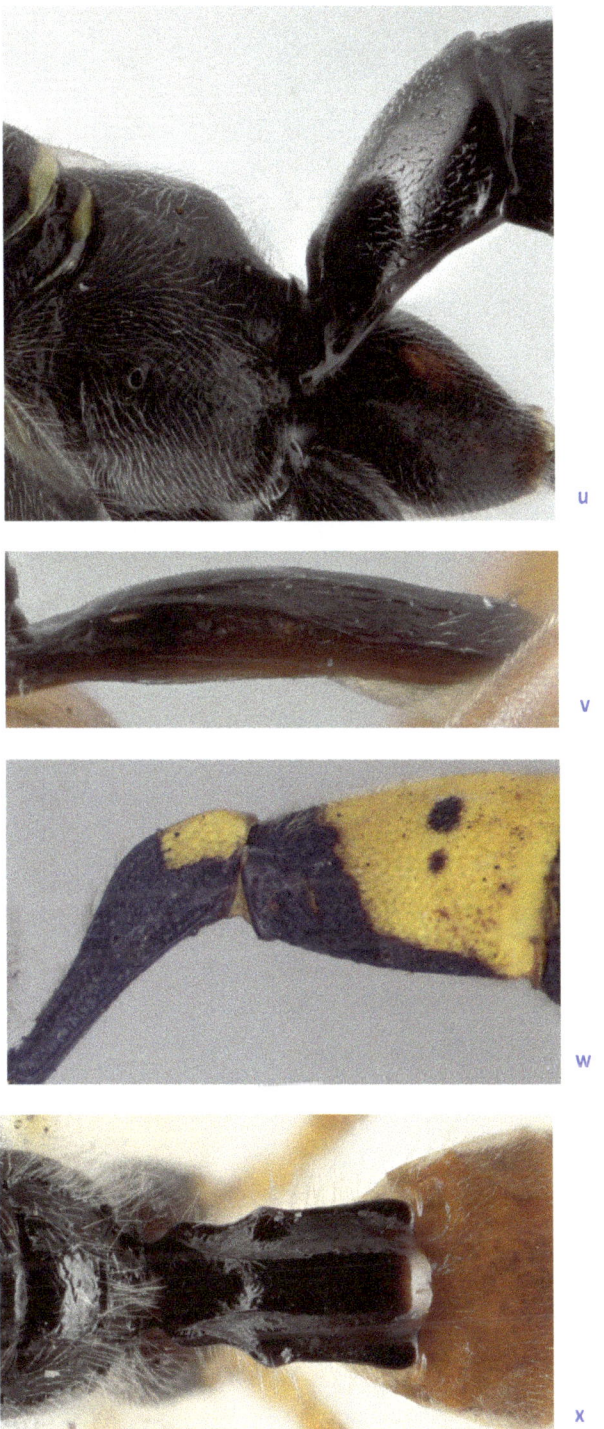

Figure 166 continued: Morphology of Tryphoninae: (u) *Phytodietus montanus* Tolkanitz (Phytodietini), propodeum and first metasomal segment; (v) *T. tener*, first metasomal segment; (w) *S. serotinus*, first and second metasomal segments; (x) *Polyblastus varitarsus* (Gravenhorst) (Tryphonini), propodeum, first and second metasomal tergites; (continued overleaf).

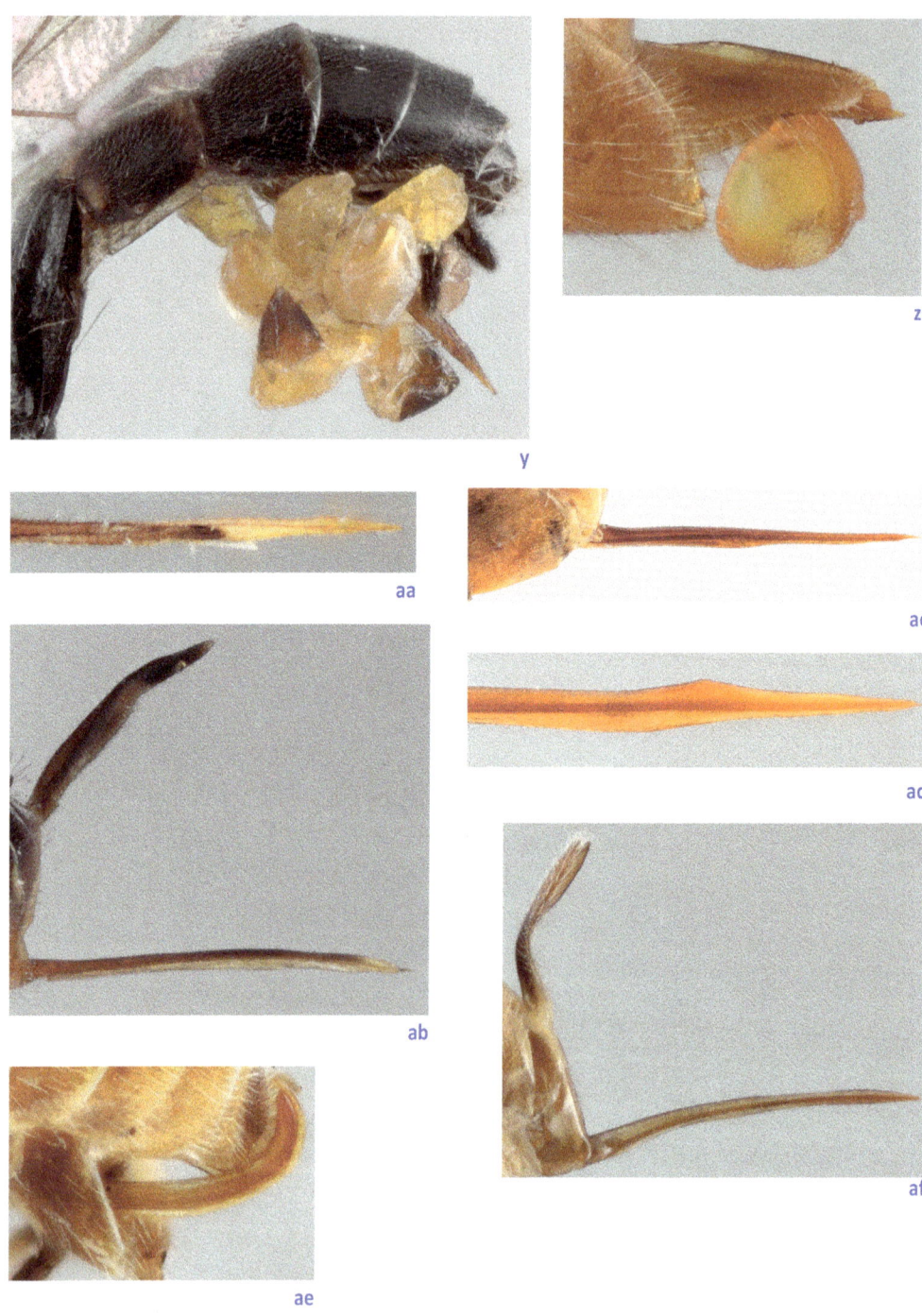

Figure 166 continued: Morphology of Tryphoninae: (y) *Polyblastus wahlbergi* Holmgren (Tryphonini), metasoma and eggs; (z) *T. latrator*, ovipositor sheaths and egg; (aa) *I. euryops*, ovipositor tip; (ab) *O. scabricula*, ovipositor and sheaths; (ac) *N. infractor*, ovipositor; (ad) *P. ornatus*, ovipositor tip; (ae) *Grypocentrus apicalis* (Thomson) (Tryphonini), ovipositor; (af) *Monoblastus brachyacanthus* (Gmelin) (Tryphonini), ovipositor and sheaths.

Subfamily **XORIDINAE**

Xoridinae is one of the smaller subfamilies of the Ichneumonidae but with representatives found throughout the world. There are about 210 described species in four genera, three of which (*Aplomerus* Provancher, *Ischnoceros* and *Odontocolon*) are mostly restricted to the north temperate zone (although the non-British *Aplomerus* reaches south to Mexico and Vietnam) while the fourth, *Xorides*, is world-wide in distribution and by far the most species-rich. Many species, particularly those in the Old World tropics, are known from very few specimens. Fourteen species representing three genera have been found in Britain.

Figure 167. *Odontocolon dentipes* (Gmelin).

Recognition. Xoridines are usually elongate, superficially resembling other ichneumonids that attack wood-boring hosts. Distinctive characters useful for recognising Xoridinae are: fore wing vein 2*rs-m* very short, 3*rs-m* absent; mesoscutum with a transverse suture anterior to the scuto-scutellar groove; first sternite and tergite fused (but not in *Aplomerus*), lacking glymmae, spiracle at or anterior to the mid-length of the tergite; ovipositor longer than the metasoma and with apical teeth on the ventral valves. Additionally, each genus in Britain has at least one distinctive character: *Ischnoceros* has a median horn on the frons, *Odontocolon* has a ventral tooth on the hind femur, and *Xorides* has unidentate mandibles and (in females) pegs projecting from the subapical flagellomeres. Xoridines are most easily confused with poemeniines, from which they can be distinguished most simply by the hind wing with the nervellus intercepted at about the middle, as opposed to intercepted far above the middle in poemeniines. The Poemeniinae with unidentate mandibles have the upper part of the vertex with small (rather inconspicuous) to large scale-like ridges and lack the epicnemial carina (this carina present in Xoridinae).

Systematics. The xoridines were for a long time included in a heterogeneous assemblage which formed part of the Pimplinae, and the traditional tribe Xoridini included genera now placed in Campopleginae, Cryptinae and Poemeniinae (Townes, 1969). The Xoridinae as now recognized was defined by Townes (1969) and forms a rather isolated subfamily within the Ichneumonidae. Quicke *et al.* (1999, 2009) found no evidence of a close relationship with any other subfamily and hypothesised that the Xoridinae may be the most basal extant subfamily of Ichneumonidae. The generic taxonomy of the group has been examined by Wahl (1997), who confirmed the monophyly of the genera but abandoned the use of the subgenera of *Xorides* that had been recognised by Townes (1969).

Biology. Xoridine larvae develop as idiobiont ectoparasitoids of immature stages of 'wood-boring' beetles, and possibly also woodwasps (Hymenoptera: Siricoidea) in the case of the the non-British *Odontocolon geniculatum* (Kriechbaumer) (Spradbery, 1970). Species of various coleopterous families have been recorded as hosts (Aubert, 1969; Gauld & Fitton, 1981; Hilszczanski, 2003; Townes *et al.*, 1960) but most authenticated host rearings have been from Cerambycidae and, to a lesser extent, Buprestidae. Some species seem largely to parasitize cambium-feeding hosts (i.e. found just below the bark) but others attack hosts living further into the wood. Exact host relations are hard to establish in such situations and comparatively little is known in detail about host ranges of individual species, their more general biology, or their importance in relation to pest species of wood borers. Our knowledge of the biology of the subfamily is based on studies of very few species, although Hilszczanski (2003) has obtained a number of rearings from known hosts in Poland of species that also occur in Britain. In many ways xoridines appear to be rather unspecialized idiobionts. Although the larval stage of the host is probably most frequently attacked they are known to consume pupae and possibly pharate adults. The majority are probably solitary but at least one of the British species, *Ischnoceros rusticus* (Geoffroy), which is often abundant as a parasitoid of the cerambycid *Rhagium bifasciatum* Fabricius, is usually gregarious (Shaw, 1999) in broods of up to about seven. One species-group of *Odontocolon* has wingless, brachypterous, and fully-winged forms in North America (Townes *et al.*, 1960) and there is also an apterous *Odontocolon* known from China (Kasparyan, 1997). However, the significance of this adaptation is unknown.

Figure 168. Cocoon and pupa (removed from cocoon) of *Odontocolon dentipes* (Gmelin) (M.R. Shaw).

Some host relations can be summarised from Gauld & Fitton (1981) and Hilszczanski (2003). The common and widespread *Odontocolon dentipes* (Gmelin) has been reared regularly from Cerambycidae (*Arhopalus rusticus* (L) and *Anastrangalia dubia* (Scopoli)) in dead coniferous trees, *Ischnoceros rusticus* (Geoffroy) has been reared from several species of Cerambycidae, especially *Rhagium* species in rotten wood of both coniferous and broadleaved trees, whilst

Xorides praecatorius (Fabricius), also common and widespread, has been recorded from a variety of cerambycids and a species of Buprestidae (*Agrilus viridis* (L.)) in deciduous trees. Cocoon shape and colour varies between species, with *Ischnoceros caligatus* (Gravenhorst) spinning a particularly distinctive, boat-shaped cocoon. Females of *Ischnoceros* and *Odontocolon* spin longer cocoons than males to accommodate the outstretched ovipositor (Hilszczanski, 2003; M.R. Shaw, pers. obs. and figured by Spradbery, 1970) that may be unique in Ichneumonidae with very long ovipositors (Fig. 164); however, where known, *Xorides* pupate with the ovipositor bent partly over the body (Obrtel, 1946; Hilszczanski, pers. comm.).

Chrystal & Skinner (1931) studied *Xorides brachylabis* (Kriechbaumer) and *Xorides irrigator* (Fabricius) parasitizing the cerambycid *Tetropium gabrieli* Weise in larch (*Larix*). The flight period of *X. brachylabis* spans the entire summer, the first adults emerging in late May and the last in mid October, with the maximum emergence in August. Courtship and copulation are relatively protracted. Males were observed searching bark using their antennae, and apparently awaiting the emergence of females. The male approaches the female with his antennae vibrating. On making contact he strokes the female's body with his antennae and slowly mounts her. The male clings to the female during copulation and if he is very small he may lie across her metasoma or hold onto her ovipositor, which points straight backwards. Pairing may last from 15 to 40 minutes, with an average of about 20 minutes.

The female *X. brachylabis* oviposits through the bark, using her ovipositor to drill into the host's gallery in the sapwood. Location of the host is much more accurate than in *Rhyssa persuasoria*, attacking similarly deeply concealed hosts, and it is thought that xoridines locate their hosts by vibrational sounding, detecting the density difference of a potential host in a gallery (Broad & Quicke, 2000). The hammer-like antennal pegs of female *Xorides* are an obvious adaptation for this, although lacking in *Ischnoceros* and *Odontocolon*. When host-searching, the female *Xorides* sweeps the antennae across the surface of the wood, rapidly tapping (presumably via the pegs) with the antennae bent near the tip (Broad, 2013; Fig. 168). After initial insertion of the ovipositor tip into the bark the sheaths are withdrawn and resume their normal position in line with the body. When the ovipositor has penetrated the host's gallery, which may be accomplished in about 15 minutes, the female stands motionless for a further period, perhaps 5 to 15 minutes. After the egg is laid the ovipositor is withdrawn by short jerks of the metasoma. The female then cleans herself and stands at the oviposition site for a few minutes, the whole oviposition sequence taking more than half an hour. Chrystal & Skinner (1931) discovered few differences between *X. irrigator* and *X. brachylabis* and showed that both had some preference for the same host in the same tree (in the laboratory they failed to attack a limited number of alternatives presented to them). The differences which were revealed include, in *X. irrigator*: later first emergence of adults (July) and a more condensed flight period; a stouter egg with little or no pedicel; and a dark brown cocoon.

In both *X. brachylabis* and *X. irrigator*, a single egg is placed on (and apparently stuck to) the cerambycid larva. Only exceptionally was an egg found near, rather than on, a host. The host is paralysed before the egg is laid, becoming immobile and having a reduced pulse rate, and it lives for several weeks in this state if the xoridine egg fails to hatch. The elongate fusiform egg has a short pedicel at one end. Incubation takes about eight days and the newly hatched larva, after spending an hour or two moving about on the body of the host, begins feeding, burying its mandibles in the integument and sucking body fluids. Chrystal & Skinner (1931) and Spradbery (1970) both recognised four larval instars. The first instar larva does not differ much from successive instars and the head capsule is not strongly sclerotized, in contrast to first instar rhyssine larvae (Spradbery, 1970). The first two instars are of 10 to 12 days duration and by the end of this period the host is undoubtedly dead. The rest of the host except the head capsule and chitinous integument is consumed over a period of about 13 days during the remaining two larval instars. The fully grown larva constructs a cocoon in the host's gallery. This takes from three to five days and the completed cocoon is thin-walled, transparent, rounded at both ends, and about twice the length and twice the diameter of the larva. The fourth instar larva overwinters in the cocoon without any obvious change until sometime between the following April and September, when it assumes a prepupal form. The pupal stage itself is very short, lasting nine or 10 days in the case of *Odontocolon*. The adult does not leave the cocoon for a further five days, before using its mandibles to gnaw through the wood and bark to escape.

Xoridines can most easily be collected around log piles and are often caught in Malaise traps sited near dead wood. Hilszczanski (2000) reports that they are readily sampled using yellow pan traps.

Identification. The British genera and species are keyed by Gauld & Fitton (1981); one additional species, *Xorides filiformis* (Gravenhorst), has been found since (Jones *et al.*, in press). Kolarov (1997) provides keys to much of the European fauna of Xoridinae and Hilszczanski (2000) revised the European species of the *Xorides rufipes* species-group.

Figure 169. *Xorides fuligator* (Thunberg) searching for hosts on dead hazel (*Corylus avellana*) (G.R. Broad). (Video available at http://youtu.be/DYnvLmm00Gc).

Figure 170. Morphology of Xoridinae: (a) *Xorides fuligator* (Thunberg), head; (b) *Ischnoceros rusticus* (Geoffroy), head, lateral; (c) *I. rusticus*, mesoscutum; (d) *X. fuligator*, antenna tip; (e) *I. rusticus*, first metasomal segment; (f) *Odontocolon dentipes* (Gmelin), hind femur.

Extralimital and fossil subfamilies

Most of the extant subfamilies of Ichneumonidae are found in Britain but there are a few, mostly small, extralimital subfamilies. With the exception of some Labeninae these are all poorly known biologically; for completeness we briefly summarise knowledge of these groups here. Athough not yet known from Britain, there is one species of Brachycyrtinae known from northern and western Europe so that subfamily is catered for in the main body of chapters.

Claseinae (Fig. 170) comprises two genera, *Ecphysis* Townes and *Clasis* Townes, known only from temperate Chile and, to a lesser extent, Argentina. Only one species of each genus has been described but there are numerous undescribed *Clasis* species in collections. Treated as a tribe of Labeninae (as Labiinae) by Townes (1969), and as a tribe of Cryptinae *s.l.* by Gauld (1983), the group was accorded subfamily status by Porter (1998), which was supported by the phylogenetic results of Quicke *et al.* (2009). In the topologies of Quicke *et al.* (2009), Claseinae form a (weakly supported) clade of presumed Gondwanan origin, along with Brachycyrtinae, Eucerotinae and Pedunculinae. *Ecphysis cyanea* Townes has been reared from a wood-boring buprestid beetle (Porter, 1998) and a *Clasis* species is reported to be a parasitoid of the prepupal stage of a pergid sawfly (Hymenoptera: Pergidae) (Bauerle *et al.*, 1997). Based on morphological correlates (Broad & Quicke, 2000), both genera are presumed to employ vibrational sounding to detect concealed hosts; by analogy and from their phylogenetic position, they are most likely to be idiobiont ectoparasitoids.

Figure 171. *Clasis* sp. (Claseinae) from Chile.

The **Labeninae** (Fig. 171) is the largest entirely extralimital subfamily, comprising just over 150 described species in 12 genera in four tribes. Gauld & Wahl (2000b) proposed relationships between the labenine genera, defined the genera cladistically, and inferred likely biogeographic and host utilisation scenarios. Thus, we know more about labenines than other extralimital ichneumonid subfamilies, although there is a paucity of host records. This is a group that seems to have radiated within the Gondwanan land mass, with most genera and species found in South America and Australia. A few species of *Grotea* Cresson and *Labena* Cresson are found in North America and a few *Certonotus* Kriechbaumer and *Labium* Brullé in New Zealand, New Guinea and adjacent islands. There is some molecular evidence that the Labeninae, together with Xoridinae, comprise one of the most basal lineages of extant ichneumonids (Quicke *et al.*, 1999), although the total evidence analyses of Bennett *et al.* (in prep.) found that labenines were placed near the base of Ichneumoniformes and not near the base of Ichneumonidae. The ancestral biology of Labeninae is thought to be idiobiont ectoparasitism of deeply concealed holometabolous insect larvae. Several species of the tribe Labenini, which generally have long ovipositors and associated adaptations for oviposition into and emergence from wood, have been reared from xylophagous beetles of the families Buprestidae, Cerambycidae and Curculionidae (all known host associations for the Labeninae were summarised by Gauld & Wahl, 2000b). The tribe Orthognatheliini (based on Orthognathellinae Szépligeti, 1908; often, but incorrectly, referred to by its junior synonym, Groteini Viereck, 1923) are, where known, parasitoids of solitary aculeate larvae; *Grotea* species access their stem-nesting hosts with their long ovipositors but the Australasian *Labium* (and presumably *Ozlabium* Gauld & Wahl) enter nests in the soil, contacting their hosts' cells directly, and thus have very short ovipositors. Orthognatheliini species act as inquilines, consuming the host's pollen store alongside the host. This phytophagous tendency is more pronounced in *Poecilocryptus* Cameron (Poecilocryptini), with at least one Australian species known to be an inquiline in galls caused by a phytophagous species of Braconidae, *Mesostoa kerri* Austin & Wharton, on *Banksia* (Austin & Dangerfield, 1998) and in galls caused by a *Fergusonina* (Diptera: Fergusoninidae) on *Melaleuca* (Goolsby *et al.*, 2001). Little is known of the biology of other Poecilocryptini, but the larval head capsule of *Poecilocryptus* has distinctive adaptations, including a uniquely (within Ichneumonidae) bidentate mandible, that are presumed to be associated with the rigours of consuming hard vegetable matter (Gauld & Wahl, 2000b). Nothing is known of the biology of *Xenothyris ferruginea* Townes, the sole member of the Xenothyrini, although it has a morphology indicative of parasitism of deeply concealed hosts.

Figure 172. *Grotea anguina* Cresson (Labeninae: Orthognatheliini) from the USA.

Nesomesochorinae (Fig. 172) was treated as a tribe of Campopleginae by Townes (1970b), comprising two genera, *Chriodes* Förster and *Nonnus* Cresson. Most recent authors have recognised a third genus, *Klutiana* Betrem, which Townes (1970) considered to be a subgenus of *Chriodes*. Miah & Bhuyia (2001) and Quicke *et al.* (2005) recognised that this grouping was not closely related to Campopleginae, a result recovered in the larger scale analysis of Quicke *et al.* (2009). There are many, mostly undescribed, Neotropical species of *Nonnus*, whilst *Chriodes* and *Klutiana* are represented by numerous species in Africa, the Oriental region and in the temperate Far East of the Palaearctic, many of which are, again, undescribed. Nothing is known of the biology of any Nesomesochorinae. The only reported host associations (of *Chriodes* having been reared from a pyralid (Lepidoptera) larva in South Africa; e.g. Conlong, 1994), were based on misidentifications of Campopleginae (Zwart, 1998). Nesomesochorines are derived ophioniformes and their phylogenetic position, together with their notched ovipositors, suggests that they will almost certainly be found to be koinobiont endoparasitoids of holometabolous insect larvae. Females have swollen, clavate fore tibiae, a phenotype associated with detection of vibrations from concealed hosts, through an expanded subgenual organ (Menzel & Tautz, 1994; Broad & Quicke, 2000).

Figure 173. *Nonnus niger* (Brullé) (Nesomesochorinae) from Brazil.

Pedunculinae (Fig. 173) comprises three genera, *Adelphion* Townes, *Monganella* Gauld and *Pedunculus* Townes, found in Chile (*Pedunculus*), Australia (*Adelphion* and *Monganella*) and just extending to New Guinea (undescribed species of *Adelphion*; Gauld, 1984). Although one species of each genus has been described, there are numerous undescribed *Adelphion* and *Pedunculus* species in the larger collections. *Monganella* is known from one Australian species, *M. variegata* Gauld (Gauld, 1984). Although they had generally been classified along with *Brachycyrtus* in a tribe of Labeninae (Townes, 1969) or in the subfamily Brachycyrtinae (Wahl, 1993a), Porter (1998) noted the dissimilarity between *Brachycyrtus* and *Pedunculus* and accordingly erected the Pedunculinae, but without considering the non-Chilean genera. Gauld & Ward (2000) formally transferred *Adelphion* and *Monganella* to the Pedunculinae and Quicke *et al.* (2009) found that the pedunculines from which they were able to obtain sequences formed a clade, apparently closely related to *Brachycyrtus* (Brachycyrtinae). There is one host record, of an *Adelphion* species reared from a spider egg sac in Australia (Chadwick & Nikitin, 1976). A few wingless and micropterous female *Pedunculus* are known, suggesting that at least some members of this genus search for hosts close to the ground (although there are other correlates of flightlessness).

Figure 174. *Pedunculus* sp. (Pedunculinae) from Chile.

Sisyrostolinae (Fig. 174) comprises six genera, four of which were included by Townes (1971) in the Phrudinae (the remaining two were described subsequently): the Afrotropical and Oriental *Brachyscleroma* Cushman, the Afrotropical and Neotropical *Erythrodolius* Seyrig, the Malagasy *Icariomimus* Seyrig, the Oriental *Laxiareola* Sheng & Sun, the Eastern Palaearctic and Oriental *Lygurus* Kasparyan, and the Afrotropical *Melanodolius* Saussure. All species have fairly long ovipositors. Most species are medium-sized, although *Melanodolius* includes some of the most massive ichneumonids known (body length, excluding ovipositor, up to 3.5 cm). Townes (1971) assembled the genera comprising his Phrudinae on the basis of several morphological trends, such as an un-notched ovipositor, first metasomal tergite with spiracles anterior to the middle, etc., but some other authors (e.g. Gupta, 1994a) suggested that these 'macrophrudines' should be separated from the 'microphrudines'. Townes, in Townes *et al.* (1961), had suggested that the rather aberrant *Brachyscleroma* is best placed in its own subfamily, Brachyscleromatinae, before transferring the genus to the Phrudinae (Townes, 1971). Although Gauld (1997) suggested that a monophyletic Phrudinae could be defined by a narrowed proboscidial fossa, Quicke *et al.* (2009) found, in their phylogenetic analyses, that the smaller phrudines belonged to an expanded Tersilochinae whereas the larger phrudines usually formed a separate clade, which they referred to as Brachyscleromatinae; however Bennett *et al.* (2013) have established that the name Sisyrostolinae has priority over Brachyscleromatinae. Bennett *et al.* (in prep.) found that their two sisyrostoline exemplars (*Brachysleroma* sp. and *Erythrodolius calamitosus* Seyrig) clustered with Tersilochinae exemplars including the 'microphrudines'; this entire grouping placed as the sister group of all Ophioniformes except Tryphoninae. Although sisyrostolines are generally rarely collected, new perspectives still regularly arise, such as the recent description of *Laxiareola* (Sheng & Sun, 2011) and the discovery of several species of *Erythrodolius* in the neotropics (Gauld, 1997; Bennett *et al.*, 2013). Virtually nothing is known of the biology of sisyrostolines, other than for

Brachyscleroma apoderi Cushman, which was described from specimens reared from *Apoderus quadripunctatus* Gyllenhal (Coleoptera: Attelabidae) (Cushman, 1940), and the description of *B. jiulongshana* He, Chen & Ma from specimens reared from another attelabine weevil, *Tomapoderus coeruleipennis* Schilsky (He *et al.*, 2000). From their ovipositor structure and phylogenetic placement, the Sisyrostolinae are presumed to be koinobiont endoparasitoids of concealed larvae.

Figure 175. *Brachyscleroma flavoabdominalis* Gupta (Sisyrostolinae) from Thailand.

The **Tatogastrinae** (Fig. 175) comprises one known species, *Tatogaster nigra* Townes. This is another Chilean endemic, part of an intriguing 'neantarctic' radiation of Ichneumonidae (Porter, 1998). Although it was included in a motley 'Microleptinae' assemblage by Townes (1971), Wahl (1991) noted that *Tatogaster* Townes has affinities with the ophioniformes clade and erected a new subfamily for this taxon, a result supported by Quicke *et al.* (2009), who found limited, mainly morphological, support for a sister-group relationship with Mesochorinae. Nothing is known about

Figure 176. *Tatogaster nigra* Townes (Tatogastrinae) from Chile.

the biology of *Tatogaster nigra*; from the short notched ovipositor, elongate metasoma and phylogenetic position, the species is presumed to be a koinobiont endoparasitoid, forcing the metasoma into weak substrate. Although rare in collections, *Tatogaster nigra* has been found in numbers, flying over low vegetation, in a very few sites in Chile (J. Heraty, pers. comm.).

Ichneumonid subfamilies known only from the fossil record

Several extant subfamilies of Ichneumonidae have described fossil representatives, from compression fossils as well as amber inclusions. However, only a tiny portion of fossil ichneumonids have been examined critically or described and when fossils have been revised, their classification has frequently changed (e.g. Spasojevic *et al.*, 2017). We are thus not in a position to say much regarding the fossil history of Ichneumonidae. We just take this opportunity to mention five subfamilies of Ichneumonidae that have been described solely from fossil material. The **Palaeoichneumoninae** and **Labenopimplinae** were described recently from Russian compression fossils dating to the Early Cretaceous (Barremian – Aptian, 113-131 ma) in the case of the Palaeoichneumoninae (Kopylov, 2009) and Late Cretaceous (Cenomanian, 93.9-100 ma) for Labenopimplinae (Kopylov, 2010a). Another Cretaceous subfamily is **Novichneumoninae**, described for two genera of Late Cretaceous (earliest Cenomanian, ~99 ma) amber fossils from Myanmar (Li *et al.*, 2016). The **Pherhombinae** and **Townesitinae** are more recent, known from Baltic and Rovno amber inclusions (i.e. Eocene, about 44 ma) (Kasparyan, 1988, 1994a; Tolkanitz *et al.*, 2005). The three described species of Pherhombinae are all included in the genus *Pherhombus* (Tolkanitz *et al.*, 2005) whereas the four genera of Townesitinae have been classified in two tribes (Kasparyan, 1994). Sharkey & Wahl (1992) and Quicke *et al.* (1999) found that the Early Cretaceous **Tanychorinae** are unplaced within Ichneumonoidea and cannot be included within Ichneumonidae, although they continue to be regarded as ichneumonids in recent works (e.g. Zhang & Rasnitsyn, 2003; Kopylov, 2010b) and are listed as a subfamily of Ichneumonidae by Yu *et al.* (2012). We regard the tanychorines as *incertae sedis* within the Ichneumonoidea. A fairly diverse array of compression fossils have been assigned to the Tanychorinae, currently comprising 23 species in eight genera (Kopylov, 2010b). Whereas some ichneumonids from Baltic amber can readily be recognised as belonging to extant subfamilies, many specimens are difficult to classify in modern subfamilies and the affinities of, for example, pherombines, are obscure. It seems that there has been much turnover of subfamilies in the past 44 million years.

Glossary

The definitions and meanings given here are those applied in the Ichneumonidae and may be applied differently in other groups of insects (including other Hymenoptera). Some definitions are taken from Shaw & Huddleston (1991). Most morphological terminology follows the system of preferred terms employed in the Hymenoptera Anatomy Ontology (HAO: Yoder et al., 2010; http://api.hymao.org/projects/32/public/ontology/), except for wing veins and cells, which are not yet included in the HAO, and a few instances in which the terminology almost universally employed in the ichneumonid literature does not match the preferred term in the HAO (e.g. 'frons'). Table 2 links morphological terms to concepts in the HAO. See also the morphology and terminology chapter.

Abscissa (pl. abscissae): a section of a wing vein between two consecutive junctions with other veins (or between its final junction and the wing margin). Measured from the mid-point of the vein junctions.

Aciculate: finely striate pattern of surface sculpture (see Fig. 24).

Aculeata (adj. aculeate): clade of Hymenoptera comprising the superfamilies Apoidea, Chrysidoidea and Vespoidea (additionally, if the Vespoidea are split, Formicoidea, Pompiloidea and Scolioidea), in which the sting has lost its oviposition function (as it has in various ectoparasitoids of exposed hosts: Shaw & Wahl, 2014).

Aedeagus: in insects, the penis.

Aestivate: undergoing a period of diapause during the summer. In some cases this may be followed by hibernation, with or without a period of renewed activity in between.

Afrotropical: an Old World tropical zoogeographical realm, also known as the Ethiopian realm, comprising sub-Saharan Africa and, often, Madagascar and associated islands (but these are sometimes accorded their own zoogeographical realm, the Malagasy realm).

Alecithal (eggs): lacking yolk. Such eggs are usually very small and absorb nutrients from the host's haemolymph. Found only in endoparasitoids.

Alitrunk: see *mesosoma*.

Allomone: a chemical produced by one species that affects the behaviour of another species.

Amblypygous: a term used of Ichneumoninae to describe a relatively shorter, less robust ovipositor nearly concealed by the hypopygium (as opposed to *oxypygous*).

Annellus (pl. annelli): the differentiated proximal section of the first segment of the *flagellum* (Fig. 9). Note that in Chalcidoidea, one or more proximal, differentiated, ring-like flagellar segments are known as anelli.

Apex (adj. apical): the part of an appendage or segment furthest from its point of attachment or from the so-called base line between the mesosoma and metasoma (see also *base, distal, proximal*).

Apocrita: the clade of Hymenoptera comprising all but the sawflies and woodwasps, i.e. Hymenoptera with the first abdominal segment fused to the thorax and forming the propodeum; parasitoid and aculeate groups.

Apolysis: the period and process during which new cuticular tissue is being secreted, prior to ecdysis. Usually applied to the resting phase between larval instars, but it would apply also between larva and pupa.

Apomorphy (n.), apomorphic (adj.): in cladistic terminology, pertaining to a character that is derived relative to the ancestral state.

Aposematic: a colour pattern that warns off potential predators.

Apparent competition: the situation in which a shared enemy of two species can adversely affect one species as a result of population growth on the other.

Apterous: entirely wingless (the tegulae also absent, see key Fig. 1aa) (see also *brachypterous, macropterous* and *micropterous*).

Area superomedia: the central area on the dorsal surface of the propodeum, delimited by carinae (see Fig. 11b) (= *areola* [not to be confused with *areolet*]).
Areola: see *area superomedia*.
Areolet: a small cell (the third submarginal in ichneumonids) in the fore wing between *RS* and *M* and bounded by cross veins 2*rs-m* and 3*rs-m* (see Figs 17a, 18a). It may be described as closed or present when both cross veins are present and as open or absent if one cross vein is absent. Its overall shape is also significant.
Arolium: pad projecting from the fifth tarsal segment, between the claws (see Fig. 12) (referred to inaccurately as the pulvillus by Fitton *et al.*, 1988).
Arrhenotoky: a form of reproduction by which males are produced from unfertilised eggs and are thus haploid. See also *thelytoky*.
Aulaciform rod: a structure in the ovipositor, interlocking the two halves of the upper valves; found in the ophioniformes (Quicke *et al.*, 1994).
Autapomorphy: a derived character unique to a species or other monophyletic group.
Australasian Realm: the zoogeographical realm comprising Australia, New Zealand, New Guinea and the parts of the Indonesian archipelago to the East of the Wallace line.
Base (adj. basal): the part of an appendage or segment nearest to its point of attachment or to the so-called base line between the mesosoma and metasoma (see also *apex*, *distal*, *proximal*).
Bidentate: having two teeth; generally used in describing mandibles (see also *unidentate*, *tridentate*) (see Fig. 10).
Bilobed: divided into two lobes.
Biocontrol/biological control: use of an organism, released outside its normal range (or its numbers augmented in its natural range) to control the population of (usually) an economically injurious species.
Bivoltine: having precisely two complete generations in a year (see also *plurivoltine*).
BMNH: the Natural History Museum, London (formerly British Museum (Natural History)).
Brachypterous: short winged, possessing only rudimentary wings, usually with some venation discernible (see key Fig. 1bb) (see also *apterous*, *macropterous* and *micropterous*).
Bracovirus: see *polydnavirus*.
Bulla (pl. bullae): a flexible, weakly sclerotised, translucent section of a wing vein where it is crossed by a wing fold/flexion line (see Fig. 18) (= *fenestra*).
Calyx (gland): an area of tissue in the lateral oviduct of the female reproductive organs, the site of production of *polydnaviruses* in some taxa.
Carina: a raised narrow ridge of the integument, in at least some cases corresponding to an underlying muscle attachment.
Caudal: posterior ('tail') end.
Cercus: a protrusion (sensory organ) situated postero-laterally on one of the posterior-most metasomal segments, attached to a large nerve cord (= *pygostyle*).
Character state: the appearance in a taxon of a character (typically, but not exclusively, morphological) that varies across taxa; in *cladistic* terminology, the coding of a character in a matrix.
Chorion: the outer layer of an insect egg.
Clade: a *monophyletic* group of taxa, i.e. exclusively sharing a common ancestor.
Cladistics (adj. cladistic): the method of phylogenetic systematics in which relationships between taxa are calculated based on shared derived characters.
Clasper: an alternative name for *paramere*, the conspicuous, outermost external part of the male genitalia.
Claval furrow: a fold in the posterior part of the wing membrane (originating in the folding of the wing in the pupa) indicated on the wing margin by the claval notch (see Fig. 18b).
Clavate: club-shaped (narrow proximally and wide apically).
Cleptoparasitism: stealing the food resource of another organism; in parasitoid Hymenoptera typically taking the form of usurping another parasitoid's host.

Clone/clonal: genetically identical individuals, the result of asexual reproduction (see *parthenogenesis*) or *polyembryony*.

Clypeal sulcus: the groove between the anterior tentorial pits defining the dorsal edge of the clypeus, delimiting it from the face (see Fig. 8) (= *epistomal sulcus*).

Clypeus: the area immediately below the face and often separated from it by a well marked clypeal sulcus (see Fig. 8). In ichneumonids the clypeus usually conceals the labrum but the labrum is sometimes visible and conspicuous.

Compressed (also laterally compressed): flattened from side to side (see also *depressed*).

Convergence: the possession by unrelated taxa of a characteristic in common. The process by which this occurs is termed convergent or parallel evolution.

Costa: the vein (*C*) along the proximal part of the leading edge of the wings (see Fig. 17). The costa of the fore wing in ichneumonoids is apposed to the second vein (*R*), more or less obliterating the costal cell (see Fig. 17a).

Coxa: the first/proximal section of the leg that articulates with the mesosoma (see Fig. 12).

Cranium: the head capsule.

Depressed (also dorso-ventrally depressed): flattened dorso-ventrally (see also *compressed*).

Diapause: a period of physiologically reduced activity; can be at any life stage or time of year.

Distal: the part of an appendage or segment furthest from its point of attachment or from the so-called base line between the mesosoma and metasoma (see also *apex, basal, proximal*).

Ditrysia: a *clade* of Lepidoptera comprising the vast majority of extant species, in which the female has two genital openings (for mating and for oviposition) (see also *Monotrysia*).

Dorsad: towards the dorsal.

Dichotomous key: an identification key that uses a sequence of choices between two alternatives.

Ecdysis (pl. ecdyses): the moulting of cuticle to allow growth, in holometabolous insects marking the transitions between larval instars. The transition between larva and pupa is usually also regarded as an ecdysis (see also *apolysis*).

Ectoparasitoid: a parasitoid which develops as a larva on the outside of the body of its host (see also *endoparasitoid*).

Ectophagous/ectoparasitic: feeding externally; the feeding mode of ectoparasitoids (see also *endophagous*).

Egg-larval parasitoid: a parasitoid that oviposits into the egg stage of its host and completes its development in the host larva.

Encapsulation: a non-specific immune response in insects whereby a foreign body in the haemocoel is surrounded by *haemocytes*, which (in the case of a parasitoid) first cut off the supply of oxygen and/or food and then melanise, to form an isolated entity that typically becomes dark and hard.

Endoparasitoid: a parasitoid which develops as a larva inside the body of its host (see also *ectoparasitoid*).

Endophagous/endoparasitic: feeding internally; the feeding mode of endoparasitoids (see also *ectophagous*).

Epicnemial carina: a carina extending along the anterior end of the mesosternum and usually extending dorsally near the anterior of the mesopleuron (see Fig. 11) (= *prepectal carina*), delimiting the epicnemium anterior of the carina.

Epistomal sulcus or suture: = *clypeal sulcus*.

Epomia: a lateral carina on the pronotum extending dorso-ventrally and sometimes continuing along the ventral edge of the pronotum (see Fig. 11a).

Exodont: denoting mandibles which have the teeth outwardly directed and not crossing when closed (see Ichneumonoidea key Fig. 6aa).

Face: the anterior area of the head between the antennal sockets, the eyes, and the clypeal suclus (see Fig. 8).

Facultative: not obligatorily so, but opportunistic. Often used to describe the role of species in a parasitoid complex that can parasitise either the host itself or its parasitoids (usually in their fully grown – e.g. cocooned – stage as a *pseudohyperparasitoid*).
Fat body: in insect larvae, the aggregations of cells where lipids are concentrated and stored.
Femur: the third (but in ichneumonoids usually the apparent fourth) section of the leg (see Fig. 12) (see also *trochantellus*), between the trochantellus (or *trochanter* if the trochantellus is not delimited) and the tibia.
Fenestra (pl. fenestrae): a flexible, weakly sclerotised, translucent section of a wing vein where it is crossed by a wing fold (see Fig. 18) (= *bulla*).
Flagellomere/flagellar segment: a repeated section of the antenna distal to the pedicel.
Flagellum: the part of the antenna comprising the flagellomeres (i.e. the antenna excluding the scape and pedicel, see also *annellus*).
Foramen magnum: the space at the back of the head where the head articulates with the mesosoma.
Frons: the area of the head between the antennal sockets and the ocelli (see Fig. 8).
Gaster: an obsolete term, at least in ichneumonoids, for the *metasoma*; frequently used in ant and chalcidoid taxonomy as the part of the metasoma posterior to the petiolar segment(s) if present, i.e. the abdomen posterior to segment 2 (or, in some ants, 3).
Gastrocoelus (pl. gastrocoeli): the lateral depression(s), sometimes coarsely sculptured, at the anterior of tergite 2 of the metasoma in some ichneumonids, immediately behind which may be the thyridium(a) (see Fig. 13) (see also *thyridium*).
Gena: the lateral surface of the head, posterior to the eye (see Fig. 8) (see also *temple*).
Genal carina: ventral portion of the *occipital carina*, where it meets the hypostomal carina (sometimes the carinae do not meet).
Glymma: a pit or groove, variable in position and size, on the lateral surface of the first segment of the metasoma (see Fig. 13a).
Grade group: a paraphyletic group left as a basal residue following the recognition of one or more monophyletic groups stemming from it.
Granulate: a weakly raised, finely reticulate form of surface sculpture (see Fig. 24).
Gregarious parasitoid (also gregarious development): a parasitoid species in which more than one larva develops on or in a single host individual.
Haemocoel: the body cavity, in which an open circulatory system for transport of biochemicals such as nutrients and gases to and from organs operates in insects.
Haemocyte: haemolymph cells in insects, which play a role in the immune system (see also *encapsulation*).
Haemolymph: the fluid within the *haemocoel*, analogous to blood.
Hamulus (pl. hamuli): one of a series of small hooks on the anterior edge of the hind wing which engage with the down-folded hind margin of the fore wing. In ichneumonids there are two groups of hamuli: costal (=proximal=basal) and radial (=distal) (see Fig. 18b).
Haplodiploidy: the mode of reproduction whereby diploid females are produced from fertilised eggs and haploid males from unfertilised ones; see *arrhenotoky*.
Hibernation: a period of diapause during the winter months.
Histolysed/histolysis: in the pupa of holometabolous insects, the process of dissolution of body tissues before their reorganisation to the adult form.
Holarctic: pertaining to or occuring in both the Palaearctic and Nearctic zoogeographical realms.
Homonym: a name that has been applied to two or more taxa (often arising when species are transferred between genera = secondary homonym).
Homoplasious/homoplastic: of a *character*, subject to reversals and or/convergence such that it is not of itself informative about evolutionary relationships.

Host-feeding: the feeding, by the adult female parasitoid, on the haemolymph and sometimes other tissues of host or potential host individuals; **concurrent host-feeding**: feeding on the same host as is parasitized; **non-concurrent host-feeding**: host-feeding on different individuals to those parasitized. Host-feeding may be destructive or non-destructive (definitions from Jervis & Kidd, 1986).

Humeral plate: the sclerite on the mesosoma that articulates with the base of the fore wing at the proximal end of the costal vein; the *tegula* covers the wing base in dorsal view (see Fig. 11a).

Hypandrium: in males, the subgenital plate, the last visible sternite (metasomal sternite 8), protecting the genitalia (see also *hypopygium)*.

Hyperparasitoid: a parasitoid which develops upon another parasitoid (see also *true hyperparasitoid*, and *pseudohyperparasitoid*).

Hypopygium: the terminal (most distal) visible metasomal sternite, in females the 7th abdominal and 6th metasomal sternite, in females, the subgenital plate, the last visible sternite (metasomal sternite 6), protecting the genitalia (see also *hypandrium*). Note that the term hypopygium has sometimes mistakenly been used for the subgenital plate in both sexes.

Hypostoma: sclerite on the posterior surface of the head delimited by the hypostomal sulcus (adjacent to the *hypostomal carina*), running from the posterior articulation of the mandibles up to the posterior tentorial pits, membranous centrally.

Hypostomal carina: *carina* on the posterior surface of the head, delimiting the hypostoma from the cranium (see Fig. 8).

Ichnovirus: see *polydnavirus*.

Idiobont (idiobiosis): a parasitoid that effectively arrests the development of its host at the time of oviposition (see p. 4) (contrast *koinobiont*).

Infuscate: darkened, suffused with black or brown.

Instar: the larval stage between *ecdyses*, although the first instar hatches from the egg.

Koinobiont (koinobiosis): a parasitoid that allows its host to continue development (and/or behaviour) for a time after oviposition (see p. 4) (contrast *idiobiont*).

Labial palp: structure forming part of the mouthparts, in ichneumonids usually composed of four palpomeres (see Fig. 8), shorter than the maxillary palp and originating more centrally, from the *labium*.

Labio-maxillary complex: part of the head, formed of the *maxilla* and *labium* and bearing the *maxillary* and *labial palps* (see Fig. 8).

Labium: appendage on the head, forming part of the mouthparts, the dorsal part of which adjoins the pharynx (mouth). The labium is attached by membranes to the maxilla to form the labio-maxillary complex, with the labial palps and maxillary palps originating from this complex (see Fig. 8).

Labrum: a sclerite attached to the underside of (and often concealed by) the clypeus (see Fig. 8).

Larval parasitoid: a parasitoid that attacks and completes its feeding within the larval period of the host.

Larva-pupal parasitoid: a parasitoid that attacks the larval stage of the host but completes its feeding and development within the host pupa.

Laterotergite: a lateral part of a tergite often demarcated by a fold or carina.

Macropterous: with wings fully formed (capable of sustained flight) (see Key Fig. 1a) (see also *apterous, brachypterous, micropterous*).

Malar space: the space (distance) between the ventral rim of the eye and the articulation of the mandible; sometimes with a well developed sulcus (*subocular sulcus*) (see Fig. 8).

Malar sulcus: see *subocular sulcus*.

Maxilla: appendage of the head, forming part of the mouthparts, connecting with the hypostoma, mandible, labrum and labium. Together with the *labium* forms the labio-maxillary complex, bearing *maxillary* and *labial palps* (see Fig. 8).

Maxillary palp: structure forming part of the mouthparts, in ichneumonids usually composed of five palpomeres (see Fig. 8), longer than the labial palp and originating more laterally, from the *maxilla*.
Meconium: stored waste voided as faeces by the parasitoid larva once it has finished feeding and is preparing to pupate; in apocritan Hymenoptera the hind gut is not connected to the mid gut until the larva has completed its feeding. Also applied to the waste material voided immediately after adult eclosion.
Melanisation: in insect immune responses, the process whereby foreign objects in the *haemocoel* are attacked by *haemocytes* that harden, blacken and smother the foreign body (in this context, an ichneumonid egg or larva).
Mesenteron: mid-gut of insects.
Mesonotum: the dorsal part of the *mesothorax*, posterior to the pronotum, divided into the (anterior) *mesoscutum* (sometimes incorrectly referred to as the mesonotum) and the (posterior) *scutellum* (properly, mesoscutellum) (see Fig. 11).
Mesopleuron: the main, lateral part of the mesothorax (see Fig. 11) (also mesopleurum).
Mesoscutellum: see *scutellum*.
Mesoscutum: see *mesonotum*, of which it is a part.
Mesosoma: the second main body division (between the head and the metasoma); in apocritan Hymenoptera it comprises the prothorax, mesothorax, metathorax and the propodeum (=first abdominal segment) (see Fig. 6) (= *alitrunk*).
Mesosternum: the ventral surface of the mesothorax (see Fig. 11).
Mesothorax: the second segment of the thorax, between the prothorax and metathorax, externally comprising the mesonotum, mesopleuron and mesosternum.
Metapleuron: lateral area of the metathorax, delimited dorsally from the propodeum by the *pleural carina* (see Fig. 11).
Metasoma: the third main body division (see also mesosoma); in apocritan Hymenoptera it comprises the abdominal segments except for the first (see Fig. 6) (= *gaster* sensu Richards, 1977).
Metathorax: the third segment of the thorax, between the mesothorax and the first abdominal segment (=propodeum in Apocrita), externally comprising the metapleuron, metasternum and metanotum and from which the hind legs articulate.
Micropterous: with only small wing remnants (the tegulae are also present) (see Key Fig. 1bb) (see also *apterous, brachypterous, macropterous*)
Monophyletic: applied to an all-inclusive group of organisms with a common ancestor (= *clade*).
Monotrysia: a paraphyletic assemblage of Lepidoptera (paraphyletic with respect to *Ditrysia*) in which the female has a single genital opening for both mating and oviposition; in Britain comprising only the Adeloidea, Tischerioidea and Nepticuloidea.
Multiparasitism: development in or on one host individual by more than one parasitoid species.
Mummy: host larval remains hardened by chemical actions of the parasitoid larva inside which the parasitoid pupates (seen in only a a few groups).
Neantarctic: the temperate South American zoogeographical realm comprising Argentina and Chile (see Porter, 1988).
Nearctic: the North American zoogeographical realm.
Nebulous: (of a wing vein), sclerotised, more-or-less pigmented, but without a lumen, viewed by transmitted light the edges less distinct because of the absence of a lumen (see also *tubular* and *spectral*).
Neotropical: the Central and South American zoogeographical realm, but sometimes excluding the more temperate South American countries Argentina and Chile, classified as *Neantarctic*.
Nervellus: in the hind wing, the combined veins *cu-a* and the first abscissa of *CU*, which is said to be intercepted (or not) by the distal abscissa of *CU* (see Figs 17, 20, 21).

Nominal: (of a taxon) having a name, irrespective of its validity.
Notaulus (pl. notauli): one of a pair of longitudinal linear depressions in the surface of the mesoscutum arising anteriorly (see Fig. 11); often sculptured and sometimes converging or coalescing posteriorly.
NMS: National Museums of Scotland (Edinburgh).
Obligate: having no alternative.
Occipital carina: a carina on the posterior surface of the head, demarking the *occiput* (see Fig. 8); the ventral portion, nearing or meeting the hypostomal carina, is often termed the *genal carina*.
Occiput: the posterior surface of the head (see Fig. 8), above the foramen magnum and delimited dorsally by the *occipital carina*.
Ocellus (pl. ocelli): a simple eye; in adult ichneumonids they are three in number and are located on the top of the head (see Fig. 8).
Oriental: an Old World tropical zoogeographical realm, also known as the Indomalayan realm, extending from Afghanistan in the West to the Wallace line in the East (which delimits the Western boundary of the Australasian realm).
overwintering: passing the winter (whether or not in diapause).
Oviposition: the laying of an egg or eggs.
Ovipositor: an anatomical cluster, formed of paired lower valves and a single upper valve (arising through fused paired valves), comprising the female egg-laying organ at the postero-ventral apex of the metasoma (see Fig. 14). The ovipositor is also used to administer *venom*.
Oxypygous: a term used of Ichneumoninae to describe a relatively longer, more robust ovipositor exposed by a short hypopygium (as opposed to *amblypygous*).
Palaearctic: the Old World north temperate zoogeographical realm, including Europe, temperate Asia and the northern parts of Africa and the Middle East.
Paramere: the main lateral component of the genital capsule of males (= *clasper*) (see Fig. 13e); note that 'paramere' refers to several different anatomical concepts in the morphological literature.
Paraphyletic: pertaining to a group of taxa derived from a single ancestral taxon, but which does not contain all the descendants of that ancestor, i.e. through separate recognition and exclusion of another monophyletic group of organisms that are descended from within the group concerned. (See also *grade group*.)
Parasitoid: an insect which is free-living as an adult but in the larval stage obligatorily feeds on a single host unit that is killed as a result (the "unit" may be an individual, or aggregated items such as the content of an egg sac). Some definitions of parasitoid would include non-insect organisms exhibiting similar behaviour.
Parthenogenesis: asexual reproduction, arising from an unfertilised egg.
Pectinate: with comb-like teeth (see Key Fig. 42g); generally applied to tarsal claws.
Pedicel: the small antennal segment between the *scape* and *flagellum*.
Petiolate: stalked; possessing a petiole. In Ichneumonidae, usually applied to the shape of the first metasomal segment (see Fig. 13), or of the areolet (see Key Fig. 60a).
Petiole: in some Ichneumonidae, the narrow anterior part of the first metasomal segment.
Pharate: a state of metamorphosis at which the instar or stage concerned is fully formed but still confined within the previous skin. For example, a pharate adult has fully formed adult cuticle but is still within its pupal skin or shell.
Phenetic: pertaining to a method of classification in which organisms are grouped according to overall similarity (= numerical taxonomy); contrast with *cladistics*.
Planidium: a type of first instar larva which is free-living and motile. Planidia are usually strongly sclerotised and highly resistant to desiccation.
Plesiomorphic: pertaining to primitive shared characters, not derived characters, so that these characters are not informative of relations between taxa.

Pleural carina: a carina at the junction of the *metapleuron* and *propodeum*.
Plurivoltine: having more than one generation in a year (see also *bivoltine*).
Polydnavirus (PDV): a double-stranded DNA virus, integrated into the parasitoid genome. They suppress the host's immune response but do not replicate within the host (see Whitfield & Asgari, 2003). Polydnaviruses in Ichneumonidae are classified as Ichnoviruses and are thought to have a separate origin from Bracoviruses, found in Braconidae.
Polyembryony: the development of more than one embryo from the cleavage, usually repetitively, of a single egg. Known to occur in Hymenoptera in just a few Braconidae, Encyrtidae, Platygastridae, and Dryinidae.
Polyphyletic: applied to a grouping of taxa sharing homoplasies, i.e. that do not share a common ancestor to the exclusion of other taxa not included in this group.
Polythetic: defined by a combination of multiple features.
Postpectal carina (=posterior transverse carina of the mesosternum): the carina at the posterior end of the *mesosternum*, immediately anterior to the mid coxae (see Fig. 11c).
Preaxillary excision: a small indentation in the fore wing margin at the end of the claval furrow (see Fig. 18b) (= *claval notch*).
Prepectal carina: see *epicnemial carina*.
Prepupa: the inactive stage of the final instar larva, before pupation. Sometimes prolonged (i.e. not necessarily simply apolysis).
Primary parasitoid: a parasitoid, the host of which is not itself a parasitoid; contrast *hyperparasitoid*.
Proboscidial fossa: the ventral space at the back of the head where the mouthparts (palpi, glossa, etc.) articulate.
Pronotum: the dorsal sclerite of the prothorax (see Fig. 11).
Pro-ovigenic: having the entire complement of eggs fully mature at or soon after eclosion (contrast *synovigenic*).
Propleuron: the latero-ventral sclerite of the prothorax (see Fig. 11).
Propodeal carinae: the *carinae* on the *propodeum*, delimiting a series of areas (see Fig. 11).
Propodeum: the first abdominal segment, forming the dorsal posterior part of the mesosoma in the Apocrita, where it is fused with the metathorax (see Fig. 11).
Proximal: the part of an appendage or segment nearest to its point of attachment or to the so-called base line between the mesosoma and metasoma (see also *apex, basal, distal*)
Pseudohyperparasitoid: a parasitoid species that parasitises other parasitoids but only once they have left the primary host; thus pseudohyperparasitoids are invariably idiobionts. Pseudohyperparasitoid species can often behave facultatively both as secondary and as primary parasitoids.
Pterostigma: a thickened area of wing venation (often more or less triangular) at the anterior edge of the fore wing (see Fig. 17).
Punctate: describing surface sculpture consisting of small pits (see Fig. 24).
Pupal parasitoid: a parasitoid attacking the pupal stage of the host. Often idiobiont endoparasitoids undergoing their entire development within the pupal stage of the host.
Radicle: the narrowed proximal part of the antennal scape, that articulates with the head by a knob.
Reticulate: describing surface sculpture comprised of small sculpticells (or scutes) (see Fig. 24).
Rugose: describing wrinkled surface sculpture (see Fig. 24).
Rugulose: a fine form of rugose sculpture (see Fig. 24).
Sagittate: spear-shaped (used in describing an ovipositor tip).
Scape: the first segment of the antenna, that articulates with the head via the *radicle* (see Fig. 9).
Sclerite: a discrete area (though sometimes fused with others) of integument that is strongly sclerotised.

Sclerotised: pertaining to an area of cuticle that has been hardened (and usually darkened) to form a tough more or less rigid structure.

Scolopendral cells: cells in the insect tibia, forming the *subgenual organ*, which detect vibrations based on their displacement in a fluid-filled cavity.

Scopa: area of short, dense setae on the ventral side of the hind coxa, particularly marked in some Ichneumoninae (see Fig. 12).

Scutellum (=mesoscutellum): the posterior part of the mesonotum, properly the mesoscutellum (see Fig. 11).

Scutoscutellar sulcus: a transverse groove between the scutellum and mesoscutum, often crossed by ridges (see Fig. 11).

Secondary host: the host of a primary parasitoid that is itself the host of another (hyper- or pseudohyper-) parasitoid.

Secondary parasitoid: an alternative term for *hyperparasitoid*.

Sensilla (sing. sensillum): sensory organs on the insect cuticle, concentrated particularly on the antennae, which detect chemical, mechanical or thermal cues.

Sequence (of DNA): the linear series of nucleotides in a (fragment of a) DNA molecule, indicated by a series of letters corresponding to the four nucleotides (bases).

Sessile: broadly attached at the base, usually applied to a metasoma in which tergite one is broad anteriorly (at the base); contrast *petiolate*.

Sex ratio: ratio of one sex to the other in a brood or population (note that some publications define the sex ratio by the number of one sex divided by the sum of both sexes).

Sister-group: the *clade* that is the closest relative to the clade under consideration (a species can also be termed a sister to another species or clade of species). Together, the two sister-groups form a monophyletic clade.

Solitary parasitoid: a parasitoid species in which normally only one larva develops on or in a host individual.

Spectral: (of a wing vein), unpigmented, its course marked only by a furrow in the wing membrane, it is invisible when viewed by transmitted light or in slide preparations (see also *nebulous* and *tubular*). Not to be confused with the independent flexion and fold lines in the wing membrane (Fig. 18b).

Spiracle: the opening affording access of the tracheal system to the atmosphere (allowing oxygen to diffuse into the body).

Spurious vein: a short vein (or line of pigmentation resembling a vein) running along the posterior edge of the second subdiscal cell in the fore wing of Ophioninae.

Sternaulus: a horizontal groove near the lower edge of the mesopleuron (see Fig. 11a). (A superficially similar structure seen in many braconids, the precoxal sulcus, is not homologous; Wharton, 2006.)

Sternite: the main ventral sclerite of a segment; when used with only a segment number, for example 'first sternite', it applies to the metasoma. In Ichneumonoidea the metasomal sternites are usually much more weakly sclerotised than the tergites (see Fig. 13).

Stigma: an abbreviation (not preferred) of *pterostigma*.

Sub- (as a prefix): somewhat less than; as in subsessile = not quite sessile.

Subalar ridge (=subtegular ridge): a longitudinal ridge along the upper edge of the mesopleuron underneath the tegula and base of the fore wing (see Fig. 11).

Subgenital plate: the terminal visible sternite of the metasoma, situated more or less ventral to the genitalia (see *hypandrium* and *hypopygium*).

Subgenual organ: a vibration-detecting organ in the tibia, comprising an aggregation of *scolopendral cells*.

Submetapleural carina: a longitudinal carina at the ventral edge of the *metapleuron*, between the mid and hind coxae (see Fig. 11a).

Subocular sulcus (=malar sulcus): a groove between the mandible base and the compound eye; if present, varying from a shallow, sculpturally differentiated strip to a narrow, sharply defined groove.
Subtegular ridge: see *subalar ridge*.
Sulcus (pl. sulci): a furrow or groove defining a secondary division of a sclerite (see, e.g. clypeal sulcus) (see also *suture*).
Superparasitism: the deposition of a clutch of eggs (which may be several or a single egg) on or in a host individual that has already been parasitised by a member of the same species (if by the same female it is termed self-superparasitism).
Suture: a groove or flexible line, usually used in the context of delimiting sclerites.
Symplesiomorphy: a shared primitive character, i.e. a character that is not informative about evolutionary relationships between taxa.
Synapomorphy: a shared derived character, i.e. a character that is informative about evolutionary relationships between taxa.
Synonym: one of multiple (two or more) names given to the same taxon.
Synovigenic: in which eggs continue to be matured during the life of the adult female (contrast *pro-ovigenic*).
Syntergite: a plate formed by the fusion of two or more tergites.
Tarsal claw: one of a pair of claws at the end of a tarsus (see Fig. 12).
Tarsomere: one of the segments of a tarsus.
Tarsus: the terminal section of the leg (see Fig. 12); divided into (usually five) segments (tarsomeres).
Taxon (pl. taxa): a division within a taxonomic classification (for example, species, genus, tribe, family).
Tegula (pl. tegulae): a small sclerite above the base of the fore wing (see Fig. 11).
Temple: the upper part of the *gena* (see Fig. 8).
Tentorial pits: small paired pits on the anterior and posterior surfaces of the head. The anterior tentorial pits are visible at the junction of face and clypeus (see Fig. 8b).
Teratocytes: discrete cells, originating from an endoparasitoid, that float in the host's haemolymph absorbing nutrients and becoming greatly enlarged. In Braconidae, at least, they originate from the *trophamnion*, which disintegrates when the larva hatches from the egg.
Tergite: the main dorsal sclerite of a segment; when used with only a segment number, for example, *first tergite*, it applies to the metasoma (see Fig. 13).
Thelytoky (or thelytokous parthenogenesis): a type of parthenogenesis in which females produce diploid female progeny asexually. See also *arrhenotoky*.
Thyridium (pl. thyridia): a weakly depressed sculptured area anterolaterally on metasomal tergite 2 behind the *gastrocoelus* (sometimes also found on tergite 3 and sometimes two pairs are present on tergite 2) (see Fig. 13).
Tibia: the section of the leg between the tarsus and femur, usually the fourth but the apparent fifth in most ichneumonoids (see *trochantellus*) (Fig. 12).
Tibial spur: an articulated spur at the distal apex of the tibia (generally single on the fore leg and paired on mid and hind legs in Ichneumonoidea) (see Fig. 12).
Tracheal: pertaining to the respiration system, whereby oxygen diffuses into the haemocoel via spiracles that open to the atmosphere, connected to a system of trachaea that permeate body tissues.
Tridentate: three-toothed; used to describe mandibles (see Fig. 10).
Trochantellus: the more-or-less differentiated proximal end of the femur, present in most Ichneumonoidea (see Fig. 12), which forms the apparent third section of the leg, between the trochanter and femur.
Trochanter: the second section of the leg, between the coxa and (usually, in Ichneumonoidea) the trochantellus (see Fig. 12).

Trophamnion: a layer of cells surrounding the developing embryo of some koinobiont endoparasitoids which, by analogy with non-ichneumonids, may take in nutrients from the host haemolymph to nourish the developing parasitoid. These cells may also have secretory properties and on dissociation from the larva they may continue to absorb nutrients and be consumed by the parasitoid larva (see *teratocyte*).

True hyperparasitoid: a hyperparasitoid that accesses its host while it is actively developing, and thus is invariably a koinobiont.

Tubular: (of a wing vein), well sclerotised, pale to almost black in colour, with a lumen filled with haemolymph, viewed by transmitted light the edges appear darker because of the cylindrical structure (see also *nebulous* and *spectral*).

Tyloids: structures, often raised and keel-like (but not homologous with placoid sensilla), on the antennal segments of males (see Fig. 9).

Unidentate: one-toothed; used to describe mandibles (see Fig. 10).

Univoltine: with only one generation in a year (see also *bivoltine* and *plurivoltine*).

Venation: the system of (usually tubular) veins that supports the wing membrane (see also *tubular*, *nebulous*, *spectral* and *bulla*).

Venom: in this context, a substance that alters the physiological state of the host when injected by a parasitoid.

Vertex: part of the top of the head, extending from the lateral ocelli to the occipital carina (see Fig. 8).

Virion: a virus particle (nucleic acid surrounded by a protein capsid).

Virus-like particles (VLP): particles that are injected into a host insect at oviposition and prevent encapsulation of the endoparasitoid egg. There are no nucleic acids but, as in polydnaviruses, they are produced in the female reproductive organs.

Wolbachia: a genus of endosymbiotic bacteria that infects many insects, with a range of outcomes. In some cases they cause sex ratio distortion, decreasing the proportion of males, or effectively eliminating males from the population, often by causing the death of male embryos or by inducing thelytokous parthenogenesis.

Xylophagous: feeding on wood.

Table 2. Morphological terms and Universal Resource Identifiers linking to concepts in the Hymenoptera Anatomy Ontology.

Label	URI
Abscissa	http://purl.obolibrary.org/obo/HAO_0000076
Aedeagus	http://purl.obolibrary.org/obo/HAO_0000091
Alitrunk	http://purl.obolibrary.org/obo/HAO_0000576
Anellus	http://purl.obolibrary.org/obo/HAO_0000287
Area superomedia	http://purl.obolibrary.org/obo/HAO_0002008
Areolet	http://purl.obolibrary.org/obo/HAO_0000147
Arolium	http://purl.obolibrary.org/obo/HAO_0000148
Bulla	http://purl.obolibrary.org/obo/HAO_0000184
Calyx	http://purl.obolibrary.org/obo/HAO_0000186
Carina	http://purl.obolibrary.org/obo/HAO_0000188
Cercus	http://purl.obolibrary.org/obo/HAO_0000191
Chorion	http://purl.obolibrary.org/obo/HAO_0001547
Claval fold	http://purl.obolibrary.org/obo/HAO_0000204
Clypeal suture	http://purl.obolibrary.org/obo/HAO_0000306
Clypeus	http://purl.obolibrary.org/obo/HAO_0000212
Costa	http://purl.obolibrary.org/obo/HAO_0000225
Coxa	http://purl.obolibrary.org/obo/HAO_0000228
Cranium	http://purl.obolibrary.org/obo/HAO_0000234
Epicnemial carina	http://purl.obolibrary.org/obo/HAO_0000292
Epomia	http://purl.obolibrary.org/obo/HAO_0000307
Face	http://purl.obolibrary.org/obo/HAO_0000502
Femur	http://purl.obolibrary.org/obo/HAO_0000327
Flagellomere	http://purl.obolibrary.org/obo/HAO_0000342
Flagellum	http://purl.obolibrary.org/obo/HAO_0000343
Foramen magnum	http://purl.obolibrary.org/obo/HAO_0000347
Frons	http://purl.obolibrary.org/obo/HAO_0001044
Gaster	http://purl.obolibrary.org/obo/HAO_0000369
Gastrocoelus	http://purl.obolibrary.org/obo/HAO_0000370
Gena	http://purl.obolibrary.org/obo/HAO_0000371
Genal carina	http://purl.obolibrary.org/obo/HAO_0001755
Glymma	http://purl.obolibrary.org/obo/HAO_0000378
Hamulus	http://purl.obolibrary.org/obo/HAO_0000394
Humeral plate	http://purl.obolibrary.org/obo/HAO_0000403
Hypopygium	http://purl.obolibrary.org/obo/HAO_0000410
Hypostoma	http://purl.obolibrary.org/obo/HAO_0000411
Hypostomal carina	http://purl.obolibrary.org/obo/HAO_0000413
Labial palp	http://purl.obolibrary.org/obo/HAO_0000450
Labium	http://purl.obolibrary.org/obo/HAO_0000453
Labrum	http://purl.obolibrary.org/obo/HAO_0000456
Laterotergite	http://purl.obolibrary.org/obo/HAO_0001861
Malar space	http://purl.obolibrary.org/obo/HAO_0001393
Maxilla	http://purl.obolibrary.org/obo/HAO_0000513
Maxillary palp	http://purl.obolibrary.org/obo/HAO_0000515
Mesonotum	http://purl.obolibrary.org/obo/HAO_0000556
Mesopleuron	http://purl.obolibrary.org/obo/HAO_0000566
Mesoscutum	http://purl.obolibrary.org/obo/HAO_0000575
Mesosoma	http://purl.obolibrary.org/obo/HAO_0000576
Mesosternum	http://purl.obolibrary.org/obo/HAO_0000580
Mesothorax	http://purl.obolibrary.org/obo/HAO_0000583
Metapleuron	http://purl.obolibrary.org/obo/HAO_0000621
Metasoma	http://purl.obolibrary.org/obo/HAO_0000626

Table 2 continued.

Label	URI
Metathorax	http://purl.obolibrary.org/obo/HAO_0000630
Notaulus	http://purl.obolibrary.org/obo/HAO_0000647
Occipital carina	http://purl.obolibrary.org/obo/HAO_0000653
Occiput	http://purl.obolibrary.org/obo/HAO_0000658
Ocellus	http://purl.obolibrary.org/obo/HAO_0000661
Ovipositor	http://purl.obolibrary.org/obo/HAO_0000679
Paramere	http://purl.obolibrary.org/obo/HAO_0000395
Pedicel	http://purl.obolibrary.org/obo/HAO_0000706
Petiole	http://purl.obolibrary.org/obo/HAO_0002411
Pleural carina	http://purl.obolibrary.org/obo/HAO_0000609
Postpectal carina	http://purl.obolibrary.org/obo/HAO_0000794
Proboscidial fossa	http://purl.obolibrary.org/obo/HAO_0000670
Pronotum	http://purl.obolibrary.org/obo/HAO_0000853
Propleuron	http://purl.obolibrary.org/obo/HAO_0000862
Propodeum	http://purl.obolibrary.org/obo/HAO_0001249
Pterostigma	http://purl.obolibrary.org/obo/HAO_0000957
Punctate	http://purl.obolibrary.org/obo/HAO_0000885
Radicle	http://purl.obolibrary.org/obo/HAO_0000889
Reticulate	http://purl.obolibrary.org/obo/HAO_0002379
Scape	http://purl.obolibrary.org/obo/HAO_0000908
Sclerite	http://purl.obolibrary.org/obo/HAO_0000909
Scutellum	http://purl.obolibrary.org/obo/HAO_0000574
Scutoscutellar sulcus	http://purl.obolibrary.org/obo/HAO_0000919
Sensillum	http://purl.obolibrary.org/obo/HAO_0000933
Spiracle	http://purl.obolibrary.org/obo/HAO_0000950
Sternaulus	http://purl.obolibrary.org/obo/HAO_0000953
Sternite	http://purl.obolibrary.org/obo/HAO_0000955
Submetapleural carina	http://purl.obolibrary.org/obo/HAO_0000974
Subalar ridge	http://purl.obolibrary.org/obo/HAO_0001221
Subgenital plate	http://purl.obolibrary.org/obo/HAO_0000410
Submetapleural carina	http://purl.obolibrary.org/obo/HAO_0000974
Subocular sulcus	http://purl.obolibrary.org/obo/HAO_0000504
Sulcus	http://purl.obolibrary.org/obo/HAO_0000978
Suture	http://purl.obolibrary.org/obo/HAO_0000982
Syntergite	http://purl.obolibrary.org/obo/HAO_0000987
Tarsal claw	http://purl.obolibrary.org/obo/HAO_0000989
Tarsomere	http://purl.obolibrary.org/obo/HAO_0000991
Tarsus	http://purl.obolibrary.org/obo/HAO_0000992
Tegula	http://purl.obolibrary.org/obo/HAO_0000993
Temple	http://purl.obolibrary.org/obo/HAO_0000955
Tentorial pit	http://purl.obolibrary.org/obo/HAO_0000999
Tergite	http://purl.obolibrary.org/obo/HAO_0001005
Thyridium	http://purl.obolibrary.org/obo/HAO_0001016
Tibia	http://purl.obolibrary.org/obo/HAO_0001017
Tibial spur	http://purl.obolibrary.org/obo/HAO_0001018
Trochantellus	http://purl.obolibrary.org/obo/HAO_0001033
Trochanter	http://purl.obolibrary.org/obo/HAO_0001033
Tyloid	http://purl.obolibrary.org/obo/HAO_0001199
Venation	http://purl.obolibrary.org/obo/HAO_0001096
Vertex	http://purl.obolibrary.org/obo/HAO_0001077

References

Achterberg, C. van 1999. The West Palaearctic species of the subfamily Paxylommatinae (Hymenoptera: Ichneumonidae), with special reference to the genus *Hybrizon* Fallén. *Zoologische Mededelingen, Leiden* **73**: 11-26.

Achterberg, C. van 2002. *Apanteles (Choeras) gielisi* spec. nov. (Hymenoptera: Braconidae: Microgastrinae) from The Netherlands and the first report of Trichoptera as hosts of Braconidae. *Zoologische Mededelingen, Leiden* **76**: 53-60.

Achterberg, C. van 2009. Can Townes type Malaise traps be improved? Some recent developments. *Entomologische Berichten* **69**: 129-135.

Achterberg, C. van and Altenhofer, E. 2013. Notes on the biology of *Seleucus cuneiformis* Holmgren (Hymenoptera, Ichneumonidae, Ctenopelmatinae). *Journal of Hymenoptera Research* **31**: 97-104.

Adam, H. 1966. Die hämocytären Abwehrreaktionen des Blutes von *Strongylogaster xanthoceros* (Stephens) und *Strongylogaster lineata* (Christ) gegen die endoparasitische Ichneumonide *Mesoleius niger* (Gravenhorst). *Beiträge zur Entomologie* **15**: 893-965.

Aeschlimann, J.-P. 1973a. Revision des espèces ouest-palearctiques du genre *Triclistus* Förster (Hymenoptera: Ichneumonidae). *Mitteilungen der Schweizerischen Entomologischen Gesellschaft* **46**: 219-252.

Aeschlimann, J.-P. 1973b. Révision des espèces ouest-paléarctiques du genre *Trieces* (Hym. Ichneumonidae). *Annales de la Société Entomologique de France* **9**: 975-987.

Aeschlimann, J.-P. 1974a. Biologie et comportement de *Chorinaeus funebris* Gravenhorst (Hymenoptera: Ichneumonidae). *Annales de Zoologie Ecologie Animale* **6**: 529-538.

Aeschlimann, J.-P. 1974b. Hibernation chez trois espèces de Métopiines: Hymenoptera, Ichneumonidae. *Entomologia Experimentalis Et Applicata* **17**: 487-492.

Aeschlimann, J.-P. 1975a. Révision des espèces ouest-paléarctiques du genre *Chorinaeus* Holmgren (Hymenoptera, Ichneumonidae). *Annales de la Société Entomologique de France* **11**: 723-744.

Aeschlimann, J.-P. 1975b. Biologie, comportement et lâcher expérimental de *Triclistus pygmaeus* Cresson (Hym., Ichn.). *Mitteilungen der Schweizerischen Entomologischen Gesellschaft* **48**: 165-171.

Aeschlimann, J.-P. 1981. Une addition et deux corrections au genre *Chorinaeus* Holmgren (Hymenoptera, Ichneumonidae). *Annales de la Société Entomologique de France* **17**: 3-6.

Aeschlimann, J.-P. 1983. Note sur les Métopiines ouest-paléarctiques, avec description de deux espèces nouvelles (Hymenoptera, Ichneumonidae). *Annales de la Société Entomologique de France* **19**: 3-6.

Aeschlimann, J.-P. 1989. Révision des espèces ouest-paléarctiques du genre *Hypsicera* Latreille (Hymenoptera : Ichneumonidae). *Annales de la Société Entomologique de France* **25**: 33-39.

Aguiar, A. P. 2005. An accurate procedure to describe colors in taxonomic works, with an example from Ichneumonidae (Hymenoptera). *Zootaxa* **1008**: 31-38.

Aguiar, A. P. and Gibson, G. A. P. 2010. The spatial complexity in describing leg surfaces of Hymenoptera (Insecta), the problem and a proposed solution. *Zootaxa* **2415**: 54-62.

Altenhofer, E. 1980. Zur Systematik und Ökologie der Larvenparasiten (Hym., Ichneumonidae, Braconidae, Eulopidae) der minierenden Blattwespen (Hym., Tenthredinidae). *Zeitschrift für Angewandte Entomologie* **89**: 250-259.

Alvarado, M., Bordera, S. and Rodríguez-Berrío, A. 2011. First record of Oxytorinae (Hymenoptera: Ichneumonidae) from South America, with description of a new species of *Oxytorus* Förster, 1869. *Biologia (Bratislava)* **66**: 866-869.

Arthur, A. P. 1966. Associative learning in *Itoplectis conquisitor* (Say) (Hymenoptera: Ichneumonidae). *Canadian Entomologist* **98**: 213-233.

Arthur, A. P. 1981. Host acceptance by parasitoids. In: Nordlund, D. A., Jones, R. L., Lewis, W. J., editors. *Semiochemicals, their role in pest control*. New York: Wiley. pp. 97-120.

Arthur, A. P. and Wylie, H. G. 1959. Effects of host size on sex ratio, development time and size of *Pimpla turionellae* (L.) (Hymenoptera: Ichneumonidae). *Entomophaga* **4**: 297-301.

Arthur, A. P., Stainer, J. E. R. and Turnbull, A. L. 1964. The interaction between *Orgilus obscurator* (Nees) (Hymenoptera: Braconidae) and *Temelucha interruptor* (Grav.) (Hymenoptera: Ichneumonidae), parasites of the pine shoot moth, *Rhyacionia buoliana* (Schiff.) (Lepidoptera: Olethreutidae). *Canadian Entomologist* **96**: 1030-1034.

Ashley, T. R., Waddill, V. H., Mitchell, E. R. and Rye, J. 1982. Impact of native parasitoids on the fall armyworm, *Spodoptera frugiperda* (Lepidoptera: Noctuidae) in south Florida, and release of the exotic parasitoid *Eiphosoma vitticolle* (Hymenoptera: Ichneumonidae). *Environmental Entomology* **11**: 833-837.

Askew, R. R. 1971. *Parasitic Insects*. London: Heinemann.

Askew, R. R. and Shaw, M. R. 1986. Parasitoid communities: their size, structure and development. In: Waage, J. and Greathead, G., editors. *Insect Parasitoids*. London: Academic Press. pp. 225-264.

Atanasov, A. Z. 1975. [New representatives and key to Palearctic species of the genus *Erigorgus* (Hymenoptera, Ichneumonidae)] (in Russian). *Zoologicheskii Zhurnal* **54**: 1480-1487.

Atanasov, A. Z. 1977. [A contribution to the system of Palaearctic Ichneumonids belonging to the genus *Habronyx* (Hymenoptera, Ichneumonidae).] (in Russian with English summary). *Acta Zoologica Bulgarica* **7**: 37-45.

Atanasov, A. Z. 1978. [On the taxonomy of Palearctic Ichneumon-wasps of the genus *Aphanistes* (Hymenoptera, Ichneumonidae).] (in Russian with English summary). *Acta Zoologica Bulgarica* **10**: 13-20.

Atanasov, A. Z. 1981. [A guide to the insects of the European part of the USSR Hymenoptera, Ichneumonidae. Subfamily Anomaloninae.] (in Russian). *Opredeliteli Faune SSSR* **129**: 432-451.

Aubert, J.-F. 1955. Un état prénymphal chez les Hyménoptères Ichneumonides. *Revue de Pathologie Végétale et d'Entomologie Agricole de France* **34**: 159-163.

Aubert, J.-F. 1959. Les hôtes et les stades immatures des Ichneumonides *Pimpla* F., *Apechthis* Först. et *Itoplectis* Först. *Bulletin Biologique de la France et de la Belgique* **93**: 235-259.

Aubert, J.-F. 1969. *Les Ichneumonides ouest-palearctiques et leurs hôtes. 1. Pimplinae, Xoridinae, Acaenitinae*. Paris: Laboratoire d'Evolution des Etres Organises.

Aubert, J.-F. 1978a. Révision préliminaire des Ichneumonides Orthocentrinae européennes (1) (Hym. Ichneumonidae). *Eos* **52(1976)**: 7-28.

Aubert, J.-F. 1978b. *Les ichneumonides ouest-palearctiques et leurs hôtes 2. Banchinae et Suppl. aux Pimplinae*. Paris: Laboratoire d'Evolution des Etres Organises, Paris & EDIFAT-OPIDA, Echauffour.

Aubert, J.-F. 1981. Révision des Ichneumonides *Stenomacrus* sensu lato. *Mitteilungen der Münchner Entomologischen Gesellschaft* **71**: 139-159.

Aubert, J.-F. 1985. Ichneumonides Scolobatinae des collections suédoises (suite) et du Musée de Léningrad. *Bulletin de la Société Entomologique de Mulhouse* **1985**: 49-58.

Aubert, J.-F. 1988. Troisème prélude à une révision des Ichneumonides Scolobatinae: les *Rhorus* Foerst., du groupe de *neustriae* Schrk. *Bulletin de la Société Entomologique de Mulhouse* **1988**: 1-10.

Aubert, J.-F. 2000. Les ichneumonides ouest-paléartiques et leurs hôtes. 3 Scolobatinae (= Ctenopelmatinae) et supplement aux volumes précédents. *Littererae Zoologicae, Actes du Musée cantonal de Zoologie, Lausanne* **5**: 1-310.

Aubert, J.-F. and Jourdheuil, P. 1959. Nouvelle description et biologie de quelques Ichneumonides appartenant aux genres *Aneuclis* Först., *Isurgus* Först. et *Thersilochus* Holm. *Revue de Pathologie Végétale et d'Entomologie Agricole de France* **37**: 175-193.

Austin, A. D. and Dangerfield, P. C. 1998. Biology of *Mesostoa kerri* (Hymenoptera: Braconidae: Mesostoinae), an endemic Australian wasp that causes stem galls on *Banksia marginata*. *Australian Journal of Botany* **46**: 559-569.

Azidah, A. A., Fitton, M. G. and Quicke, D. L. J. 2000. Identification of the *Diadegma* species (Hymenoptera: Ichneumonidae, Campopleginae) attacking the diamondback moth, *Plutella xylostella* (Lepidoptera : Plutellidae). *Bulletin of Entomological Research* **90**: 375-389.

Baarlen, P. van, Topping, C. J. and Sunderland, K. D. 1996. Host location by *Gelis festinans*, an eggsac parasitoid of the linyphiid spider *Erigone atra*. *Entomologia Experimentalis Et Applicata* **81**: 155-163.

Bahena, F., Budia, F., Adan, A., DelEstal, P. and Vinuela, E. 1999. Scanning electron microscopy of *Hyposoter didymator* (Hymenoptera: Ichneumonidae) in host *Mythimna umbrigera* (Lepidoptera: Noctuidae) larvae. *Annals of the Entomological Society of America* **92**: 144-152.

Baker, C. H., Graham, G. C., Scott, K. D., Cameron, S. L., Yeates, D. K. and Merritt, D. J. 2008. Distribution and phylogenetic relationships of Australian glow-worms *Arachnocampa* (Diptera, Keroplatidae). *Molecular Phylogenetics and Evolution* **48**: 506-514.

Baltensweiler, W. 1958. Zur Kenntnis der Parasiten des Grauen Lärchenwicklers (*Zeiraphera griseana* Hübner) in Oberengadin. *Mitteilungen der Schweizerischen Anstalt für das Forstliche Versuchwesen* **34**: 399–478.

Barlow, N. D., Moller, H. and Beggs, J. R. 1996. A model for the effect of *Sphecophaga vesparum vesparum* as a biological control agent of the common wasp in New Zealand. *Journal of Applied Ecology* **33**: 31-44.

Barron, J. R. 1976. Systematics of Nearctic *Euceros* (Hymenoptera: Ichneumonidae: Eucerotinae). *Le Naturaliste Canadien* **103**: 285-375.

Barron, J. R. 1977. The Nearctic species of *Orthopelma* (Hymenoptera: Ichneumonidae). *Systematic Entomology* **2**: 283-299.

Barron, J. R. 1978. Systematics of the world Eucerotinae (Hymenoptera, Ichneumonidae). Part II: non-nearctic species. *Le Naturaliste Canadien* **105**: 327-374.

Barron, J. R. 1981. The Nearctic species of *Ctenopelma* (Hymenoptera, Ichneumonidae, Ctenopelmatinae). *Naturaliste Canadien* **108**: 17-56.

Barron, J. R. 1994. The Nearctic species of *Lathrolestes* (Hymenoptera, Ichneumonidae, Ctenopelmatinae). *Contributions of the American Entomological Institute* **28**: 1-135.

Barron, J. R. and Walley, G. S. 1983. Revision of the Holarctic genus *Pyracmon* (Hymenoptera: Ichneumonidae). *Canadian Entomologist* **115**: 227-241.

Bartell, D. P. and Pass, B. C. 1978. Morphology, development, and behavior of the immature stages of the parasite, *Bathyplectes curculionis*. *Annals of the Entomological Society of America* **71**: 23-30.

Bartell, D. P. and Pass, B. C. 1980. Morphology, development, and behaviour of the immature stages of the parasite *Bathyplectes anurus* (Hymenoptera: Ichneumonidae). *Canadian Entomologist* **112**: 481-487.

Barthélémy, C. and Broad, G. R. 2012. A new species of *Hadrocryptus* (Hymenoptera, Ichneumonidae, Cryptinae), with the first account of the biology for the genus. *Journal of Hymenoptera Research* **24**: 47-57. doi: 10.3897/JHR.24.1888

Basibuyuk, H. H. and Quicke, D. L. J. 1994. Evolution of antennal cleaner structure in the Hymenoptera (Insecta). *Norwegian Journal of Agricultural Sciences* **Supplement 16**: 199-206.

Basibuyuk, H. H. and Quicke, D. L. J. 1995. Morphology of the antenna cleaner in the Hymenoptera with particular reference to non-aculeate families (Insecta). *Zoologica Scripta* **24**: 157-177.

Basibuyuk, H. H. and Quicke, D. L. J. 1997. Hamuli in the Hymenoptera (Insecta) and their phylogenetic implications. *Journal of Natural History* **31**: 1563-1585.

Basibuyuk, H. H. and Quicke, D. L. J. 1999. Grooming behaviours in the Hymenoptera (Insecta): potential phylogenetic significance. *Zoological Journal of the Linnean Society* **125**: 349-382.

Basibuyuk, H. H., Quicke, D. L. J., Rasnitsyn, A. P. and Fitton, M. G. 2000. Morphology and sensilla of the orbicula, a sclerite between the tarsal claws, in the Hymenoptera. *Annals of the Entomological Society of America* **93**: 625-636.

Bauer, R. 1961. Ichneumoniden aus Franken, Teil II (Hymenoptera, Ichneumonidae). *Beiträge zur Entomologie* **11**: 732-792.

Bauerle, P., Rutherford, P. and Lanfranco, D. 1997. Defoliadores de roble (*Nothofagus obliqua*), raulí (*N. alpina*), coigüe (*N. dombeyi*) y lenga (*N. pumilio*). *Bosque* **18**: 97-107.

Beaver, R. A. 1972. Ecological studies on Diptera breeding in dead snails. I. Biology of the species found in *Cepaea nemoralis* (L.). *Entomologist* **105**: 41-52.

Beckage, N. E. & Templeton, T. J. 1985. Temporal synchronization of emergence of *Hyposoter exiguae* and *H. fugitivus* (Hymenoptera: Ichneumonidae) with apolysis preceding larval molting in *Manduca sexta* (Lepidoptera: Sphingidae). *Annals of the Entomological Society of America* **78**: 775-782.

Beirne, B. P. 1941a. A consideration of the cephalic structures and spiracles of the final instar larvae of the Ichneumonidae (Hym.). *Transactions of the Society for British Entomology* **7**: 123-190.

Beirne, B. P. 1941b. British species of Diplazonini (Bassini auctt.) with a study of the genital and postgenital abdominal sclerites in the male (Hym.: Ichneum.). *Transactions of the Royal Entomological Society of London* **91**: 661-712.

Belshaw, R., Fitton, M. G., Herniou, E., Gimeno, C. and Quicke, D. L. J. 1998. A phylogenetic reconstruction of the Ichneumonoidea (Hymenoptera) based on the D2 variable region of 28S ribosomal RNA. *Systematic Entomology* **23**: 109-123.

Benjamin, D. M. and Demba, W. M. 1969. Notes on the biologies of three parasites of the eggs of *Entypotrachelus meyeri* Kolbe and *E. micans* Hust. laid on tea. *Bulletin of Entomological Research* **59 (1968)**: 473-478.

Bennett, A. M. R. 2001. Phylogeny of the Agriotypinae (Hymenoptera: Ichneumonidae), with comments on the subfamily relationships of the basal Ichneumonidae. *Systematic Entomology* **26**: 329-356.

Bennett, A. M. R. 2002. *Cladistics of the Tryphoninae (Hymenoptera: Ichneumonidae) with a discussion of host use and the evolution of parasitism in the Ichneumonidae.* PhD thesis: University of Toronto.

Bennett, A. M. R. 2011. Review of the aquatic Hymenoptera of the world. *Entomological Society of America (ESA) 59th Annual Meeting*; November 13-16, 2011; Reno-Sparks Convention Center, Reno, NV, USA.

Bennett, A. M. R. 2015. Revision of the world genera of Tryphoninae (Hymenoptera: Ichneumonidae). *Memoirs of the American Entomological Institute* **86**: viii+1-387.

Bennett, A. M. R., Sääksjärvi, I. E. and Broad, G. R. 2013. Revision of the New World species of *Erythrodolius* (Hymenoptera: Ichneumonidae: Sisyrostolinae), with a key to the world species. *Zootaxa* **3702**: 425-436.

Bennett, A.M.R., Cardinal, S., Gauld, I.D. and Wahl, D.B. in prep. Phylogeny of the subfamilies of Ichneumonidae. *Journal of Hymenoptera Research* (to be submitted).

Bennett, F. D., Askew, R. R. and Shaw, M. R. 2002. A second rearing of *Telepsogina adelognathi* Hedqvist, 1958 (Hym., Pteromalidae, Miscogastrinae). *Entomologist's Monthly Magazine* **138**: 59-61.

Berberet, R. C. 1986. Relationship of temperature to embryogenesis and encapsulation of eggs of *Bathyplectes curculionis* (Hymenoptera: Ichneumonidae) in larvae of *Hypera postica* (Coleoptera: Curculionidae). *Annals of the Entomological Society of America* **79**: 985-988.

Berland, D. 1947. Hyménopteres, Tenthredoides. *Faune de France* **47**: 1-496.

Bigot, Y., Rabouille, A., Sizart, P. Y., Hamelin, M. H. and Périquet, G. 1997. Particle and genomic characteristics of a new member of the Ascoviridae: *Diadromus pulchellus* ascovirus. *Journal of General Virology* **78**: 1139-1147.

Blair, K. G. 1945. Notes on the economy of the rose galls formed by *Rhodites* (Hymenoptera, Cynipidae). *Proceedings of the Royal Entomological Society of London (A)* **20**: 26-31.

Bland, K. P. 1989. Notes of *Lampronia fuscatella* (Teng.) (Lep.: Incurvariidae). *Entomologist's Record and Journal of Variation* **101**: 249-253.

Bledowski, R. and Krainska, M. K. 1926. Die Entwicklung von *Banchus femoralis* Thoms. (Hymenoptera, Ichneumonidae). *Bibliotheca Universitatis Liberae Polonae* **16**: 1-49+viii.

Blunck, H. and Janssen, M. 1957. Zur Kenntnis von *Hemiteles melanarius* Grav. (Ichn.) Ein Fall des Übergangs vom Ekto- zum Endoparasitismus. *Zeitschrift für Pflanzenkrankheiten und Pflanzenschutz* **64**: 600-606.

Bogush, P. P. 1959. [Materials on parasitic insects of Turkmenia] (in Russian with English summary). *Zoologicheskii Zhurnal* **38**: 189-195.

Bordera, S. and Sääksjärvi, I. E. 2012. Western Amazonian Ateleutina (Hymenoptera, Ichneumonidae, Cryptinae). *Journal of Hymenoptera Research* **29**: 83-118.

Bordera, S., Kolarov, J. and Mazón, M. 2007. An illustrated species key of *Enclisis* Townes including descriptions of two new species (Hymenoptera: Ichneumonidae). *Insect Systematics & Evolution* **38**: 293-310.

Bradley, G.A. 1974. Parasites of forest Lepidoptera in Canada. Subfamily Metopiinae and Pimplinae (Hymenoptera: Ichneumonidae). Part 1. Publication No.1336, Environment Canada, Canadian Forestry Service, 99pp.

Bradley, W. G. and Burgess, E. D. 1934. The biology of *Cremastus flavoorbitalis* (Cameron), an ichneumonid parasite of the European Corn Borer. *United States Department of Agriculture. Technical Bulletin* **441**: 1-15.

Broad, G. R. 2010. Status of *Batakomacrus* Kolarov (Hymenoptera: Ichneumonidae: Orthocentrinae), with new generic combinations and description of a new species. *Zootaxa* **2394**: 51-68.

Broad, G. R. 2012. Keys for the identification of British and Irish nocturnal Ichneumonidae [Internet]. 2012. Available from: http://nocturnalichs.myspecies.info/

Broad, G. R. 2013. Wasps on the phone. *Hamuli* **4(2)**: 12-14.

Broad, G. R. 2016. Checklist of British and Irish Hymenoptera - Ichneumonidae. *Biodiversity Data Journal* **4**: e9042. doi: 10.3897/BDJ.4.e9042

Broad, G. R. and Davis, A. M. 2015. *Lymantrichneumon disparis* (Poda, 1761) (Hymenoptera: Ichneumonidae), a genus and species new to Britain. *Entomologist's Gazette* **66**: 216-218.

Broad, G. R. and Quicke, D. L. J. 2000. The adaptive significance of host location by vibrational sounding in parasitoid wasps. *Proceedings of the Royal Society of London B* **267**: 2403-2409.

Broad, G. R. and Shaw, M. R. 2005. The species of four genera of Metopiinae (Hymenoptera: Ichneumonidae) in Britain, with new host records and descriptions of four new species. *Journal of Natural History* **39**: 2389-2407.

Broad, G. R. and Shaw, M. R. 2016. The British species of *Enicospilus* (Hymenoptera: Ichneumonidae: Ophioninae). *European Journal of Taxonomy* **187**: 1-31. doi: http://dx.doi.org/10.5852/ejt.2016.187

Broad, G. R., Sääksjärvi, I. E., Veijalainen, A. and Notton, D. G. 2011. Three new genera of Banchinae (Hymenoptera: Ichneumonidae) from Central and South America. *Journal of Natural History* **45**: 1311–1329.

Broad, G. R., Schnee, H. and Shaw, M. R. 2015. The hosts of *Ophion luteus* (Linnaeus) (Hymenoptera, Ichneumonidae, Ophioninae) in Europe. *Journal of Hymenoptera Research* **46**: 115-125.

Brock, J. P. 1982. A systematic study of the genus *Ophion* in Britain (Hymenoptera, Ichneumonidae). *Tijdschrift voor Entomologie* **125**: 57-97.

Brock, J. P. 2017. The banchine wasps (Ichneumonidae: Banchinae) of the British Isles. *Handbooks for the Identification of British Insects* **7(4)**: 1-150+vi.

Brock, J. P. and Shaw, M. R. 1997. *Perithous albicinctus* (Gravenhorst), a large pimpline ichneumon-wasp new to Britain (Hymenoptera: Ichneumonidae). *Entomologist's Gazette* **48**: 49-50.

Burdick, D. J. 1961. A taxonomic and biological study of the genus *Xyela* Dalman in North America. *University of California Publications in Entomology* **17**: 285-355.

Callan, E. M. 1943. A note on *Orthopelma luteolator* Grav. and *O. brevicornis* Morl. (Hymenoptera, Ichneumonidae). *Proceedings of the Royal Entomological Society of London (A)* **18**: 30-32.

Campadelli, G. and Scaramozzino, E. P. L. 1994. Imenotteris parassitoidi di insetti xilofagi in Romagna [Parasitoid Hymenoptera of xylophagous insects in Romagna.]. *Bollettino dell'Istituto di Entomologia della Universita degli Studi di Bologna* **48**: 115-121.

Campbell, R. W. 1963. Some ichneumonid sarcophagid interactions in the gypsy moth *Porthetria dispar* (L.) (Lepidoptera: Lymantriidae). *Canadian Entomologist* **95**: 337-345.

Carl, K. P. 1968. *Thymelicus lineola* (Lepidoptera: Hesperidae) and its parasites in Europe. *Canadian Entomologist* **100**: 785-801.

Carleton, M. 1939. The biology of *Pontania proxima* Lep., the bean gall sawfly of willows. *Journal of the Linnean Society of London (Zoology)* **40**: 575-624.

Carlson, R. W. 1979. Family Ichneumonidae. In: Krombein, K. V., Hurt, P. D., Smith, D. R., Burks, B. D., editors. *Catalog of Hymenoptera in America north of Mexico*. Washington. pp. 315-740.

Carton, Y. 1978. Biologie de *Pimpla instigator* (Hym.: Ichneumonidae) IV. Modalités du développement larvaire en fonction du site de ponte; rôle des réactions hémocytaires de l'hôte. *Entomophaga* **23**: 249-259.

Casas, J., Driessen, G., Mandon, N., Wielaard, S., Desouhant, E., Van Alphen, J., Lapchin, L., Rivero, A., Christides, J. P. and Bernstein, C. 2003. Energy dynamics in a parasitoid foraging in the wild. *Journal of Animal Ecology* **72**: 691-697.

Casiraghi, M., Andrietti, F., Bonasoro, I. A. and Martinoli, A. 2001. A note on host detection by *Buathra tarsoleuca* (Schrank) (Hymenoptera: Ichneumonidae), a parasite of *Ammophila sabulosa* (L.) and *Podalonia affinis* (Kirby) (Hymenoptera: Sphecidae). *Journal of Insect Behavior* **14**: 299-312.

Castillo, C., Sääksjärvi, I. E., Bennett, A. M. R. and Broad, G. R. 2011. First record of Acaenitinae (Hymenoptera, Ichneumonidae) from South America with description of a new species and a key to the world species of *Arotes* Gravenhorst. *ZooKeys* **137**: 77-88.

Ceballos, G. 1942. *Las Tribus de los Himenópteros de España*. Madrid (1941): Instituto Español de Entomologia.

Chadwick, C. E. and Nikitin, M. I. 1976. Records of parasitism in the families Ichneumonidae, Braconidae and Aulacidae (Hymenoptera). *The Journal of the Entomological Society of Australia* **9**: 28-38.

Chrystal, R. N. and Skinner, E. R. 1931. Studies in the biology of *Xylonomus brachylabris* Kr., and *X. irrigator* F., parasites of the larch longhorn beetle, *Tetropium gabriele* Weise. *Forestry* **5**: 21-33 + plates I, II.

Chrystal, R. N. and Skinner, E. R. 1932. Studies on the biology of the woodwasp *Xiphydria prolongata* Geoffr. and its parasite *Thalessa curvipes* Grav. *Scottish Forestry Journal* **46**: 36-51.

Claret, J. 1973. La diapause facultative de *Pimpla instigator* (Hym.: Ichneumonidae) I. Rôle de la photopériode. *Entomophaga* **18**: 409-418.

Claret, J. 1978. La diapause facultative de *Pimpla instigator* (Hym.: Ichneumonidae) II. Rôle de la température. *Entomophaga* **23**: 411-415.

Claret, J. and Carton, Y. 1975. Influence de l'espèce hôte sur la diapause de *Pimpla instigator* F. (Hyménoptère, Ichneumonidae). *Compte Rendu de l'Academie des Sciences. Paris (D)* **281**: 279-282.

Clausen, C. P. 1931. Biological observations on *Agriotypus* (Hymenoptera). *Proceedings of the Entomological Society of Washington* **33**: 29-37.

Clausen, C. P. 1932. The early stages of some Tryphonine Hymenoptera parasitic on sawfly larvae. *Proceedings of the Entomological Society of Washington* **34**: 49-60.

Clausen, C. P. 1940. *Entomophagous insects*. New York and London: McGraw-Hill.

Clément, E. 1930. Opuscula Hymenopterologica III. Die paläarktischen *Metopius*-Arten (Hym., Ichneumon.). *Konowia* **8 [1929]**: 325-437.

Clemons, L. 2009. *Anomalon cruentatum* (Goeffroy [sic]) (Hym.: Ichneumonidae) in north Kent. *Entomologist's Record and Journal of Variation* **121**: 241-242.

Cobelli, R. 1906. *Pachylomma Cremieri* de Romand ed il *Lasius fuliginosus* Latr. *Verhandlungen der k.k. zoologisch botanischen Gesellschaft in Wien* **56**: 475-477.

Cole, L. R. 1967. A study of the life-cycles and hosts of some Ichneumonidae attacking pupae of the green oak-leaf roller moth, *Tortrix viridana* (L.) (Lepidoptera: Tortricidae) in England. *Transactions of the Royal Entomological Society of London* **119**: 267-281.

Cole, L. R. 1970. Observations on the finding of mates by male *Phaeogenes invisor* and *Apanteles medicaginis* (Hym., Ichneumonoidea). *Animal Behaviour* **18**: 184-189.

Cole, L. R. 1979. Notes on the biology of *Ischnus inquisitorius* (Müll.) (Hym., Ichneumonidae), an ectoparasitoid of tortricid pupae. *Entomologist's Monthly Magazine* **114**: 115-118.

Cole, L. R. 1981. A visible sign of a fertilization action during oviposition by an ichneumonid wasp, *Itoplectis maculator*. *Animal Behaviour* **29**: 299-300.

Conlong, D. E. 1994. A review and perspectives for the biological control of the African sugarcane stalkborer *Eldana saccharina* Walker (Lepidoptera: Pyralidae). *Agriculture, Ecosystems and Environment* **48**: 9-17.

Constantineanu, I. and Constantineanu, R. 1994. Contributions of parasitoid Hymenoptera to limiting the outbreak of some defoliator Lepidoptera populations in the oak woods. *Revue Roumaine De Biologie Serie De Biologie Animale* **39**: 151-157.

Constantineanu, M. I. 1961. Vorschlag für eine neue Einteilung der Ichneumoniden in Unterfamilien (Hymenoptera: Ichneumonidae). *Beiträge zur Entomologie* **11**: 685-792.

Corbet, S. A. and Rotheram, S. 1965. The life history of the Ichneumonid *Nemeritis* (*Devorgilla*) *canescens* (Gravenhorst) as a parasite of the Mediterranean flour moth *Ephestia* (*Anagasta*) *kuehniella* Zeller, under laboratory conditions. *Proceedings of the Royal Entomological Society of London (A)* **40**: 67-72.

Coronado-Rivera, J., González-Herrera, A., Gauld, I. D. and Hanson, P. 2004. The enigmatic biology of the Ichneumonid subfamily Lycorininae. *Journal of Hymenoptera Research* **13**: 223-227.

Coseglia, A. F., Simpson, R. G. and Eklund, L. R. 1977. Biology of *Mesochorus nigripes*: a hyperparasite of *Bathyplectes* spp. *Annals of the Entomological Society of America* **70**: 695-698.

Criddle, N. 1928. The introduction and establishment of the larch sawfly parasite *Mesoleius tenthredinis* Morley into southern Manitoba. *Canadian Entomologist* **60**: 51-53.

Cross, J. E. and Simpson, R. G. 1972. Cocoon construction by *Bathyplectes curculionis* and attack by secondary parasites. *Environmental Entomology* **1**: 631-633.

Cummins, H. M., Wharton, R. A. and Colvin, A. M. 2011. Eggs and egg loads of field-collected Ctenoplematinae [sic] (Hymenoptera: Ichneumonidae): Evidence for phylogenetic constraints and life-history trade-offs. *Annals of the Entomological Society of America* **104**: 465-475.

Cushman, R. A. 1916. *Thersilochus conotracheli*, a parasite of the plum curculio. *Journal of Agricultural Research* **6**: 847-855.8

Cushman, R. A. 1926. Address of the retiring President. *Proceedings of the Entomological Society of Washington* **28**: 25-51.

Cushman, R. A. 1937a. Revision of the North American species of Ichneumon-flies of the genus *Exetastes* Gravenhorst. *Proceedings of the United States National Museum* **84**. 243-312.

Cushman, R. A. 1937b. The genus *Lysiognatha* Ashmead. *Journal of the Washington Academy of Sciences* **27**: 438-444.

Cushman, R. A. 1940. New genera and species of Ichneumon-flies with taxonomic notes. *Proceedings of the United States National Museum* **88**: 355-372.

Cushman, R. A. 1947. A generic revision of the Ichneumon-flies of the tribe Ophionini. *Proceedings of the United States National Museum* **96**: 417-482.

Cusson, M., Lucarroti, C., Stoltz, D., Krell, P. and Doucet, D. 1998. A polydnavirus from the Spruce Budworm parasitoid, *Tranosema rostrale* (Ichneumonidae). *Journal of Invertebrate Pathology* **72**: 50-56.

Daly, H. V. 1983. Taxonomy and ecology of Ceratinini of North Africa and the Iberian Peninsula (Hymenoptera: Apoidea). *Systematic Entomology* **8**: 29-62.

Danks, H. V. 1971. Biology of some stem-nesting aculeate Hymenoptera. *Transactions of the Royal Entomological Society of London* **122**: 323-399.

Dasch, C. E. 1964. Ichneumon-flies of America north of Mexico: 5. Subfamily Diplazontinae. *Memoirs of the American Entomological Institute* **3**: 1-304.

Dasch, C. E. 1971. Ichneumon-flies of America north of Mexico: 6. Subfamily Mesochorinae. *Memoirs of the American Entomological Institute* **16**: 1-376.

Dasch, C. E. 1984. Ichneumon-flies of America north of Mexico: 9. Subfamilies Theriinae and Anomaloninae. *Memoirs of the American Entomological Institute* **36**: 1-610.

Dasch, C. E. 1988. Ichneumon-flies of America north of Mexico: 10. Subfamily Banchinae, tribe Glyptini. *Memoirs of the American Entomological Institute* **43**: 1-644.

Dasch, C.[E.] 1992. The ichneumon-flies of America north of Mexico. Pt. 12. Subfamilies Microleptinae, Helictinae, Cylloceriinae and Oxytorinae (Hymenoptera: Ichneumonidae). *Memoirs of the American Entomological Institute* **52**: 1-470.

Day, W. H. 1970. The survival value of its jumping cocoons to *Bathyplectes anurus*, a parasite of the alfalfa weevil. *Journal of Economic Entomology* **63**: 586-589.

Day, W. H. and Hedlund, R. C. 1988. Biological comparisons between arrhenotokous and thelytokous biotypes of *Mesochorus nigripes* (Hym.,: Ichneumonidae). *Entomophaga* **33**: 201-210.

Dbar, R. S. 1984. [Review of the Palaearctic species of the genus *Cymodusa* Holmgren (Hymenoptera, Ichneumonidae)] (in Russian). *Entomologicheskoe Obozrenie* **63**: 802-812.

Dbar, R. S. 1985. Revision of the Palaearctic species of *Cymodusa* Holmgren (Hymenoptera, Ichneumonidae). II. *Entomological Review* **65(1986)**: 5-17.

Dell, D. and Burckhardt, D. 2004. *Psilomastax pyramidalis* (Hymenoptera, Ichneumonidae), ein Parasitoid von *Apatura iris* (Lepidoptera, Nymphalidae): Beobachtungen aus der Region Basel (CH, F) aus den Jahren 1982-2002. *Mitteilungen der Entomologischen Gesellschaft Basel* **54**: 83-87.

Delrio, G. 1975. Révision des espèces ouest-paléarctiques du genre *Netelia* Gray (Hym., Ichneumonidae). *Studi Sassaresi sez. III.- Annali della Facolta di Agraria dell'Università di Sassari* **23**: 1-126.

Dijkerman, H. J. 1988. Notes on the parasitization behaviour and larval development of *Trieces tricarinatus* and *Triclistus yponomeutae* (Hymenoptera, Ichneumonidae), endoparasitoids of the genus *Yponomeuta* (Lepidoptera, Yponomeutidae). *Proceedings of the Koninklijke Nederlandse Akademie van Wetenschappen Series C Biological and Medical Sciences* **91**: 19-30.

Dijkerman, H. J. and Koenders, J. T. H. 1988. Competition between *Trieces tricarinatus* and *Triclistus yponomeutae* in multiparasitized hosts (Hymenoptera, Ichneumonidae). *Entomologia Experimentalis Et Applicata* **47**: 289-295.

Diller, E. 1981. Die Arten der Gattung *Dilleritomus* Aubert, 1979 und *Epitomus* Foerster, 1868 (Ichneumonidae, Phaeogenini). *Entomofauna* **1**: 1-10.

Diller, E. 2006. Comments on the genera of Phaeogenini and new descriptions of Palaearctic species (Insecta: Ichneumonidae, Ichneumoninae, Phaeogenini). *Linzer Biologische Beitraege* **38**: 1255-1268.

Diller, E. and Schönitzer, K. 2003. Revision einiger westpaläarktischer und nearktischer Arten der Gattung *Colpognathus* Wesmael, [1845] (Hymenoptera, Ichneumonidae, Ichneumoninae, Alomyini). *Entomofauna* **24**: 333-344.

Diller, E. and Shaw, M. R. 2014. Western Palaearctic Oedicephalini and Phaeogenini (Hymenoptera: Ichneumonidae, Ichneumoninae) in the National Museums of Scotland, with distributional data including 28 species new to Britain, rearing records, and descriptions of two new species of *Aethecerus* Wesmael and one of *Diadromus* Wesmael. *Entomologist's Gazette* **65**: 109-129.

Djoumad, A., Stoltz, D., Béliveau, C., Boyle, B., Kuhn, L. and Cusson, M. 2013. Ultrastructural and genomic characterization of a second banchine polydnavirus confirms the existence of shared features within this ichnovirus lineage. *Journal of General Virology* **94**: 1888-1895.

Donisthorpe, H. 1912. Myrmecophilous notes for 1911. *Entomologist's Record* **24**: 34-40.

Donisthorpe, H. S. J. K. and Wilkinson, D. S. 1930. Notes on the genus *Paxylomma* (Hym. Brac.), with the description of a new species taken in Britain. *Transactions of the Entomological Society of London* **78**: 87-93.

Donovan, B. J. 1991. Life cycle of *Sphecophaga vesparum* (Curtis) (Hymenoptera: Ichneumonidae), a parasitoid of some vespid wasps. *New Zealand Journal of Zoology* **18**: 181-192.

Doutt, R. L. 1964. Biological characteristics of entomophagous adults. In: DeBach, P., editor. *Biological Control of Insect Pests and Weeds*. London: Reinhold. pp. 145-167.

Duff, A. G. 2012. *Checklist of Beetles of the British Isles*. United Kingdom: Pemberley Books.

Dupuis, J. R., Mori, B. A. and Sperling, F. A. H. 2016. *Trogus* parasitoids of *Papilio* butterflies undergo extended diapause in western Canada (Hymenoptera, Ichneumonidae). *Journal of Hymenoptera Research* **50**: 179-190.

Dušek, J., Láska, P. and Šedivý, J. 1979. Parasitization of aphidophagous Syrphidae (Diptera) by Ichneumonidae (Hymenoptera) in the Palaearctic region. *Acta entomologica Bohemoslovaca* **76**: 366-378.

Dysart, R. J., Maltby, H. L. and Brunson, M. H. 1973. Larval parasites of *Oulema melanopus* in Europe and their colonization in the United States. *Entomophaga* **18**: 133-167.

Eady, R. D. 1968. Some illustrations of microsculpture in the Hymenoptera. *Proceedings of the Royal Entomological Society of London (A)* **43**: 66-72.

Eady, R. D. 1974. The present state of nomenclature of wing venation in the Braconidae (Hymenoptera); its origins and comparison with related groups. *Journal of Entomology (B)* **43**: 63-72.

Eberhard, W. G. 2000. The natural history and behavior of *Hymenoepimecis argyraphaga* (Hymenoptera: Ichneumonidae) a parasitoid of *Plesiometa argyra* (Araneae: Tetragnathidae). *Journal of Hymenoptera Research* **9**: 220-240.

Eberhard, W. G. 2001. Under the influence: webs and building behavior of *Plesiometa argyra* (Araneae, Tetragnathidae) when parasitized by *Hymenoepimecis agyraphaga* (Hymenoptera, Ichneumonidae). *The Journal of Arachnology* **29**: 354-366.

Eberhard, W. G. 2013. The polysphinctine wasps *Acrotaphus tibialis*, *Eruga* ca. *gutfreundi*, and *Hymenoepimecis tedfordi* (Hymenoptera, Ichneumonidae, Pimplinae) induce their host spiders to build modified webs. *Annals of the Entomological Society of America* **106**: 652-660.

Eggleton, P. 1989. *The Phylogeny and Evolutionary Biology of the Pimplinae (Hymenoptera: Ichneumonidae)*. PhD thesis: University of London.

Eggleton, P. 1990. Male reproductive behaviour of the parasitoid wasp *Lytarmes maculipennis* (Hymenoptera: Ichneumonidae). *Ecological Entomology* **15**: 357-360.

Eggleton, P. 1991. Patterns in male mating strategies of the Rhyssini: a holophyletic group of parasitoid wasps (Hymenoptera: Ichneumonidae). *Animal Behaviour* **41**: 829-838.

Eichhorn, O. 1988. Untersuchungen über die Fichtengespinstblattwespen *Cephalcia* spp. Panz. (Hym., Pamphiliidae). II. Die Larven- und Nymphenparasiten. *Journal of Applied Entomology* **105**: 105-140.

Eichhorn, O. and Pschorn-Walcher, H. 1973. The parasites of the birch leaf-mining sawfly (*Fenusa pusilla* (Lep.), Hym.: Tenthredinidae) in central Europe. *Technical Bulletin of the Commonwealth Institute of Biological Control* **16**: 79-104.

Elliott, J. M. 1982. The life cycle and spatial distribution of the aquatic parasitoid *Agriotypus armatus* (Hymenoptera: Agriotypidae) and its caddis host *Silo pallipes* (Trichoptera: Goeridae). *Journal of Animal Ecology* **51**: 923-941.

Elliott, J. M. 1983. The responses of the aquatic parasitoid *Agriotypus armatus* (Hymenoptera: Agriotypidae) to the spatial distribution and density of its caddis host *Silo pallipes* (Trichoptera: Goeridae). *Journal of Animal Ecology* **52**: 315-330.

Ellsbury, M. M. and Simpson, R. G. 1978. Biology of *Mesochorus agilis*, an indirect hyperparasite of *Bathyplectes curculionis*. *Annals of the Entomological Society of America* **71**: 865-868.

Elzinga, J. A., Zwakhals, K., Harvey, J. A. and Biere, A. 2007. The parasitoid complex associated with the herbivore *Hadena bicruris* (Lepidoptera: Noctuidae) on *Silene latifolia* (Caryophyllaceae) in the Netherlands. *Journal of Natural History* **41**: 101-123. doi: 10.1080/00222930601121668|issn 0022-2933

Federici, B. A. and Bigot, Y. 2003. Origin and evolution of polydnaviruses by symbiogenesis of insect DNA viruses in endoparasitic wasps. *Journal of Insect Physiology* **49**: 419-432.

Felland, C. M. 1990. Habitat-specific parasitism of the Stalk Borer (Lepidoptera: Noctuidae) in Northern Ohio. *Environmental Entomology* **19**: 162-166.

Filipy, F. L., Burbutis, P. P. and Fuester, R. W. 1985. Biological control of the European wheat stem sawfly in Delaware (Hymenoptera: Cephidae). *Environmental Entomology* **14**: 665-668.

Finch, O. D. 2005. The parasitoid complex and parasitoid-induced mortality of spiders (Araneae) in a Central European woodland. *Journal of Natural History* **39**: 2339-2354.

Finlayson, T. 1960. Taxonomy of cocoons and puparia, and their contents, of Canadian parasites of *Neodiprion sertifer* (Geoff.) (Hymenoptera: Diprionidae). *Canadian Entomologist* **92**: 20-47.

Finlayson, T. 1964. The caudal appendage of final instar larvae of some Porizontinae (Hymenoptera: Ichneumonidae). *Canadian Entomologist* **96**: 1155-1158.

Finlayson, T. 1966. The false cocoon of *Hyposoter parorgyiae* (Vier.) (Hymenoptera: Ichneumonidae). *Canadian Entomologist* **98**: 139.

Finlayson, T. 1975. The cephalic structures and spiracles of final-instar larvae of the Campopleginae, tribe Campoplegini (Hymenoptera: Ichneumonidae). *Memoirs of the Entomological Society of Canada* **94**: 1-137.

Finlayson, T. 1976. Cephalic structures and spiracles of final-instar larvae of the genus *Toxophoroides* Hymenoptera: Ichneumonidae: Lycorininae). *Canadian Entomologist* **108**: 981-984.

Fiori, G. 1947. Contributo alla conoscenza degli insetti del "*Daucus carota*" L. *Bollettino dell'Istituto di Entomologia della Universita degli Studi di Bologna* **16**: 291-314.

Fischer, S., Samietz, J., Wackers, F. L. and Dorn, S. 2004. Perception of chromatic cues during host location by the pupal parasitoid *Pimpla turionellae* (L.) (Hymenoptera: Ichneumonidae). *Environmental Entomology* **33**: 81-87.

Fisher, K. 1932. *Agriotypus armatus* (Walk.) (Hymenoptera) and its relations with its hosts. *Proceedings of the Zoological Society of London* **1932**: 451-461.

Fisher, R. C. 1959. Life history and ecology of *Horogenes chrysostictos* Gmelin (Hymenoptera, Ichneumonidae), a parasite of *Ephestia sericarium* Scott (Lepidoptera, Phycitidae). *Canadian Journal of Zoology* **37**: 429-446.

Fitton, M. G. 1975. A review of the British species of *Tryphon* Fallén (Hym.,Ichneumonidae). *Entomologist's Monthly Magazine* **110**: 153-171.

Fitton, M. G. 1981. The British Acaenitinae (Hymenoptera: Ichneumonidae). *Entomologist's Gazette* **32**: 185-192.

Fitton, M. G. 1984. A review of the British Collyriinae, Eucerotinae, Stilbopinae and Neorhacodinae (Hymenoptera: Ichneumonidae). *Entomologist's Gazette* **35**: 185-195.

Fitton, M. G. 1985a. The ichneumon-fly genus *Banchus* (Hymenoptera) in the Old World. *Bulletin of the British Museum (Natural History) (Entomology)* **51**: 1-60.

Fitton, M. G. 1985b. The British species of *Cidaphus* (Hymenoptera: Ichneumonidae). *Entomologist's Gazette* **36**: 293-297.

Fitton, M. G. 1987. A review of the *Banchus*-group of ichneumon-flies, with a revision of the Australian genus *Philogalleria* (Hymenoptera: Ichneumonidae). *Systematic Entomology* **12**: 33-45.

Fitton, M. G. and Boston, M. 1988. The British species of *Phthorima* (Hymenoptera: Ichneumonidae). *Entomologist's Gazette* **39**: 165-170.

Fitton, M. G. and Ficken, L. 1990. British ichneumon-flies of the tribe Oedemopsini (Hymenoptera: Ichneumonidae). *The Entomologist* **109**: 200-214.

Fitton, M. G. and Gauld, I. D. 1976. The family-group names of the Ichneumonidae (excluding Ichneumoninae) (Hymenoptera). *Systematic Entomology* **1**: 247-258.

Fitton, M. G. and Gauld, I. D. 1980. A review of the British Cremastinae (Hymenoptera: Ichneumonidae), with keys to the species. *Entomologist's Gazette* **31**: 63-71.

Fitton, M. G. and Rotheray, G. E. 1982. A key to the European genera of diplazontine ichneumon-flies, with notes on the British fauna. *Systematic Entomology* **7**: 311-320.

Fitton, M. G., Gauld, I. D. and Shaw, M. R. 1982. The taxonomy and biology of the British Adelognathinae (Hymenoptera: Ichneumonidae). *Journal of Natural History* **16**: 275-283.

Fitton, M. G., Shaw, M. R. and Austin, A. D. 1987. The Hymenoptera associated with spiders in Europe. *Zoological Journal of the Linnean Society* **90**: 65-93.

Fitton, M. G., Shaw, M. R. and Gauld, I. D. 1988. Pimpline Ichneumon-flies. Hymenoptera, Ichneumonidae (Pimplinae). *Handbooks for the Identification of British Insects* **7(1)**: 1-110.

Forbes, A. A., Bagley, R. K., Beer, M. A., Hippee, A. C. and Widmayer, H. A. 2018. Quantifying the unquantifiable: why Hymenoptera - not Coleoptera - is the most speciose animal order. *bioRxiv* preprint doi: https://doi.org/10.1101/274431

Force, D. C. 1989. Observations on the parasitoids of *Mesepiola specca* Davis (Lepidoptera: Incurvariidae). *Pan-Pacific Entomologist* **65**: 436-439.

Fox, J. H. 1927. The life history of *Exeristes roborator* Fab., a parasite of the European Corn Borer. *National Research Council of Canada Report* **No. 21**: 1-58+14 plates.

Franz, J. M. 1958. Studies on *Laricobius erichsonii* Rosenh. (Coleoptera: Derodontidae) a predator on chermesids. Part I, distribution, life history and ecology. *Entomophaga*: 109-164.

Fraser, S. E. M., Dytham, C. and Mayhew, P. J. 2007. Determinants of parasitoid abundance and diversity in woodland habitats. *Journal of Applied Ecology* **44**: 352-361.

Fraser, S. E. M., Dytham, C. and Mayhew, P. J. 2008a. Patterns in the abundance and distribution of ichneumonid parasitoids within and across habitat patches. *Ecological Entomology* **33**: 473-483.

Fraser, S. E. M., Dytham, C., Mayhew, P. J., Mouillot, D. and Anderson, B. J. 2008b. Community structure in ichneumonid parasitoids at different spatial scales. *Oecologia* **157**: 521-530.

Fritzén, N. and Sääksjärvi, I. E. 2016. Spider silk felting - functional morphology of the ovipositor tip of *Clistopyga* sp. (Ichneumonidae) reveals a novel use of the hymenopteran ovipositor.*Biology Letters* **12**: e20160350. DOI: 10.1098/rsbl.2016.0350

Fritzén, N. R. and Shaw, M. R. 2014. On the spider parasitoids *Polysphincta longa* Kasparyan and *P. boops* Tschek (Hymenoptera, Ichneumonidae, Pimplinae), with the first host records of *P. longa*. *Journal of Hymenoptera Research* **39**: 71-82.

Fritzsche, R. 1957. Beeinflussung der Populationsdichte verschiedener *Meligethes*-Arten von gleichen Wirtspflanzen durch Parasiten. In. *Bericht über die Hundertjahrfeier der Deutschen Entomologischen Gesellschaft, Berlin [1956]*. pp. 141-145.

Führer, E. 1975. Über die physiologische Spezifität des polyphagen Puppenparasiten *Pimpla turionellae* L. (Hym., Ichneumonidae) und ihre ökologischen Folgen. *Zentralblatt für das gesamte Forstwesen* **92**: 218-227.

Führer, E. and Kilincer, N. 1972. Die motorische aktivität der endoparasitischen larven von *Pimpla turionellae* L. und *Pimpla flavicoxis* Ths. [Hym,. Ichneum.] in der wirstpuppe. *Entomophaga* **17**: 149-163.

Führer, E. and Willers, D. 1986. The anal secretion of the endoparasitic larva *Pimpla turionellae*: sites of production and effects. *Journal of Insect Physiology* **32**: 361-367.

Gaston, K. J. and Gauld, I. D. 1993. How many species of pimplines (Hymenoptera, Ichneumonidae) are there in Costa Rica? *Journal of Tropical Ecology* **9**: 491-499.

Gauld, I. D. 1973. Notes on the British Ophionini (Hym., Ichneumonidae) including a provisional key to species. *Entomologist's Gazette* **24**: 55-65.

Gauld, I. D. 1976a. The classification of the Anomaloninae (Hymenoptera: Ichneumonidae). *Bulletin of the British Museum (Natural History), Entomology* **33**: 1-135.

Gauld, I. D. 1976b. The taxonomy of the genus *Heteropelma* Wesmael (Hymenoptera: Ichneumonidae). *Bulletin of the British Museum (Natural History), Entomology* **34**: 153-219.

Gauld, I. D. 1977. A revision of the Ophioninae (Hymenoptera: Ichneumonidae) of Australia. *Australian Journal of Zoology* **Supplementary Series No. 49**: 1-112.

Gauld, I. D. 1978. A revision of the Anomaloninae (Hymenoptera: Ichneumonidae) of Melanesia I. The genera *Anomalon* Panzer to *Aphanistes* Förster. *Bulletin of Entomological Research* **68**: 501-519.

Gauld, I. D. 1979. An analysis of the classification of the *Ophion* genus-group (Ichneumonidae). *Systematic Entomology* **5 [1980]**: 59-82.

Gauld, I. D. 1983. The classification, evolution and distribution of the Labeninae, an ancient southern group of Ichneumonidae (Hymenoptera). *Systematic Entomology* **8**: 167-178.

Gauld, I. D. 1984. *An introduction to the Ichneumonidae of Australia*: British Museum (Natural History).

Gauld, I. D. 1985. The phylogeny, classification and evolution of parasitic wasps of the subfamily Ophioninae (Ichneumonidae). *Bulletin of the British Museum (Natural History), Entomology* **51**: 61-185.

Gauld, I. D. 1987. Some factors affecting the composition of tropical ichneumonid faunas. *Biological Journal of the Linnean Society* **30**: 299-312.

Gauld, I. D. 1988a. A survey of the Ophioninae (Hymenoptera: Ichneumonidae) of tropical Mesoamerica with special reference to the fauna of Costa Rica. *Bulletin of the British Museum (Natural History), Entomology* **57**: 1-309.

Gauld, I. D. 1988b. Evolutionary patterns of host utilization by ichneumonoid parasitoids (Hymenoptera: Ichneumonidae and Braconidae). *Biological Journal of the Linnean Society* **35**: 351-377.

Gauld, I. D. 1991. The Ichneumonidae of Costa Rica, 1. *Memoirs of the American Entomological Institute* **47**: 1-589.

Gauld, I. D. 1997. The Ichneumonidae of Costa Rica, 2. *Memoirs of the American Entomological Institute* **57**: 1-485.

Gauld, I. D. 2000. The Ichneumonidae of Costa Rica, 3. *Memoirs of the American Entomological Institute* **63**: 1-453.

Gauld, I. D. and Bolton, B., eds. 1988. *The Hymenoptera*. Oxford: British Museum (Natural History) and Oxford University Press.

Gauld, I. D. and Dubois, J. 2006. Phylogeny of the *Polysphincta* group of genera (Hymenoptera: Ichneumonidae; Pimplinae): a taxonomic revision of spider ectoparasitoids. *Systematic Entomology* **31**: 529-564.

Gauld, I. D. and Fitton, M. G. 1980. The British species of Phrudinae (Hym., Ichneumonidae). *Entomologist's Monthly Magazine* **115**: 197-199.

Gauld, I. D. and Fitton, M. G. 1981. Keys to the British xoridine parasitoids of the wood-boring beetles (Hymenoptera: Ichneumonidae). *Entomologist's Gazette* **32**: 259-267.

Gauld, I. D. and Huddleston, T. 1976. The nocturnal Ichneumonoidea of the British Isles, including a key to genera. *Entomologist's Gazette* **27**: 35-49.

Gauld, I. D. and Janzen, D. H. 2004. The systematics and biology of the Costa Rican species of parasitic wasps in the *Thyreodon* genus-group (Hymenoptera: Ichneumonidae). *Zoological Journal of the Linnean Society* **141**: 297-351.

Gauld, I. D. and Mitchell, P. A. 1977a. Nocturnal Ichneumonidae of the British Isles: the genus *Alexeter* Foerster. *Entomologist's Gazette* **28**: 51-55.

Gauld, I. D. and Mitchell, P. A. 1977b. Ichneumonidae. Orthopelmatinae and Anomaloninae. *Handbooks for the Identification of British Insects* **VII(2)**: 1-32.

Gauld, I. D. and Mitchell, P. A. 1978. *The taxonomy, distribution and host preferences of African parasitic wasps of the subfamily Ophioninae*. Slough: Commonwealth Institute of Entomology, London.

Gauld, I. D. and Mitchell, P. A. 1981. *The taxonomy, distribution and host preferences of Indo-Papuan parasitic wasps of the subfamily Ophioninae.* Slough: CAB.

Gauld, I. D. and Sithole, R. 2002. Subfamily Metopiinae. In: Gauld, I. D., Godoy, C., Sithole, R., Ugalde Gómez, J. A., editors. *Ichneumonidae of Costa Rica, 4.*: Memoirs of the American Entomological Society **66**. pp. 11-262.

Gauld, I. D. and Wahl, D. B. 2000a. The Townesioninae: a distinct subfamily of Ichneumonidae (Hymenoptera) or a clade of the Banchinae? *Transactions of the American Entomological Society* **126**: 279-292.

Gauld, I. D. and Wahl, D. B. 2000b. The Labeninae (Hymenoptera: Ichneumonidae): a study in phylogenetic reconstruction and evolutionary biology. *Zoological Journal of the Linnean Society* **129**: 271-347.

Gauld, I. D. and Wahl, D. B. 2002. The Eucerotinae: a Gondwanan origin for a cosmopolitan group of Ichneumonidae? *Journal of Natural History* **36**: 2229-2248.

Gauld, I. D. and Wahl, D. B. 2006. The relationship and taxonomic position of the genera *Apolophus* and *Scolomus* (Hymenoptera: Ichneumonidae). *Zootaxa* **1130**: 35-41.

Gauld, I. D. and Ward, S. 2000. Subfamily Brachycyrtinae. In: Gauld, I. D., editor. *The Ichneumonidae of Costa Rica, 3.* Memoirs of the American Entomological Institute **63**. pp. 13-35.

Gauld, I. D., Godoy, C., Sithole, R. and Ugalde Gómez, J. A. 2002a. The Ichneumonidae of Costa Rica, 4. Memoirs of the American Entomological Institute **66**: vi+1-768.

Gauld, I. D., Wahl, D. B. and Broad, G. R. 2002b. The suprageneric groups of the Pimplinae (Hymenoptera: Ichneumonidae): a cladistic re-evaluation and evolutionary biological study. *Zoological Journal of the Linnean Society* **136**: 421-485.

Gaulle, J. de 1908. Catalogue systématique et biologique des Hyménoptères de France. (Extrait de la Feuille des Juenes Naturalistes, années 1906, 1907, 1908). *Feuille des Juenes Naturalistes*.

George, A. M. 1978. A new Ichneumon parasite of the purple hair streak (*Quercusia quercus* L.). *Entomologist's Record* **90**: 275.

Gerig, L. 1960. Zur Morphologie der Larvenstadien einiger parasitischer Hymenopteren des Grauen Lärchenwicklers (*Zeiraphera griseana* Hübner). *Zeitschrift für Angewandte Entomologie* **46**: 121-177.

Gibson, G. A. P. 1985. Some pro- and mesothoracic characters important for phylogenetic analysis of Hymenoptera, with a review of terms used for structures. *Canadian Entomologist* **117**: 1395-1443.

Gillespie, J. J., Yoder, M. J. and Wharton, R. A. 2005. Predicted secondary structure for 28S and 18S rRNA from Ichneumonoidea (Insecta: Hymenoptera: Apocrita): Impact on sequence alignment and phylogeny estimation. *Journal of Molecular Evolution* **61**: 114-137.

Given, B. B. 1944. Notes on the anatomy of *Diadromus* (*Thyraeella*) *collaris* Grav. (Hymenoptera, Ichneumonidae). *Transactions of the Royal Society of New Zealand* **74**: 154-164.

Godfray, H. C. J. 1994. *Parasitoids, behavioural and evolutionary ecology.* Princeton N.J.: Princeton University Press.

Gokhman, V. E. 1992. On the origin of endoparasitism in the subfamily Ichneumoninae (Hymenoptera, Ichneumonidae). *Zhurnal Obschei Biologii* **53**: 600-608.

Gómez Durán, J.-M. and Achterberg, C. van 2011. Oviposition behaviour of four ant parasitoids (Hymenoptera, Braconidae, Euphorinae, Neoneurini and Ichneumonidae, Hybrizontinae), with the description of three new European species. *ZooKeys* **125**: 59-106.

Gómez, I. C., Sääksjärvi, I. E., Mayhew P. J., Pollet, M., Rey del Castillo, C., Nieves-Aldrey, J.-L., Broad, G. R., Roininen, H. and Tuomisto, H. 2017. Variation in the species richness of parasitoid wasps (Ichneumonidae: Pimplinae and Rhyssinae) across sites on different continents. *Insect Conservation and Diversity* online early. DOI: 10.1111/icad.12281

Gómez, I. C., Sääksjärvi, I. E., Puhakka, L., Castillo, C. and Bordera, S. 2015. The Peruvian Amazonian species of *Epirhyssa* Cresson (Hymenoptera: Ichneumonidae: Rhyssinae), with notes on tropical species richness. *Zootaxa* **3937**: 311-336.

Gonzaga, M. O. and Sobczak, J. F. 2007. Parasitoid-induced mortality of *Araneus omnicolor* (Araneae, Araneidae) by *Hymenoepimecis* sp (Hymenoptera, Ichneumonidae) in southeastern Brazil. *Naturwissenschaften* **94**: 223-227. doi: 10.1007/s00114-006-0177-z | ISSN 0028-1042

Gonzaga, M. O., Sobczak, J. F., Penteado-Dias, A. M. and Eberhard, W. G. 2010. Modification of *Nephila clavipes* (Araneae Nephilidae) webs induced by the parasitoids *Hymenoepimecis bicolor* and *H. robertsae* (Hymenoptera Ichneumonidae). *Ethology Ecology & Evolution* **22**: 151-165.

Goolsby, J. A., Burwell, C. J., Makinson, J. and Driver, F. 2001. Investigation of the biology of Hymenoptera associated with *Fergusonina* sp. (Diptera: Fergusoninidae), a gall fly of *Melaleuca quinquenervia*, integrating molecular techniques. *Journal of Hymenoptera Research* **10**: 163-182.

Gordh, G. and Hendrickson, R. 1976. Courtship behavior in *Bathyplectes anurus* (Thomson) (Hymenoptera: Ichneumonidae). *Entomological News* **87**: 271-274.

Goulet, H. and Huber, J. T., eds. 1993. *Hymenoptera of the world: An identification guide to families*. Ottawa: Agriculture Canada.

Graham, A. R. 1953. Biology and establishment in Canada of *Mesoleius tenthredinis* Morley a parasite of the larch sawfly, *Pristiphora erichsonii* (Hartig) (Hymenoptera, Tenthredinidae). *Report of the Quebec Society for the Protection of Plants* **35**: 61-75.

Grenier, S. 1970. Biologie d'*Agriotypus armatus* Curtis (Hymenoptera: Agriotypidae), parasite de nymphes de Trichoptères. *Annales de Limnologie* **6**: 317-361.

Griffiths, G. C. D. 1964. The Alysiinae (Hym. Braconidae) parasites of the Agromyzidae (Diptera). I. General questions of taxonomy, biology and evolution. *Beiträge zur Entomologie* **14**: 823-914.

Grissell, E. E. 1999. Hymenoptera biodiversity: some alien notions. *American Entomologist* **45**: 235-244.

Günthart, E. 1949. Beiträge zur Lebensweise und Bekämfung von *Ceuthorrhynchus quadridens* Panz. und *Ceuthorrhynchus napi* Gyll. mit Beobachtungen an weiteren Kohl- und Rapsschädlingen. *Mitteilungen der Schweizerischen Entomologischen Gesellschaft* **22**: 441-591.

Gupta, V. K. 1970. *Ichneumon hunting in India*. Delhi: Department of Zoology, University of Delhi. v+80 pp.

Gupta, V. K. 1984. A revision of the world species of *Hercus* (Hymenoptera: Ichneumonidae). *International Journal of Entomology* **26**: 222-234.

Gupta, V. K. 1988. Relationships of the genera of the tryphonine tribe Oedemopsini and a revision of *Acaenitellus* Morley (Hymenoptera: Ichneumonidae: Tryphoninae). In: Gupta, V. K., editor. *Advances in Parasitic Hymenoptera Research*. Leiden, New York, København, Köln: E.J.Brill. pp. 243-258.

Gupta, V. K. 1990. The taxonomy of the *Kristotomus*-complex of genera and a revision of *Kristotomus* (Hymenoptera: Ichneumonidae: Tryphoninae). *Contributions of the American Entomological Institute* **25**: 1-88.

Gupta, V. K. 1991. A review of the Exenterine genus *Acrotomus* Holmgren, 1855 (Hymenoptera: Ichneumonidae). *Entomofauna* **12**: 33-48.

Gupta, V. K. 1994a. A review of the genus *Brachyscleroma* with descriptions of new species from Africa and the Orient (Hymenoptera: Ichneumonidae: Phrudinae). *Oriental Insects* **28**: 353-382.

Gupta, V. K. 1994b. A review of the world species of *Orthomiscus* Mason (Hymenoptera: Ichneumonidae: Tryphoninae). *Journal of Hymenoptera Research* **3**: 157-173.

Gürbüz, M. F. and Kolarov, J. 2006. A review of the Collyriinae (Hymenoptera: Ichneumonidae). *Entomologica Fennica* **17**: 118-122.

Haeselbarth, E. 1979. Zur Parasitierung der Puppen von Forleule (*Panolis flammea* (Schiff.)), Kiefernspanner (*Bupalus piniarius* (L.)) und Heidelbeerspanner (*Boarmia bistortata* (Goerze)) in bayerischen Kiefernwäldern. Teil 1. *Zeitschrift für Angewandte Entomologie* **87**: 186-202.

Haeselbarth, E. 1983. Determination list of entomophagous insects. 9. International Union of Biological Sciences. International Organization for Biological Control of noxious animals and plants. WPRS Bulletin.

Haeselbarth, E. 1989. Determination list of entomophagous insects. 11. International Union of Biological Sciences. International Organization for Biological Control of noxious animals and plants. WPRS Bulletin.

Halstead, A. J. 1987. Unusual behaviour by *Pimpla instigator* (F.) (Hym., Ichneumonidae). *Entomologist's Monthly Magazine* **123**: 189.

Hamilton, K. G. A. 1972. The insect wing, part III. Venation of the orders. *Journal of the Kansas Entomological Society* **45**: 145-162.

Hancock, G. L. R. 1925. Notes on the hibernation of Ichneumonidae and on some parasites of *Tortrix viridana* L. *Entomologist's Monthly Magazine* **61**: 23-28.

Hanson, H. S. 1939. Ecological notes on the *Sirex* wood wasps and their parasites. *Bulletin of Entomological Research* **30**: 27-76.

Hanson, P. E. and Gauld, I. D. 1995. *The Hymenoptera of Costa Rica*. Oxford: Oxford University Press and The Natural History Museum.

Hanson, P. E. and Gauld, I. D. 2006. Hymenoptera de la Region Neotropical. *Memoirs of the American Entomological Institute* **77**: 1-994.

Harvey, J. A. 1996. *Venturia canescens* parasitizing *Galleria mellonella* and *Anagasta kuehniella*: is the parasitoid a conformer or regulator? *Journal of Insect Physiology* **42**: 1017-1025.

Harvey, J. A. and Witjes, L. M. A. 2005. Comparing and contrasting life history and development strategies in the pupal hyperparasitoids *Lysibia nana* and *Gelis agilis* (Hymenoptera: Ichneumonidae). *Applied Entomology and Zoology* **40**: 309-316.

Harvey, J. A., Witjes, L. M. A. and Wagenaar, R. 2004. Development of hyperparasitoid wasp *Lysibia nana* (Hymenoptera: Ichneumonidae) in a multitrophic framework. *Environmental Entomology* **33**: 1488-1496.

Harvey, J. A., Wagenaar, R. and Bezemer, T. M. 2009. Life-history traits in closely related secondary parasitoids sharing the same primary parasitoid host: evolutionary opportunities and constraints. *Entomologia Experimentalis Et Applicata* **132**: 155-164.

Haye, T. and Kenis, M. 2004. Biology of *Lilioceris* spp. (Coleoptera: Chrysomelidae) and their parasitoids in Europe. *Biological Control* **29**: 399-408.

He, J., Chen, X. and Ma, Y. 2000. Revision of the genus *Grachyscleroma* [*Brachyscleroma*] Cushman (Hymenoptera: Ichneumonidae) from China with a key to the known species of the world. In: Zhang, Y., editor. *Systematic and faunistic research on Chinese insects. Proceedings of the 5th National Congress of Insect Taxonomy*. China Agriculture Press. pp. 235-245.

He, J.-h., Chen, X.-x. and Ma, Y. 1996. [*Hymenoptera: Ichneumonidae*] (in Chinese). Beijing, China: Science Press.

Heath, J. 1961. Some parasites of Eriocraniidae (Lep.). *Entomologist's Monthly Magazine* **92**: 163.

Hedwig, K. 1939. Verzeichnis der bisher in Schlesien aufgefundenen Hymenopteren. V. Ichneumonidae. *Zeitschrift für Entomologie. Breslau* **18**: 12-28.

Hedwig, K. 1950. Beiträge zur Kenntnis europäischer Schlupfwespen. 1. *Hemiteles difficilis* sp.n. 2. Mitteleuropäische Schlupfwespen und ihre Wirte. *Nachrichten des Naturwissenschaftlichen Museums der Stadt Aschaffenburg* **29**: 17-42.

Heinrich, G. H. 1934. Die Ichneumoninae von Celebes: bearbeitet auf Grund der Ausbeute der Celebes-expedition G. Heinrich 1930-32. *Mitteilungen aus dem Zoologischen Museum in Berlin* **20**: 1-263.

Heinrich, G. H. 1961a. Synopsis of the Nearctic Ichneumoninae Stenopneusticae with particular reference to the northeastern region (Hymenoptera). Part I. Introduction, key to Nearctic genera of Ichneumoninae Stenopneusticae, and synopsis of the Protichneumonini north of Mexico. *The Canadian Entomologist* **Supplement 15**: 1-88.

Heinrich, G. H. 1961b. Synopsis of the Nearctic Ichneumoninae Stenopneusticae with particular reference to the northeastern region (Hymenoptera). Part II. Synopsis of the Ichneumonini: genera *Orgichneumon, Cratichneumon, Homotherus, Aculichneumon, Spilichneumon*. *The Canadian Entomologist* **Supplement 18**: 89-206.

Heinrich, G. H. 1961c. Synopsis of the Nearctic Ichneumoninae Stenopneusticae with particular reference to the northeastern region (Hymenoptera). Part III. Synopsis of the Ichneumonini: genera *Ichneumon* and *Thyrateles*. *The Canadian Entomologist* **Supplement 21**: 207-368.

Heinrich, G. H. 1961d. Synopsis of the Nearctic Ichneumoninae Stenopneusticae with particular reference to the northeastern region (Hymenoptera). Part IV. Synopsis of the Ichneumonini: genera *Chasmias, Neamblymorpha, Anisopygus, Limerodops, Eupalamus, Tricholabus, Pseudamblyteles, Eutanyacra, Ctenichneumon, Exephanes, Ectopimorpha, Pseudoamblyteles*. *The Canadian Entomologist* **Supplement 23**: 369-506.

Heinrich, G. H. 1962a. Synopsis of the Nearctic Ichneumoninae Stenopneusticae with particular reference to the northeastern region (Hymenoptera). Part V. Synopsis of the Ichneumonini: genera *Protopelmus, Patrocloides, Probolus, Stenichneumon, Aoplus, Limonethe, Hybophorellus, Rubicundiella, Melanichneumon, Stenobarichneumon, Platylabops, Hoplismenus, Hemihoplis, Trogomorpha*. *The Canadian Entomologist* **Supplement 26**: 507-672.

Heinrich, G. H. 1962b. Synopsis of the Nearctic Ichneumoninae Stenopneusticae with particular reference to the northeastern region (Hymenoptera). Part VI. Synopsis of the Ichneumonini (genus *Plagiotrypes*), Acanthojoppini, Listrodromini and Platylabini. *The Canadian Entomologist* **Supplement 27**: 675-802.

Heinrich, G. H. 1962c. Synopsis of the Nearctic Ichneumoninae Stenopneusticae with particular reference to the northeastern region (Hymenoptera). Part VII. Synopsis of the Trogini. Addenda and Corrigenda. *The Canadian Entomologist* **Supplement 29**: 803-886.

Heinrich, G. H. 1965a. Burmesische Ichneumoninae I. *Entomologisk Tidskrift* **86**: 74-130.

Heinrich, G. H. 1965b. Burmesische Ichneumoninae II. *Entomologisk Tidskrift* **86**: 133-177.

Heinrich, G. H. 1966. Burmesische Ichneumoninae III. *Entomologisk Tidskrift* **87**: 184-247.

Heinrich, G. H. 1967a. *Synopsis and reclassification of the Ichneumoninae Stenopneusticae of Africa south of Sahara (Hym.).Vol.I*: Farmington State College Press.

Heinrich, G. H. 1967b. *Synopsis and reclassification of the Ichneumoninae Stenopneusticae of Africa south of Sahara (Hym.).Vol.II*: Farmington State College Press.

Heinrich, G. H. 1967c. *Synopsis and reclassification of the Ichneumoninae Stenopneusticae of Africa south of Sahara (Hym.).Vol.III*: Farmington State College Press.

Heinrich, G. H. 1968a. *Synopsis and reclassification of the Ichneumoninae Stenopneusticae of Africa south of Sahara (Hym.).Vol.IV*: Farmington State College Press.

Heinrich, G. H. 1968b. *Synopsis and reclassification of the Ichneumoninae Stenopneusticae of Africa south of Sahara (Hym.).Vol.V*: Farmington State College Press.

Heinrich, G. H. 1968c. Burmesische Ichneumoninae IV. *Entomologisk Tidskrift* **89**: 77-106.

Heinrich, G. H. 1968d. Burmesische Ichneumoninae V. *Entomologisk Tidskrift* **89**: 197-228.

Heinrich, G. H. 1968e. Contributions to the knowledge of North America Ichneumoninae. *Naturaliste Canadien* **95**: 703-722.

Heinrich, G. H. 1969a. Burmesische Ichneumoninae VI. *Entomologisk Tidskrift* **90**: 100-130.

Heinrich, G. H. 1969b. Synopsis of Nearctic Ichneumoninae Stenopneustcae with particular reference to the northeastern region (Hymenoptera) - Supplement 1. *Naturaliste Canadien* **96**: 935-963.

Heinrich, G. H. 1970. Burmesische Ichneumoninae VII. *Entomologisk Tidskrift* **91**: 68-102.

Heinrich, G. H. 1971. Synopsis of Nearctic Ichneumoninae Stenopneusticae with particular reference to the northeastern region (Hymenoptera) Supplement 2. *Naturaliste Canadien* **98**: 959-1026.

Heinrich, G. H. 1972. Zur Systematik der Ichneumoninae Stenopneusticae IX. Eine Spätlese (Hymenoptera, Ichneumonidae). *Mitteilungen der Münchner Entomologischen Gesellschaft* **60 [1970]**: 80-101.

Heinrich, G. H. 1973a. Synopsis of Nearctic Ichneumoninae Stenopneusticae with particular reference to the northeastern region (Hymenoptera). Supplement 3. *Naturaliste Canadien* **99**: 173-211.

Heinrich, G. H. 1973b. Synopsis of Nearctic Ichneumoninae Stenopneusticae with particular reference to the northeastern region (Hymenoptera). Supplement 4. *Naturaliste Canadien* **100**: 461-465.

Heinrich, G. H. 1974a. Burmesische Ichneumoninae VIII. *Annales Zoologici* **31**: 407-457.

Heinrich, G. H. 1974b. Burmesische Ichneumoninae IX. *Annales Zoologici* **32**: 103-197.

Heinrich, G. H. 1975a. Burmesische Ichneumoninae X. *Annales Zoologici* **32**: 441-514.

Heinrich, G. H. 1975b. Synopsis of Nearctic Ichneumoninae Stenopneusticae with particular reference to the northeastern region (Hymenoptera). Supplement 5: Ichneumoninae of the island of Newfoundland. *Naturaliste Canadien* **102**: 753-782.

Heinrich, G. H. 1978a. Synopsis of Nearctic Ichneumoninae Stenopneusticae, with particular reference to the northeastern region (Hymenoptera). Supplement 6. *Naturaliste Canadien* **105**: 159-168.

Heinrich, G. H. 1978b. *[Eastern Palearctic Hymenopterous insects of the subfamily Ichneumoninae.] (in Russian)*. Leningrad.

Heitland, W. and Pschorn-Walcher, H. 1992. A brief note on unusually high superparasitism in some sawfly parasitoids (Hym., Ichneumonidae: Euryproctini). *Mitteilungen der Schweizerischen Entomologischen Gesellschaft* **65**: 75-79.

Heitland, W. and Pschorn-Walcher, H. 2005. Biology and parasitoids of the peculiar alder sawfly, *Platycampus luridiventris* (Fallen) (Insecta, Hymenoptera, Tenthredinidae). *Senckenbergiana Biologica* **85**: 215-231.

Henaut, A. and Guerdoux, J. 1982. Location of a lure by the drumming insect *Pimpla instigator* (Hymenoptera, Ichneumonidae). *Experientia* **38**: 346-347.

Heraty, J. M., Burks, R. A., Cruaud, A., Gibson, G. A. P., Liljeblad, J., Munro, J., Rasplus, J.-Y., Delvare, G., Jansta, P., Gumovsky, A., Huber, J., Woolley, J. B., Krogmann, L., Heydon, S., Polaszek, A., Schmidt, S., Darling, D. C., Gates, M. W., Mottern, J., Murray, E., Dal Molin, A., Triapitsyn, S., Baur, H., Pinto, J. D., Van Noort, S., George, J. and Yoder, M. J. 2013. A phylogenetic analysis of the megadiverse Chalcidoidea (Hymenoptera). *Cladistics* **29**: 466-542.

Hilpert, H. 1992. Zur Systematik der Gattung *Ichneumon* Linnaeus in der Westpalaearktis (Hymenoptera, Ichneumonidae, Ichneumoninae). *Entomofauna* **Supplement 6**: 1-389.

Hilszczanski, J. 2000. European species of subgenus *Moerophora* Foerster of *Xorides* Latreille (Hymenoptera: Ichneumonidae: Xoridinae), with descriptions of two new species. *Insect Systematics and Evolution* **31**: 247-255.

Hilszczanski, J. 2003. Polish xoridines and their host associations. In: Melika, G., Thuróczy, C., editors. *Parasitic wasps: Evolution, Systematics, Biodiversity and Biological Control*. Budapest, Hungary: Agroinform. pp. 294-298.

Hinz, R. 1961. Über Blattwespenparasiten (Hym. und Dipt.). *Mitteilungen der Schweizerischen Entomologischen Gesellschaft* **34**: 1-29.

Hinz, R. 1968. Die Untersuchung der Lebensweise der Ichneumoniden (Hym.) mit Anhang: Bemerkungen zur Praeparation von Ichneumoniden. *Entomologische Nachrichten* **12**: 73-81.

Hinz, R. 1975. Die Arten der Gattung *Glyptorhaestus* Thomson (Hymenoptera, Ichneumonidae). *Zeitschrift der Arbeitsgemeinschaft Österreich Entomologen* **27**: 39-46.

Hinz, R. 1976. Zur Systematik und Ökologie der Ichneumoniden V. *Deutsche Entomologische Zeitschrift* **23**: 99-105.

Hinz, R. 1981. Die europäischen Arten der Gattung *Stilbops* Förster (Hymenoptera, Ichneumonidae). *Nachrichtenblatt der Bayerischen Entomologen* **30**: 62-64.

Hinz, R. 1983. The biology of the European species of the genus *Ichneumon* and related species (Hymenoptera : Ichneumonidae). *Contributions of the American Entomological Institute* **20**: 151-152.

Hinz, R. 1986. Die paläarktischen Arten der Gattung *Trematopygus* Holmgren (Hymenoptera, Ichneumonidae). *Spixiana* **8 [1985]**: 265-276.

Hinz, R. 1991. Die palaearktischen Arten der Gattung *Sympherta* Förster (Hymenoptera, Ichneumonidae). *Spixiana* **14**: 27-43.

Hinz, R. 1996. Übersicht über die europäischen Arten von *Lethades* Davis (Insecta, Hymenoptera, Ichneumonidae, Ctenopelmatinae). *Spixiana* **19**: 271-279.

Hinz, R. and Horstmann, K. 1998. Holarctic species of *Trematopygodes* Aubert (Insecta Hymenoptera, Ichneumonidae, Ctenopelmatinae). *Spixiana* **21**: 241-251.

Hinz, R. and Horstmann, K. 1999. Zur Lebensweise der europäischen Arten von *Chasmias* Ashmead, 1900 und *Limerodops* Heinrich, 1949 (Hymenoptera, Ichneumonidae, Ichneumoninae). *Entomofauna* **20**: 301-305.

Hinz, R. and Horstmann, K. 2000. Die westpaläarktischen Arten von *Exephanes* Wesmael (Insecta, Hymenoptera, Ichneumonidae, Ichneumoninae). *Spixiana* **23**: 15-32.

Hinz, R. and Horstmann, K. 2007. Über Wirstbeziehungen europäischer *Ichneumon*-Arten (Insecta, Hymenoptera, Ichneumonidae, Ichneumoninae). *Spixiana* **30**: 39-63.

Hinz, R. and Short, J. 1983. Life-history and systematic position of the European *Alomya* species (Hymenoptera: Ichneumonidae). *Entomologica Scandinavica* **14**: 462-466.

Hisasue, Y., Konishi, K. and Takashino, T. 2018. Oviposition behavior of *Ghilaromma orientalis* (Ichneumonidae, Hybrizontinae). Poster presentation at *Congress of the International Society of Hymenopterists*, Matsuyama, July 2018.

Hoffmann, V. 2014. Die Schlupfwespe *Alomya* spec. bestäubt *Coeloglossum viride*. *Journal Europäischer Orchideen* **46**: 730-735.

Holy, K., Bezdečková, K. and Bezdečka, P. 2017. Occurrence of *Ogkosoma cremieri* Romand (Ichneumonidae: Hybrizontinae) in the Czech Republic with notes on adult behaviour. *Acta Musei Moraviae, Scientiae biologicae* **102**: 139-143.

Horstmann, K. 1968. Zur Systematik und Biologie von *Neorhacodes enslini* (Ruschka) (Hymenoptera, Ichneumonidae). *Entomologische Nachrichten* **12**: 33-36.

Horstmann, K. 1969. Typenrevision der europäischen Arten der Gattung *Diadegma* Foerster (syn. *Angitia* Holmgren) (Hymenoptera: Ichneumonidae). *Beiträge zur Entomologie* **19**: 413-472.

Horstmann, K. 1970. Bemerkungen zur Systematik einiger Gattungen der Campopleginae (Hymenoptera, Ichneumonidae). *Nachrichtenblatt der Bayerischen Entomologen* **19**: 77-84.

Horstmann, K. 1971a. Revision der europäischen Arten der Gattung *Lathrostizus* Förster (Hymenoptera, Ichneumonidae). *Mitteilungen der Deutschen Entomologischen Gesellschaft* **30**: 8-12, 16-18.

Horstmann, K. 1971b. Revision der europäischen Tersilochinen 1 (Hymenoptera: Ichneumonidae). *Veröffentlichungen der Zoologischen Staatssammlung (München)* **15**: 47-138.

Horstmann, K. 1973a. Übersicht über die europäischen Arten der Gattung *Venturia* Schrottky (Hymenoptera, Ichneumonidae). *Mitteilungen der Deutschen Entomologischen Gesellschaft* **32 [1972]**: 7-12.

Horstmann, K. 1973b. Revision der Gattung *Nepiesta* Förster (mit einer Übersicht über die Arten der Gattung *Leptoperilissus* Schmiedeknecht) (Hymenoptera, Ichneumonidae). *Polskie Pismo Entomologiczne* **43**: 729-741.

Horstmann, K. 1974. Revision der westpaläarktischen Arten der Schlupfwespen-Gattungen *Bathyplectes* und *Biolysia* (Hymenoptera: Ichneumonidae). *Entomologica Germanica* **1**: 58-81.

Horstmann, K. 1978a. Bemerkungen zur Systematik einiger Gattungen der Campopleginae II (Hymenoptera, Ichneumonidae). *Mitteilungen der Münchner Entomologischen Gesellschaft* **67**: 65-83.

Horstmann, K. 1978b. Revision der gattungen der Mastrina Townes (Hymenoptera, Ichneumonidae, Hemitelinae). *Zeitschrift der Arbeitsgemeinschaft Österreichischer Entomologen* **30**: 65-70.

Horstmann, K. 1979. Revision der von Kokujev beschriebenen Campopleginae-Arten (mit Teiltabellen der Gattungen *Venturia* Schrottky, *Campoletis* Förster und *Diadegma* Förster) (Hymenoptera: Ichneumonidae). *Beiträge zur Entomologie* **29**: 195-199.

Horstmann, K. 1980a. Neue westpaläarktische Campopleginen-Arten (Hymenoptera, Ichneumonidae). *Mitteilungen der Münchner Entomologischen Gesellschaft* **69**: 117-132.

Horstmann, K. 1980b. Revision der europäischen Arten der Gattung *Aclastus* Förster (Hymenoptera, Ichneumonidae). *Polskie Pismo Entomologiczne* **50**: 133-158.

Horstmann, K. 1980c. Revision der europäischen Arten der Gattung *Rhimphoctona* Förster (Hymenoptera, Ichneumonidae). *Nachrichtenblatt der Bayerischen Entomologen* **29**: 17-24.

Horstmann, K. 1981. Revision der europäischen Tersilochinen II (Hymenoptera, Ichneumonidae). *Spixiana* **Supplement 4 (1980)**: 1-76.

Horstmann, K. 1983. Die westpaläarktischen Arten der Gattung *Chirotica* Förster, 1869 (Hymenoptera Ichneumonidae). *Entomofauna* **4**: 1-33.

Horstmann, K. 1985. Revision der mit *difformis* (Gmelin, 1790) verwandten westpaläarktischen Arten der Gattung *Campoplex* Gravenhorst, 1829 (Hymenoptera, Ichneumonidae). *Entomofauna* **6**: 129-163.

Horstmann, K. 1986. Die westpaläarktischen Arten der Gattung *Gelis* Thunberg, 1827, mit macropteren oder brachypteren Weibchen (Hymenoptera, Ichneumonidae). *Entomofauna* **7**: 389-424.

Horstmann, K. 1987a. Bemerkungen zur Systematik einiger Gattungen der Campopleginae. III (Hymenoptera, Ichneumonidae). *Mitteilungen der Münchner Entomologischen Gesellschaft* **76 [1986]**: 143-164.

Horstmann, K. 1987b. Die europäischen Arten der Gattungen *Echthronomas* Förster und *Eriborus* Förster (Hymenoptera, Ichneumonidae). *Nachrichtenblatt der Bayerischen Entomologen* **36**: 57-67.

Horstmann, K. 1990a. Die westpaläarktischen Arten einiger Gattungen der Cryptini (Hymenoptera, Ichneumonidae). *Mitteilungen der Münchner Entomologischen Gesellschaft* **79 [1989]**: 65-89.

Horstmann, K. 1990b. Neubeschreibungen einiger Schlupfwespen-Arten aus den Gattungen *Mastrus* Förster, *Odontoneura* Förster und *Zoophthorus* Förster (Hymenoptera, Ichneumonidae, Cryptinae). *Zeitschrift der Arbeitsgemeinschaft Österreichischer Entomologen* **42**: 1-14.

Horstmann, K. 1990c. Die westpaläarktischen Arten der Gattung *Pristomerus* Curtis, 1836 (Hymenoptera, Ichneumonidae). *Entomofauna* **11**: 9-44.

Horstmann, K. 1992. Revision einiger Gattungen und Arten der Phygadeuontini (Hymenoptera, Ichneumonidae). *Mitteilungen der Münchner Entomologischen Gesellschaft* **81 [1991]**: 229-254.

Horstmann, K. 1993a. Revision der brachypteren Weibchen der wespaläarktischen Cryptinae (Hymenoptera, Ichneumonidae). *Entomofauna* **14**: 85-148.

Horstmann, K. 1993b. Nachträge zu Revisionen der Gattungen *Aclastus* Förster, *Ceratophygadeuon* Viereck, *Chirotica* Förster und *Gelis* Thunberg (Hymenoptera, Ichneumonidae, Cryptinae). *Nachrichtenblatt der Bayerischen Entomologen* **42**: 7-15.

Horstmann, K. 1993c. Die europäischen Arten von *Gnotus* Förster und *Uchidella* Townes (Hymenoptera, Ichneumonidae, Cryptinae). *Zeitschrift der Arbeitsgemeinschaft Österreichischer Entomologen* **45**: 35-45.

Horstmann, K. 1993d. Revision der Gattung *Leptoperilissus* Schmiedeknecht (Hymenoptera, Ichneumonidae, Campopleginae). *Zeitschrift der Arbeitsgemeinschaft Österreich Entomologen* **45**: 87-98.

Horstmann, K. 1994a. Nachtrag zur Revision der wespaläarktischen *Nemeritis*-Arten (Hymenoptera, Ichneumonidae, Campopleginae). *Mitteilungen der Münchner Entomologischen Gesellschaft* **84**: 79-90.

Horstmann, K. 1994b. Die europäischen Arten von *Picrostigeus* Förster (Hymenoptera, Ichneumonidae, Orthocentrinae). *Zeitschrift der Arbeitsgemeinschaft Österreichischer Entomologen* **46**: 111-120.

Horstmann, K. 1995. Die europäischen Arten von *Arotrephes* Townes, 1970 und *Pleurogyrus* Townes, 1970 (Hymenoptera, Ichneumonidae, Cryptinae). *Entomofauna* **16**: 261-275.

Horstmann, K. 1998a. Revisionen einiger Gattungen und Arten der Phygadeuontini II (Hymenoptera, Ichneumonidae, Cryptinae). *Entomofauna* **19**: 433-460.

Horstmann, K. 1998b. Revisionen von Schlupfwespen-Arten II (Hymenoptera: Ichneumonidae, Braconidae). *Mitteilungen der Münchner Entomologischen Gesellschaft* **88**: 3-12.

Horstmann, K. 1998c. Die europäischen Arten von *Megarhyssa* Ashmead, 1900 (Hymenoptera, Ichneumonidae). *Entomofauna* **19**: 337-350.

Horstmann, K. 1999. Revisionen von Schlupfwespen-Arten III (Hymenoptera: Ichneumonidae). *Mitteilungen der Münchner Entomologischen Gesellschaft* **89**: 47-57.

Horstmann, K. 2000a. Die westpaläarktischen Arten von *Ethelurgus* Förster, 1869 und *Rhembobius* Förster, 1869 (Hymenoptera, Ichneumonidae, Cryptinae). *Entomofauna* **21**: 65-76.

Horstmann, K. 2000b. Die europäischen Arten von *Probolus* Wesmael, 1845 (Hymenoptera: Ichneumonidae). *Entomofauna* **21**: 293-300.

Horstmann, K. 2001a. Revision der bisher zu *Iselix* Förster gestellten westpaläarktischen Arten von *Phygadeuon* Gravenhorst (Insecta, Hymenoptera, Ichneumonidae, Cryptinae). *SPIXIANA - Zeitschrift für Zoologie* **24**: 207-229.

Horstmann, K. 2001b. Revisionen von Schlupfwespen-Arten V (Hymenoptera: Ichneumonidae). *Mitteilungen der Münchner Entomologischen Gesellschaft* **91**: 77-86.

Horstmann, K. 2002a. Revisionen von Schlupfwespen-Arten VI (Hymenoptera: Ichneumonidae) . *Mitteilungen der Münchner Entomologischen Gesellschaft* **92**: 79-91.

Horstmann, K. 2002b. Über einige mit *Coelichneumon orbitator* (Thunberg, 1824) nah verwandte Arten (Hymenoptera, Ichneumonidae, Ichneumoninae). *Entomofauna* **23**: 73-84.

Horstmann, K. 2004a. Bemerkungen zur Systematik einiger Gattungen der Campopleginae IV (Hymenoptera, Ichneumonidae). *Zeitschrift der Arbeitsgemeinschaft Österreichischer Entomologen* **56**: 13-35.

Horstmann, K. 2005. Über einige mit *Scambus inanis* (Schrank, 1802) nah verwandte Arten (Hymenoptera, Ichneumonidae, Pimplinae). *Entomofauna* **26**: 101-116.

Horstmann, K. 2006a. Revision einiger europäischer Mesochorinae (Hymenoptera, Ichneumonidae). *Linzer Biologische Beiträge* **38**: 1449-1492.

Horstmann, K. 2006b. Revisionen von Schlupfwespen-Arten IX (Hymenoptera, Ichneumonidae). *Mitteilungen der Münchner Entomologischen Gesellschaft* **95**: 75-86.

Horstmann, K. 2006c. Revisionen der von Kriechbaumer aus der Westpaläarktis und Zentralasien beschriebenen Ichneumonidae (Insecta, Hymenoptera). *Spixiana* **29**: 1-30.

Horstmann, K. 2008a. Neue westpaläarktische Arten der Campopleginae (Hymenoptera: Ichneumonidae). *Zeitschrift der Arbeitsgemeinschaft Österreichischer Entomologen* **60**: 3-27.

Horstmann, K. 2008b. Revision der europäischen Arten von *Ephialtes* Gravenhorst, 1829, mit Bemerkungen zu weiteren holarktischen Arten (Hymenoptera, Ichneumonidae, Pimplinae). *Entomofauna* **29**: 145-168.

Horstmann, K. 2009a. Revision of the western Palearctic species of *Dusona* Cameron (Hymenoptera, Ichneumonidae, Campopleginae). *Spixiana* **32**: 45-110.

Horstmann, K. 2009b. Revision der europäischen Arten von *Isadelphus* Förster, 1869 (Hymenoptera, Ichneumonidae, Cryptinae). *Entomofauna* **30**: 473-492.

Horstmann, K. 2009c. Revisionen von Schlupfwespen-Arten XIII (Hymenoptera: Ichneumonidae). *Mitteilungen der Münchner Entomologischen Gesellschaft* **99**: 37-44.

Horstmann, K. 2010a. Revision der europäischen Arten von *Stibeutes* Förster, 1850 (Hymenoptera, Ichneumonidae, Cryptinae). *Entomofauna* **31**: 229-264.

Horstmann, K. 2010b. Revisions of Nearctic Tersilochinae II. Genera *Allophrys* Förster, *Barycnemis* Förster, *Ctenophion* gen. nov., *Sathropterus* Förster, *Spinolochus* Horstmann and *Stethantyx* Townes (Hymenoptera, Ichneumonidae). *Spixiana* **33**: 73-109.

Horstmann, K. 2011a. Verbreitung und Wirte der *Dusona*-Arten in der Westpalaearktis (Hymenoptera. Ichneumonidae, Campopleginae). [Distribution and hosts of *Dusona* species in the West Palaearctic (Hymenoptera. Ichneumonidae, Campopleginae).]. *Linzer Biologische Beiträge* **43**: 1295-1330.

Horstmann, K. 2011b. Revisionen von Schlupfwespen-Arten XV (Hymenoptera: Ichneumonidae). *Mitteilungen der Münchner Entomologischen Gesellschaft* **101**: 5-13.

Horstmann, K. 2012a. Revisionen einiger Gattungen und Arten der Phygadeuontini III (Hymenoptera, Ichneumonidae, Cryptinae). *Entomofauna* **33**: 397-422.

Horstmann, K. 2012b. Revisions of Nearctic Tersilochinae III. Genera *Aneuclis* Förster and *Diaparsis* Förster (Hymenoptera, Ichneumonidae). *Spixiana* **35**: 117-142.

Horstmann, K. 2013a. Revisions of Nearctic Tersilochinae IV. Genus *Phradis* Förster. *Spixiana* **36**: 67-92.

Horstmann, K. 2013b. Revisions of Nearctic Tersilochinae V. Genera *Allophroides* Horstmann and *Gelanes* Horstmann (partim). *Spixiana* **36**: 227-261.

Horstmann, K. 2013c. Revisionen von Schlupfwespen-Arten XVII (Hymenoptera: Ichneumonidae). *Mitteilungen der Münchner Entomologischen Gesellschaft* **103**: 1-14.

Horstmann, K. and Shaw, M. R. 1984. The taxonomy and biology of *Diadegma chrysostictos* (Gmelin) and *Diadegma fabricianae* sp.n. (Hymenoptera: Ichneumonidae). *Systematic Entomology* **9**: 329-337.

House, H. L. 1978. An artificial host: encapsulated synthetic medium for in vitro oviposition and rearing the endoparasitoid *Itoplectis conquisitor* (Hymenoptera: Ichneumonidae). *Canadian Entomologist* **110**: 331-333.

Huang, F., Shi, M., Chen, Y. F., Cao, T. T. and Chen, X. X. 2008. Oogenesis of *Diadegma semiclausum* (Hymenoptera: Ichneumonidae) and its associated polydnavirus. *Microscopy Research and Technique* **71**: 676-683.

Huddleston, T. and Gauld, I. D. 1988. Parasitic wasps (Ichneumonoidea) in British light-traps. *The Entomologist* **107**: 134-154.

Humala, A. E. 2002. A Review of Parasitic Wasps of the Genera *Cylloceria* Schiodte, 1838 and *Allomacrus* Forster, 1868 (Hymenoptera, Ichneumonidae) of the Fauna of Russia. *Entomological Review* **82**: 301-313.

Humala, A. E. 2003. [*The Ichneumonid Wasps in the Fauna of Russia and Adjacent Countries: Microleptinae and Oxytorinae (Hymenoptera: Ichneumonidae)*] (In Russian). Moscow: Nauka.

Humala, A. E. 2007a. To the system of the subfamily Orthocentrinae s. l. (Hymenoptera, Ichneumonidae). In: Rasnitsyn, A. P., Gokhman, V. E., editors. [*Studies on hymenopterous insects. Collection of scientific papers.*] (In Russian with English summaries). Moscow: KMK Scientific Press Ltd. pp. 76-85.

Humala, A. E. 2007b. Orthocentrinae (Microleptinae + Orthocentrinae sensu Townes). In: Ler, P. A., editor. *Key to the insects of Russian Far East. Vol. IV. Neuropteroidea, Mecoptera, Hymenoptera. Pt 5 [in Russian].* Vladivostok: Dal'nauka. p. 680-718.

Humala, A. E. 2010. Review of the genus *Batakomacrus* Kolarov, 1986 (Hymenoptera: Ichneumonidae, Orthocentrinae) with description of new species. *Proceedings of the Russian Entomological Society. St. Petersburg* **81**: 29-38.

Humala, A. E., Kruidhof, H. M. and Woelke, J. B. 2016. New species of *Megastylus* (Hymenoptera: Ichneumonidae: Orthocentrinae) reared from larvae of Keroplatidae fungus gnats (Diptera) in a Dutch orchid greenhouse. *Journal of Natural History* **51**: 83-95. DOI: 10.1080/00222933.2016.1257074

Hung, A. C. F., Day, W. H. and Hedlund, R. C. 1988. Genetic variability in arrhenotokous and thelytokous forms of *Mesochorus nigripes* (Hym.: Ichneumonidae). *Entomophaga* **33**: 7-15.

I.C.Z.N. 1945. Opinion 159. On the status of the names *Ephialtes* Schrank, 1802, *Ichneumon* Linnaeus, 1758, *Pimpla* Fabricius, [1804-1805], and *Ephialtes* Gravenhorst, 1829 (Class Insecta, Order Hymenoptera). *Opinions of the International Commission on Zoological Nomenclature* **2**: 275-290.

Idar, M. 1975. Redescriptions of *Hadrodactylus fugax* (Gr.), *H. confusus* (Hlgr.), *H. genalis* Th. and *H. larvatus* (Krb.) (Hym.: Ichneumonidae). *Entomologica Scandinavica* **6**: 286-296.

Idar, M. 1979. Revision of the European species of the genus *Hadrodactylus* Förster (Hymenoptera: Ichneumonidae). Part 1. *Entomologica Scandinavica* **10**: 303-313.

Idar, M. 1981. Revision of the European species of the genus *Hadrodactylus* Förster (Hymenoptera: Ichneumonidae). Part 2. *Entomologica Scandinavica* **12**: 231-239.

Idar, M. 1983. Revision of European *Synomelix* Förster (Hymenoptera: Ichneumonidae) with description of *S. faciator* n. sp. *Entomologica Scandinavica* **14**: 168-172.

Isenhour, D. J. 1986. Developmental time, adult reproductive capability, and longevity of *Campoletis sonorensis* (Hymenoptera: Ichneumonidae) as a parasitoid of Fall Armyworm, *Spodoptera frugiperda* (Lepidoptera: Noctuidae). *Annals of the Entomological Society of America* **79**: 893-897.

Ives, W. G. H. and Muldrew, J. A. 1984. *Pristiphora erichsonii* (Hartig), Larch Sawfly (Hymenoptera: Tenthredinidae). In: Kelleher, J. S. and Hulme, M. A., editors. *Biological Control Programmes against Insects and Weeds in Canada 1969-1980*. Slough, Commonwealth Agricultural Bureaux. pp. 369-380.

Iwata, K. 1958. Ovarian eggs of 223 species of the Japanese Ichneumonidae (Hymenoptera). *Acta Hymenopterologica* **1**: 63-74.

Iwata, K. 1960. The comparative anatomy of the ovary in Hymenoptera, Part V. Ichneumonidae. *Acta Hymenopterologia* **1**: 115-169.

Jackson, D. J. 1937. Host-selection in *Pimpla examinator* F. (Hymenoptera). *Proceedings of the Royal Entomological Society of London - series A* **12**: 81-91 + 81 plate.

Jamieson, W. and Resh, V. H. 1998. Biology of *Tanychela pilosa* (Hymenoptera: Ichneumonidae), a parasitoid of the aquatic moth *Petrophila confusalis* (Lepidoptera: Pyralidae). *Entomological News* **109**: 329-338.

Jenner, W. H., Kuhlmann, U., Cappuccino, N. and Mason, P. G. 2010. Pre-release analysis of the overwintering capacity of a classical biological control agent supporting prediction of establishment. *BioControl* **55**: 351-362.

Jerman, E. J. and Gauld, I. D. 1988. *Casinaria*, a paraphyletic ichneumonid genus (Hymenoptera), and a review of the Australian species. *Journal of Natural History* **22**: 589-609.

Jervis, M. A. and Kidd, N. A. C. 1986. Host-feeding strategies in hymenopteran parasitoids. *Biological Reviews* **61**: 395-434.

Jervis, M. A., Ellers, J. and Harvey, J. A. 2008. Resource acquisition, allocation, and utilization in parasitoid reproductive strategies. *Annual Review of Entomology* **53**: 361-385.

Jolivet, P. 1950. Les parasites, prédateurs et phorétiques des Chrysomeloidea (Coleoptera) de la faune Franco-belge. *Bulletin de l'Institut Royal des Sciences Naturelles de Belgique* **26(34)**: 39pp.

Jolivet, P. and Théodoridès, J. 1952. Les parasites, phorétiques et prédateurs des Chrysomeloidea (Coleoptera). *Bulletin de l'Institut Royal des Sciences Naturelles de Belgique* **28(20)**: 19 pp.

Jones, R. A. 2001. *Ctenochares bicolorus* (L.), an African ichneumonid (Hymenoptera) found in Britain. *British Journal of Entomology and Natural History* **14**: 96-99.

Jordan, T. 1998a. *Tersilochus curvator* Horstmann und *Tersilochus* sp. n. (Ichneumonidae, Tersilochinae), neue Parasitoiden der an Birken minierenden Trugmotten (Lepidoptera, Eriocraniidae) [*Tersilochus curvator* Horstmann and *Tersilochus* sp. n. (Ichneumonidae, Tersilochinae), new parasitoids of birch leaf mining *Eriocrania* species (Lepidoptera, Eriocraniidae)]. *Bonner Zoologische Beitraege* **47**: 411-419.

Jordan, T. 1998b. Eiformen und Lebenweise zweier Artengruppen der Gattung *Grypocentrus* (Ichneumonidae: Tryphoninae), spezifische Parasitoiden der im Frühjahr an Birken minierenden Trugmotten-Arten (Lepidoptera: Eriocraniidae). *Entomologia Generalis* **23**: 223-231.

Jørgensen, O. F. 1975. Competition among larvae of *Pimplopterus dubius* Hgn. (Hymenoptera: Ichneumonidae), a parasitoid of *Epinotia tedella* Cl. (Lepidoptera: Tortricidae). *Zeitschrift für Angewandte Entomologie* **79**: 301-309.

Jourdheuil, P. 1960. Influence de quelques facteurs écologiques sur les fluctuations de population d'une biocénose parasitaire: étude relative a quelques Hyménoptères (Ophioninae, Diospilinae, Euphorinae) parasites de divers Coléoptères inféodés aux Crucifères. *Annales des Épiphyties (C)* **11**: 445-658.

Juillet, J. A. 1959. Morphology of immature stages, life-history, and behaviour of three hymenopterous parasites of the European Pine Shoot Moth, *Rhyacionia buoliana* (Schiff.) (Lepidoptera: Olethreutidae). *The Canadian Entomologist* **91**: 709-719.

Jumean, Z., Unruh, T., Gries, R. and Gries, G. 2005. *Mastrus ridibundus* parasitoids eavesdrop on cocoon-spinning codling moth, *Cydia pomonella*, larvae. *Naturwissenschaften* **92**: 20-25.

Jussila, R. 1975. Ichneumonological (Hym.) reports from Finland III. *Annales Entomologici Fennici* **41**: 49-55.

Jussila, R. 1979. A revision of the genus *Atractodes* (Hymenoptera, Ichneumonidae) in the Western Palaearctic region. *Acta Entomologica Fennica* **34**: 1-44.

Jussila, R. 1987. Revision of the genus *Stilpnus* (Hymenoptera, Ichneumonidae) of the western Palaearctic Region. *Annales Entomologici Fennici* **53**: 1-16.

Jussila, R. 1994. *Aperileptus rossemi* sp. n., *Aniseres lapponicus* sp. n., and additions to descriptions of other Oxytorinae species (Hymenoptera, Ichneumonidae). *Entomologica Fennica* **5**: 115-118.

Jussila, R. 1999. Additions to the revisions of the genus *Stilpnus* (Hymenoptera, Ichneumonidae) of the Palaearctic Region. I. *Entomologica Fennica* **10**: 107-112.

Jussila, R. 2001. Additions to the revision of the genus *Atractodes* (Hymenoptera: Ichneumonidae) of the Palaearctic Region. III. *Entomologica Fennica* **12**: 193-216.

Jussila, R., Sääksjärvi, I. E. and Bordera, S. 2010. Revision of the western Palaearctic *Mesoleptus* (Hymenoptera: Ichneumonidae). *Annales de la Société entomologique du France (n.s.)* **46**: 499-518.

Kamath, M. K. and Gupta, V. K. 1972. Ichneumologia Orientalis. Part II. The tribe Rhyssini (Hymenoptera : Ichneumonidae). *Oriental Insects Monograph* **2**: 1-300+viii.

Karimpour, Y., Fathipour, Y., Talebi, A. A. and Moharramipour, S. 2007. Biology of *Chamaesphecia schizoceriformis* (Lep.: Sesiidae), a biocontrol agent of *Euphorbia boissieriana* (Euphorbiales: Euphorbiaceae) in north west of Iran. *Journal of Entomological Society of Iran* **26**: 35-45.

Karlsson, D. and Ronquist, F. 2012. Skeletal morphology of *Opius dissitus* and *Biosteres carbonarius* (Hymenoptera: Braconidae), with a discussion of terminology. *PLoS ONE* **7**: e32573 (32538 pp.). doi: 10.1371/journal.pone.0032573

Kasparyan, D. R. 1973a. Review of Palearctic ichneumonids of the tribe Pimplini (Hymenoptera, Ichneumonidae). The genera *Itoplectis* Först. and *Apechthis* Först. *Entomological Review* **52**: 445-455.

Kasparyan, D. R. 1973b. *Fauna of the USSR Hymenoptera Vol.III Number 1. Ichneumonidae (Subfamily Tryphoninae) Tribe Tryphonini*. Leningrad: Nauka Publishers [translated from Russian. Amerind Publishing Co. Ltd., New Delhi, 1981. 414pp.].

Kasparyan, D. R. 1974. A review of Palearctic species of the tribe Pimplini (Hymenoptera, Ichneumonidae). The genus *Pimpla* Fabricius. *Entomological Review* **53**: 102-117.

Kasparyan, D. R. 1976. New species of the tribe Cteniscini (Hymenoptera, Ichneumonidae) from East Asia. The genera *Cycasis* Townes, *Orthomiscus* Mason and *Kristotomus* Mason. *Entomological Review* **55**: 99-108.

Kasparyan, D. R. 1977. A revision of the genus *Eclytus* Holmgren (Hymenoptera, Ichneumonidae). *Entomological Review* **56**: 116-129.

Kasparyan, D. R. 1981. [*Hymenoptera, Ichneumonidae. Keys to the insects of the European part of the U.S.S.R.*] (in Russian). Opredeliteli Faune SSSR 3(3), 688pp.

Kasparyan, D. R. 1982. East-Palaearctic ichneumonids of the genera *Monoblastus* Htg. and *Otoblastus* Förster (Hymenoptera, Ichneumonidae). *Entomological Review* **61**: 137-144.

Kasparyan, D. R. 1984. [New species of Ichneumonidae (Hymenoptera) of the genus *Euceros* Grav. from the Far East]. (in Russian). In: Ler, P. A., editor. *Systematics of insects from the Far East. Collected scientific papers*. SSR, Vladivostok: Akademii Nauk. p. 78-83.

Kasparyan, D. R. 1986. [Towards a revision of the ichneumonids genus *Adelognathus* Holmgren (Hymenoptera, Ichneumonidae)] (in Russian). *Proceedings of the Zoological Institute, Leningrad* **159**: 38-56.

Kasparyan, D. R. 1987. [Structure of head capsule of larva of *Atractogaster semisculptus* Kriechb. & taxonomic notes on the tribe Delomeristini (Hymenoptera, Symphyta).] (in Russian). *Morphology and Phylogeny of Insects* **69**: 219-222.

Kasparyan, D. R. 1988. [New taxa of fam. Paxylommatidae (Hymenoptera, Ichneumonoidea) from the Baltic amber] (in Russian). *Trudy Vsesoyuznogo Entomologicheskogo Obshchestva* **70**: 125-131.

Kasparyan, D. R. 1990. [*Fauna of USSR. Insecta Hymenoptera. Vol.III(2). Ichneumonidae. Subfamily Tryphoninae: Tribe Exenterini. Subfamily Adelognathinae*] (in Russian). Leningrad: Nauka Publishing House.

Kasparyan, D. R. 1992. [New east Palaearctic species of the ichneumonid genera *Idiogramma* Först., *Sphinctus* Grav. and *Euceros* Grav. (Hymenoptera: Ichneumonidae)]. *Entomologicheskoye Obozreniye* **71**: 887-899.

Kasparyan, D. R. 1993a. Townesioninae, a new ichneumonid subfamily from the Eastern Palearctic (Hymenoptera: Ichneumonidae). *Zoosystematica Rossica* **2**: 155-159.

Kasparyan, D. R. 1993b. [Revision of the genus *Thymaris* (Hymenoptera, 1. Ichneumonidae)]. *Zoologichesky Zhurnal* **72**: 105-117.

Kasparyan, D. R. 1994a. [Review of Ichneumon flies of the Townesitinae subfam. nov. (Hymenoptera, Ichneumonidae) from the Baltic ambers] (in Russian with English summary). *Paleontologicheskiy Zhurnal* **4**: 86-96.

Kasparyan, D. R. 1994b. Review of Palearctic species of wasps of the genus *Phytodietus* Grav. (Hymenoptera, Ichneumonidae). *Entomological Review* **73**: 56-79.

Kasparyan, D. R. 1996. The main trends in evolution of parasitism in Hymenoptera. *Entomological Review* **76**: 1107-1136.

Kasparyan, D. R. 1997. A new apterous xoridine species from China and notes on the status of *Aderaeon* Townes (Hymenoptera: Ichneumonidae). *Zoosystematica Rossica* **5**: 295-296.

Kasparyan, D. R. 1998. Taxonomic notes on the species of *Mesoleius* s.l., *Hyperbatus* and *Phaestus* in the museums of Stokholm, Lund and Munich (Hymenoptera: Ichneumonidae, Ctenopelmatinae). *Zoosystematica Rossica* **7**: 181-183.

Kasparyan, D. R. 1999. New species of ichneumonid wasps of the subfamily Stilbopinae (Hymenoptera, Ichneumonidae) from the old world. *Entomological Review* **79**: 376-385.

Kasparyan, D. R. 2000. Palaearctic ichneumonid wasps of the genus *Mesoleius* (s. str.) Holmgren (Hymenoptera, Ichneumonidae): I. *Entomological Review* **80**: 144-168.

Kasparyan, D. R. 2001. Palaearctic ichneumonid wasps of the genus *Mesoleius* (s. str.) Holmgren (Hymenoptera, Ichneumonidae): II. *Entomological Review* **81**: 642-665.

Kasparyan, D. R. 2002. [Analysis of the Fauna of Parasitoids (Diptera et Hymenoptera) of Sawflies of the Family Pamphiliidae (Hymenoptera). A Review of the Palaearctic Ichneumonid Genus *Notopygus* Holmg. (Hymenoptera, Ichneumonidae)]. *Entomologicheskoe Obozrenie* **81**: 890-917.

Kasparyan, D. R. 2003. [Palaearctic species of the ichneumonid-wasp genus *Campodorus* Foerster (s. str.) (Hymenoptera, Ichneumonidae) with pectinate claws]. *Entomologicheskoe Obozrenie* **82**: 758-766.

Kasparyan, D. R. 2004a. [A review of Palaearctic species of tribe Ctenopelmatini (Hymenoptera, Ichneumonidae). The genera *Ctenopelma* Holmgren and *Homaspis* Foerster]. *Entomologicheskoe Obozrenie* **83**: 437-467.

Kasparyan, D. R. 2004b. Nomenclatural notes on some Ctenopelmatinae from Dutch and Hungarian museums (Hymenoptera: Ichneumonidae). *Zoosystematica Rossica* **13**: 47-48.

Kasparyan, D. R. 2005. Palaearctic Ichneumonid Wasps of the Genus *Campodorus* Foerster (Hymenoptera, Ichneumonidae): II. Species with red mesothorax and species with yellow face. *Entomological Review* **85**: 177-192.

Kasparyan, D. R. 2006. Palaearctic species of the ichneumon-fly genus *Campodorus* Forster (Hymenoptera; Ichneumonidae). III. Species with long-haired ovipositor sheath, species with uniformly rufous hind tibiae, and species with white-banded tibiae. *Entomologicheskoe Obozrenie* **85**: 632-661.

Kasparyan, D. R. 2011. A review of the Palaearctic species of the genus *Hadrodactylus* Förster (Hymenoptera, Ichneumonidae, Ctenopelmatinae) with a description of five new species. *Entomological Review* **91**: 866-888.

Kasparyan, D. R. 2012. Review of the ichneumon-flies of the genus *Rhorus* Förster, 1869 (Hymenoptera, Ichneumonidae: Ctenopelmatinae): I. The species from the Far East (with description of 24 new species and with a key). *Entomological Review* **92**: 650-687.

Kasparyan, D. R. 2014. Review of the Western Palaearctic ichneumon-flies of the genus *Rhorus* Förster, 1869 (Hymenoptera, Ichneumonidae: Ctenopelmatinae). II. The species of the *punctus, longicornis, chrysopygus,* and *substitutor* groups, the species with the black metasoma and some others. *Entomological Review* **94**: 712-755.

Kasparyan, D. R. 2015. Review of the Western Palaearctic ichneumon-flies of the genus *Rhorus* Förster, 1869 (Hymenoptera, Ichneumonidae: Ctenopelmatinae). Part III. The species with the reddish metasoma and black face. *Entomological Review* **95**: 1257-1291.

Kasparyan, D. R. 2017. Review of the Western Palaearctic ichneumon-flies of the genus *Rhorus* Förster, 1869 (Hymenoptera, Ichneumonidae: Ctenopelmatinae). Part IV. The species with the reddish metasoma and black face (addendum). *Entomological Review* **97**: 116-131.

Kasparyan, D. R. and Khalaim, A. I. 2007. Subfamily Tryphoninae. In: Ler, P. A., editor. *Key to the insects of Russian Far East. Vol. IV. Neuropteroidea, Mecoptera, Hymenoptera. Pt 5* [in Russian]. Vladivostok: Dal'nauka. pp. 333-404.

Kasparyan, D. R. and Kopelke, J.-P. 2009. Taxonomic review and key to European ichneumon flies (Hymenoptera, Ichneumonidae), parasitoids of gall-forming sawflies of the genera *Pontania* Costa, *Phyllocolpa* Benson, and *Euura* Newman (Hymenoptera, Tenthredinidae) on willows: part I. *Entomological Review* **89**: 933-957.

Kasparyan, D. R. and Kopelke, J.-P. 2010. A taxonomic review of ichneumon-flies (Hymenoptera, Ichneumonidae), parasitoids of gall-forming sawflies (Hymenoptera, Tenthredinidae) on *Salix*. Part II. Review of the Palaearctic species of the genus *Saotis* Förster with description of four new species. *Entomological Review* **90**: 71-98.

Kasparyan, D. R. and Manukyan, A. R. 1987. A new genus of parasitic wasps (Hymenoptera: Ichneumonidae Diplazontinae) from the eastern Palearctic region. *Entomological Review* **67**: 14-17.

Kasparyan, D. R. and Shaw, M. R. 2003. A preliminary key to the European species of the genus *Saotis* Förster, 1869, with a list of British species (Ichneumonidae: Ctenopelmatinae: Mesoleiini). *Zoosystematica Rossica* **11**: 351-355.

Kasparyan, D. R. and Shaw, M. R. 2008. British and European *Phytodietus* Gravenhorst (Hymenoptera: Ichneumonidae,Tryphoninae) in the National Museums of Scotland, with a key to European species of the subgenus *Neuchorus* Uchida and descriptions of three new species. *Entomologist's Gazette* **59**: 184-198.

Kasparyan, D. R. and Shaw, M. R. 2009. A new species of *Hadrodactylus* Foerster (Hymenoptera: Ichneumonidae, Ctenopelmatinae, Euryproctini) from Britain and mainland Europe, with a review of material of the genus in the National Museums of Scotland. *Entomologist's Gazette* **60**: 251-258.

Kasparyan, D. R. and Tolkanitz, V. I. 1999. [*Ichneumonidae subfamily Tryphoninae: tribes Sphinctini, Phytodietini, Oedemopsini, Tryphonini (Addendum), Idiogrammatini. Subfamilies Eucerotinae, Adelognathinae (addendum), Townesioninae*] (in Russian). Saint Petersburg: Nauka.

Kasparyan, D. R., Oh, S.-H., Choi, G. W. and Lee, J.-W. 2014. First record of the subfamily Oxytorinae (Hymenoptera: Ichneumonidae) from Korea with descriptions of two new species. *Entomological Research* **44**: 315-322.

Kelleher, J. S. and Hulme, M. A. 1984. *Biological Control Programmes against Insects and Weeds in Canada 1969-1980*. Slough: Commonwealth Agricultural Bureaux. 410 pp.

Kerrich, G. J. 1936. Notes on larviposition in *Polyblastus* (Hym. Ichn. Tryphoninae). *Proceedings of the Royal Entomological Society of London (A)* **11**: 6-12.

Kerrich, G. J. 1939. Systematic notes on the Oxytorina (Hym., Ichneumonidae, Mesoleptini Auctt.). *Opuscula Entomologica* **4**: 126-128.

Kerrich, G. J. 1952. A review, and a revision in greater part, of the Cteniscini of the old world (Hym., Ichneumonidae). *Bulletin of the British Museum (Natural History) (Entomology)* **2**: 305-460.

Kerrich, G. J. 1953. A preliminary study of the European species of the genus *Eudiaborus* mihi (Hym. Ichneumonidae). *Opuscula Entomologica* **18**: 151-159.

Kerrich, G. J. 1969. Description of an ichneumonid (Hym.) that preys on egg-masses of weevils harmful to tea culture in Kenya. *Bulletin of Entomological Research* **59**: 469-472.

Khalaim, A. I. 2002a. A review of the subgenera *Nanodiaparsis, Ischnobatis*, and *Lanugopasrsis* subgen. n. of the genus *Diaparsis* Forster (Hymenoptera, Ichneumonidae) with descriptions of new species. *Entomological Review* **82**: 76-82.

Khalaim, A. I. 2002b. A review of the species of the genus *Gelanes* (Hymenoptera, Ichneumonidae, Tersilochinae) of the Palaearctic Region. *Vestnik Zoologii* **36**: 3-12.

Khalaim, A. I. 2003. Review of the Palaearctic subgenus *Rugodiaparsis* Horstmann, 1971 of the genus *Probles* Förster, 1869 (Hymenoptera: Ichneumonidae: Tersilochinae). *Russian Entomological Journal* **12**: 75-78.

Khalaim, A. I. 2004a. A review of the Palaearctic species of the genera *Barycnemis* Först., *Epistathmus* Först. and *Spinolochus* Horstm. (Hymenoptera: Ichneumonidae, Tersilochinae). *Proceedings of the Russian Entomological Society* **75**: 46-63.

Khalaim, A. I. 2004b. A review of the genera *Aneuclis* Förster and *Sathropterus* Förster (Hymenoptera, Ichneumonidae, Tersilochinae). *Entomological Review* **83**: 664-678.

Khalaim, A. I. 2005. A review of the subgenera *Diaparsis* s. str. and *Pectinoparsis* subgen. n. of the genus *Diaparsis* Förster (Hymenoptera, Ichneumonidae, Tersilochinae). *Entomological Review* **85**: 538-554.

Khalaim, A. I. 2009. South African species of *Aneuclis* Förster, 1869 (Hymenoptera: Ichneumonidae: Tersilochinae). *African Invertebrates* **50**: 123-136.

Khalaim, A. I. 2011. Tersilochinae of South, Southeast and East Asia, excluding Mongolia and Japan (Hymenoptera: Ichneumonidae). *Zoosystematica Rossica* **20**: 96-148.

Khalaim, A. I. and Broad, G. R. 2012. Tersilochinae (Hymenoptera: Ichneumonidae) of Costa Rica, part 1. Genera *Allophrys* Förster, *Barycnemis* Förster and *Meggoleus* Townes. *Zootaxa* **3185**: 36-52.

Khalaim, A. I. and Broad, G. R. 2013. Tersilochinae (Hymenoptera: Ichneumonidae) of Costa Rica, part 2. Genera *Megalochus* gen. nov. and *Stethantyx* Townes. *Zootaxa* **3693**: 221-266.

Khalaim, A. I. and Ruíz Cancino, E. 2017. Ichneumonidae (Hymenoptera) associated with xyelid sawflies (Hymenoptera, Xyelidae) in Mexico. *Journal of Hymenoptera Research* **58**: 17-27. DOI: 10.3897/jhr.58.12919

Khalaim, A. I. and Sheng, M.-L. 2009. Review of Tersilochinae (Hymenoptera, Ichneumonidae) of China, with descriptions of four new species. *ZooKeys* **14**: 67-81.

Khalaim, A. I., Bordera, S. and Rodríguez-Berrío, A. 2009. A review of the European species of *Phradis* (Hymenoptera: Ichneumonidae: Tersilochinae), with description of a new species from Spain. *European Journal of Entomology* **106**: 107-118.

Khalaim, A. I., Ruíz-Cancino, E. and Coronado-Blanco, J. M. 2012. A new genus and species of Metopiinae (Hymenoptera, Ichneumonidae) from Mexico. *ZooKeys* **207**: 1-10.

Khalaim, A. I., Sääksjärvi, I. E. and Bordera, S. 2013. Tersilochinae of Western Amazonia (Hymenoptera: Ichneumonidae). Genus *Stethantyx* Townes, part 1. *Zootaxa* **3741**: 301-326.

Khalaim, A. I., Ruíz-Cancino, E. and Coronado-Blanco, J. M. 2017. *Labilochus brevipalpis*, a new genus and species with extremely long mouthparts (Hymenoptera, Ichneumonidae, Tersiochinae) from Mexico. *Journal of Hymenoptera Research* **55**: 121-127. DOI: 10.3897/jhr.55.11452

Kishi, Y. 1970. Difference in the sex ratio of the pine bark weevil parasite, *Dolichomitus* sp. (Hymenoptera: Ichneumonidae), emerging from different host species. *Applied Entomology and Zoology* **5**: 126-132.

Klomp, H. and Teerink, B. J. 1978. The epithelium of the gut as a barrier against encapsulation by blood cells in three species of parasitoids of *Bupalus piniarius* (Lep., Geometridae). *Netherlands Journal of Zoology* **28**: 132-138.

Klopfstein, S. 2014. Revision of the Western Palaearctic Diplazontinae (Hymenoptera, Ichneumonidae). *Zootaxa* **3801**: 1-143.

Klopfstein, S. 2016. Nine new species of *Dimophora* from Australia (Hymenoptera: Ichneumonidae): new insights on the distribution of a poorly known genus of parasitoid wasps. *Austral Entomology* **55**: 185-207.

Klopfstein, S., Kropf, C. and Quicke, D. L. J. 2010a. An evaluation of phylogenetic informativeness profiles and the molecular phylogeny of Diplazontinae (Hymenoptera, Ichneumonidae). *Systematic Biology* **59**: 226-241.

Klopfstein, S., Quicke, D. L. J. and Kropf, C. 2010b. The evolution of antennal courtship in diplazontine parasitoid wasps (Hymenoptera, Ichneumonidae, Diplazontinae). *BMC Evolutionary Biology* **10**: 12 pp.

Klopfstein, S., Quicke, D. L. J., Kropf, C. and Frick, H. 2011. Molecular and morphological phylogeny of Diplazontinae (Hymenoptera, Ichneumonidae). *Zoologica Scripta* **40**: 379-402.

Klopfstein, S., Vilhelmsen, L., Heraty, J. M., Sharkey, M. and Ronquist, F. 2013. The hymenopteran tree of life: evidence from protein-coding genes and objectively aligned ribosomal data. *PLoS ONE* **8**: 23pp. doi: 10.1371/journal.pone.0069344

Klopfstein, S., Langille, B., Spasojevic, T., Broad, G. R., Cooper, S. J. B., Austin, A. D. and Niehuis, O. In press. Hybrid capture data unravels a rapid radiation in pimpliform parasitoid wasps (Hymenoptera: Ichneumonidae: Pimpliformes). *Systematic Entomology*.

Kolarov, J. A. 1987. A new Ctenopelmatinae genus and species from Bulgaria (Hymenoptera, Ichneumonidae). *Entomofauna* **8(6)**: 69-76.

Kolarov, J. A. 1997. Hymenoptera, Ichneumonidae. Part 1. Pimplinae, Xoridinae, Acaenitinae, Collyriinae. *Fauna Bulgarica* **25**: 1-326.

Kolarov, J. and Bechev, D. 1995. Hymenopterenparasiten (Hymenoptera) auf Pilzmücken (Mycetophiloidea, Diptera). *Acta Entomologica Bulgarica* **2**: 18-20.

Komatsu, T. and Konishi, K. 2010. Parasitic behaviors of two ant parasitoid wasps (Ichneumonidae: Hybrizontinae). *Sociobiology* **56**: 575-584.

Konishi, K. 1986. A new subgenus and species of the genus *Netelia* (Hymenoptera, Ichneumonidae) from New Guinea. *Kontyu* **54**: 415-419.

Konishi, K. 1991. A revision of the subgenus *Prosthodocis* Enderlein of the genus *Netelia* Gray of Japan (Hymenoptera, Ichneumonidae), I. *Japanese Journal of Entomology* **59**: 775-788.

Konishi, K. 1992. A revision of the subgenus *Prosthodocis* Enderlein of the genus *Netelia* Gray of Japan (Hymenoptera, Ichneumonidae), II. *Japanese Journal of Entomology* **60**: 39-53.

Kopelke, J.-P. 1987. *Adelognathus cubiceps* Roman 1924 (Ichneumonidae: Adelognathinae) - ein ungewöhnlicher Parasitoid der gallenbildenden *Pontania*-Arten (Tenthredinidae: Nematinae) (Insecta: Hymenoptera). *Senckenbergiana Biologica* **67**: 253-259.

Kopelke, J.-P. 2003. Natural enemies of gall-forming sawflies on willows (*Salix* spp) (Hymenoptera: Tenthredinidae: Euura, Phyllocolpa, Pontania). *Entomologia Generalis* **26**: 277-312.

Koponen, M., Jussila, R. and Vikberg, V. 2000. Suomen loispistiäisluettelo (Hymenoptera, Parasitica) Osa 4. heimo Ichneumonidae, alaheimot Lycorininae, Neorhacodinae, Stilbopinae, Banchinae ja Ctenopelmatinae. *Sahlbergia* **5**: 51-82.

Kopylov, D. S. 2009. A new subfamily of ichneumons from the Lower Cretaceous of Transbaikalia and Mongolia (Insecta: Hymenoptera: Ichneumonidae). *Paleontological Journal* **43**: 83-93.

Kopylov, D. S. 2010a. A new subfamily of ichneumon wasps (Insecta: Hymenoptera: Ichneumonidae) from the Upper Cretaceous of the Russian Far East. *Paleontological Journal* **44**: 422-433.

Kopylov, D. S. 2010b. Ichneumonids of the subfamily Tanychorinae (Insecta: Hymenoptera: Ichneumonidae) from the Lower Cretaceous of Transbaikalia and Mongolia. *Paleontological Journal* **44**: 179-186.

Kovács, G., Kaasik, R., Metspalu, L. and Veromann, E. 2013. The attractiveness of wild cruciferous plants on the key parasitoids of *Meligethes aeneus*. *Integrated Control in Oilseed Crops IOBC-WPRS Bulletin* **96**: 81-92.

Kuhlmann, U. and Mills, N. J. 1999. Exploring the biodiversity of Central Asia to assess specialized parasitoids for biological control of apple pests in Europe and North America. *Integrated Plant Protection in Orchards. IOBC/wprs Bulletin* **22**: 1-6.

Kusigemati, K. 1981. New host records of Ichneumonidae from Japan (IV). *Memoirs of the Faculty of Agriculture, Kagoshima University* **17**: 135-138.

Kuslitzky, W. S. 1981. [*A guide to the insects of the European part of the USSR. Hymenoptera, Ichneumonidae. Subfamily Banchinae*] Opredeliteli Faune SSSR **3**: 276-316.

Kuslitzky, V. S. and Kasparyan, D. R. 2011. A new genus of ichneumonid flies of the subfamily Collyriinae (Hymenoptera: Ichneumonidae) from Syria and Israel. *Zoosystematica Rossica* **20**: 319-324.

Lapointe, R., Tanaka, K., Barney, W. E., Whitfield, J. B., Banks, J. C., Béliveau, C., Stoltz, D., Webb, B. A. and Cusson, M. 2007. Genomic and morphological features of a banchine polydnavirus: comparison with bracoviruses and ichnoviruses. *Journal of Virology* **81**: 6491-6501.

LaSalle, J. and Gauld, I. D. 1993. *Hymenoptera and Biodiversity*. Wallingford: C.A.B. International.

Laurenne, N. M., Broad, G. R. and Quicke, D. L. J. 2006. Direct optimization and multiple alignment of 28S D2–D3 rDNA sequences: problems with indels on the way to a molecular phylogeny of the cryptine ichneumon wasps (Insecta: Hymenoptera). *Cladistics* **22**: 442-473.

Lawton, F. D. 1981. An introduction to the Mesochorinae (Hymenoptera, Ichneumonidae). *Proceedings and Transactions of the British Entomological and Natural History Society* **14**: 93-97.

Le Lannic, J. and Nénon, J.-P. 1999. Functional morphology of the ovipositor in *Megarhyssa atrata* (Hymenoptera, Ichneumonidae) and its penetration into wood. *Zoomorphology* **119**: 73-79.

Leblanc, L. 1999. The Nearctic species of *Protarchus* Foerster (Hymenoptera: Ichneumonidae: Ctenopelmatinae). *Journal of Hymenoptera Research* **8**: 251-267.

Legner, E. F. 1995. Biological control of Diptera of medical and veterinary importance. *Journal of Vector Ecology* **20**: 59-120.

Lehmann, W. 1969. Beiträge zur Parasitenfauna in Rapsbeständen. *Deutsche Akademie der Landwirtschaftswissenschaften zu Berlin. Tagungsberichte* **80**: 513-528.

Leius, K. 1960. Attractiveness of different foods and flowers to the adults of some hymenopterous parasites. *Canadian Entomologist* **92**: 369-376.

Leius, K. 1961a. Influence of food on fecundity and longevity of adults of *Itoplectis conquisitor* (Say) (Hymenoptera: Ichneumonidae). *Canadian Entomologist* **93**: 771-780.

Leius, K. 1961b. Influence of various foods on fecundity and longevity of adults of *Scambus buolianae* (Htg.) (Hymenoptera: Ichneumonidae). *Canadian Entomologist* **93**: 1079-1084.

Li, L., Kopylov, D. S., Shih, C. and Ren, D. 2016. The first record of Ichneumonidae (Insecta: Hymenoptera) from the Upper Cretaceous of Myanmar. *Cretaceous Research* **70**: 152-162.

Lloyd, D. C. 1940. Host selection by Hymenopterous parasites of the moth *Plutella maculipennis* Curtis. *Proceedings of the Royal Society of London Series B - Biological Sciences* **128**: 451-484.

Luhman, J. C. 1991. A revision of the world *Amphibulus* Kriechbaumer (Hymenoptera: Ichneumonidae, Phygadeuontinae). *Insecta Mundi* **5**: 129-152.

Mansbridge, G. H. 1933. On the biology of some Ceroplatinae and Macrocerinae (Diptera, Mycetophilidae). *Transactions of the Royal Entomological Society of London* **81**: 75-92.

Marsh, P. M. 1989. Notes on the genus *Hybrizon* in North America (Hymenoptera: Paxylommatidae). *Proceedings of the Entomological Society of Washington* **91 (1988)**: 29-34.

Mason, W. R. M. 1967. Specialization in the egg structure of *Exenterus* (Hymenoptera: Ichneumonidae) in relation to distribution and abundance. *Canadian Entomologist* **99**: 375-384.

Mason, W. R. M. 1971. An Indian *Agriotypus* (Hymenoptera: Agriotypidae). *Canadian Entomologist* **103**: 1521-1524.

Mason, W. R. M. 1981. Paxylommatidae: the correct family-group name for *Hybrizon* Fallén (Hymenoptera: Ichneumonoidea), with figures of unusual antennal sensilla. *Canadian Entomologist* **113**: 433-439.

Mason, W. R. M. 1990. Cubitus posterior in Hymenoptera. *Proceedings of the Entomological Society of Washington* **92**: 93-97.

Matsumoto, R. 2016. Molecular phylogeny and systematics of the *Polysphincta* group of genera (Hymenoptera, Ichneumonidae, Pimplinae). *Systematic Entomology* **41**: 854-864. doi: 10.1111/syen.12196

Matsumoto, R. and Broad, G. R. 2011. Discovery of *Rodrigama* Gauld in the Old World, with description of two new species (Hymenoptera, Ichneumonidae, Poemeniinae). *Journal of Hymenoptera Research* **20**: 65-75.

Matsumoto, R. and Saigusa, T. 2001. The biology and immature stages of *Thrybius togashii* Kusigemati (Hymenoptera: Ichneumonidae: Cryptinae), with a description of the male. *Journal of Natural History* **35**: 1507-1516.

Mayhew, P. J., Dytham, C., Shaw, M. R. and Fraser, S. E. M. 2009. Collections of ichneumonids wasps (subfamilies Diacritinae, Diplazontinae, Pimplinae and Poemeniinae) from woodlands near York and their implications for conservation planning. *Naturalist* **134**: 3-24.

Mazanec, Z. 1990. Immature stages and life history of *Enytus* sp. (Hymenoptera: Ichneumonidae), a parasitoid of *Perthida glyphopa* Common (Lepidoptera: Incurvariidae). *Journal of the Australian Entomological Society* **29**: 57-66.

McConnell, H. S. 1938. Additional notes on *Oocenteter tomostethae*. *Proceedings of the Entomological Society of Washington* **40**: 23-24.

Menzel, J. G. and Tautz, J. 1994. Functional morphology of the subgenual organ of the Carpenter Ant. *Tissue and Cell* **26**: 735-746.

Messner, B. and Taschenberger, D. 1981. Zur Funktionsmorphologie des Atembandes von *Agriotypus armatus* Walk. (Hym., Agriotypidae). *Deutsche Entomologische Zeitschrift* **28**: 7-9 + 1 plate.

Miah, M. I. and Bhuyia, B. A. 2001. The relationships of the subfamily Campopleginae (Hymenoptera, Ichneumonidae) with its related subfamilies in cladistic assessment. *Proceedings of the Zoological Society (Calcutta)* **54**: 27-37.

Miller, D. and Clark, A. F. 1935. *Sirex noctilio* (Hym.) and its parasites in New Zealand. *Bulletin of Entomological Research* **26**: 149-154.

Miller, D. and Clark, A. F. 1937. The establishment of *Rhyssa persuasoria* in New Zealand. *New Zealand Journal of Science and Technology* **19**: 63-64.

Mills, N. 2005. Selecting effective parasitoids for biological control introductions: Codling moth as a case study. *Biological Control* **34**: 274-282.

Momoi, S., Kishitani, Y. and Iwata, K. 1965. Studies on an ichneumonid parasite (Hymenoptera) of *Clania minuscula* (Lepidoptera). *Science Reports of the Hyogo University of Agriculture. Series: Plant Protection* **7**: 25-31.

Montgomery, V. E. and DeWitt, P. R. 1975. Morphological differences among immature stages of three genera of exotic larval parasitoids attacking the cereal leaf beetle in the United States. *Annals of the Entomological Society of America* **68**: 574-578.

Morley, C. 1903. *Ichneumonologia Britannica. Ichneumoninae*. Plymouth.

Morley, C. 1907. *Ichneumonologica Britannica, ii. The ichneumons of Great Britain. Cryptinae*. Plymouth.

Morley, C. 1908. *Ichneumonologica Britannica, iii. The ichneumons of Great Britain. Pimplinae*. London.

Morley, C. 1911. *Ichneumonologia Britannica, iv. Tryphoninae*. London.

Morley, C. 1915 [1914]. *Ichneumonologia Britannica, V. The Ichneumons of Great Britain. Ophioninae*. London.

Morris, K. R. S. 1937. The prepupal stage in Ichneumonidae, illustrated by the life-history of *Exenterus abruptorius*, Thb. *Bulletin of Entomological Research* **28**: 525-534.

Morris, K. R. S., Cameron, E. and Jepson, W. F. 1937. The insect parasites of the Spruce sawfly (*Diprion polytomum* Thg.) in Europe. *Bulletin of Entomological Research* **28**: 341-393.

Moutia, L. A. and Courtois, C. M. 1952. Parasites of the moth-borers of sugar-cane in Mauritius. *Bulletin of Entomological Research* **43**: 325-359.

Muldrew, J. A. 1967. Biology and initial dispersal of *Olesicampe* (*Holocremnus*) sp. nr. *nematorum* (Hymenoptera: Ichneumonidae), a parasite of the Larch Sawfly recently established in Manitoba. *Canadian Entomologist* **99**: 312-321.

Munro, V. M. W. and Henderson, I. M. 2002. Nontarget effect of entomophagous biocontrol: shared parasitism between native lepidopteran parasitoids and the biocontrol agent *Trigonospila brevifacies* (Diptera: Tachinidae) in forest habitats. *Environmental Entomology* **31**: 388-396.

Münster-Swendsen, M. 1979. The parasitoid complex of *Epinotia tedella* (Cl.) (Lepidoptera: Tortricidae). *Entomologiske Meddelelser* **47**: 63-71.

Narolsky, N. B. 1990. New data on the taxonomy of Cremastine wasps (Hymenoptera, Ichneumonidae) from Kazakhstan and central Asia. *Entomological Review* **69**: 66-71.

Narolsky, N. B. 1994. [A new Palaearctic genus of the Ichneumonid wasp subfamily Cremastinae (Hymenoptera, Ichneumonidae.] (in Ukrainian with English summary). *Zhurnal Ukrayins'kogo Entmologichnogo Tovaristva* **1**: 41-58.

Narolsky, N. B. 2002. *Tersoakus* gen. nov., a new genus of cremastine wasps from the Russian Far East (Hymenoptera: Ichneumonidae: Cremastinae). *Zoologische Mededelingen* **76**: 97-102.

Narolsky, N. B. and Schönitzer, K. 2001. A new Palearctic genus of cremastine wasps from the flats of the Danube delta. *Entomofauna* **22**: 197-204.

Nielsen, E. 1923. Contributions to the life history of the pimpline spider parasites (*Polysphincta, Zaglyptus Tromatobia*) (Hym. Ichneum.). *Entomologiske Meddelelser* **14**: 137-205.

Nielsen, E. 1928. A supplementary note upon the life histories of the Polysphinctas (Hym. Ichneum.). *Entomologiske Meddelelser* **16**: 152-155.

Nielsen, E. 1929. A second supplementary note upon the life histories of the Polysphinctas (Hym. Ichneum.). *Entomologiske Meddelelser* **16**: 366-368.

Nielsen, E. 1935. A third supplementary note upon the life histories of the Polysphinctas (Hym. Ichneum.). *Entomologiske Meddelelser* **19**: 191-215.

Nielsen, E. 1937. A fourth supplementary note upon the life histories of the Polysphinctas (Hym. Ichneum.). *Entomologiske Meddelelser* **20**: 25-28.

Nieves Aldrey, J. L. 1984. Contribution to the knowledge of the hymenopterous inquiline fauna and parasites in the galls of *Diplolepis mayri* (Schlectendal) and *Diplolepis eglanteriae* (Hartig) (Hym., Cynipidae). *Graellsia* **39**: 93-102.

Notton, D. G. and Shaw, M. R. 1998. A review of the Palaearctic Neorhacodinae (Hymenoptera, Ichneumonidae) with *Eremura* Kasparyan, 1995 new to the west Palaearctic. *Bulletin of the British Museum (Natural History) (Entomology)* **67**: 209-218.

Nouhuys, S. van 2005. Effects of habitat fragmentation at different trophic levels in insect communities. *Annales Zoologici Fennici* **42**: 433-447.

Nouhuys, S. van and Hanski, I. 2000. Apparent competition between parasitoids mediated by a shared hyperparasitoid. *Ecology Letters* **3**: 82-84.

Nouhuys, S. van and Kaartinen, R. 2008. A parasitoid wasp uses landmarks while monitoring potential resources. *Proceedings of the Royal Society of London Series B - Biological Sciences* **275**: 377-385.

Noyes, J. S. 1994. The reliability of published host-parasitoid records: a taxonomist's view. *Norwegian Journal of Agricultural Sciences* **16**: 59-69.

Noyes, J. S. 2012. An inordinate fondness of beetles, but seemingly even more fond of microhymenoptera! *Hamuli* **3**: 5-8.

Nuzhna, A. D. 2013. Овариальные яйца наездников-ихневмонид подсемейства Anomaloninae (Hymenoptera, Ichneumonidae) фауны Украины [Ovarian eggs of the Anomaloninae wasps (Hymenoptera, Ichneumonidae) from Ukraine] [in Ukrainian with English summary]. *Український Ентомологічний Журнал [Ukrainian Entomological Magazine]* **2(7)**: 60-63.

Obrtel, R. 1946. *Xylonominus gracilicornis* Grav., parasit krasců *Anthaxia* Eschsch. (Hym. Ichn.) [*Xylonominus gracilicornis* Grav. as a parasite of the genus *Anthaxia* Eschsch. (Hym. Ichn.)]. *Folia Entomologica* **9**: 94-98.

Oehlke, J. 1966. Die in europäischen Kiefernbuschhornblattwespen (Diprionidae) parasitierenden Ichneumonidae. *Beiträge zur Entomologie* **15 [1965]**: 761-879.

Oehlke, J. 1967. Westpaläarktische Ichneumonidae I: Ephialtinae. *Hymenopterorum Catalogus (nova editio)* **2**: 1-49.

Oldroyd, H. 1969. Tabanoidea and Asiloidea. *Handbooks for the Identification of British Insects* **9(4)**: 1-132.

Osborne, P. 1960. Observations on the natural enemies of *Meligethes aeneus* (F.) and *M. viridescens* (F.) [Coleoptera: Nitidulidae]. *Parasitology* **50**: 91-110.

Osman, S. E. 1978. Der Einfluss der Imaginalernährung und der Begattung auf die Sekretproduktion der weiblichen Genitalanhangdrüsen und auf die Eireifung von *Pimpla turionellae* L. (Hym., Ichneumonidae). *Zeitschrift für Angewandte Entomologie* **85**: 113-122.

Pagliano, G., Santini, L. and Strumia, F. 1997. Descrizione delle carte del manoscritto. In: Bernardi, W., Pagliano, G., Santini, L., Strumia, F., Tongiorgi Tomasi, L., Tongiorgi, P., editors. *Natura e Immagine*. Pisa: ETS ed. pp. 95-246.

Pair, S. D., Raulston, J. R., Sparks, A. N. and Martin, P. B. 1986. Fall armyworm (Lepidoptera: Noctuidae) parasitoids: Differential spring distribution and incidence on corn and sorghum in the southern United States and northeastern Mexico. *Environmental Entomology* **2**: 342-348.

Pampel, W. 1914. Die weiblichen Geschlechtsorgane der Ichneumoniden. *Zeitschrift für Wissenschaftliche Zoologie* **108**: 290-357.

Parker, H. L., Berry, P. A. and Silveira, A. 1950. Vegetable weevils and their natural enemies in Argentina and Uruguay. *United States Department of Agriculture, Technical Bulletin* **1016**: 1-28.

Parkinson, N., Smith, I., Weaver, R. and Edwards, J. P. 2001. A new form of arthropod phenoloxidase is abundant in venom of the parasitoid wasp *Pimpla hypochondriaca*. *Insect Biochemistry and Molecular Biology* **31**: 57-63.

Parkinson, N., Richards, E. H., Conyers, C., Smith, I. and Edwards, J. P. 2002. Analysis of venom consitituents from the parasitoid wasp *Pimpla hypochondriaca* and cloning of cDNA encoding a venom protein. *Insect Biochemistry and Molecular Biology* **32**: 729-735.

Paull, C. and Austin, A. D. 2006. The hymenopteran parasitoids of light brown apple moth, *Epiphyas postvittana* (Walker) (Lepidoptera: Tortricidae) in Australia. *Australian Journal of Entomology* **45**: 142-156.

Perkins, J. F. 1936. Notes on British Metopiini (Hym. Ichneumonidae). *Entomologist's Monthly Magazine* **72**: 83-86.

Perkins, J. F. 1940. Notes on the synonymy of some genera of European Pimplinae (s.l.) (Hym. Ichneumonidae). *The Entomologist* **63**: 54-56.

Perkins, J. F. 1943. Notes on the British species of Adelognathini Roman, with descriptions of two new species (Hym. Ichneumonidae). *Transactions of the Royal Entomological Society of London* **93**: 95-114.

Perkins, J. F. 1952. On some British species of *Ichneumon* and *Alomyia* (Hym., Ichneumonidae). *Bulletin of Entomological Research* **43**: 361-363.

Perkins, J. F. 1959. Hymenoptera. Ichneumonoidea. Ichneumonidae, key to subfamilies and Ichneumoninae - 1. *Handbooks for the Identification of British Insects* **7(2ai)**: 1-116.

Perkins, J. F. 1960. Hymenoptera. Ichneumonoidea. Ichneumonidae, subfamilies Ichneumoninae II, Alomyinae, Agriotypinae and Lycorininae. *Handbooks for the Identification of British Insects* **7(2aii)**: 117-213.

Perkins, J. F. 1962. On the type species of Foerster's genera (Hymenoptera: Ichneumonidae). *Bulletin of the British Museum (Natural History) (Entomology)* **11**: 385-483.

Peters, R. S., Krogmann, L., Mayer, C., Donath, A., Gunkel, S., Meusemann, K., Kozlov, A., Podsiadlowski, L., Petersen, M., Lanfear, R., Diez, P. A., Heraty, J., Kjer, K. M., Klopfstein, S., Meier, R., Polidori, C., Schmitt, T., Liu, S., Zhou, X., Wappler, T., Rust, J., Misof, B. and Niehuis, O. 2017. Evolutionary history of the Hymenoptera. *Current Biology* **27**: 1-6. DOI: 10.1016/j.cub.2017.01.027

Plotnikov, V. 1914. Contribution à la biologie de *Bupalus piniarius* L. et de quelques uns de ses parasites. *Revue Russe d'Entomologie* **14**: 23-43.

Porter, C. C. 1978. A revision of the genus *Epirhyssa* (Hymenoptera, Ichneumonidae). *Studia Entomologia* **20**: 297-412.

Porter, C. C. 1998. Guía de los géneros de Ichneumonidae en la región neantárctica del sur de Sudamérica. *Opera Lilloana* **42**: 1-234.

Price, P. W. 1975. Reproductive strategies of parasitoids. In: Price, P. W., editor. *Evolutionary strategies of parasitic insects and mites*. New York and London: Springer. pp. 87-111.

Prous, M., Blank, S. M., Goulet, H., Heibo, E., Liston, A., Malm, T., Nyman, T., Schmidt, S., Smith, D. R., Vårdal, H., Viitasaari, M., Vikberg, V. and Taeger, A. 2014. The genera of Nematinae (Hymenoptera,Tenthredinidae). *Journal of Hymenoptera Research* **40**: 1-69.

Pschorn-Walcher, H. 1967. Biology of the ichneumonid parasites of *Neodiprion sertifer* (Geoffroy) (Hym.: Ichneumonidae) in Europe. *Commonwealth Institute of Biological Control Technical Bulletin* **8**: 7-51.

Pschorn-Walcher, H. 1973. Die Parasiten der gesellig lebenden Kiefern-Buschhornblattwespen (Familie Diprionidae) als Beispiel für Koexistenz und Konkurrenz in multiplen Parasit-Wirt-Komplexen. *Verhandlungen der Deutschen Zoologischen Gesellschaft* **66**: 136-145.

Pschorn-Walcher, H. 1987. Interspecific competition between the principal larval parasitoids of the pine sawfly *Neodiprion sertifer* (Geoff.) (Hym. Diprionidae). *Ecologia* **73**: 621-625.

Pschorn-Walcher, H. 1988. Die Parasitenkomplexe europäischer Diprionidae in ökologisch-evolutionsbiologischer Sicht. *Zeitschrift für Zoologische Systematik und Evolutionsforschung* **26**: 89-103.

Pschorn-Walcher, H. 1990. A brief note on the biology and larvae of *Megalodontes klugi* Leach (Hymenoptera: Megalodontidae). *Mitteilungen der Schweizerischen Entomologischen Gesellschaft* **63**: 303-308.

Pschorn-Walcher, H. and Altenhofer, E. 1989. The parasitoid community of leaf-mining sawflies Fenusini and Heterarthrini, a comparative analysis. *Zoologische Annalen* **222**: 37-56.

Pschorn-Walcher, H. and Altenhofer, E. 1999. Auftreten und Parasitierung von Fichten-Buschhornblattwespen in Osterreich (Hymenoptera: Diprionidae). *Linzer Biologische Beiträge* **31**: 83-91.

Pschorn-Walcher, H. and Kriegl, M. 1965. Zur Kenntnis der Parasiten der Bürsthorn-Blattwespen der Gattung *Arge* Schrank (Hymenoptera: Argidae). *Zeitschrift für Angewandte Entomologie* **56**: 263-275.

Pschorn-Walcher, H. and Zinnert, K. D. 1971. Investigations on the ecology and natural control of the Larch Sawfly (*Pristiphora erichsonii* Htg., Hym.: Tenthredinidae) in Central Europe. Part II: natural enemies: their biology and ecology, and their role as mortality factors in *P. erichsonii*. *Commonwealth Institute of Biological Control Technical Bulletin* **14**: 1-50.

Puttler, B. 1967. Interrelationship of *Hypera postica* (Coleoptera: Curculionidae) and *Bathyplectes curculionis* (Hymenoptera: Ichneumonidae) in the eastern United States with particular reference to encapsulation of the parasite eggs by the weevil larvae. *Annals of the Entomological Society of America* **60**: 1031-1038.

Quednau, F. W. and Guevremont, H. 1975. Observations on mating and oviposition behaviour of *Priopoda nigricollis* (Hymenoptera: Ichneumonidae), a parasite of the birch leaf-miner, *Fenusa pusilla* (Hymenoptera: Tenthredinidae). *Canadian Entomologist* **107**: 1199-1204.

Quicke, D. L. J. 1991. Ovipositor mechanics of the braconine wasp genus *Zaglyptogastra* and the ichneumonid genus *Pristomerus*. *Journal of Natural History* **25**: 971-977.

Quicke, D. L. J. 1997. *Parasitic Wasps*. London: Chapman and Hall.

Quicke, D.[L.]J. 2005. Biology and immature stages of *Panteles schnetzeanus* [sic] (Hymenoptera: Ichneumonidae), a parasitoid of *Lampronia fuscatella* (Lepidoptera: Incurvariidae). *Journal of Natural History* **39**: 431-443.

Quicke, D. L. J. 2015. *The Braconid and Ichneumonid Parasitoid Wasps: Biology, Systematics, Evolution and Ecology*. Oxford: Wiley-Blackwell.

Quicke, D. L. J., Fitton, M. G. and Ingram, S. 1992a. Phylogenetic implications of the structure and distribution of ovipositor valvilli in the Hymenoptera (Insecta). *Journal of Natural History* **26**: 587-608.

Quicke, D. L. J., Ingram, S. N., Baillie, H. S. and Gaitens, P. V. 1992b. Sperm structure and ultrastructure in the Hymenoptera (Insecta). *Zoologica Scripta* **21**: 381-402.

Quicke, D. L. J., Fitton, M. G., Tunstead, J. R., Ingram, S. N. and Gaitens, P. V. 1994. Ovipositor structure and function within the Hymenoptera, with special reference to the Ichneumonoidea. *Journal of Natural History* **28**: 635-682.

Quicke, D. L. J., Wyeth, P., Fawke, J. D., Basibuyuk, H. H. and Vincent, J. F. 1998. Manganese and zinc in the ovipositors and mandibles of hymenopterous insects. *Zoological Journal of the Linnean Society* **124**: 387-396.

Quicke, D. L. J., Basibuyuk, H. H., Fitton, M. G. and Rasnitsyn, A. P. 1999. Morphological, palaeontological and molecular aspects of ichneumonoid phylogeny (Hymenoptera, Insecta). *Zoologica Scripta* **28**: 175-202.

Quicke, D. L. J., Fitton, M. G., Notton, D. G., Broad, G. R. and Dolphin, K. 2000. Phylogeny of the Ichneumonidae (Hymenoptera): a simultaneous molecular and morphological analysis. In: Austin, A. D., Dowton, M., editors. *Hymenoptera: Evolution, Biodiversity and Biological Control*. Canberra: CSIRO Publishing. pp. 74-83.

Quicke, D. L. J., Fitton, M. G., Broad, G. R., Crocker, B., Laurenne, N. M. and Miah, M. I. 2005. The parasitic wasp genera *Skiapus*, *Hellwigia*, *Nonnus*, *Chriodes*, and *Klutiana* (Hymenoptera, Ichneumonidae): Recognition of the Nesomesochorinae stat. rev. and Nonninae stat. nov. and transfer of *Skiapus* and *Hellwigia* to the Ophioninae. *Journal of Natural History* **39**: 2559-2578.

Quicke, D. L. J., Laurenne, N. M., Fitton, M. G. and Broad, G. R. 2009. A thousand and one wasps: a 28S rDNA and morphological phylogeny of the Ichneumonidae (Insecta: Hymenoptera) with an investigation into alignment parameter space and elision. *Journal of Natural History* **43**: 1305–1421.

Radcliffe, E. B. and Flanders, K. L. 1998. Biological control of alfalfa weevil in North America. *Integrated Pest Management Reviews* **3**: 225-242.

Rahoo, G. M. and Luff, M. L. 1987. The biology of *Adelognathus granulatus* Perkins (Hym., Ichneumonidae) a parasitoid of the small goosberry sawfly, *Pristiphora pallipes* (Lep.) (Hym., Tenthredinidae). *Journal of Applied Entomology* **104**: 480-484.

Ranin, O. 1983. Über die Artengruppe *Tycherus elongatus* (Thomson) und nahe Verwandte (Hymenoptera, Ichneumonidae). *Annales Entomologici Fennici* **49**: 33-44.

Rappaport, N. and Page, M. 1985. Rearing *Glypta fumiferanae* (Hym.: Ichneumonidae) on a multivioltine laboratory colony of the Western Spruce Budworm (*Choristoneura occidentalis*) (Lep.: Tortricidae). *Entomophaga* **30**: 347-352.

Rasnitsyn, A. P. 1964. Overwintering of Ichneumon-flies (Hymenoptera, Ichneumonidae). *Entomologicheskoye Obozreniye* **43**: 46-51.

Rasnitsyn, A. P. 1980. Origin and evolution of Hymenoptera [Translated in 1984 by Biosystematics Research Centre, Ottawa]. *Transactions of the Paleontological Institute of the Academy of Sciences of the USSR* **174**: 1-192.

Ratzeburg, J. T. C. 1852. *Die Ichneumonen der Forstinsecten in forstlicher und entomologischer Beziehung. Dritter Band*. Berlin.

Resh, V. H. and Jamieson, W. 1988. Parasitism of the aquatic moth *Petrophila confusalis* (Lepidoptera, Pyralidae) by the aquatic wasp *Tanychela pilosa* (Hymenoptera, Ichneumonidae). *Entomological News* **99**: 185-188.

Reshchikov, A. V. 2013. Two new species of *Lathrolestes* (Hymenoptera, Ichneumonidae) from Norway, northern Russia and Finland with a key to western Palaearctic species. *Zootaxa* **3681**: 59-72.

Reshchikov, A. V. 2016. A revision of the genus *Rhinotorus* Förster, 1869 (Hymenoptera, Ichneumonidae, Ctenopelmatinae), with descriptions of three new species and an illustrated identification key. *European Journal of Taxonomy* **235**: 1-40. doi: 10.5852/ejt.2016.235

Revels, R. 2004. The rise and fall of the holly blue butterfly. *British Wildlife* **5**: 236-239.

Richards, E. H. and Parkinson, N. M. 2000. Venom from the endoparasitic wasp *Pimpla hypochondriaca* adversely affects the morphology, viability, and immune function of hemocytes from larvae of the tomato moth, *Lacanobia oleracea*. *Journal of Invertebrate Pathology* **76**: 33-42.

Richards, O. W. 1956. Hymenoptera. Introduction and keys to families. *Handbooks for the Identification of British Insects* **6(i)**: 1-94.

Richards, O. W. 1977. Hymenoptera. Introduction and key to families. Second edition. *Handbooks for the Identification of British Insects* **6(i)**: 1-100.

Riedel, M. 2008. Revision der westpaläarktischen Platylabini 1. Die Gattung *Platylabus* Wesmael, 1845 (Hymenoptera, Ichneumonidae, Ichneumoninae). *Spixiana* **31**: 105-172.

Riedel, M. 2012. Revision der westpaläarktischen Arten der Gattung *Coelichneumon* Thomson (Hymenoptera: Ichneumonidae: Ichneumoninae). *Linzer Biologische Beiträge* **44**: 1477-1611.

Riedel, M. 2015. Revision of the European species of the genus *Astiphromma* Förster, 1869. *Spixiana* **38**: 85-132.

Riedel, M. 2017. Die westpaläarktischen Arten der Gattung *Campoletis* Förster (Hymenoptera, Ichneumonidae, Campopleginae). *Spixiana* **40**: 95-137.

Riedel, M. 2018a. Contribution to the Palaearctic species of *Mesochorus* Gravenhorst (Hymenoptera, Ichneumonidae, Mesochorinae): 1. The *M. fulvus* group. *Linzer Biologische Beiträge* **50**: 687-716.

Riedel, M. 2018b. Revision of the Western Palaearctic species of the genus *Casinaria* Holmgren (Hymenoptera, Ichneumonidae, Campopleginae). *Linzer Biologische Beiträge* **50**: 723-763.

Rietra, E. 1932. *Lets over den bouw en de levenswijze van Nemeritis canescens (Gravenhorst) als intern parasiet van de larve van Ephestia kuehniella Zeller*. PhD thesis: Rijksuniversiteit te Leiden.

Rizzo, M. C. and Massa, B. 2006. Parasitism and sex ratio of the bedeguar gall wasp *Diplolepis rosae* (L.) (Hymenoptera: Cynipidae) in Sicily (Italy). *Journal of Hymenoptera Research* **15**: 277-285.

Rohwer, S. A. 1914. Descriptions of two parasitic Hymenoptera. *Proceedings of the Entomological Society of Washington* **16**: 141-142.

Rojas-Rousse, D. and Benoit, M. 1977. Morphology and biometry of larval instars of *Pimpla instigator* (F.) (Hymenoptera: Ichneumonidae). *Bulletin of Entomological Research* **67**: 129-141.

Roman, A. 1923. Ichneumonids reared from Diptera Nematocera. *Entomologist's Monthly Magazine* **59**: 71-76.

Roman, A. 1925. Schwedische Schlupfwespen, alte und neue. *Arkiv för Zoologi* **17A**: 1-34.

Roman, A. 1939. Nordische Ichneumoniden - und einige andere. *Entomologisk Tidskrift* **60**: 176-205.

Rosenberg, H. T. 1934. The biology and distribution in France of the larval parasites of *Cydia pomonella*, L. *Bulletin of Entomological Research* **25**: 201-256.

Ross, H. H. 1936. The ancestry and wing venation of the Hymenoptera. *Annals of the Entomological Society of America* **29**: 99-111.

Rossem, G. van 1966. A study of the genus *Trychosis* Förster in Europe (Hymenoptera, Ichneumonidae, Cryptinae). *Zoologische Verhandelingen* **79**: 1-40.

Rossem, G. van 1969a. A revision of the genus *Cryptus* Fabricius s. str. in the western Palearctic region, with keys to genera of Cryptina and species of *Cryptus* (Hymenoptera, Ichneumonidae). *Tijdschrift voor Entomologie* **112**: 299-374.

Rossem, G. van 1969b. A study of the genus *Meringopus* Förster in Europe and of some related species from Asia (Hymenoptera, Ichneumonidae, Cryptinae). *Tijdschrift voor Entomologie*: 112.

Rossem, G. van 1971. The genus *Buathra* Cameron in Europe (Hymenoptera, Ichneumonidae). *Tijdschrift voor Entomologie* **114**: 201-208.

Rossem, G. van 1981. A revision of some western Palaearctic oxytorine genera (Hymenoptera, Ichneumonidae). *Spixiana* **Supplement 4(1980)**: 79-135.

Rossem, G. van 1982. A revision of some western Palaearctic oxytorine genera. Part II. Genus *Eusterinx* (Hymenoptera, Ichneumonidae). *Spixiana* **5**: 149-170.

Rossem, G. van 1983a. A revision of Western Palearctic oxytorine genera. Part III. Genus *Proclitus* (Hymenoptera, Ichneumonidae). *Contributions of the American Entomological Institute* **20**: 153-165.

Rossem, G. van 1983b. A revision of Western Palaearctic oxytorine genera. Part IV. Genus *Megastylus*. *Entomofauna* **4**: 121-132.

Rossem, G. van 1985. A revision of Western Palaearctic oxytorine genera. Part V. Genus *Aperileptus* (Hymenoptera, Ichneumonidae). *Spixiana* **8**: 145-152.

Rossem, G. van 1987. A revision of Western Palaearctic oxytorine genera. Part VI. (Hymenoptera, Ichneumonidae). *Tijdschrift voor Entomologie* **130**: 49-108.

Rossem, G. van 1988. A revision of Palaearctic oxytorine genera. Part VII. (Hymenoptera, Ichneumonidae). *Tijdschrift voor Entomologie* **131**: 103-112.

Rossem, G. van 1990. Key to the genera of the Palaearctic Oxytorinae, with the description of three new genera (Hymenoptera: Ichneumonidae). *Zoologische Mededelingen* **63**: 309-323.

Rossem, G. van 1991. New Oxytorinae from Siberia, with revised keys to *Plectiscidea* Viereck and *Eusterinx* Förster s.l. (Hymenoptera: Ichneumonidae). *Zoologische Mededelingen* **65**: 25-38.

Rotheray, G. E. 1979. *Biological Studies on some Diplazontinae, parasitoids of aphidophagous Syrphidae (Diptera)*. PhD thesis: University of Wales.

Rotheray, G. E. 1981a. Host searching and oviposition behaviour of some parasitoids of aphidophagous Syrphidae. *Ecological Entomology* **6**: 79-87.

Rotheray, G. E. 1981b. Emergence from the host puparium by *Diplazon pectoratorius* (Gravenhorst) (Hymenoptera: Ichneumonidae), a parasitoid of aphidophagous syrphid larvae. *Entomologist's Gazette* **32**: 39-41.

Rotheray, G. E. 1981c. Courtship, male swarms and a sex pheromone of *Diplazon pectoratorius* (Thunberg) (Hymenoptera: Ichneumonidae). *Entomologist's Gazette* **32**: 193-196.

Rotheray, G. E. 1984. Host relations, life cycles and multiparasitism in some parasitoids of aphidophagous Syrphidae (Diptera). *Ecological Entomology* **9**: 303–310. DOI: 10.1111/j.1365-2311.1984.tb00853.x

Rotheray, G. E. 1988. Larval morphology and feeding patterns of four *Cheilosia* species (Diptera: Syrphidae) associated with *Cirsium palustre* L. Scopoli (Compositae) in Scotland. *Journal of Natural History* **22**: 17-25.

Rotheray, G. E. 1990. A new species of *Bioblapsis* (Hymenoptera: Ichneumonidae) from Scotland parasitising a mycophagous hoverfly, *Cheilosia longula* (Diptera: Syrphidae). *Entomologica Scandinavica* **21**: 277-280.

Rouleux-Bonnin, F., Renault, S., Rabouille, A., Periquet, G. and Bigot, Y. 1999. Free serosal cells originating from the embryo of the wasp *Diadromus pulchellus* in the pupal body of parasitized leek-moth, *Acrolepiosis assectella*. Are these cells teratocyte-like? *Journal of Insect Physiology* **45**: 479-484.

Rousse, P. and van Noort, S. 2014a. Revision of the Afrotropical Lycorininae (Ichneumonidae; Hymenoptera) II. Three new *Lycorina* species and additional distribution records. *Zootaxa* **3884**: 222-234.

Rousse, P. and van Noort, S. 2014b. A review of the Afrotropical Rhyssinae (Hymenoptera: Ichneumonidae) with the descriptions of five new species. *European Journal of Taxonomy* **91**: 1-42.

Rousse, P., Villemant, C. and Seyrig, A. 2011. Ichneumonid wasps from Madagascar. V. Ichneumonidae Cremastinae. *Zootaxa* **3118**: 1-30.

Rousse, P., Quicke, D. L. J., Matthee, C. A., Lefeuvre, P. and van Noort, S. 2016. A molecular and morphological reassessment of the phylogeny of the subfamily Ophioninae (Hymenoptera: Ichneumonidae). *Zoological Journal of the Linnean Society* **178**: 128-148.

Russ, K. and Rupf, O. 1974. Influence of parasites and pathogens on the hibernating population of codling moth (*Laspeyresia pomonella* L.) in Austria. In: IAEA, editors. *Sterility principle for insect control*. Vienna: International Atomic Energy Agency. pp. 557-563.

Sääksjärvi, I. E., Haataja, S., Neuvonen, S., Gauld, I. D., Jussila, R., Salo, J. and Marmol Burgos, A. 2004. High local species richness of parasitoid wasps (Hymenoptera: Ichneumonidae; Pimplinae and Rhyssinae) from the lowland rain forests of Peruvian Amazonia. *Ecological Entomology* **29**: 735-743.

Salisbury, A. 2003. Two parasitoids of the lily beetle, *Lilioceris lilii* (Scopoli) (Coleoptera: Chrysomelidae), in Britain, including the first record of *Lemophagus errabundus* Gravenhorst (Hymenoptera: Ichneumonidae). *British Journal of Entomology and Natural History* **16**: 103-104.

Salt, G. 1931. Parasites of the wheat-stem sawfly, *Cephus pygmaeus*, Linnaeus, in England. *Bulletin of Entomological Research* **22**: 479-545.

Salt, G. 1932. Superparasitism by *Collyria calcitrator*, Grav. *Bulletin of Entomological Research* **23**: 211-215.

Salt, G. 1952. Trimorphism in the Ichneumonid parasite *Gelis corruptor*. *Quarterly Journal of Microscopical Science* **93**: 453-474.

Salt, G. 1957. Experimental studies in insect parasitism X. The reactions of some endopterygote insects to an alien parasite. *Proceedings of the Royal Society, series B- Biological Sciences* **147**: 167-184.

Salt, G. 1964. The ichneumonid parasite *Nemeritis canescens* (Gravenhorst) in relation to the wax moth *Galleria mellonella* (L.). *Transactions of the Royal Entomological Society of London* **116**: 1-14.

Salt, G. 1965. Experimental studies in insect parasitism XIII. The haemolytic reaction of a caterpillar to eggs of its habitual parasite. *Proceedings of the Royal Society of London Series B - Biological Sciences* **162**: 303-318.

Salt, G. 1973. Experimental studies in insect parasitism. XVI. The mechanism of the resistance of *Nemeritis* to defence reactions. *Proceedings of the Royal Society of London Series B - Biological Sciences* **183**: 337-350.

Salt, G. 1976. The hosts of *Nemeritis canescens*, a problem in the host specificity of insect parasitoids. *Ecological Entomology* **1**: 63-67.

Salt, G. 1977. Problems of orientation associated with cocoon-spinning by *Nemeritis*. *Ecological Entomology* **2**: 171-177.

Sanborne, M. 1984. A revision of the world species of *Sinophorus* (Ichneumonidae). *Memoirs of the American Entomological Institute* **38**: 1-403.

Sanborne, M. 1986. Revision of the Nearctic species of *Xylophylax* Kriechbaumer (Hymenoptera: Ichneumonidae). *Contributions of the American Entomological Institute* **22(8)**: 1-45.

Sandlan, K. 1979a. Host feeding and its effects on the physiology and behaviour of the ichneumonid parasitoid, *Coccygomimus turionellae*. *Physiological Entomology* **4**: 383-392.

Sandlan, K. 1979b. Sex ratio regulation in *Coccygomimus turionella* [sic] Linnaeus (Hymenoptera: Ichneumonidae) and its ecological implications. *Ecological Entomology* **4**: 365-378.

Sandlan, K. 1980. Host location by *Coccygomimus turionellae* (Hymenoptera: Ichneumonidae). *Entomologia Experimentalis Et Applicata* **27**: 233-245.

Sandlan, K. P. 1982. Host suitability and its effects on parasitoid biology in *Coccygomimus turionellae* (Hymenoptera: Ichneumonidae). *Annals of the Entomological Society of America* **75**: 217-221.

Santos, B. F. 2017. Phylogeny and reclassification of Cryptini (Hymenoptera, Ichneumonidae, Cryptinae), with implications for ichneumonid higher-level classification. *Systematic Entomology* **42**: 650-676.

Sato, M. and Takasu, K. 2000. Food odor learning by both sexes of the pupal parasitoid *Pimpla alboannulatus* Uchida (Hymenoptera : Ichneumonidae). *Journal of Insect Behavior* **13**: 263-272.

Sawoniewicz, J. 1980. Revision of European species of the genus *Bathythrix* Foerster (Hymenoptera, Ichneumonidae). *Annales Zoologici* **35**: 1-47.

Sawoniewicz, J. 1985. Revision of European species of the subtribe Endaseina (Hymenoptera, Ichneumonidae), I. *Annales Zoologici* **39**: 131-145.

Sawoniewicz, J. 1990. Revision of European species of the subtribe Endaseina (Hymenoptera, Ichneumonidae), II. Genus *Amphibulus* Kriechbaumer, 1893. *Annales Zoologici* **43**: 287-291.

Sawoniewicz, J. 2008. *Hosts of the world Aptesini (Hymenoptera, Ichneumonidae, Cryptinae)*. Olsztyn: Mantis.

Sawoniewicz, J. and Luhman, J. C. 1992. Revision of European species of the subtribe Endaseina, III genus: *Endasys* Foerster, 1868. *Entomofauna* **13**: 1-94.

Scaramozzino, P. L. 1982. Note su alcuni Ichneumonidae italiani (Hymenoptera). *Rivista Piemontese Storia Naturale* **3**: 141-144.

Schaffner, U. and Müller, C. 2001. Exploitation of the fecal shield of the lily leaf beetle, *Lilioceris lilii* (Coleoptera: Chrysomelidae), by the specialist parasitoid *Lemophagus pulcher* (Hymenoptera: Ichneumonidae). *Journal of Insect Behavior* **14**: 739-757.

Schedl, W. 1997. [Contribution to the morphology and biology of *Xyela curva* Benson, 1938 (Hymenoptera: Symphyta, Xyelidae).] (in German). *Zeitschrift der Arbeitsgemeinschaft Österreichischer Entomologen* **49**: 37-40.

Schiff, N., Goulet, H., Smith, D. R., Boudreault, C., Wilson, A. D. and Scheffler, B. E. 2012. Siricidae (Hymenoptera: Symphyta: Siricoidea) of the Western Hemisphere. *Canadian Journal of Arthropod Identification* **No.21**: 305 pp.

Schmiedeknecht, O. 1902. *Opuscula Ichneumonologica. I. Band. (Fasc. I.) Allgemeine Eintheilung. Die Gattungen der Joppinen, Ichneumoninen, Listrodrominen, Heresiarchinen, Gyrodontinen und Alomyinen. Bestimmungstabelle der paläarktischen Arten der Gattung Ichneumon.*: Blankenburg in Thüringen.

Schmiedeknecht, O. 1903. *Opuscula Ichneumonologica. I. Band. (Fasc. II-IV.) Bestimmungstabelle der paläarktischen Arten der Gattung Ichneumon. Bestimmungstabelle und Beschreibungen weiter Gattungen der Ichneumoninae*: Blankenburg in Thüringen.

Schmiedeknecht, O. 1904a. *Opuscula Ichneumonologica. I. Band. (Fasc. V.) Bestimmungstabelle und Beschreibungen weiter Gattungen der Ichneumoninae*: Blankenburg in Thüringen.

Schmiedeknecht, O. 1904b. *Opuscula Ichneumonologica. II. Band. (Fasc. VI-VII.) Cryptinae*: Blankenburg in Thüringen.

Schmiedeknecht, O. 1905. *Opuscula Ichneumonologica. II. Band. (Fasc. VIII-XI.) Cryptinae*: Blankenburg in Thüringen.

Schmiedeknecht, O. 1906a. *Opuscula Ichneumonologica. II. Band. (Fasc. XII-XIII.) Cryptinae*: Blankenburg in Thüringen.

Schmiedeknecht, O. 1906b. *Opuscula Ichneumonologica. III. Band. (Fasc. XIII-XIV.) Pimplinae*: Blankenburg in Thüringen.

Schmiedeknecht, O. 1907. *Opuscula Ichneumonologica. III. Band. (Fasc. XV-XVII.) Pimplinae*: Blankenburg in Thüringen.

Schmiedeknecht, O. 1908. *Opuscula Ichneumonologica. IV. Band. (Fasc. XVIII-XX.) Ophioninae*: Blankenburg in Thüringen.

Schmiedeknecht, O. 1909. *Opuscula Ichneumonologica. IV. Band. (Fasc. XXI-XXIII.) Ophioninae*: Blankenburg in Thüringen.

Schmiedeknecht, O. 1910. *Opuscula Ichneumonologica. IV. Band. (Fasc. XXIV-XXVI.) Ophioninae*: Blankenburg in Thüringen.

Schmiedeknecht, O. 1911a. *Opuscula Ichneumonologica. IV. Band. (Fasc. XXVII-XXIX.) Ophioninae*: Blankenburg in Thüringen.

Schmiedeknecht, O. 1911b. *Opuscula Ichneumonologica. V. Band. (Fasc. XXIX.) Tryphoninae*: Blankenburg in Thüringen.

Schmiedeknecht, O. 1912. *Opuscula Ichneumonologica. V. Band. (Fasc. XXX-XXXII.) Tryphoninae*: Blankenburg in Thüringen.

Schmiedeknecht, O. 1913. *Opuscula Ichneumonologica. V. Band. (Fasc. XXXIII-XXXV.) Tryphoninae*: Blankenburg in Thüringen.

Schmiedeknecht, O. 1914. *Opuscula Ichneumonologica. V. Band. (Fasc. XXXVI-XXXVII.) Tryphoninae*: Blankenburg in Thüringen.

Schnee, H. 1978. Revision der palaearktischen Arten der Gattung *Perisphincter* Townes (Hym. Ichneumonidae). *Entomologische Nachrichten* **22**: 129-143.

Schnee, H. 1989. Revision der von Gravenhorst beschriebenen und redeskribierten Anomaloninae mit Beschreibung zweier neuer Arten (Hymenoptera, Ichneumonidae). *Deutsche Entomologische Zeitschrift* **36**: 241-266.

Schnee, H. 2008. Die Anomaloninae der Sammlung Arnold Förster - Typenrevision und faunistische Anmerkungen (Hymenoptera, Ichneumonidae) [The Anomaloninae of the Arnold Förster collection - type revision and faunistic remarks (Hymenoptera, Ichneumonidae)]. *Beiträge zur Entomologie* **58**: 249-266.

Schnee, H. 2018. Typenrevision der von Hellén beschriebenen Anomaloninae (Hymenoptera, Ichneumonidae) und Übersicht über die finnischen Arten. *Beiträge zur Entomologie* **68**: 151-175.

Schneider, F. 1950. Die Entwicklung des Syrphidenparasiten *Diplazon fissorius* Grav. (Hym. Ichneum.) in uni-, oligo-, und polyvoltinen Wirten und sein Verhalten bei parasitaerer Aktivierung der Diapauselarven durch *Diplazon pectoratorius* Grav. *Mitteilungen der Schweizerischen Entomologischen Gesellschaft* **23**: 155-194.

Schneider, F. 1951. Einige physiologische Beziehungen zwischen Syrphidenlarven und ihren Parasiten. *Zeitschrift für Angewandte Entomologie* **33**: 150-162.

Schneider, M. V., Driessen, G., Beukeboom, L. W., Boll, R., van Eunen, K., Selzner, A., Talsma, J. and Lapchin, L. 2003. Gene flow between arrhenotokous and thelytokous populations of *Venturia canescens* (Hymenoptera). *Heredity* **90**: 260-267.

Scholler, M. 1999. Field studies of Cryptocephalinae biology. In: Cox, M. L., editor. *Advances in Chrysomelidae biology. 1.* Leiden: Backhuys. pp. 421-436.

Schröder, D. 1967. *Diplolepsi* (=*Rhodites*) *rosae* (L.) (Hym: Cynipidae) and a review of its parasite complex in Europe. *Commonwealth Institute of Biological Control Technical Bulletin* **9**: 93-131.

Schröder, D. 1974. A study of the interactions between the internal larval parasites of *Rhyacionia buoliana* (Lepidoptera: Olethreutidae). *Entomophaga* **19**: 145-171.

Schulmeister, S. 2003. Genitalia and terminal abdominal segments of male basal Hymenoptera (Insecta): morphology and evolution. *Organisms Diversity & Evolution* **3**: 253-279.

Schwarz, M. 1988. Die europäischen Arten der Gattung *Idiolispa* Förster (Hymenoptera, Ichneumonidae). *Linzer Biologische Beiträge* **20**: 37-66.

Schwarz, M. 1989. Revision der Gattung *Enclisis* Townes (Ichneumonidae, Hymenoptera). *Linzer Biologische Beiträge* **21**: 479-522.

Schwarz, M. 1990. Bemerkungen zur Systematik und Taxonomie westpaläarktischer Schlupfwespen (Ichneumonidae, Hymenoptera). *Linzer Biologische Beiträge* **22**: 59-67.

Schwarz, M. 1991a. Revision der westpaläarktischen Arten der Schlupfwespen-Gattungen *Caenocryptus* Thomson 1873 und *Nippocryptus* Uchida 1936 (Ichneumonidae, Hymenoptera). *Linzer Biologische Beiträge* **22**: 359-380.

Schwarz, M. 1991b. Eine neue Art der Gattung *Microleptes* Gravenhorst (Ichneumonidae, Hymenoptera) aus Österreich. *Linzer Biologische Beiträge* **23**: 399-405.

Schwarz, M. 1994. Beitrag zur Systematik und Taxonomie europäischer *Gelis*-Arten mit macropteren oder brachypteren Weibchen (Hymenoptera, Ichneumonidae). *Linzer Biologische Beiträge* **26**: 381-391.

Schwarz, M. 1995. Revision der westpaläarktischen Arten der Gattungen *Gelis* Thunberg mit apteren Weibchen und *Thaumatogelis* Schmiedeknecht (Hymenoptera, Ichneumonidae). Teil 1. *Linzer Biologische Beiträge* **27**: 5-105.

Schwarz, M. 2001. Revision der westpaläarktischen Arten der Gattungen *Gelis* Thunberg mit apteren Weibchen und *Thaumatogelis* Schwarz (Hymenoptera, Ichneumonidae). Teil 4. *Linzer Biologische Beiträge* **33**: 1111-1155.

Schwarz, M. 2002a. Schlupfwespen (Insecta, Hymenoptera, Ichneumonidae) in den Hochlagen der Hohen Tauern (Österreich). Teil 1: Überblick. *Mitteilungen aus dem Haus der Natur* **15**: 45-52.

Schwarz, M. 2002b. Revision der westpaläarktischen Arten der Gattungen *Gelis* Thunberg mit apteren Weibchen und *Thaumatogelis* Schwarz (Hymenoptera, Ichneumonidae). Teil 3. *Linzer Biologische Beiträge* **34**: 1293-1392.

Schwarz, M. 2005. Revisionen und Neubeschreibungen von Cryptinae (Hymenoptera, Ichneumonidae) 1. *Linzer Biologische Beiträge* **37**: 1641-1710.

Schwarz, M. 2007. Revision der westpaläarktischen Arten der Gattung *Hoplocryptus* Thomson (Hymenoptera, Ichneumonidae). *Linzer Biologische Beiträge* **39**: 1161-1219.

Schwarz, M. 2015. Zur Kenntnis paläarktischer *Cryptus*-Arten (Hymenoptera, Ichneumonidae, Cryptinae). *Linzer Biologische Beiträge* **47**: 749-896.

Schwarz, M. 2016. Die Schlupfwespengattung *Gelis* (Hymenoptera, Ichneumonidae, Cryptinae) mit macropteren Weibchen in der Westpaläarktis [The ichneumonid genus *Gelis* (Hymenoptera, Ichneumonidae, Cryptinae) with macropterous females in the western Palaearctic region]. *Linzer Biologische Beiträge* **48**: 1677-1752.

Schwarz, M. and Shaw, M. R. 1998. Western Palaearctic Cryptinae (Hymenoptera: Ichneumonidae) in the National Museums of Scotland, with nomenclatural changes, taxonomic notes, rearing records and special reference to the British check list. Part 1. Tribe Cryptini. *Entomologist's Gazette* **49**: 101-127.

Schwarz, M. and Shaw, M. R. 1999. Western Palaearctic Cryptinae (Hymenoptera: Ichneumonidae) in the National Museums of Scotland, with nomenclatural changes, taxonomic notes, rearing records and special reference to the British check list. Part 2. Genus *Gelis* Thunberg (Phygadeuontini: Gelina). *Entomologist's Gazette* **50**: 117-142.

Schwarz, M. and Shaw, M. R. 2000. Western Palaearctic Cryptinae (Hymenoptera: Ichneumonidae) in the National Museums of Scotland, with nomenclatural changes, taxonomic notes, rearing records and special reference to the British check list. Part 3. Tribe Phygadeuontini, subtribes Chiroticina, Acrolytina, Hemitelina and Gelina (excluding *Gelis*), with descriptions of new species. *Entomologist's Gazette* **51**: 147-186.

Schwarz, M. and Shaw, M. R. 2010. Western Palaearctic Cryptinae (Hymenoptera: Ichneumonidae) in the National Museums of Scotland, with nomenclatural changes, taxonomic notes, rearing records and special reference to the British check list. Part 4.Tribe Phygadeuontini, subtribes Mastrina, Ethelurgina, Endaseina (excluding *Endasys*), Bathythrichina and Cremnodina. *Entomologist's Gazette* **61**: 187-206.

Schwarz, M. and Shaw, M. R. 2011. Western Palaearctic Cryptinae (Hymenoptera: Ichneumonidae) in the National Museums of Scotland, with nomenclatural changes, taxonomic notes, rearing records and special reference to the British check list. Part 5.Tribe Phygadeuontini, subtribe Phygadeuontina, with descriptions of new species. *Entomologist's Gazette* **62**: 175-210.

Schwarz, M., Hilszczanski, J. and Shaw, M. R. 2013. *Cryptus genalis* Tschek, 1872 (Hymenoptera: Ichneumonidae, Cryptinae), a gregarious ectoparasitoid in scarabaeid pupal chambers. *Entomologist's Gazette* **64**: 131-134.

Schwarzfeld, M. D. and Sperling, F. A. H. 2015. Comparison of five methods for delimitating species in *Ophion* Fabricius, a diverse genus of parasitoid wasps (Hymenoptera, Ichneumonidae). *Molecular Phylogenetics and Evolution* **93**: 234-248.

Schwarzfeld, M. D., Broad, G. R. and Sperling, F. A. H. 2016. Molecular phylogeny of the diverse parasitoid wasp genus *Ophion* Fabricius (Hymenoptera: Ichneumonidae: Ophioninae). *Systematic Entomology* **41**: 191-206.

Schwenke, W. 1999. Revision der europäischen Mesochorinae (Hymenoptera, Ichneumonoidea, Ichneumonidae). *Spixiana* **Supplement**: 1-124.

Schwenke, W. 2002. Neue europäische Mesochorinae-Arten (Hymenoptera, Ichneumonidae, Mesochorinae). *Entomofauna* **23**: 85-92.

Schwenke, W. 2004. Eine neue Gattung und 19 neue Arten und Geschlechter europäischer Mesochorinae (Hymenoptera, Ichneumonidae). *Entomofauna* **25**: 81-88.

Scott, E. I. 1939. An account of the developmental stages of some aphidophagous Syrphidae (Dipt.) and their parasites (Hymenopt.). *Annals of Applied Biology* **26**: 509-532.

Šedivý, J. 1970. Westpaläarktische Arten der Gattung *Dimophora*, *Pristomerus*, *Eucremastus* und *Cremastus* (Hym., Icheumonidae). *Prirodovedne Prace Ustavu Ceskoslovenske Akademie Ved v Brne* **(N.S.). 4(11)**: 38 pp.

Šedivý, J. 1971. Revision der europäischen *Temelucha*-Arten (Hym., Ichneumonidae). *Prirodovedne Prace Ustavu Ceskoslovenske Akademie Ved v Brne* **(N.S.). 5(1)**: 34 pp.

Šedivý, J. 1983. Tersilochinae as parasitoids of insect pests of winter rape (Hymenoptera: Ichneumonidae). *Contributions of the American Entomological Institute* **20**: 266-276.

Šedivý, J. 2004. European species of the genus *Phobocampe* Förster (Hymenoptera: Ichneumonidae). *Acta Universitatis Carolinae Biologica* **48**: 203-235.

Šedivý, J. and Ševčík, J. 2003. Ichneumonid (Hymenoptera: Ichneumonidae) parasitoids of fungus gnats (Diptera: Mycetophilidae): rearing records from the Czech Republic. *Studia Dipterologica* **10**: 153-158.

Selfa, J. and Diller, E. 1994. Illustrated key to the Western Palearctic genera of Phaeogenini (Hymenoptera, Ichneumonidae, Ichneumoninae). *Entomofauna* **15**: 237-251.

Seltmann, K., Yoder, M. J., Mikó, I., Forshage, M., Bertone, M. A., Agosti, D., Austin, A. D., Balhoff, J. P., Borowiec, M. L., Brady, S. G., Broad, G. R., Brothers, D. J., Burks, R. A., Buffington, M. L., Campbell, H. M., Dew, K. J., Ernst, A. F., Fernández-Triana, J. L., Gates, M. W., Gibson, G. A. P., Jennings, J. T., Johnson, N. F., Karlsson, D., Kawada, R., Krogmann, L., Kula, R. R., Mullins, P. L., Ohl, M., Rasmussen, C., Ronquist, F., Schulmeister, S., Sharkey, M. J., Talamas, E., Tucker, E., Vilhelmsen, L., Ward, P. S., Wharton, R. A. and Deans, A. R. 2012. A hymenopterists' guide to the Hymenoptera Anatomy Ontology: utility, clarification, and future directions. *Journal of Hymenoptera Research* **27**: 67-88.

Seyrig, A. 1952. Les Ichneumonides de Madagascar. IV Ichneumonidae Cryptinae. *Mémoires de l'Académie Malgache* **Fascicule XIX**: 213 pp.

Sharkey, M. J. 2007. Phylogeny and classification of Hymenoptera. *Zootaxa* **1668**: 521-548.

Sharkey, M. J. and Wahl, D. B. 1992. Cladistics of the Ichneumonoidea (Hymenoptera). *Journal of Hymenoptera Research* **1**: 15-24.

Shaw, M. R. 1975. A rationale for abnormal, male-dominated sex-ratios in adult populations of *Zygaena* (Lep.: Zygaenidae). *Entomologist's Record and Journal of Variation* **87**: 52-54.

Shaw, M. R. 1977. On the distribution of some Satyrid (Lep.) larvae at a coastal site in relation to their Ichneumonid (Hym.) parasite. *Entomologist's Gazette* **28**: 133-134.

Shaw, M. R. 1986. *Coleocentrus excitator* (Poda) (Hymenoptera: Ichneumonidae) new to Britain. *Entomologist's Gazette* **37**: 221-224.

Shaw, M. R. 1989. A host record for *Stilbops limneriaeformis* Schmiedeknecht (Hymenoptera: Ichneumonidae, Stilbopinae) in Scotland. *Entomologist's Gazette* **40**: 5-6.

Shaw, M. R. 1991. *Phrudus badensis* Hilpert (Hym., Ichneumonidae) new to Britain. *Entomologist's Monthly Magazine* **127**: 157-158.

Shaw, M. R. 1993. Species of *Mesochorus* (Hymenoptera: Ichneumonidae) reared as hyperparasitoids of Lepidoptera via koinobiont ectoparasitoid Tryphoninae (Ichneumonidae). *Entomologist's Gazette* **44**: 181-182.

Shaw, M. R. 1994. Parasitoid host ranges. In: Hawkins, B. A., Sheehan, W., editors. *Parasitoid Community Ecology*. Oxford, New York, Tokyo: Oxford University Press. pp. 111-144.

Shaw, M. R. 1997. *Rearing Parasitic Hymenoptera*. Orpington: The Amateur Entomologist.

Shaw, M. R. 1999. Gregarious development in endoparasitic koinobiont Ichneumonidae (Hymenoptera). *Entomologist's Gazette* **50**: 55-56.

Shaw, M. R. 2001. Interactions between adults of some species of *Netelia* Gray (Hymenoptera: Ichneumonidae: Tryphoninae) and their caterpillar hosts (Lepidoptera). *Journal of Hymenoptera Research* **10**: 101-111.

Shaw, M. R. 2003[2002]. Host ranges of *Aleiodes* species (Hymenoptera: Braconidae), and an evolutionary hypothesis. In: Melika, G., Thuróczy, C., editors. *Parasitic wasps: Evolution, Systematics, Biodiversity and Biological Control*. Budapest, Hungary: Agroinform. pp. 321-327.

Shaw, M. R. 2004. Notes on the biology of *Lycorina triangulifera* Holmgren (Hymenoptera: Ichneumonidae: Lycorininae). *Journal of Hymenoptera Research* **13**: 302-308.

Shaw, M. R. 2006a. Habitat considerations for parasitic wasps (Hymenoptera). *Journal of Insect Conservation* **10**: 117-127.

Shaw, M. R. 2006b. Notes on British Pimplinae and Poemeniinae (Hymenoptera, Ichneumonidae), with additions to the British list. *British Journal of Entomology and Natural History* **19**: 217-238.

Shaw, M. R. 2009. Notes on the host-feeding and hyperparasitic behaviours of *Itoplectis* species (Hymenoptera: Ichneumonidae, Pimplinae). *Entomologist's Gazette* **60**: 113-116.

Shaw, M. R. 2014. Illustrated notes on the biology of two European species of *Euceros* Gravenhorst (Hymenoptera: Ichneumonidae: Eucerotinae) *Proceedings of the Russian Entomological Society* **85**: 122-132.

Shaw, M. R. 2017a. A few recommendations on recording host information for reared parasitoids. *Hamuli* **8**: 7-9.

Shaw, M. R. 2017b. Anatomy, reach and classification of the parasitoid complex of a common British moth, *Anthophila fabriciana* (L.) (Choreutidae). *Journal of Natural History* **51**: 1119-1149.

Shaw, M. R. and Aeschliman, J.-P. 1994. Host ranges of parasitoids (Hymenoptera: Braconidae and Ichneumonidae) reared from *Epermenia chaerophyllella* (Goeze) (Lepidoptera: Epermeniidae) in Britain, with description of a new species of *Triclistus* (ichneumonidae). *Journal of Natural History* **28**: 619-629.

Shaw, M. R. and Askew, R. R. 1979. Hymenopterous parasites of Diptera (Hymenoptera parasitica). *Amateur Entomologist* **15**: 164-171.

Shaw, M. R. and Askew, R. R. 2010. Hymenopterous parasitoids of Diptera. In: Chandler, P. J., editor. *A Dipterist's Handbook*: Amateur Entomologists' Society. pp. 347-361.

Shaw, M. R. and Bennett, F. D. 2001. Host prepupal mummification by *Colpognathus* Wesmael (Hymenoptera: Ichneumonidae, Ichneumoninae). *Entomologist's Gazette* **52**: 201-203.

Shaw, M. R. and Hochberg, M. E. 2001. The neglect of parasitic Hymenoptera in insect conservation strategies: the British fauna as a prime example. *Journal of Insect Conservation* **5**: 253-263.

Shaw, M. R. and Horstmann, K. 1997. An analysis of host range in the *Diadegma nanus* group of parasitoids in Western Europe, with a key to species (Hymenoptera: Ichneumonidae: Campopleginae). *Journal of Hymenoptera Research* **6**: 273-296.

Shaw, M. R. and Huddleston, T. 1991. Classification and biology of braconid wasps (Hymenoptera: Braconidae). *Handbooks for the Identification of British Insects* **7(11)**: 1-126.

Shaw, M. R. and Kasparyan, D. R. 2002. Some British records of *Olethrodotis modestus* (Gravenhorst) (Hymenoptera: Ichneumonidae: Ctenopelmatinae). *Entomologist's Record* **114**: 137-139.

Shaw, M. R. and Kasparyan, D. R. 2003. Some genera of British and European Mesoleiini (Hym., Ichneumonidae: Ctenopelmatinae) in the National Museums of Scotland, including a new species of *Mesoleius* and a further twenty species new to Britain. *Entomologist's Monthly Magazine* **139**: 17-28.

Shaw, M. R. and Kasparyan, D. R. 2005. British and European Tryphonini, Exenterini, Eclytini and Idiogrammatini (Hym., Ichneumonidae: Tryphoninae) in the National Museums of Scotland, including 19 species new to Britain. *Entomologist's Monthly Magazine* **141**: 1-14.

Shaw, M. R. and Quicke, D. L. J. 2000. The biology and early stages of *Acampsis alternipes* (Nees), with comments on the relationships of the Sigalphinae (Hymenoptera: Braconidea). *Journal of Natural History* **34**: 611-628.

Shaw, M. R. and Voogd, J. 2016. Illustrated notes on the biology of *Sphinctus serotinus* Gravenhorst (Hymenoptera, Tryphoninae, Sphinctini). *Journal of Hymenoptera Research* **49**: 81-93. doi: 10.3897/JHR.49.7705

Shaw, M. R. and Wahl, D. B. 1989. The biology, egg and larvae of *Acaenitus dubitator* (Panzer) (Hymenoptera, Ichneumonidae: Acaenitinae). *Systematic Entomology* **14**: 117-125.

Shaw, M. R. and Wahl, D. B. 2014. Biology, early stages and description of a new species of *Adelognathus* Holmgren (Hymenoptera: Ichneumonidae: Adelognathinae). *Zootaxa* **3884**: 235-252.

Shaw, M. R., Kasparyan, D. R. and Fitton, M. G. 2003. Revision of the British checklist of Ctenopelmatini (Hymenoptera: Ichneumonidae, Ctenopelmatinae). *Entomologist's Gazette* **54**: 137-141.

Shaw, M. R., Stefanescu, C. and van Nouhuys, S. 2009. Parasitoids of European butterflies. In: Settele, J., Shreeve, T., Konvicka, M., Van Dyck, H., editors. *Ecology of Butterflies in Europe*. Cambridge: Cambridge University Press. pp. 130-156.

Shaw, M. R., Jennings, M. T. and Quicke, D. L. J. 2011. The identity of *Scambus planatus* (Hartig, 1838) and *Scambus ventricosus* (Tschek, 1871) as seasonal forms of *Scambus calobatus* (Gravenhorst, 1829) in Europe (Hymenoptera, Ichneumonidae, Pimplinae, Ephialtini). *Journal of Hymenoptera Research* **23**: 55-64.

Shaw, M. R., Kan, P. and Kan-van Limburg Stirum, B. 2015. Emergence behaviour of adult *Trogus lapidator* (Fabricius) (Hymenoptera, Ichneumonidae, Ichneumoninae, Heresiarchini) from pupa of its host *Papilio machaon* L. (Lepidoptera, Papilionidae), with a comparative overview of emergence of Ichneumonidae from Lepidoptera pupae in Europe. *Journal of Hymenoptera Research* **47**: 65-85.

Shaw, M. R., Horstmann, K. and Whiffin, A. L. 2016. Two hundred and twenty-five species of reared western Palaearctic Campopleginae (Hymenoptera: Ichneumonidae) in the National Museums of Scotland, with descriptions of new species of *Campoplex* and *Diadegma*, and records of fifty-four species new to Britain. *Entomologist's Gazette* **67**: 177-222.

Sheng, M.-L. and Sun, S.-P. 2010a. A new genus and species of Acaenitini (Hymenoptera, Ichneumonidae, Acaenitinae) from China. *ZooKeys* **49**: 87-93.

Sheng, M.-L. and Sun, S.-P. 2010b. [*Parasitic Ichneumonids on Woodborers in China*] (in Chinese with English summary). Beijing: Science Press.

Sheng, M.-L. and Sun, S.-P. 2011. A new genus and species of Brachyscleromatinae (Hymenoptera: Ichneumonidae) from China, *Laxiareola ochracea*. *Journal of Insect Science* **11**: 6 pp. doi: insectscience.org/11.27

Sheng, M.-L. and Sun, S.-P. 2014. *Ichneumonid Fauna of Liaoning*. Beijing: Science Press.

Sheng, M.-L., Broad, G. R. and Sun, S.-P. 2012. A new genus and species of Collyriinae (Hymenoptera, Ichneumonidae). *Journal of Hymenoptera Research* **25**: 103-125. doi: 10.3897/JHR.25.2319

Sheng, M.-L., Sun, S-P., Ding, D.-S. and Luo, J.-G. 2013. *Ichneumonid Fauna of Jiangxi (Hymenoptera: Ichneumonidae)*. Beijing: Science Press.

Shevyrev, I. J. 1912. *Parazity i iz mira nasekomykh [Parasites and hyperparasites of the insect world]*. St Petersburg.

Shimizu, S. 2018. First record of the genus *Seleucus* Holmgren, 1860 (Hymenoptera: Ichneumonidae: Ctenopelmatinae) from the Oriental region, with description of a new species. *Oriental Insects* **52**: 88-95. DOI: 10.1080/00305316.2017.1357054

Short, J. R. T. 1957. On the final instar larva of *Stilbops* (*Aphanoroptrum*) *abdominale* (Grav.) (Hymenoptera: Ichneumonidae). *Proceedings of the Royal Entomological Society of London (B)* **26**: 175-176.

Short, J. R. T. 1978. The final larval instars of the Ichneumonidae. *Memoirs of the American Entomological Institute* **25**: 1-508.

Sime, K. R. and Wahl, D. B. 2002. The cladistics and biology of the *Callajoppa* genus-group (Hymenoptera: Ichneumonidae, Ichneumoninae). *Zoological Journal of the Linnean Society* **134**: 1-56.

Simmonds, F. J. 1947. Biology of *Phytodietus pulcherrimus* (Cress.), parasite of *Loxostege sticticalis* L. in North America. *Parasitology* **38**: 150-156.

Sivaraman, A. and Manickavasagam, S. 2011. Record of *Nipponaetus* [sic] *haeussleri* (Uchida) (Hymenoptera: Ichneumonidae) as an egg parasitoid of sugarcane top shoot borer, *Scirpophaga excerptalis* Walker from India. *Journal of Biological Control* **25**: 62-63.

Skinner, E. R. and Thompson, G. H. 1960. *The alder woodwasp and its insect enemies*. Film produced by University of Oxford Department of Forestry, distributed by World Educational Films.

Slovák, M. 1984. New data about developmental stages of *Exetastes cinctipes* (Hym., Ichneumonidae). *Biologia (Bratislava)* **39**: 611-616.

Slovák, M. 1986. Longevity of adults and fecundity in *Exetastes cinctipes* Retz. (Hym., Ichneumonidae). *Biologia (Bratislava)* **41**: 105-114.

Smilowitz, Z. & Iwantsch, G. F. 1973. Relationships between the parasitoid *Hyposoter exiguae* and the cabbage looper, *Trichoplusia ni*: effects of host age on development rate of the parasitoid. *Environmental Entomology* **2**: 759–763.

Smilowitz, Z. & Iwantsch, G. F. 1975. Relationships between the parasitoid *Hyposoter exiguae* and the cabbage looper, *Trichoplusia ni*: the effect of host age on oviposition rate of the parasitoid and successful parasitism. *Canadian Entomologist* **107**: 689–694.

Smith, C. W. 1931. Colonisation in Canada of *Collyria calcitrator* (Hym. Ichn.), a parasite of the wheat sawfly. *Bulletin of Entomological Research* **22**: 547-550.

Smith, E. L. 1970. Evolutionary morphology of external insect genitalia. 2. *Annals of the Entomological Society of America* **63**: 1-27.

Smithers, C. N. 1956. On *Philopsyche abdominalis* Morley (Hym.: Ichneumonidae), a parasite of *Acanthopsyche junodi* Heylaerts (Lep.: Psychidae). *Journal of the Entomological Society of Southern Africa* **19**: 225-249.

Snodgrass, R. E. 1956. *The anatomy of the honey bee*. Ithaca, New York: Comstock.

Sonan, J. 1933. On the tribe Rhyssini Cush. and Rohw. from Formosa (Hym. Ichneumonidae). *Transactions of the Natural History Society of Formosa. Taihoku* **23**: 242-248.

Spasojevic, T., Broad, G. R., Bennett, A. M. R. and Klopfstein, S. 2017. Ichneumonid parasitoid wasps from the Early Eocene Green River Formation: five new species and a revision of the known fauna (Hymenoptera, Ichneumonidae). *Paläontologische Zeitschrift* **92**: 35-63. DOI: 10.1007/s12542-017-0365-5

Spradbery, J. P. 1968. The biology of *Pseudorhyssa sternata* Merrill (Hym., Ichneumonidae), a cleptoparasite of Siricid woodwasps. *Bulletin of Entomological Research* **59**: 291-297.

Spradbery, J. P. 1970. Host finding by *Rhyssa persuasoria* (L.), an Ichneumonid parasite of Siricid woodwasps. *Animal Behaviour* **18**: 103-114.

Spradbery, J. P. and Kirk, A. A. 1978. Aspects of the ecology of siricid woodwasps (Hymenoptera: Siricidae) in Europe, North Africa and Turkey with special reference to the biological control of *Sirex noctilio* F. in Australia. *Bulletin of Entomological Research* **68**: 341-359.

Spradbery, J. P. and Ratkowsky, D. A. 1974. An analysis of geographical variation in the parasitoid *Rhyssa persuasoria* (L.) (Hymenoptera, Ichneumonidae). *Bulletin of Entomological Research* **64**: 653-668.

Steiner, S. M., Kropf, C., Graber, W., Nentwig, W. and Klopfstein, S. 2010. Antennal courtship and functional morphology of tyloids in the parasitoid wasp *Syrphoctonus tarsatorius* (Hymenoptera: Ichneumonidae: Diplazontinae). *Arthropod Structure & Development* **39**: 33-40.

Stelfox, A. W. 1929. On the distinction of *Pimpla clavicornis* Thoms. and *P. curticauda* Kriech. *Entomologist's Monthly Magazine* **65**: 17-18.

Stelfox, A. W. 1961. *Gnathoniella egregia* Schmiedeknecht (Hym., Ichneumonidae) new to the British Isles. *Entomologist's Monthly Magazine* **97**: 181-182.

Stelfox, A. W. 1966. The species of the genus *Eclytus* so far found in Ireland. *Proceedings of the Royal Irish Academy* **64B**: 509-511.

Stenton, R. 1926. [Untitled]. *Entomologist's Monthly Magazine* **62**: 98.

Stoltz, D. B. 1981. A putative baculovirus in the Ichneumonid parasitoid, *Mesoleius tenthredinis*. *Canadian Journal of Microbiology* **27**: 116-122.

Stoltz, D. B. 1986. Interactions between parasitoid-derived products and host insects: an overview. *Journal of Insect Physiology* **32**: 347-350.

Stoltz, D.[B.] and Vinson, S. B. 1979. Viruses and parasitism in insects. *Advances in Virus Research* **24**: 125-171.

Stoltz, D. B. and Whitfield, J. B. 1992. Viruses and virus-like entities in the parasitic Hymenoptera. *Journal of Hymenoptera Research* **1**: 125-139.

Stoltz, D. B., Krell, P. J. and Vinson, S. B. 1981. Polydisperse viral DNAs in ichneumonid ovaries: a survey. *Canadian Journal of Microbiology* **27**: 123-130.

Subinprasert, S. 1987. Natural enemies and their impact on overwintering codling moth populations (*Laspeyresia pornonella* L.) (Lep., Tortricidae) in South Sweden. *Journal of Applied Entomology* **103**: 46-55.

Takasuka, K. and Matsumoto, R. 2011a. Lying on the dorsum: unique host-attacking behaviour of *Zatypota albicoxa* (Hymenoptera, Ichneumonidae). *Journal of Ethology* **29**: 203-207. doi: 10.1007/s10164-010-0263-8

Takasuka, K. and Matsumoto, R. 2011b. Infanticide by a solitary koinobiont ichneumonid ectoparasitoid of spiders. *Naturwissenschaften* **98**: 529-536. doi: 10.1007/s00114-011-0797-9

Takasuka, K., Matsumoto, R. and Ohbayashi, N. 2009. Oviposition behavior of *Zatypota albicoxa* (Hymenoptera, Ichneumonidae), an ectoparasitoid of *Achaearanea tepidariorum* (Araneae, Theridiidae). *Entomological Science* **12**: 232-237.

Takasuka, K., Yasui, T., Ishigami, T., Nakata, K., Matsumoto, R., Ikeda, K. and Maeto, K. 2015. Host manipulation by an ichneumonid spider ectoparasitoid that takes advantage of preprogrammed web-building behaviour for its cocoon protection. *Journal of Experimental Biology* **218**: 2326-2332.

Takasuka, K., Fritzén, N., Tanaka, Y., Matsumoto, R., Maeto, K. and Shaw, M. R. 2018. The changing use of the ovipositor in host shifts by ichneumonid ectoparasitoids of spiders (Hymenoptera, Ichneumonidae, Pimplinae). *Parasite* **25**: 17. DOI: 10.1051/parasite/2018011

Taylor, K. L. 1977. The introduction and establishment of insect parasitoids to control *Sirex noctilio* in Australia. *Entomophaga* **21(1976)**: 429-440.

Tereshkin, A. M. 2009. Illustrated key to the tribes of subfamilia Ichneumoninae and genera of the tribe Platylabini of world fauna (Hymenoptera, Ichneumonidae). *Linzer Biologische Beiträge* **41**: 1317-1608.

Tereshkin, A. M. 2011. Illustrated key to the genera of the subtribe Amblytelina of Palaearctic (Hymenoptera, Ichneumonidae, Ichneumoninae, Ichneumonini). *Linzer Biologische Beiträge* **43**: 597-711.

Teunissen, H. G. M. 1945. Ueber die Gattung *Mesoleius* (Hymenoptera Ichneumonidae). *Zoologische Mededelingen* **25**: 200-238.

Thomas, J. A., Knapp, J. J., Akino, T., Gerty, S., Wakamura, S., Simcox, D. J., Wardlaw, J. C. and Elmes, G. W. 2002. Parasitoid secretions provoke ant warfare. *Nature* **417**: 505-506.

Thorpe, W. H. 1939. On the occurrence of the male of *Nemeritis canescens* (Grav.) in Britain. *Proceedings of the Royal Entomological Society of London (A)* **14**: 47.

Thorpe, W. H. and Caudle, H. B. 1938. A study of the olfactory responses of insect parasites to the food plant of their host. *Parasitology* **30**: 523-528.

Thorpe, W. H. and Jones, F. G. W. 1937. Olfactory conditioning in a parasitic insect and its relation to the problem of host selection. *Proceedings of the Royal Society of London series B – Biological Sciences* **124**: 56-81.

Tobias, V. I., editor. 1968. [Problems of classification and phylogeny of the family Braconidae (Hymenoptera).] (In Russian). *Reports on the 20th annual lectures in memory of N.A. Kholodkovskii*. pp. 3-43.

Tobias, V. I. 1988a. [The family Paxylommatidae (Hymenoptera) in the fauna of the USSR] (in Russian). *Trudy Vsesoyuznogo Entomologicheskogo Obshchestva* **70**: 131-143.

Tobias, V. I. 1988b. Paxylommatidae. In: Tobias, V. I. and Zinovjev, A. G., editors. [*Hymenoptera, XX. Keys to the Insects of the European part of the U.S.S.R.*] (In Russian). pp. 237-242.

Tolkanitz, V. I. 2007. Ichneumon flies of the genus *Exochus* Gravenhorst (Hymenoptera: Ichneumonidae: Metopiinae) of the fauna of Palaearctic region. *Russian Entomological Journal* **16**: 339-358

Tolkanitz, V. I. 2011. Ichneumon-Flies of the Genus *Hypsicera* (Hymenoptera, Ichneumonidae, Metopiinae) of the Fauna of Palaearctic Region. *Vestnik Zoologii* **45**: 277-282.

Tolkanitz, V. I. 2015. Ichneumon flies of the genus *Metopius* Panzer (Hymenoptera, Ichneumonidae, Metopiinae) of the Palaearctic fauna. *Entomological Review* **95**: 647-665.

Tolkanitz, V. I., Narolsky, N. B. and Perkovsky, E. E. 2005. A new species of parasitic wasp of the genus *Pherhombus* (Hymenoptera, Ichneumonidae, Pherhombinae) from the Rovno amber. *Paleontological Journal* **39**: 511-513.

Tothill, J. D. 1922. The natural control of the Fall Webworm (*Hyphantria cunea* Drury). With an account of its several parasites. *Canadian Department of Agriculture Bulletin* **Entomological Bulletin 19**: 1-107.

Townes, H. K. 1939a. Protective odors among the Ichneumonidae (Hymenoptera). *Bulletin of the Brooklyn Entomological Society* **34**: 29-30.

Townes, H. K. 1939b. The Nearctic species of *Netelia* (*Paniscus*) of authors and a revision of genera of Neteliini (Hymenoptera, Ichneumonidae). *Lloydia* **1(1938)**: 168-231.

Townes, H. K. 1944-1945. A catalogue and reclassification of the Nearctic Ichneumonidae (Hymenoptera). *Memoirs of the American Entomological Society* **11**: 1-925.

Townes, H. K. 1967. A new *Ateleute* from the United States. *Proceedings of the Entomological Society of Washington* **69**: 181-182.

Townes, H. K. 1969. The genera of Ichneumonidae, Part 1. *Memoirs of the American Entomological Institute* **11**: 1-300.

Townes, H. K. 1970a. The genera of Ichneumonidae, Part 2. *Memoirs of the American Entomological Institute* **12**: 1-537.

Townes, H. K. 1970b. The genera of Ichneumonidae, Part 3. *Memoirs of the American Entomological Institute* **13**: 1-307.

Townes, H. K. 1971. The genera of Ichneumonidae, Part 4. *Memoirs of the American Entomological Institute* **17**: 1-372.

Townes, H. K. 1983. Revisions of twenty genera of Gelini (Ichneumonidae). *Memoirs of the American Entomological Institute* **35**: 1-281.

Townes, H. K. and Townes, M. 1951. Family Ichneumonidae. In: Muesebeck, C. F. W., Krombein K. V, and Townes, H. K., editors. *Hymenoptera of America north of Mexico - Synoptic catalog*: USDA. Agriculture Monograph. pp. 184-409.

Townes, H. K. and Townes, M. 1959. Ichneumon-flies of America north of Mexico: 1. Subfamily Metopiinae. *United States National Museum Bulletin* **216**: i-ix+1-318.

Townes, H. K. and Townes, M. 1966. A catalogue and reclassification of the Neotropic Ichneumonidae. *Memoirs of the American Entomological Institute* **8**: 1-367.

Townes, H. K. and Townes, M. 1973. A catalogue and reclassification of the Ethiopian Ichneumonidae. *Memoirs of the American Entomological Institute* **19**: 1-416.

Townes, H. K. and Townes, M. 1978. Ichneumon-flies of America north of Mexico: 7. Subfamily Banchinae, tribes Lissonotini and Banchini. *Memoirs of the American Entomological Institute* **26**: 1-614.

Townes, H. K., Townes, M., Walley, G. S., Walkley, L., Habeck, D. and Townes, G. 1960. Ichneumon-flies of American north of Mexico: 2 Subfamily Ephialtinae, Xoridinae, Acaenitinae. *United States National Museum Bulletin* **216**: 1-676.

Townes, H. K., Townes, M. and Gupta, V. K. 1961. A catalogue and reclassification of the Indo-Australian Ichneumonidae. *Memoirs of the American Entomological Institute* **1**: 1-522.

Townes, H. K., Momoi, S. and Townes, M. 1965. A catalogue and reclassification of the eastern Palearctic Ichneumonidae. *Memoirs of the American Entomological Institute* **5**: 1-661.

Tripp, H. A. 1961. The biology of a hyperparasite, *Euceros frigidus* Cress. (Ichneumonidae) and a description of the planidial stage. *Canadian Entomologist* **93**: 40-58.

Tsankov, G. 1988. The ichneumonoid parasites of the European Pine Shoot Moth caterpillars in Bulgaria (Hymenoptera: Ichneumonoidea). In: Gupta, V. K., editor. *Advances in Parasitic Hymenoptera Research*. Leiden: E.J. Brill. pp. 391-393.

Tschopp, A., Riedel, M., Kropf, C., Nentwig, W. and Klopfstein, S. 2013. The evolution of host associations in the parasitic wasp genus *Ichneumon* (Hymenoptera: Ichneumonidae): convergent adaptations to host pupation sites. *BMC Evolutionary Biology* **13**: 13 pp. doi: http://www.biomedcentral.com/1471-2148/13/74

Uçkan, F., Sinan, S., Savaşçi, Ş. and Ergin, E. 2004. Determination of venom components from the endoparasitoid wasp *Pimpla turionellae* L. (Hymenoptera: Ichneumonidae). *Annals of the Entomological Society of America* **97**: 775-780.

Ueno, T. 2004. Offspring performance in the pupal parasitoid *Pimpla* (=*Coccygomimus*) *luctuosa* (Hymenoptera: Ichneumonidae) as influenced by host age and size. *Journal of the Faculty of Agriculture Kyushu University* **49**: 321-329.

Vance, A. M. 1927. On the biology of some Ichneumonids of the genus *Paniscus* Schrank. *Annals of the Entomological Society of America* **20**: 405-415.

Várkonyi, G. 1998. Notes on *Leptocampoplex cremastoides* (Hymenoptera, Ichneumonidae, Campopleginae), a new genus and species to the Finnish fauna. *Entomologica Fennica* **9**: 215-216.

Varley, G. C. 1965. A note on the life-history of the Ichneumon fly *Euceros unifasciatus* Voll. with a description of its planidium larva. *Entomologist's Monthly Magazine* **100**: 113-116.

Varley, G. C., Gradwell, G. R. and Hassell, M. P. 1973. *Insect Population Ecology [:] an analytical approach*. Oxford: Blackwell.

Vas, Z. 2016. A new species of *Temelucha* Förster from Malta with an updated and revised identification key to the Western Palaearctic *Temelucha* species (Hymenoptera, Ichneumonidae, Cremastinae). *Journal of Hymenoptera Research* **48**: 67-84.

Vas, Z. and Kutasi, C. 2016. Hymenoptera from caves of Bakony Mountains, Hungary - an overlooked taxon in hypogean research. *Subterranean Biology* **19**: 31-39.

Veen, J. C. van 1981. The biology of *Poecilostictus cothurnatus* (Hymenoptera, Ichneumonidae) an endoparasite of *Bupalus piniarius* (Lepidoptera, Geometridae). *Annales Entomologici Fennici* **47**: 77-93.

Veen, J. C. van 1982. Notes on the biology of *Banchus femoralis* Thomson (Hym., Ichneumonidae) an endoparasitoid of *Panolis flammea* (D. & S.) (Lep., Noctuidae). *Zeitschrift für Angewandte Entomologie* **94**: 300-311.

Veijalainen, A., Sääksjärvi, I. E., Erwin, T. L., Gómez, I. C. and Longino, J. T. 2012a. Subfamily composition of Ichneumonidae (Hymenoptera) from western Amazonia: Insights into diversity of tropical parasitoid wasp. *Insect Conservation and Diversity* **6**: 28-37. doi: 10.1111/j.1752-4598.2012.00185.x

Veijalainen, A., Wahlberg, N., Broad, G. R., Erwin, T. L., Longino, J. T. and Sääksjärvi, I. E. 2012b. Unprecedented ichneumonid parasitoid wasp diversity in tropical forests. *Proceedings of the Royal Society of London Series B - Biological Sciences* **279**: 4694–4698.

Vickery, R. A. 1929. Studies on the fall armyworm in the Gulf Coast district of Texas. *United States Department of Agriculture. Technical Bulletin* **No.138**: 63 pp.

Viereck, H. L. 1923. Groteinae - a new subfamily. *Proceedings of the Biological Society of Washington* **36**: 201.

Viitasaari, M. 1979. A study on the Palaearctic species of the genus *Protarchus* Förster (Hymenoptera, Ichneumonidae). *Notulae Entomologicae* **59**: 33-39.

Vikberg, V. and Koponen, M. 2000. On the taxonomy of *Seleucus* Holmgren and the European species of Phrudinae (Hymenoptera: Ichneumonidae). *Entomologica Fennica* **11(4)**: 195-228.

Viktorov, G. A. 1968. The position of the tribe Therionini (Hymenoptera, Ichneumonidae) in the classification of the family and the composition of some of its genera. *Entomological Review*.

Vilhelmsen, L. 1996. The preoral cavity of lower Hymenoptera (Insecta): Comparative morphology and phylogenetic significance. *Zoologica Scripta* **25**: 143-170.

Vilhelmsen, L. 1997a. The phylogeny of lower Hymenoptera (Insecta), with a summary of the early evolutionary history of the order. *Journal of Zoological Systematics and Evolutionary Research* **35**: 49-70.

Vilhelmsen, L. 1997b. Head capsule concavities accommodating the antennal bases in Hymenoptera pupating in wood: possible emergence-facilitating adaptations. *International Journal of Insect Morphology and Embryology* **26**: 129-138.

Vilhelmsen, L. 1999. The occipital region in the basal Hymenoptera (Insecta): a reappraisal. *Zoologica Scripta* **28**: 75-85.

Vilhelmsen, L. 2001. Phylogeny and classification of the extant basal lineages of the Hymenoptera (Insecta). *Zoological Journal of the Linnean Society* **131**: 393-442.

Vilhelmsen, L. 2003. Flexible ovipositor sheaths in parasitoid Hymenoptera (Insecta). *Arthropod Structure & Development* **32**: 277-287.

Vilkamaa, P. and Komonen, A. 2001. Redescription and biology of *Trichosia* (*Baeosciara*) *sinuata* Menzel & Mohrig (Diptera: Sciaridae). *Entomologica Fennica* **12**: 46-49.

Villemant, C., Jingxian, L. and Rousse, P. 2016. Deep into the head of a remarkable new genus of Orthocentrinae (Hymenoptera: Ichneumonidae) from the highest peak of Papua New Guinea. In: Robillard, T., F., L., Villemant, C., Leponce, M., editors. *Insects of Mount Wilhelm, Papua New Guinea*. Paris: Muséum national d'Histoire naturelle. pp. 375-392.

Vinson, S. B. 1976. Host selection by insect parasitoids. *Annual Review of Entomology* **21**: 109-133.

Vinson, S. B. 1981. Habitat location. In: Nordlund, D. A., Jones, R. L., Lewis, W. J., editors. *Semiochemicals, their role in pest control*. New York. pp. 51-77.

Vinson, S. B. & Iwantsch, G. F. 1980a. Host suitability for insect parasitoids. *Annual Review of Entomology* **25**: 397–419.

Vinson, S. B. & Iwantsch, G. F. 1980b. Host regulation by insect parasitoids. *Quarterly Review of Biology* **55**: 143–165.

Visser, B., Le Lann, C., den Blanken, F. J., Harvey, J. A., van Alphen, J. J. M. and Ellers, J. 2010. Loss of lipid synthesis as an evolutionary consequence of a parasitic lifestyle. *Proceedings of the National Academy of Sciences of the USA* **107**: 8677-8682. doi: 10.1073/pnas.1001744107

Wäckers, F. L., Mitter, E. and Dorn, S. 1998. Vibrational sounding by the pupal parasitoid *Pimpla* (*Coccygomimus*) *turionellae*: an additional solution to the reliability-detectability problem. *Biological Control* **11**: 141-146.

Wahl, D. B. 1986. Larval structures of oxytorines and their significance for the higher classification of some Ichneumonidae (Hymenoptera). *Systematic Entomology* **11**: 117-127.

Wahl, D. B. 1988. A review of the mature larvae of the Banchini and their phylogenetic significance, with comments on the Stilbopinae (Hymenoptera: Ichneumonidae). In: Gupta, V. K., editor. *Advances in Parasitic Hymenoptera Research*. Leiden, New York, København, Köln: E.J.Brill. pp. 147-161.

Wahl, D. B. 1990. A review of the mature larvae of Diplazontinae, with notes on larvae of Acaenitinae and Orthocentrinae and proposal of two new subfamilies (Insecta: Hymenoptera, Ichneumonidae). *Journal of Natural History* **24**: 27-52.

Wahl, D. B. 1991. The status of *Rhimphoctona*, with special reference to the higher categories within Campopleginae and the relationships of the subfamily (Hymenoptera: Ichneumonidae). *Transactions of the American Entomological Society* **117**: 193-213.

Wahl, D. B. 1993a. Cladistics of the ichneumonid subfamily Labeninae (Hymenoptera: Ichneumonidae). *Entomologia Generalis* **18**: 91-105.

Wahl, D. B. 1993b. Cladistics of the genera of Mesochorinae (Hymenoptera: Ichneumonidae). *Systematic Entomology* **18**: 371-387.

Wahl, D. B. 1993c. Key to subfamilies of Holarctic and Neotropical Ichneumonidae. In: Goulet, H., Huber, J. T., editors. *Hymenoptera of the world: An identification guide to families*. Ottawa: Agriculture Canada. pp. 396-509.

Wahl, D. B. 1996. Two new species of *Megastylus* from the New World (Hymenoptera: Ichneumonidae; Orthocentrinae). *Journal of the New York Entomological Society* **104**: 221-225.

Wahl, D. B. 1997. The cladistics of the genera and subgenera of Xoridinae. In: Gauld, I. D., editor. *Ichneumonidae of Costa Rica, 2*. Memoirs of the American Entomological Institute **57**. pp. 454-460.

Wahl, D. B. and Gauld, I. D. 1998. The cladistics and higher classification of the Pimpliformes (Hymenoptera: Ichneumonidae). *Systematic Entomology* **23**: 299-303.

Wahl, D. B. and Mason, W. R. M. 1995. The family-group names of the Ichneumoninae (Hymenoptera: Ichneumonidae). *Journal of Hymenoptera Research* **4**: 285-293.

Wahl, D. B., Shanower, T. G. and Hoelmer, K. A. 2007. A new species of *Collyria* Schiødte (Hymenoptera: Ichneumonidae: Collyriinae), a parasitoid of *Cephus fumipennis* (Hymenoptera: Cephidae) in China, and potential biological control agent for *Cephus cinctus* in North America. *Journal of the Kansas Entomological Society* **80**: 43-50.

Wallace, I. D. 2003. An unusual ichneumon. *The Vasculum* **88(3)**: 11.

Waloff, N. 1967. Biology of three species of *Leiophron* (Hymenoptera: Braconidae, Euphorinae) parasitic on Miridae on broom. *Transactions of the Royal Entomological Society of London* **119**: 187-213.

Wardle, A. R. and Borden, J. H. 1985. Age-dependent associative learning by *Exeristes roborator* (F.) (Hymenoptera: Ichneumonidae). *Canadian Entomologist* **117**: 605-616.

Watanabe, C. 1984. Notes on Paxylommatinae with review of Japanese species (Hymenoptera, Braconidae). *Kontyu* **52**: 553-556.

Watanabe, K. and Maeto, K. 2012. Taxonomic study of the genus *Stilbops* Förster from Japan (Hymenoptera: Ichneumonidae: Stilbopinae). *Zootaxa* **3456**: 51-81.

Watanabe, K. and Maeto, K. 2013. Review of the subgenus *Diblastomorpha* Förster, 1869, of the genus *Glypta* Gravenhorst, 1829 (Hymenoptera, Ichneumonidae, Banchinae). *Journal of Japanese Systematic Entomology* **19**: 129-137.

Watanabe, K., Tanikawi, T. and Kasparyan, D. R. 2015. *Tanzawana flavomaculata* (Hymenoptera, Ichneumonidae, Ctenopelmatinae), a new genus and species of parasitoid of *Fagineura crenativora* (Tenthredinidae, Nematinae), a serious pest of beech tree. *Zootaxa* **4040**: 236-242.

Waterston, J. 1929a. Additional notes on *Belesica braconoides*, Waterst. *Annals and Magazine of Natural History* **3**: 628-632.

Waterston, J. 1929b. A new fungus gnat-parasite. *Annals and Magazine of Natural History* **3**: 632-636.

Weiffenbach, H. 1988. Über einige aus Blattwespenlarven (Hymenoptera, Symphyta) gezogene Ichneumoniden (Hymenoptera, Ichneumonidae). *Nachrichtenblatt der Bayerischen Entomologen* **37**: 103-107.

Weseloh, R. M. 1981. Host location by parasitoids. In: Nordlund, D. A., Jones, R. L., Lewis, W. J., editors. *Semiochemicals, their role in pest control*. New York. pp. 79-95.

Wharton, R. A. 2006. The species of *Sternaulopius* Fischer (Hymenoptera: Braconidae, Opiinae) and the braconid sternaulus. *Journal of Hymenoptera Research* **15**: 316-347.

Whitfield, J. B. and Asgari, S. 2003. Virus or not? Phylogenetics of polydnaviruses and their wasp carriers. *Journal of Insect Physiology* **49**: 397-405.

Winfield, A. L. 1963. A study on the effects of insecticides on parasites of larvae of blossom beetles (*Meligethes aeneus* F., Coleoptera: Nitidulidae). *Entomologia Experimentalis Et Applicata* **6**: 306-318.

Wisseman, R. W. and Anderson, N. H. 1984. Mortality factors affecting Trichoptera eggs and pupae in an Oregon coast range watershed. In: *Proceedings of the International Symposium on Trichoptera* **4(1983)**: 455-460.

Wootton, R. J. 1978. Function, homology and terminology in insect wings. *Systematic Entomology* **4 [1979]**: 81-93.

Wyatt, T. D. and Foster, W. A. 1989. Parental care in the subsocial intertidal beetle, *Bledius spectabilis*, in relation to parasitism by the ichneumonid wasp, *Barycnemis blediator*. *Behaviour* **110**: 76-92.

Wyrostkiewicz, K. and Blazejewska, A. 1985. *Isurgus heterocerus* Thoms. and *I. morionellus* Holmgr. (Hym. Ichneumonidae) parasitoids of *Meligethes aeneus* F. larvae (Col., Nitidulidae). *Polskie Pismo Entomologiczne* **55**: 391-404.

Yeargan, K. V. and Braman, S. K. 1989. Life history of the hyperparasitoid *Mesochorus discitergus* (Hymenoptera: Ichneumonidae) and tactics used to overcome the defensive behavior of the green cloverworm (Lepidoptera: Noctuidae). *Annals of the Entomological Society of America* **82**: 393-398.

Yoder, M. J., Mikó, I., Seltmann, K. C., Bertone, M. A. and Deans, A. R. 2010. A gross anatomy ontology for Hymenoptera. *PLoS ONE* **5**: e15991. doi: 10.1371/journal.pone.0015991

Yoshida, T., Nagasaki, O. and Hirayama, T. 2011. A new species of the genus *Apsilops* Förster (Hymenoptera: Ichneumonidae: Cryptinae) from Japan; parasitoid of an aquatic crambid moth. *Zootaxa* **2916**: 41-50.

Yu, D. S., Achterberg, C. van and Horstmann, K. 2012. *Taxapad 2012, Ichneumonoidea 2011*. Database on flash-drive. www.taxapad.com, Ottawa, Ontario, Canada.

Yu, D. S., Achterberg, C. van and Horstmann, K. 2016. *Taxapad 2016, Ichneumonoidea 2015*. Database on flash-drive. www.taxapad.com, Nepean, Ontario, Canada.

Zhang, C., Stadler, T., Klopfstein, S., Heath, T. A. and Ronquist, F. 2016. Total-evidence dating under the fossilized birth–death process. *Systematic Biology* **65**: 228-249.

Zhang, H. C. and Rasnitsyn, A. P. 2003. Some ichneumonids (Insecta, Hymenoptera, Ichneumonoidea) from the Upper Mesozoic of China and Mongolia. *Cretaceous Research* **24**: 193-202.

Zhaurova, K. and Wharton, R. A. 2009. Recognition of Scolobatini and Westwoodiini (Hymenoptera, Ctenopelmatinae) and revision of the component genera. *Contributions of the American Entomological Institute* **35(5)**: 1-77.

Zijp, J. P. and Blommers, L. 1993. *Lathrolestes ensator*, a parasitoid of the apple sawfly. *Proceedings of the Section Experimental and Applied Entomology of the Netherlands Entomological Society* **4**: 237-242.

Zijp, J.-P. and Blommers, L. H. M. 2002. Survival mode between the yearly reproduction periods, and reproductive biology of *Scambus pomorum* (Hymenoptera: Ichneumonidae: Pimplinae), a parasitoid of the apple blossom weevil *Anthonomus pomorum* (Coleoptera: Curculionidae). *Entomologia Generalis* **26**: 29-46.

Zinnert, K. D. 1969a. Vergleichende Untersuchungen zur Morphologie und Biologie der Larvenparasiten (Hymenoptera Ichneumonidae und Braconidae) mitteleuropäischer Blattwespen aus der Subfamily Nematinae (Hymenoptera: Tenthredinidae). Teil I. *Zeitschrift für Angewandte Entomologie* **64**: 180-217.

Zinnert, K. D. 1969b. Vergleichende Untersuchungen zur Morphologie und Biologie der Larvenparasiten (Hymenoptera Ichneumonidae und Braconidae) mitteleuropäischer Blattwespen aus der Subfamily Nematinae (Hymenoptera: Tenthredinidae). Teil II. *Zeitschrift für Angewandte Entomologie* **64**: 277-306.

Zúñiga Ramírez, R. J. 2004. The taxonomy and biology of the *Polycyrtus* species (Hymenoptera: Ichneumonidae, Cryptinae) of Costa Rica. *Contributions of the American Entomological Institute* **33(4)**: 1-159.

Zwakhals, C. J. 2006. The European species of the genera *Zatypota* and *Sinarachna* (Hymenoptera: Ichneumonidae, Pimplinae, Polysphinctini). *Entomologische Berichten* **66**: 34-37.

Zwakhals, C. J. 2010. Identification of Western Palearctic *Dolichomitus* species (Hymenoptera: Ichneumonidae: Pimplinae). *Entomologische Berichten* **70**: 111-127.

Zwart, K. W. R. 1998. Ichneumonidae. In: Polaszek, A., editor. *African Cereal Stem Borers: economic importance, taxonomy, natural enemies and control*. Wallingford: CAB International. pp. 205-258.

Zwölfer, H. 1961. A comparative analysis of the parasite complexes of the European fir budworm, *Choristoneura murinana* (Hüb.) and the North American spruce budworm, *C. fumiferana* (Clem.). *Commonwealth Institute of Biological Control Technical Bulletin* **1**: 1-162.

Index to Ichneumonoidea

Main entries are given in **bold**. Illustrations in *italics*

abbreviatus, Agrothereutes *106, 110, 168, 172*
abruptorius, Exenterus 182, *311*, 312-314
Absyrtus 18, 179
Absyrtus vernalis 179
Absyrtus vicinator 179
Acaenitellus 257, 307
Acaenitinae 8, 12, 17, 64, 81, *107 108,* **111**, 112-114, 158, 270, 280
Acaenitus 17, 113
Acaenitus dubitator *108, 111,* 113, *114*
Achaius 19
Achaius oratorius *15*, 210
aciculatus, Diacritus 77, *106- 108, 189, 190*
Aclastus 20, 249, 261, 265, 269
Aclastus solutus *106*
Aconias 17, 169
Acrodactyla 21, 269, *275,* 280
Acrodactyla degener 280
Acrolyta 20, 262, 263- 265, *266*
Acrolyta nens 265
Acropimpla 21, 272
Acropimpla didyma 272
Acroricnus 18, 171
Acrotomus 22, 316
aculeata, Diadegma 148
Adelognathinae 5, 6, 8, 11, 17, 56, 63, *107,* **115**, *116,* 117, *118,* 166, 167, 261, 271, 307
Adelognathus 17, 115-117
Adelognathus brevicornis *107, 118*
Adelognathus chrysopygus 116, 117, *118*
Adelognathus cubiceps 117
Adelognathus difformis 116, 117
Adelognathus dorsalis 56, *107,* 115, 117, *118*
Adelognathus laevicollis 116
Adelognathus leucotrochi *116,* 117
Adelognathus pallipes *107,* 115, *118*
Adelphion 328
admontina, Dusona 148
adustus, Enicospilus *106*
Aethecerus 19
aethiops, Pimpla 280
Agasthenes 20, 265
Agathilla 136
agilis, Gelis 264, 265
agilis, Mesochorus 153, 223, 224
Agriotypidae 8
Agriotypinae 8, 11, 17, 76, *110,* **119**, *121,* 135, 167,
Agriotypus 11, 17, 38, 119-121, 135
Agriotypus armatus 65, 76, *108, 119,* 120, *121*
Agrothereutes 18,168
Agrothereutes abbreviatus *106, 110, 168, 172*
Agrothereutes leucorhaeus 168
Agrothereutes mandator 168
Agrothereutes saturniae *30,* 168
Agrypon 17, 129
Agrypon anomelas 129, *130*
Agrypon batis *106, 126,* 129, *130*
Agrypon canaliculatum *106*
Agrypon flaveolatum *127,* 128, 129

alacer, Eridolius 314
albicinctus, Arotes 113
albicinctus, Perithous 280
albicoxa, Zatypota *108, 279, 283*
albiditarsus, Zele *106*
albilineatus, Virgichneumon *106, 107*
albitarsus, Euceros *108,* 198, 199, *200*
aloguttata, Goedartia *205,* 207
albopictus, Hypamblys 178
Alcima 17, 107, 155
Alcima orbitale 107, *114,* 156, *157*
Alexeter 18, 183
Alexeter multicolor 109, *184, 185*
Alexeter niger 182
Allomacrus 19, 99, 101, 186, 187
Allomacrus arcticus 99, *110, 187, 188*
Allophroides 21
Alloplasta 17, 137
Alloplasta piceator 137
Alomya 17, 122, 123, 205, 210
Alomya debellator *106, 107, 122, 124*
Alomya semiflava *106,* 123, *124*
Alomyinae 7, 8, 11, 17, 44, 60, *106,* **122**, 123, *124,* 167, 201, 205, 210,
alpestris, Pseudorhyssa *106,* 269, 271, *282*
alter, Bathythrix 262
Amblyjoppa 19, 208
Amblyjoppa fuscipennis 208
Amblyjoppa proteus 208
Amblyteles 19, 211
Amblyteles armatorius *24,* 210, *213*
americanus, Enicospilus 245, 243
amictorius, Exenterus 313
amoenus, Arotes 112
Amphibulus 20
analis, Homaspis *107, 110, 184*
anatorius, Cubocephalus *106, 107, 173*
Aneuclis 21, 301, 302
Aneuclis melanarius 301
angulna, Grotea *327*
Anisobas 19, 50
Anisobas cingulatellus 207
Anisobas platystylus 208
Anisotacrus 18, 178
Ankylophon 308
Anykylophon obligatus 307
Ankylophonini 305, 307, 308
Anomalinae 9
Anomalon 17, 126
Anomalon cruentatum *106, 125, 130*
Anomaloninae 5, 9, 13,17, 41, 47, 48, *106,* **125**, 126, 128, 129, *130,* 135, 199, 201, 229
Anomalonini 125
anomelas, Agrypon 129, *130*
Anoncus 18
Anoplectes 308
anthophilae, Triclistus 230
anurus, Bathyplectes 152, 153
Aoplus 19, 209

Apaeleticus 20
Apechthis 21, 270, 274, 275, 280
Apechthis quadridentata *30*
Apechthis rufata *109, 283*
Aperileptus 20, 251
Aperileptus vanus 251
Aphanistes 17, 129
Aphanistes ruficornis *128*
Aphidius colemani *106*
apiarius, Excavarus *107, 108, 318*
apicalis, Grypocentrus 314, *320*
Aplomerus 321
Apoclima 20, 186, 187, 250, 251
apoderi, Brachyscleroma 330
Apolophus 229
Apophua 17, 137, 140
Apophua simplicipes 140
Apsilops 18, 169, 171
apterus, Chasmodon *106*
Aptesini 5, 7, 106, 166-171
Aptesis 18, 169
aquisgranensis, Microleptes *108, 109*
Arbelus 18
arctica, Pimpla 280
arcticus, Allomacrus 99, *110, 187, 188*
areator, Gelis 264
arenarius, Mesochorus 223, 224
Arenetra 17
Arenetra pilosella 137, 140
areolatus, Cidaphus 225
areolatus, Triclistus 230
argyraphaga, Hymenoepimecis 279
Aritranis 18, 171
armator, Cryptus *107, 172*
armatorius, Amblyteles *24*, 210, *213*
armatus, Agriotypus 65, 76, *108, 119-121*
armatus, Oxytorus 258
armatus, Pristomerus 164
Arotes 17, 111
Arotes albicinctus 113
Arotes amoenus 112
Arotrephes 20, 265
Asthenara 18
Asthenolabus 20
Astiphromma 20, 221, 224, 225
Astiphromma dorsale 222, 225
Astiphromma nigripes 225
Astiphromma splenium 225
Astrenis 21, 54
Astrenis paradoxus *304*
atalantae, Theronia *273*, 275, 280
Ateleute 11, 17 131
Ateleute linearis 62, *107, 131-133*
Ateleute minusculae 132
Ateleutinae 11, 62, **131**, 133,166
aterrima, Itoplectis *272*
atomator, Phobetes 182
Atractodes 20, 265
Atractogaster semisculptus 272
atrata, Megarhyssa 288
atrator, Exetastes 140
atricillus, Cidaphus *107, 108*, 225, *226, 227*
atriventris, Mesochorus 221

Atrometus 17
Atrophini 85, 134, 136, 137, 139-141, 294
Aubertiella 158
Aubertiella nigricator 159
auriculatus, Scolobates 176, 181, *183, 184*
auscultator, Periope 55, *107, 108*, 230, *233, 234*
Azelus 18

Baeosemus 19, 213
Banchinae 9, 12, 75, 80, 84, 85, **134**, 135, 138, 140, 141, *142*, 182, 239, 280, 306
Banchini 134-137, 140, 141, 192
Banchus 17, *29*, 80, 135-141, 179, 191, 192, 225
Banchus hastator 138-140
Banchus volutatorius *107- 109, 134, 141, 142*
Baranisobas 19
Barichneumon 19, 209
Barichneumon heracliana 210
Barronia 197
Barycnemis 21, 299, 302
Barycnemis blediator 302
Barycnemis gravipes *107, 300, 303*
Barycnemis harpura *30*
Barylypa 17
Barylypa delictor 129
Barytarbes 18, 182
basalis, Grypocentrus 313
Batakomacrus 20
Batakomacrus caudatus *108, 249, 252*
Bathyplectes 17, 146, 147, 149, 152, 153, 155, 224
Bathyplectes anurus 152, 153
Bathyplectes curculionis 151- 153, 223
Bathyplectes stenostigma 152
Bathythrix 20, 262, 265
Bathythrix alter 262
Bathythrix formosa 262
Bathythrix fragilis 262
batis, Agrypon *106*, 126, 129, *130*
Belesica 162
Belesica pictipennis 163
benefactor, Olesicampe 152
Bessobates 310, 311, 312
Bessobates cristata 310
bicingulata, Phobocampe 149
bicoloripes, Dusona *106, 107, 156, 157*
Bicurta 158
bifoveolatus, Ophion 243
biguttata, Lissonota 137
bilineatus, Ophion 243
Bioblapsis 19, 192, 193
Bioblapsis cultiformis 193
Bioblapsis mallochi 193
Bioblapsis polita 193
Blapsidotes vicinus 263
blediator, Barycnemis 302
Boethus 310
bohemani, Notosemus *207*, 208
bombycivorus, Stauropoctonus 245
borealis, Delomerista *110, 281*
borealis, Scolomus 68, *108*, 230, *233-235*
brachyacanthus, Monoblastus *108- 110, 320*
Brachycyrtinae 8, 12, 49, *107*, 144, 198, **143**, 326, 328
Brachycyrtini 8, 143

Brachycyrtus 143, *144*, 328
Brachycyrtus ornatus *107*, *143*, *144*
brachylabis, Xorides 323
Brachyscleroma 329
Brachyscleroma apoderi 330
Brachyscleroma flavoabdominalis 330
Brachyscleroma jiulongshana 330
Brachyscleromatinae 9, 299, 329, 330
Braconidae1, 5-7, 9, 11, 12, 33, 38-41, *106*, 137, 140, 154, 201, 301, 307
Bremiella 229
brevicornis, Adelognathus *107*, *118*
brevicornis, Campoplex 148
brevicornis, Ophion 245
brevicornis, Orthopelma 255
brevicornis, Scambus *109*, *110*, *282*, *283*
brevispina, Thrybius 169
breviuscula, Symplecis *107*, *252*
brischkei, Cidaphus *225*
brischkei, Teleutaea 137, 138
brunnea, Lissonota 138
brunnicornis, Herpestomus 210
Buathra 18, 171
Buathra laborator *24*, *25*, *106*, *107*
Buathra tarsoleuca 169
buccatus, Hybrizon *35*, *106*, *201-203*

Caenocryptus 18, 171
calamitosus, Erythrodolius 329
calcator, Erromenus 313
calcitrator, Collyria 158
caligata, Cylloceria *109*, *187*, *188*
caligatus, Ischnoceros 323
Callajoppa 19, 208, 209, 211, 213
callicerus, Virgichneumon 210
Callidora 17, 149
Calliephialtes 276
calobatus, Scambus 273
caloscelis, Ichneumon 211
camelinus, Thyrateles 210
Campocraspedon 19
Campodorus 18, 178, 183
Campodorus holmgreni *109*, *110*,
Campoletis 17, 148, 149, 151, 153, 155
Campoletis sonorensis 153
Campoletis varians *107*, 157
Campopleginae 3, 5, 9, 11, 13, 17, 58, *106*, *107*, *126*, *135*, 140, **145**, 146, 147, *148*, *149*,150-154, *156*, *157*, 161, 162, 164, 176, 179, 182, 199, 223, 224, 229, 242, *262*, 263, *272*, 299, 316, 321, 328
Campoplegini 147
Campoplex 17, 148
Campoplex brevicornis 148
Campoplex eudoniae *145*
canadensis, Scirtetes 151
canaliculatum, Agrypon *106*
canescens, Nemeritis 154
canescens, Venturia 150-152, 154
cantator, Charops 150
capitator, Helotorus 139
carbonarius, Hyposoter 149
carinifer, Diaparsis 302
Carria 20, 230

Carria paradoxa 230
Casinaria 17, 150, 155
Casinaria mesozosta *148*
Catastenus 20, 251
catoptron, Collyria 159
caudatus Batakomacrus *108*, *249*, *252*
caudatus, Monoblastus *317*
celerator, Colpognathus 210
Centeterus 19
Cephalobaris 20, 265
Cephaloglypta 137
Cephaloglypta murinanae 138-140
cephalotes, Cremastus 162
cephalotes, Phygadeuon *266*, *267*
Ceratojoppini 204
Ceratophygadeuon 20, 265
Certonotus 327
Charitopes 20, 262, 265
Charops 17, 150, 155
Charops cantator 150
Chasmias 19
Chasmias paludator 212
Chasmodon apterus *106*
Chirotica 20, 265
Chorinaeus 20, 229-232
Chorinaeus funebris 230, 231, 313
Chriodes 146, 328
Chrionotini 18, 175-178, 183, *185*
chrysopus, Rhorus *184*
chrysopygus, Adelognathus 116, 117, *118*
chrysostictos, Diadegma 150-153
Cidaphus 20, 220, 221, 224, 225
Cidaphus areolatus 225
Cidaphus atricillus *107*, *108*, *225*, *226*, *227*
Cidaphus brischkei 225
cincta, Colpotrochia 230
cinctellus, Grypocentrus *109*, *110*, *317*
cingulata, Mesoleptidea *179*
cingulatellus, Anisobas 207
circumflexum, Therion 127
circumflexum, Trichomma 129
citratus, Metopius *108*, *233*
citrofrontalis, Tranosemella 156
Cladeutes 22
clandestina, Olesicampe 151
Claseinae 8, 12, 198, *326*
Clasis *326*
clavicornis, Itoplectis 276
Clistopyga 21, 273, 274, 276, 278
Clistopyga incitator *108*- *110*, 274, *282*
clypeatus, Lathrolestes 177
Clypeodromini 204
Clypeoplex 17, 155
Clypeoteles 20, 265
Coelichneumon 19, 206, 209, 213
Coelichneumon orbitator 213
Coelichneumonops 19
cognata, Sussaba *109*,*196*
colemani, Aphidius *106*
Coleocentrus 17, 81, 112
Coleocentrus croceicornis *109*, 113, *114*
Coleocentrus excitator *108*, *109*, *112*, 113, *114*
collaris, Diadromus 210, 212

collaris, Plectiscidea 251
collaris, Poemenia 286
Collyria 17, 112, 158, 159
Collyria calcitrator 158
Collyria catoptron 159
Collyria coxator *30*, 158, 159
Collyria trichophthalma *108*, *109*, *158*, 159, *160*
Collyriinae 5, 8, 12, 17, 30, 82, *108*, **158**, 159, 160, 280
Colocnema 19
Colpognathus 11, 19, 205, 210, 213
Colpognathus celerator 210
Colpotrochia 20
Colpotrochia cincta 230
combustus, Enicospilus *106*
compressa, Phthorima *196*
Compsophorini 204
confector, Hoplocryptus *166*
confluens, Temelucha 163
confusa, Dusona *107*, *157*
conotracheli, Tersilochus 300, 301
contiguus, Polytribax 170
corruptor, Demopheles 169
Cosmoconus 22, 305
costatus, Ophion 245
cothurnatus, Poecilostictus 212
cothurnatus, Polyblastus *312*
Cotiheresiarches 20, 208
Cotiheresiarches dirus 208, *209*, *216*
coxator, Collyria *30*, 158, 159
crassicornis, Hyperacmus 72, *108*, *187*, *188*
crassula, Gnathochorisis *110*, *252*
Cratichneumon 19, 209
Cratichneumon culex 209
Cratocryptus 18
Creagrura 163
Cremastinae 2, 9, 11, 13, 17, 58, *107*, 126, 135, 145, 146, **161**, 162, 164, 229, 299
cremastoides, Leptocampoplex 155
Cremastus 17, 162, 164
Cremastus cephalotes 162
Cremastus geminus *107*, *164*
Cremastus kratochvili 162
Cremastus spectator *107*, *164*
cremieri, Ogkosoma 202
Cremnodes 20, 265
cristata, Bessobates 310
cristata, Netelia (Bessobates) 312
cristata, Netelia *108*, *312*
croceicornis, Coleocentrus *109*, 113, *114*
cruentator, Megastylus *110*, *248*, 251, *252*
cruentatum, Anomalon *106*, *125*, *130*
Crypteffigies 19, 61, 165, 205
Crypteffigies lanius *107*, *214*, *216*,
Cryptinae 3, 5-11, 18, 30, 52, 66, 91-93, 105, *106*- *108*, *110*, 115, 123, 131, 132, **165**, 166, 167, 170, *172*-*174*, 199, 205, 222, 254, 257, 260-263, 321, 326
Cryptini *110*, 165, *166*, 167- 171, 263, *172*, *174*
Cryptopimpla 17, *29*, 137, 138
Cryptus 18, 168- 171
Cryptus armator *107*, *172*
Cryptus dianae *107*, *172*, *173*
Cryptus genalis 169
Crytea 19

Ctenichneumon 19
Cteniscus 22, 316
Ctenocalini 204
Ctenochares 19
Ctenochira 22, 310, 311
Ctenopelma 18, 178, 183
Ctenopelma ruficorne 178
Ctenopelma tomentosum *109*
Ctenopelmatinae 9, 13, 18, 60, 72, 83, 88, 89, 92, 93, 93, 98, 103, 128, 152, **175**, 176-178, 181, 182, *185*, 197, 198, 199, 221, 229, 257, 258, 280, 299, 306, 310, 316
Ctenopelmatini 18, 60, *106*- *109*, *175*-*177*, 178, 180, 183
cubiceps, Adelognathus 116, 117
Cubocephalus 19, 169
Cubocephalus anatorius *106*, *107*, *173*
Cubocephalus distinctor *165*
culex, Cratichneumon 209
cultiformis, Bioblapsis 193
cuneiformis, Seleucus 181, 183
curculionis, Bathyplectes 151- 153
curvator, Hypsicera *228*
curvator, Tersilochus 299
curvinervis, Eremotylus 245
curvulus, Mesochorus 222
Cushmania 236
cyanea, Ecphysis 326
Cycasis 22
Cyclolabus 20
Cyclopneusticae 206
cylindrator, Diblastomorpha *30*
Cylloceria 19, 79, 186, 187, 236
Cylloceria caligata *109*, *186*, *187*
Cylloceria ?melancholica 187
Cylloceriinae 8, 19, 72, 79, 99, 101 *108*- *110*, **186**-*188*, 192, 193, 236, 249-251, 257, 280
Cylloceriini 186
Cymodusa 17, 155

debellator, Alomya *106*, *107*, *122*, *124*
debilis, Nematopodius 169
decimator, Procinetus 112
defectus, Phrudus 35 *107*, *298*, 301, 302
degener, Acrodactyla 280
delictor, Barylypa 129
deliratorius, Ichneumon 206
Delomerista 21, 271
Delomerista borealis *110*, *281*
Delomeristini 7, 12, 21, 95, 106, *110*, 268-271, 274, 275, *281*, *282*, 286
Demopheles 18
Demopheles corruptor 169
dentatus, Metopius *232*
dentifer, Gnathochorisis *110*, *252*
Dentilabus 20
dentipes, Odontocolon *110*, 321, *322*, *325*
detrita, Endromopoda *270*
Deuterolabops 19
Deuteroxorides 21, 285
Deuteroxorides elevator *284*, 286
Diacritinae 8, 13, 19, 77, *106*, 111, **189**, *190*, 250, 269, 285
Diacritus 8, 19
Diacritus aciculatus 77, *106*- *108*, *189*, *190*
Diadegma 17, 147, 148, 150, 151, 155, 156

Diadegma aculeata 148
Diadegma chrysostictos 150-153
Diadegma fabricianae *147*, 148
Diadegma nanus 150
Diadegma scotiae 148
Diadegma semiclausum *107*, 153, *156*, *157*,
Diadromus 19
Diadromus collaris 210, 212
Diadromus pulchellus 210, 212
Diaglyptellana 265
Diaglyptidea 20
Dialipsis 20, 251
dianae, Cryptus *107*, *172*, *173*
Diaparsis 21, 300-302
Diaparsis carinifer 302
Diaparsis jucunda 224
Diaparsis stramineipes 301
Diaparsis temporalis 301
Diblastomorpha 17, 137
Diblastomorpha cylindrator *30*
Dicaelotus 19, 51, 166, 205, 261
Dicaelotus pumilus *107*, *214- 216*
Dichrogaster 20, 262, 265
Dichrogaster liostylus *107*, *267*
didyma, Acropimpla 272
didymator, Hyposoter 153
difformis, Adelognathus 116, 117
digestor, Lissonota 112
Dilleritomus 19, 213
diluta, Fredegunda 272, 273
dimidiator, Mesochorus 225
dimidiatus, Mesochorus 224, 225
Dimophora 17, 161, 163, 164
Dimophora evanialis 163
Diphyus 19, 211
Diphyus quadripunctorius *107*, 210, *214*, *215*,
Diplazon 19, 194, 195
Diplazon laetatorius 193- 195, *196*
Diplazon pectoratorius 194, 195
Diplazon tetragonus 195
Diplazontinae 8, 12, 19, 84, *106*, *109*, **191**, 192, 194- 196, 236, 250, 280
Diradops 137
Dirophanes 19
Dirophanes invisor 210
dirus, Cotiheresiarches 208, *209*, *216*
discitergus, Mesochorus 223, 224
disparis, Lymantrichneumon 213
distinctor, Cubocephalus *165*
diversicostae, Dolichomitus 280
Dolichochorus 20, 220, 221, 225
Dolichochorus longiceps 225
Dolichomitus 21, 197, 268, 272, *274*, 280
Dolichomitus diversicostae 280
Dolichomitus sericeus 197
Dolophron 17, 146, 156
dorsale, Astiphromma 222, 225
dorsalis, Adelognathus 56, *107*, 115, 117, *118*
Dreisbachia 279
dubia, Lissonota 138-140
dubitator, Acaenitus *108*, 113, *114*
Dusona 17, 147, 148, 156,199, 225
Dusona admontina 148

Dusona bicoloripes *106*, *107*, *156*, *157*
Dusona confusa *107*, *157*
Dusona erythrogaster *107*, *157*
Dusona leptogaster 148
Dusona stragifex *106*, *146*
Dusona terebrator *107*, *157*
Dyspetes 22, 311, 316
Dyspetes luteomarginatus *106*, *109*, *110*, 313, *317*, *318*

Earobia 301
Earobia paradoxa 302
Echthrini 9, 167
Echthronomas 17,156
Echthrus 18, 66, 166, 170
Echthrus relucatator *108*, *172*, *173*
Eclytini 22, 103, *109*, *305*, 306- 308, 312, 316
Eclytus 22, 176, 306-308, 313, 316
Eclytus exornatus *305*
Eclytus multicolor *108*, *109*, 308, *317*
Eclytus ornatus 313
Ecphysis 326
Ecphysis cyanea 326
Ectopius 20
edwardsi, Proclitus 250
Eiphosoma 162, 163
elegans, Neliopisthus 309
elevator, Deuteroxorides *284*, 286
elongatus, Limerodops 212
elongatus, Tycherus 213
elongator, Glypta *135*
emarginatus, Notopygus *185*
Enclisis 18, 171
Encrateola 21, 265
Endasys 21, 262, 265
Endromopoda 21, 272, 273
Endromopoda detrita *270*
enecator, Trichomma 128
Enicospilini 243
Enicospilus 20, 225, 242, 243, 245, 246
Enicospilus adustus *106*
Enicospilus americanus 243, 245
Enicospilus combustus *106*
Enicospilus inflexus 244, 245
Enicospilus merdarius 245
Enicospilus ramidulus *245*, *247*
Enicospilus repentinus 245
Enicospilus sesamiae 243
Enicospilus tournieri 245
Enicospilus undulatus 245
Enizemum 19, 194, 195
Enizemum ornatum *106*, *108*, *109*, 193-*196*
Enizemum scutellare 194
ensata, Episemura 192
ensator, Lathrolestes 180
enslini, Neorhacodes *35*,43, *106*, *239*, 240, *241*, 299
Entypoma 20, 186, 187, 250, 251
Enytus 17, 148, 151, 152, 156
Enytus montanus 147
Eparces 19
epermeniae, Triclistus 230
Ephialtes 21, 270, 272, 280
Ephialtes manifestator *108*, *110*, 278, *281*
Ephialtinae 8, 12

Ephialtini 12, 21, 95, 97, *108- 110*, 268, 269, *270, 271, 272, 274*, 276, 278, *281, 282*
Epirhyssa 291
Episemura ensata 192
Epistathmus 21, *302*
Epitomus 19, 213
Epitomus proximus 210
Epitropus 251
eques, Lamachus *30*, 182
Eremotylus 20
Eremotylus curvinervis 245
Eremotylus marginatus *29*, 243, 245
Eremura 240
Eremura perepetshaenkoi 239
Eriborus 17, *156*
Eridolius 22, 316
Eridolius alacer 314
Eridolius pachysoma *108, 317*
erigator, Sussaba 193
Erigorgus 17, 128, 129
Eriplatys 19
Eristicus 19
errabundus, Lemophagus 153
erratum, Hemiphanes *110, 174*
Erromenus 22, 310
Erromenus calcator 313
Erythrodolius 329
Erythrodolius calamitosus 329
erythrogaster, Dusona *107, 157*
erythrurus, Monoblastus 314
Ethelurgus 21, 26, 265
Euceratinae 9
Eucerinae 9
Euceros 11, 12, 19, 66, 197, 198, 199
Euceros albitarsus *108*, 198, 199, *200*
Euceros frigidus 198, 199
Euceros pruinosus *108*, 198, 199, *200*
Euceros serricornis *108*, *197*, 199, *200*
Euceros superbus 199
Euceros unispina 199
Eucerotinae 9, 19, 66, *107, 108*, **197**, 198-200, 221, 307, 326
Eucerotini 197
Eudelus 21, 265
Eudiaborus 316
eudoniae, Campoplex *145*
eugracilis, Aclastus *267*
eumerus, Ichneumon 211
Eupalamus 19
Eurygenys 162, 163
Eurylabini 19, *204*, 207
Eurylabus 19
Eurylabus larvatus 206
Eurylabus torvus *204*, 207, *214*
Eurylabus tristis 207
Euryophion 245
euryops, Idiogramma 96, *110, 308, 317, 320*
Euryproctini 18, 60, 83, 98, 103, *107*, 128 175- *179*, 182- *184*, 257, 258
Euryproctus 18, 182
Euryproctus ratzeburgi 178
Eurypterna 202
Eurytyloides 19, 195

Eusterinx 20, 251
Eutanyacra 19, 211
Eutanyacra glaucatoria 210
evanialis, Dimophora 162
exareolatus, Gelis *110*
exareolatus, Horogenes 147
Excavarus 22
Excavarus apiarius *107, 108, 318*
excitator, Coleocentrus *108, 109, 112*, 113, *114*
Exenterini 307
Exenterus 22, 64, 125, 176, 306, 307, 310, 311, 313, 316
Exenterus abruptorius 182, *311*, 312-314
Exenterus amictorius 313
Exephanes 19, 211-213
Exeristes 21, 272
Exeristes roborator 276
Exeristes ruficollis 278
Exetastes 17, *30*, 84, 136, 139
Exetastes atrator 140
Exetastes illusor *109, 141, 142*
exhortator, Linycus *208*
Exochus 20, 179, 230-232
Exochus tibialis 140, 230, 231
exornatus, Eclytus *305*
Exyston 22
Exyston pratorum 311

fabricianae, Diadegma *147*, 148
facialis, Perispuda *109, 110, 185*
femorella, Olesicampe 156
ferruginea, Xenothyris *327*
festinans, Gelis 264
Fetialis 251
Fianoniella 21
filiformis, Xorides 324
fissorius, Syrphoctonus 194
flaveolatum, Agrypon *127*, 128, 129
flavimanus, Phygadeuon *107, 267*
flavipes, Sussaba 195
flavoabdominalis, Brachyscleroma 330
flavoorbitalis, Trathala 163
flavopictus, Rhynchobanchus 137
Flavopimpla 21, 272
fontinalis, Hercus *110*, 309, 312, *317*
formosa, Bathythrix 262
forticornis, Ophion 245
fortipes, Pion *108, 109, 110, 184, 185*
Fossatyloides 19, 195
fragilis, Bathythrix 262
Fredegunda 21, 272, 273
Fredegunda diluta 272, 273
frigida, Schizopyga *108, 271, 281*
frigidus, Euceros 198, 199
fulgurans, Mesochorus *220*, 221
fuligator, Xorides *110, 324, 325*
fuliginosi, Ghilaromma 201, 202
fulvidens, Trichomma 129
fulvipes, Orthocentrus *108*
fulvipes, Xenoschesis *30, 31*
fulvus, Mesochorus 225
fumiferanae, Glypta 139, 140
funebris, Chorinaeus 230, 231, 313
fuscicornis, Netelia 312

fuscipennis, Amblyjoppa 208

Gabunia 170
Gahus 162
galloisi, Triancyra 291
Gambrus 18, 171
Gambrus incubitor 171
Gambrus ornatus 168, 171
Gareila 19
Gelanes 21, 299, 302
Gelinae 8, 9, 167, 261
Gelini 161, 167, 170, 261
Gelis 21 26, 40, 61, 104, 105, *107*, 261-265, *266, 267*
Gelis areator 264
Gelis agilis *264*, 265
Gelis exareolatus *110*
Gelis festinans 264
Gelis melanocephalus *106, 110, 262*
Gelis rufogaster *267*
Gelis vicinus 263, 265
geminus, Cremastus *107, 164*
genalis, Cryptus 169
geniculatae, Olesicampe 152
geniculatum, Odontocolon 322
geniculatus, Phytodietus *109*
Ghilaromma 19
Ghilaromma fuliginosi 201, 202
Ghilaromma orientalis 202
Ghilarovitini 201
gilvipes, Probles 301
Giraudia 18
glaucatoria, Eutanyacra 210
glaucopterus, Opheltes 179
globulipes, Triclistus *108, 234, 235*
Glyphicnemis 21, 265
Glypta 17, 134, 136, 137, *138*, 140
Glypta elongator *135*
Glypta fumiferanae 139, 140
Glypta mensurator 137
Glypta similis *108, 142*
Glyptini 17, 75, *108*, 134-137, 140, *142*, 217
Glyptorhaestus 18, 181, 183
Glyptorhaestus tomostethae 180
gnathaulax, Paraperithous 272
Gnathochorisis 20, 251
Gnathochorisis crassula *110, 253*
Gnathochorisis dentifer *110, 252*
Gnotus 21, 265
Gnypetomorpha 21, 43
Goedartia 19, 208
Goedartia alboguttata *205*, 207
Goedartiini 19, 204, *205*, 207
goesi, Ischyrocnemis 55
Gonotypus 17, 156
Goryphus 170
gracilenta, Lissonota 137
Grasseiteles 21, 265
gravator, Hemiphanes 168
Gravenhorstia 17, 47
Gravenhorstia picta *106*, 129, *130*
Gravenhorstiini 17, *106*, 125, *126*, 127, *130*
graviceps,Orthizema 41
gravipes, Barycnemis *107*, *300*, 303

Gregopimpla 21, 272
Gregopimpla inquisitor 274
griseanae, Phytodietus *309*
Grotea 327
Grotea anguina *327*
Grypocentrus 22, 310
Grypocentrus apicalis 314, *320*
Grypocentrus basalis 313
Grypocentrus cinctellus *109, 110, 317*
Gunomeria 18

Habrocampulum 17
Habronyx 17, 129
Hadrodactylus 18, 178, 183
Hadrodactylus indefessus *107, 109, 110, 184, 185*
harpura, Barycnemis 30,
hastator, Banchus 138- 140
hectica, Poemenia *109, 286, 287*
Helcostizus 18, 92, 166, 170, 171
Helcostizus restaurator 92, *110, 172, 173*
Helictes 20, 249, 250, 251
Helictinae 8
Hellwigia 146, 242, 243
Helotorus capitator 139
Hemichneumon 19, 210
Hemigaster 166
Hemigasterinae 167
Hemigastrini 166, 167
Hemiphanes 7, 18, 91, 92, 166, 167, 171
Hemiphanes erratum *110, 174*
Hemiphanes gravator 168
Hemiteles 21, 263, 265
Hemitelinae 9, 167, 261
Hemitelini 167
Hepiopelmus 19
Hepiopelmus melanogaster 210
Hepiopelmus variegatorius *206*
heracliana, Barichneumon 210
Hercus 22, 312, 314, 316
Hercus fontinalis *110*, 309, 312, *317*
Heresiarches 19
Heresiarchini 19, 204, *205*, 206, 208, 213
Herpestomus 20
Herpestomus brunnicornis 210
Heterischnus 20, 210
heterocerus, Tersilochus 301, 302
Heterocola 21, 47
Heterocola similis *106, 303, 304*
heteromallus, Platylabus *107, 214, 216*
Heteropelma 17
Heteropelma megarthrum 127-129
Hidryta 18, 169, 171
Himerta 18, 178, 183
Himerta sepulchralis *180*
histrio, Platylabus 208
Holcomastrus 21, 265
holmgreni, Campodorus *109, 110*
Homaspis 18, 178, 183
Homaspis analis *107, 110, 184*
Homotherus 19
Homotropus 19, 192, 194, 195
Homotropus pallipes *191*
Homotropus pictus 193

Hoplismenus 19, 210
Hoplocryptus 18, 169, 171
Hoplocryptus confector *166*
Horogenes exareolatus 147
horticola, Hyposoter 152
Hybrizon 19, 201, 202
Hybrizon buccatus 35, *106, 201* 202, *203*
Hybrizontinae 5, 7, 9, 13, 19, 35, 43, *106*, **201**, 202, 203, 205
Hylesicida 137
Hymenoepimecis argyraphaga 279
Hypamblys 18
Hypamblys albopictus 178
Hyperacmus 19, 72, 186, 187, 236, 249, 250
Hyperacmus crassicornis 72, *108, 187, 188*
Hyperbatus 18, 183
Hypomecus 20
Hyposoter 17, 146-149, 151, 155
Hyposoter carbonarius 149
Hyposoter didymator 153
Hyposster horticola 152
Hyposoter notatus *149*
Hyposoter rhodocerae *149*
Hyposoter tricolor *148*
Hypsantyx 18
Hypsicera 20, 229, 230, 232
Hypsicera curvator *228*

Iania 21, 279
Icariomimus 329
Ichneumon 19, 206, 207, 209, 211, 213
Ichneumon caloscelis 211
Ichneumon deliratorius 206
Ichneumon eumerus 211
Ichneumon oblongus *110*, 204, *215*, 216
Ichneumonidae 1, 3, 5, 7, *10*-12, *16*, 17, 23, **38**, 39-41, *106*, 119, 128, 135, 137, 140, 145, 147, 155, 162, 163, 166, 175, 179, 191, 193, 201, 205, 212, 220, 221, 223-225, 228, 243, 262, 263, 269-271, *272*, 278, 294, 299, 301, 306, 321, 323, 326, 327, 330, 331
Ichneumoninae 3, 5-11, 19, *28*, 29, 50, 51, 61, 104, *106, 107, 110*, 115, 122, 123, 131, 145, 165-167, *204*-206, 210, *211*, 212, *214-216*, 254, 261
Ichneumonini 19, *106, 107, 110, 204*, 205, *206*, 208, 209, 211-213
Idiogramma 22, 96, 306, 308, 309, 311, 316
Idiogramma euryops 96, *110, 308, 317, 320*
Idiogrammatini 22, 96, *110*, 305, 306, *308*, 311, 316
Idiolispa 18, 169, 171
illusor, Exetastes *109, 141, 142*
impressor, Lissonota *108, 109, 141*
inanis, Scambus 273
incitator, Clistopyga *108- 110*, 274, *282*
incubitor, Gambrus 171
indefessus, Hadrodactylus *107, 109, 110, 184, 185*
inflexus, Enicospilus 244, 245
infractor, Netelia *108, 109, 318, 320*
iniquus, Mesochorus 224
inquisitor, Gregopimpla 274
inquisitorius, Ischnus 170
interruptor, Temelucha 163
invisor, Dirophanes 210
irrigator, Xorides 323
Isadelphus 21, 263, 265

Ischnoceros 22, 321, 323
Ischnoceros caligatus 323
Ischnoceros rusticus *110*, 322, *325*
Ischnojoppini 204
Ischnus 18, 168, 169
Ischnus inquisitorius 170
Ischnus migrator 170
Ischyrocnemis 20, 55, *107*, 175, 228-230, 232, *233-235*, 306
Ischyrocnemis goesi 55
Iselix 265
Iseropus 21, 272, 280
Iseropus stercorator 274
Itoplectis 21, 274-276, 280
Itoplectis aterrima *272*
Itoplectis clavicornis 276
Itoplectis maculator *271*, 278
Itoplectis tunetana 212

Javra 18
jiulongshana, Brachyscleroma 330
Joppocryptini 204
jucunda, Diaparsis 224

kerichoensis, Sozites 263
Klutiana 146, 328
kratochvili, Cremastus 162
Kristotomus 22, 316

Labena 327
Labeninae 7, 8, 143, 198, 326-328
Labenopimplinae 331
Labiinae 8, 326
Labium 327
laborator, Buathra 24, 25, *106, 107*
Labrossyta 18
laetatorius, Diplazon 193- 195, *196*
laevicollis, Adelognathus 116
laevigatus, Mesoleptus 260
Lagarotis 18
Lamachus 18, 178, 183, 192
Lamachus eques *30*, 182
lanius, Crypteffigies *107, 214, 216,*
lapidator, Trogus 208, *211*
Lapton 229
laricis, Lethades 225
larvatus, Eurylabus 206
Lathiponus 18
Lathrolestes 18, 177, 179, 180, 183, 310
Lathrolestes clypeatus 177
Lathrolestes ensator 180
Lathrolestes mnemonicae 178
Lathrolestes nigricollis 179, 181, 182
Lathrolestes zeugophorae 178
Lathroplex 17, 146, 155
Lathrostizus 17, 146, 156
lativentris, Triclistus *108*
latrator, Tryphon *107- 110, 317, 318, 320*
Laxiareola 329
Lemophagus 17, 146, 153, 156
Lemophagus errabundus 153
Lemophagus pulcher 153
Leptacoenites 17, 81, 111
Leptacoenites notabilis *107, 109*, 113, *114*

Leptixys 308
Leptocampoplex 17, 155
Leptocampoplex cremastoides 155
Leptocryptoides 21, 263
leptogaster, Dusona 148
Leptoperilissus 147
Lethades 18, 183
Lethades laricis 225
leucorhaeus, Agrothereutes 168
leucotrochi, Adelognathus *116*, 117
lilioceriphilus, Mesochorus 224
Limerodes 19
Limerodops 19
Limerodops elongatus 212
limneriaeformis, Stilbops 294
linearis, Ateleute 62, *107*, *131*-*133*
lineator, Sypasis 209, 212
lineatoria, Tromatobia *277*
lineolaris, Lissonota *138*,
Linycus 20
Linycus exhortator *208*
liostylus, Dichrogaster *107*, 267
Liotryphon 21, 272
Lissonota 17, 134, 136-141, 187
Lissonota biguttata 137
Lissonota brunnea 138
Lissonota digestor 112
Lissonota dubia 138-140
Lissonota gracilenta 137
Lissonota impressor *108*, *109*, *141*
Lissonota lineolaris *138*
Lissonota luffiator *134*, 138
Lissonota mutator 136
Lissonota saturator 137
Lissonota setosa 137, 138
Lissonota stigmator 139
Lissonotinae 9
Listrocryptus 18
Listrodromini 19, 50, *107*, 166, 204, *207*, 261
Listrodromus 19, 166, 261
Listrodromus nycthemerus *107*, *207*, 208, *211*, 212, 214, 215
Listrognathus 18, 169, 171
Listrognathus obnoxious *168*
Lochetica 21, 265
longa, Polysphincta *110*, *281*
longiceps, Dolichochorus 225
longicornis, Rhorus *108*, 181, *185*
longigena, Ophion *106*, *242*, 245, *244*
longissima, Rodrigama 286
Lophyroplectus 18
Lophyroplectus oblongopunctatus 179, 182, 314
lubricus, Pantisarthrus *253*
luffiator, Lissonota *134*, 138
luridator, Oxytorus *107*, *108*, *110*, *257-259*
luteomarginatus, Dyspetes *106*, *109*, *110*, 313, *317*, *318*
luteus, Ophion 244, 245, *246*
Lycorina 20, 75, 217, 218, 313
Lycorina triangulifera 75, *106*, *108*, *217*, 218, *219*
Lycorininae 9, 12, 20, 75, *106*, 135, **217**, 219, 271
Lygurus 329
Lymantrichneumon 19
Lymantrichneumon disparis 213
Lysibia 21, 265

Lysibia nanus 265
Lytarmes maculipennis 291

Macrocentrinae 6
Macrus 17, 155
maculator, Itoplectis *271*, 278
maculipennis, Lytarmes 291
maculosa, Theronia 275
mallochi, Bioblapsis 193
mandator, Agrothereutes 168
manifestator, Ephialtes *108*, *110*, 278, *281*
marginatus, Eremotylus 243, 245
marshalli, Mastrulus 262
Mastrina 263, 265
Mastrulus 21, 265
Mastrulus marshalli 262
Mastrus 21, 263, 265
Mastrus ridibundus 264, 265
mediator, Orthopelma *106*, *108*, *109*, *254*, *255*, *256*
Medophron 21, 263, 265
Megacara 21, 265
Megaetaira 21, 269
Megaplectes 18, 169
Megarhyssa 288, 290, 291
Megarhyssa atrata 288
megarthrum, Heteropelma 127-129
Megastylus 20, 250, 251
Megastylus cruentator *110*, *248*, 251, *252*
Megastylus orbitator 251
melanarius, Aneuclis 301
Melanichneumon 19
melanocephalus, Gelis *106*, *110*, 262
?melancholica, Cylloceria 187
Melanodolius 329
melanogaster, Hepiopelmus 210
melanogaster, Olesicampe 152, 179
Melanoplex 17, 155
Meloboris 17, 151, 156
mensurator, Glypta 137
merdarius, Enicospilus 245
Meringopus 18, 171
Mesochorinae 5, 9, 12, 20, 67, *107*, *108*, 176, 177, **220**, 221-223, *226*, *227*, 330
Mesochorus 20, 151, 220-225
Mesochorus agilis 153, 223, 224
Mesochorus arenarius 223, 224
Mesochorus atriventris 221
Mesochorus curvulus 222
Mesochorus dimidiator 225
Mesochorus dimidiatus 224, 225
Mesochorus discitergus 223, 224
Mesochorus fulgurans *220*, 221
Mesochorus iniquus 224
Mesochorus lilioceriphilus 224
Mesochorus nigripes 224
Mesochorus olerum 225
Mesochorus pallipes *223*, 225
Mesochorus politus *108*, 221, 222, *226*, *227*
Mesochorus punctipleuris *108*, 226
Mesochorus temporalis 221
Mesochorus fulvus 225
Mesoleiini 18, 98, 103, *109*, 175, 176, 178, *180*, 182, 183, *184*
Mesoleius 18, 178, 183

Mesoleius tenthredinis 152, 178, 179, 182
Mesoleptidea 18
Mesoleptidea cingulata *179*
Mesoleptidea prosoleuca 178
Mesoleptus 21, 263, 264, 265
Mesoleptus laevigatus 260
Mesostenidea 169
Mesosteninae 9, 167
Mesostenini 9, 167
Mesostenus 18
mesozosta, Casinaria *148*
Metopiinae 5, 9, 11, 12, 13, 20, 55, 63, 64, 68, 70, *107*, *108*, 125, 126 140, 176, 177, 179, 186, **228**, 229-231, *233-235*, 280, 306, 313
Metopius 20, 63, 125, 228-230, 232
Metopius citratus *108, 233*
Metopius dentatus *232*
Mevesia 20
Microleptes 20, 236, 237
Microleptes aquisgranensis *108, 109*
Microleptes obenbergeri 237
Microleptes rectangulus *108, 109, 236-238*
Microleptes splendidulus 237
Microleptinae 8, 9, 11, 20, 78, *108*, 167, 186, **236**, *237*, 249, 250, 251, 257, 261, 307, 330
Micromonodon 21, 265
migrator, Ischnus 170
minusculae, Ateleute 132
minutus, Ophion 245
Misetus 20
mnemonicae, Lathrolestes 178
mocsaryi, Ophion *106, 247*
modesta, Olethrodotis *175*, 176, 178, *183, 184*
Monganella 328
Monganella variegata 328
monilicornis, Phrudus *107, 303, 304*
Monoblastus 22, 316
Monoblastus brachyacanthus *108-* 110, *320*
Monoblastus caudatus *317*
Monoblastus erythrurus 314
monopicida, Stethoncus *108*, 230, *233*
montanus, Enytus 147
montanus, Phytodietus *109, 319*
morio, Therion 127, 128
morionellus, Phradis 301
multicolor, Alexeter *109, 184, 185*
multicolor, Eclytus *109*, 308, *317*
multicolor, Zaglyptus *15*
murinanae, Cephaloglypta 138-140
mutator, Lissonota 136

naevius, Oetophorus *180*
nanus, Diadegma 150
nanus, Lysibia 264, 265
Neleges 22
Neliopisthus 22
Neliopisthus elegans 309
Nematomicrus 20
Nematomicrus tenellus *208*
Nematopodius 18, 169, 171
Nematopodius debilis 169
Nemeritis 17, 147, 156
Nemeritis canescens 154

Neopimpla 21, 265
Neorhacodes 20, 136, 239, 240, 299
Neorhacodes enslini *35*, 43, *106*, *239*, 240, *241*, 299
Neorhacodinae 5, 7, 9, 12, 20, *35*, 43, *106*, 136, **239**, 240, *241*
Neotheronia 280
Neotheronia tacubaya 275
Neotypus 19
Neotypus nobilitator 208
Neoxorides 21, 285
Neoxorides nitens 286
Neoxoridini 8
Nepiesta 17, 146, 147, 156
Nesomesochorinae 9, 126, 146, 162, 229, 328
Netelia 15, 22, 83, 195, 242, 243, 245, 305- 307, 309-316
Netelia (Bessobates) cristata 312
Netelia (Prosthodocis) 313
Netelia cristata *108, 312*
Netelia fuscicornis 312
Netelia infractor *108, 109, 318, 320*
Netelia nigricarpa *310*
Netelia vinulae 312, *318, 319*
Neuchorus 316
Neurateles 20, 251
Neurateles papyraceus *16*, 250
niger, Alexeter 182
niger, Nonnus *328*
niger, Stethantyx *107*
nigra, Tatogaster *330*, 331
nigricarpa, Netelia *310*
nigricator, Aubertiella 159
nigricollis, Lathrolestes 179, 181, 182
nigricornis, Pseudorhyssa 271, 272
nigripes, Astiphromma 225
nigripes, Mesochorus 224
nigripes, Pion *181*
nigrovarius, Ophion 243
nitens, Neoxorides 286
nitidus, Pygmaeolus 89, *109*, 302, *303*
nobilitator, Neotypus 208
Nonnini 9
Nonnus 146, 328
Nonnus niger *328*
notabilis, Leptacoenites *107, 109*, 113, *114*
notatus, Hyposoter *149*
Notopygus 18, 183
Notopygus emarginatus *185*
Notosemus 19
Notosemus bohemani *207*, 208
Notostilbops 293, 294
Novichneumoninae 331
nycthemerus, Listrodromus *107, 207*, 208, *211*, 212, *214*, 215

obenbergeri, Microleptes 237
Obisiphaga 21
obligatus, Anykylophon 307
oblongopunctatus, Lophyroplectus 179, 182, 314
oblongus, Ichneumon *110*, 204, *215*, *216*
obnoxius, Listrognathus 168
obscuratus, Ophion *242*, *244*, 245
obscurus pulcherrimus, Phytodietus *311*
Obtusodonta 19
Occapes 18

ocellaris, Ophion 244, 246
Odontocolon 22, 321, 322-324
Odontocolon dentipes *110*, 321, *322*, *325*
Odontocolon geniculatum 322
Odontoneura 21, 265
Oecotelma 21
Oedemopsini 22, 66, 98, 103, *106*, *110*, 222, 305-*308*, 309, 311, 312, 314, 316
Oedemopsis 22, 312
Oedemopsis scabricula *106*, *110*, *308*, *318*, *320*
Oedicephalini 19, 204, *207*, 208, 213
Oetophorus 18
Oetophorus naevius *180*
Ogkosoma cremieri 202
Oiorhinus 20
Oiorhinus pallipalpis 210
Ojuelos 228
olerum, Mesochorus 225
Olesicampe 17, 146-148, 150-152, 155, 224, 225
Olesicampe benefactor 152
Olesicampe clandestina 151
Olesicampe femorella 156
Olesicampe geniculatae 152
Olesicampe melanogaster 152, 179, 225
Olesicampe vexata 151
Olethrodotini 175, 177
Olethrodotis 18
Olethrodotis modesta *175*, *176*, *178*, 183, *184*
Opheltes 18
Opheltes glaucopterus 179
Ophion 20, 242-246
Ophion bifoveolatus 243
Ophion bilineatus 243
Ophion brevicornis 245
Ophion costatus 245
Ophion forticornis 245
Ophion longigena *106*, *242*, 245, *244*
Ophion luteus 14, 244, 245, *246*
Ophion minutus 245
Ophion mocsaryi *106*, *247*
Ophion nigrovarius 243
Ophion obscuratus *242*, *244*, 245
Ophion ocellaris 244
Ophion parvulus 244
Ophion pteridis 245, *247*
Ophion scutellaris 245
Ophion ventricosus 245
Ophioninae 9, 15, 20, 46, 99, *106*, 125, 126, 135, 146, 162, 195, 197, 199, 229, **242**, 243, 245, *247*, 299, 310
Ophionini 243
oratorius, Achaius *15*, 210
orbitale, Alcima *107*, *114*, 156, *157*
orbitator, Coelichneumon 213
orbitator, Megastylus 251
Oresbius 18, 171
orientalis, Ghilaromma 202
ornatum, Enizemum *106*, *108*, *109*, 193-*196*
ornatus, Brachycyrtus *107*, *143*, 144
ornatus, Eclytus 313
ornatus, Gambrus 168, 171
ornatus, Phytodietus *109*, *317*, *320*
Oronotus 20
Orotylus 20

Orthizema 21, 265
Orthizema graviceps 41
Orthocentrinae 7- 9, 16, 20, 59, 71, 99, 101, 104, *106-108*, 111, 160, 167, 186, 187, 192, 193, 228, 236, **248**, 249, *252*, *253*, 257, 269, 280
Orthocentrus 20, 71, 228, 249-251
Orthocentrus fulvipes *108*
Orthognatheliini 327
Orthognathellinae 327
Orthomiscus 22, 316
Orthomiscus pectoralis 307
Orthomiscus unicinctus 307, 311
Orthopelma 20, 254, 255
Orthopelma brevicornis 255
Orthopelma mediator *106*, *108*, *109*, *254*, *255*, *256*
Orthopelmatinae 9, 20, 78, *106*, **254**, 255, 256.
Otlophorus 18, 178
Otoblastus 22, 316
Oxyrrhexis 21
Oxytorinae 8, 9, 13, 20, 57, 90, 93, *107*, 176, 250, **257**, *259*
Oxytorus 20, 93, 257, 258
Oxytorus armatus 258
Oxytorus luridator *107*, *108*, *110*, *257*-259
Ozlabium 327

pachysoma, Eridolius *108*, *317*
Palaeoichneumoninae 331
pallipalps, Oiorhinus 210
pallipes, Adelognathus *107*, *115*, 118
pallipes, Homotropus *191*
pallipes, Mesochorus *223*, 225
Palpostilpnus 263
paludator, Chasmias 212
Panteles 21, 87, 136, 293-295
Panteles schuetzeanus 87, *106*, *109*, 294, *296*, *297*
Pantisarthrus 20, 251
Pantisarthrus lubricus *253*
Pantomima 251
Pantorhaestes 18
papyraceus, Neurateles *16*, 250, 263
Parabates 310
paradoxa, Carria 230
paradoxa, Earobia 302
paradoxus, Astrenis *304*
paradoxus, Polyaulon *266*
Paraethecerus 20
Parania 17
Paraperithous 21
Paraperithous gnathaulax 272
parkeri, Stethantyx 301
Parmortha 18
Paropheltes 310
parvulus, Ophion 244
Paxyllomatidae 9
Paxylommatinae 43, 201
pectoralis, Orthomiscus 307
pectoratorius, Diplazon 194, 195
pedestris, Spathius *106*
pedestris, Stenomacrus *106*, *110*
Pedunculinae 8, 12, 143, 198, 326, 328, *329*
Pedunculus 328, *329*
perepetshaenkoi, Eremura 239
Perilissini 18, 175, 176, 179, *180*, 182, 183, 220, 221, 229

Perilissus 18, 179, 180, 183
Perilissus variator 179, *184*
Periope 20, 55, 64, 125, 230, 232
Periope auscultator 55, *107, 108*, 230, *233, 234*
Perisphincter 129
Perispuda 18, 178
Perispuda facialis *109, 110, 185*
Perithoini 269
Perithous 21, 271, 278
Perithous albicinctus 280
Perithous scurra *281*
Perithous septemcinctorius 268
persuasoria, Rhyssa *106*, 271, 278, 288-290, *290-292*, 323
Peucobius 302
Phaenolobus 17, 112
Phaenolobus terebrator *108, 113, 114*
Phaeogenes 20, 210, 213
Phaeogenini 8, 19, 112, 122, 166, 204, 205, 206, *208*, 210, 213, 260, 261
Phaestus 18, 183
Pherhombinae 331
Philogalleria 137
Phobetes 18, 182, 183
Phobetes atomator 182
Phobocampe 17, 148, 149, 151, 152, 153, 156
Phobocampe bicingulata 149
Phosphoriana 251
Phradis 21, 47, 301, 302
Phradis morionellus 301
Phrudinae 9, 54, 177, 299, 329
Phrudus 21, 54, 240, 298, 299, 301, 302
Phrudus defectus 35, *107*, 298, 301, 302
Phrudus monilicornis *107*, 301 *303, 304*
Phthorima 19, 192, 193, 195
Phthorima compressa *196*
Phygadeuon 21, 222, 263, 265
Phygadeuon cephalotes *266, 267*
Phygadeuon flavimanus *107, 267*
Phygadeuon vexator 262
Phygadeuontinae 9, 11, 20, 40, 42, 52, 61, 105, *106, 107, 110*, 115, 131, 145, 162, 165-167, 170, 204, 205, 222, 236, 254, 256, **260**, 261-264, *266, 267*, 269, 270
Phygadeuontini 7, 167, 266
Phytodietini 20, 83, 86, *108, 109*, 199, 305-*309*, 311, 312, 314, 316
Phytodietus 22, 86, 221, 306, 309, 312, 315, 316
Phytodietus geniculatus *109*
Phytodietus griseanae *309*
Phytodietus montanus *109, 319*
Phytodietus obscurus pulcherrimus *311*
Phytodietus ornatus *109, 317, 320*
Picardiella 18
piceator, Alloplasta 137
Picrostigeus 20, 251
picta, Gravenhorstia *106*, 129, *130*
pictipennis, Belesica 163
pictus, Homotropus 193
pilosa, Tanychela 163
pilosella, Arenetra 140
Pimpla 21, 179, 270, 275, 276, 277, 280
Pimpla aethiops 280
Pimpla arctica 280
Pimpla rufipes *108- 110*, 268, 275-278, *281-283*

Pimpla turionellae *109*, 277, 278, *281*
Pimplinae 7, 8,12, 21, *32*, 69, 74, 84, 95, 97, 100, 106, *108-110*, 111,113, 134, 135, 159, 179, 189, 217, 222, 254, 257, 262, 263, **268**, 269-271, 278-280, *281-283*, 285, 286, 289, 312, 321
Pimplini 12, 21, *108, 109*, 268, 269, *271*, 274-277, *281*
Piogaster 21, 2274
Pion 18
Pion fortipes *108, 109, 110, 184, 185*
Pion nigripes *181*
Pionini 13, 18, *108*, 175, 176, 180, 183, 221, 228, 229
pisorius, Protichneumon *205,*
Platophion 244, 245
Platylabini 20, *108, 204, 208*
Platylabops 19
Platylabus 20, 213
Platylabus heteromallus *107, 214, 216*
Platylabus histrio 208
Platylabus tenuicornis 208
Platymischos 20
Platyrhabdus 21, 265
platystylus, Anisobas 208
Plectiscidea 20, 250, 251
Plectiscidea collaris 251
Plectiscinae 8, 9, 186, 250
Plectiscus 20,250, 251
Plectocryptus 18
Pleolophus 18, 171
Pleurogyrus 21, 265
podagricus, Triclistus 231
Podoschistus 21, 285
Podoschistus scutellaris *108, 109, 285*, 286, *287*
Poecilocryptini 327
Poecilocryptus 327
Poecilostictus 20
Poecilostictus cothurnatus 212
Poemenia 21, 285, 286
Poemenia collaris 286
Poemenia hectica *109*, 286, *287*
Poemeniinae 7, 8, 21, 44, 80, 270, **284**, 285, 286, *287*, 289, 321
Poemeniini 21, *108, 109*, 284-286 *287*
polita, Bioblapsis 193
politus, Mesochorus *108*, 221, 222, *226, 227*
Polyaulon 21, 105
Polyaulon paradoxus *266*
Polyblastus 22, 306, 310, 312, 314
Polyblastus cothurnatus *312*
Polyblastus tener 315
Polyblastus varitarsus *109, 110, 319*
Polyblastus wahlbergi *107, 320*
Polycyrtus 11, 169
Polysphincta 12, 21, 268-271, 273, 274, 278-280, 307
Polysphincta longa *110, 281*
Polysphinctini 269
Polytribax 18, 170
Polytribax contiguus 170
pomorum, Scambus 273, 276, 278
Porizon 17, 155
Porizontinae 9, 146
Porizontini 147
praecatorius, Xorides 323
praedator, Thrybius 169

praetor, Proclitus *108, 110, 252, 253*
pratorum, Exyston 311
Priopoda 183
Pristicerops 20
Pristiceros 20
Pristomerus 17, 162-164
Pristomerus armatus 164
Pristomerus vulnerator *107, 161, 162-164*
Probles 21, 302
Probles gilvipes 301
Probolus 19, 213
Procinetus 11, 112
Procinetus decimator 112
Proclitus 20, 250, 251
Proclitus edwardsi 250
Proclitus praetor *108, 110, 252, 253*
Proeliator 20, 251
Promethes 19
Promethes sulcator 195
prosoleuca, Mesoleptidea 178
Prosthodocis 310, 313
Protarchus 18, 178, 179, 183
Protarchus testatorius *109*, 185
proteus, Amblyjoppa 208
Protichneumon 19, 208
Protichneumon pisorius *205*
Protonetelia 309
proximus, Epitomus 210
pruinosus, Euceros *108*, 198, 199, *200*
Pseudorhyssa 7, 21, 268, 270, 271, 286
Pseudorhyssa alpestris *106, 269, 271, 282*
Pseudorhyssa nigricornis 271, 272
Pseudorhyssa sternata 271
Psilomastax 19
Psilomastax pyramidalis 208
pteridis, Ophion 245, *247*
pulchellus, Diadromus 210, 212
pulcher, Lemophagus 153
pumilus, Dicaelotus *107, 214- 216*
pumilus, Sathropterus *106, 107, 304*
punctipleuris, Mesochorus *108, 226*
Pygmaeolus 21, 299
Pygmaeolus nitidus 89, 302, *303*
pygmaeus, Triclistus 231
Pygocryptus 21, 265
Pyracmon 17, 146, 147, 156

quadridentata, Apechthis *30,*
quadripunctorius, Diphyus *107*, 210, *214, 215,*
Quillonota 138

ramidulus, Enicospilus 245, *247*
ratzeburgi, Euryproctus 178
Reclinervellus 21
rectangulus, Microleptes *108, 109, 236-238*
relator, Tryphon *110*
reluctator, Echthrus *108, 172, 173*
repentinus, Enicospilus 245
resplendens, Xenoschesis *177*
restaurator, Helcostizus 92, *110, 172, 173*
Rhaestus 18
Rhembobius 18, 169, 170, 171
Rhimphoctona 17, 146, 147, 150, 156, 176

Rhinotorus 18,183
rhodocerae, Hyposoter 1*49*
Rhorus 18, 72, 180-183, 229
Rhorus longicornis *108, 181, 185*
Rhorus chrysopus *184*
Rhynchobanchus 17, 80, 135, 141, 192
Rhynchobanchus flavopictus 137
Rhyssa 21, 271-272, 278
Rhyssa persuasoria *106, 271, 278, 288-290, 290-292,* 323
Rhyssinae 8, 21, 44, *106,* 268-270, 278, 285, 286, **288**, 289-*292*
Rictichneumon 19
ridibundus, Mastrus 264, 265
roborator, Exeristes 276
robustus, Scirtetes *107,* 149, 262, 276
Rodrigama longissima 286
Rodrigamini 284, 286
Romaniella 240
Rossemia 186
rostrale, Tranosema 153-155
rufata, Apechthis *109, 283*
ruficollis, Exeristes 278
ruficorne, Ctenopelma 178
ruficornis, Aphanistes 128
ruficornis, Stilbops *108, 110,* 294, *295- 297*
rufipes, Xorides 324
rufipes, Pimpla *108- 110,* 268, 275-278, *281-283*
rufogaster, Gelis *267*
Rugodiaparsis 302
rusticus, Ischnoceros *110,* 322, *325*

Sachtlebenia 136
Saotis 18, 178, 183
Sathropterus 21, 302
Sathropterus pumilus *106, 107, 304*
saturator, Lissonota 137
saturniae, Agrothereutes *30,* 168
scabricula, Oedemopsis *106, 110, 308, 318, 320*
Scambus 21, 272, 273, 276, 280
Scambus brevicornis *109, 110, 282, 283*
Scambus calobatus 273
Scambus inanis 273
Scambus pomorum 273, 276, 278
Scambus signatus 273
Scambus tenthredinum 273
Scambus vesicarius 276
Schenkia 18, 169
Schizopyga 21, 69, 279
Schizopyga frigida *108, 271, 281*
schuetzeanus, Panteles 87, *106, 109,* 294, *296, 297*
Scirtetes 17, 149, 151, 153, 156
Scirtetes canadensis 151
Scirtetes robustus 149, 262, 276
Scolobates 19
Scolobates auriculatus 176, 181, *183, 184*
Scolobatinae 9
Scolobatini 19, 176, 177, 181, *183, 184*
Scolomus 20, 68, 175, 228, 229, 232
Scolomus borealis 68, *108,* 230, *233-235*
Scopesis 18
scotiae, Diadegma 148
scurra, Perithous *281*
scutellare, Enizemum 194

scutellaris, Ophion 245
scutellaris, Podoschistus *108, 109, 285*, 286, *287*
scutellator, Syspasis 209
scutulata, Zemiophora 178
Seleucini 175, 177, 181
Seleucus 177, 229
Seleucus cuneiformis 181, 183
semiclausum, Diadegma *107*, 153, *156, 157*
semiflava, Alomya *106*, 123, *124*
Semimesoleius 18
semisculptus, Atractogaster 272
septemcinctorius, Perithous 268
sepulchralis, Himerta *180*
sericeus, Dolichomitus 197
Sericopimpla 272-274, 279
serotinus, Sphinctus 55, *107, 310 313*- 315, *317-319*
serricornis, Euceros *108*, *198*, 199, *200*
sesamiae, Enicospilus 243
setosa, Lissonota 137, 138
signatus, Scambus 273
similis, Glypta *108, 142*
similis, Heterocola *106, 303,304*
simplicipes, Apophua 140
Sinarachna 21, 280
Sinophorus 17, 147, 156
Sisyrostolinae 9, 12, 13, 135, 176, 299, 329
Skiapus 146, 242
Smicrolius 18
Smicroplectus 22
solutus, Aclastus *106, 266*
sonorensis, Campopletis 153
Sozites kerichoensis 263
Spathius pedestris *106*
spectabilis, Barycnemis 302
spectator, Cremastus *107, 164*
Sphecophaga 18, 91, 93,166, 169, 170
Sphecophaga vesparum 39, 41, 91, *108*- *110*, 166, *168*, 169 170, *174*
Sphinctini 22, 55, *107*, 305-308, 310, 316
Sphinctus 22, 176, 306, 310, 311, 313, 314, 316
Sphinctus serotinus 55, *107, 310, 313*, 314, 315, *317-319*
Spilichneumon 19
Spilothyrateles 19
Spinolochus 22, 302
splendidulus, Microleptes 237
Stauropoctonus 20
Stauropoctonus bombycivorus 245
Stenaoplus 19
Stenichneumon 19
Stenobarichneumon 19
Stenodontus 20
Stenomacrus 20,104, 250, 251
Stenomacrus pedestris *106, 110*
Stenopneusticae 206
stenostigma, Bathyplectes 152
stercorator, Iseropus 274
sternata, Pseudorhyssa 271
Stethantyx 301
Stethantyx niger *107*
Stethantyx parkeri 301
Stethoncus 20, 229, 232
Stethoncus monopicida *108*, 230, *233*
Stibeutes 21, 265

Stictopisthus 221, 222
stigmator, Lissonota 139
Stilbopinae 9, 13, 21, 87, 98, 102, *106*, *108*, 136, **293**, 294, 296, 297
Stilbopini 294
Stilbops 21, 98, 102, 136, 293-295
Stilbops limneriaeformis 294
Stilbops ruficornis *108, 110*, 294, *295*- *297*
Stilbops vetula *108*- *110*, *293*, 294, *296, 297*
Stilpnus 21, 263-265
stragifex, Dusona *106, 146*
stramineipes, Diaparsis 301
strigatorius, Tricholabus 210
suavis, Vulgichneumon 210
Sulcarius 21, 263, 265
sulcator, Promethes 195
superbus, Euceros 199
Sussaba 19, 192
Sussaba cognata *109,196*
Sussaba erigator 193
Sussaba flavipes 195
Sweaterella 186
Sycaonia 19
Sympherta 18, 183
Symplecis 20, 59, 251
Symplecis breviuscula *107, 252*
Syndipnus 18
Synetaeris 17, 156
Synodites 18
Synoecetes 18
Synomelix 18, 183
Synosis 20, 230, 232
Syntactus 18
Syrphoctonus 19, 192, 195
Syrphoctonus fissorius 194
Syrphoctonus tarsatorius 193, 195
Syrphophilus 19, 194
Syrphophilus tricinctorius *109*, 195, *196*
Syspasis 19, 209
Syspasis lineator 209, 212
Syspasis scutellator 209
Syzeuctus 17, 137

tacubaya, Neotheronia 275
Tamaulipeca 131
Tanychela pilosa 163
Tanychorinae 331
Tanzawana 175
tarsatorius, Syrphoctonus 193, 195
tarsoleuca, Buathra 169
Tatogaster 330
Tatogaster nigra *330*, 331
Tatogastrinae 9, 176, 257, *330*
Teleutaea 17, 137
Teleutaea brischkei 137, 138
Temelucha 17, 162-64
Temelucha confluens 163
Temelucha interruptor 163
temporalis, Diaparsis 301
temporalis, Mesochorus 221
tenellus, Nematomicrus *208*
tener, Polyblastus 315
tener, Thymaris *106, 108, 318, 319*

tenthredinis, Mesoleius 152, 178, 179, 182
tenthredinum, Scambus 273
tenuicornis, Platylabus 208
terebrator, Dusona *107, 157*
terebrator, Phaenolobus *108*, 113, 114
tergenus, Virgichneumon 210
Tersilochinae 9, 12, 13, 21, *30*, *35*, 43, 47, 54, 89, *106-109*, 126, 135, 136, 162, 240, **298**, 299, *302-304*, 306, 329
Tersilochus 22, 298, 299, 301, 302
Tersilochus conotracheli 300, 301
Tersilochus curvator 299
Tersilochus heterocerus 301
Tersilochus tripartitus 301
testatorius, Protarchus *109,* 185
tetragonus, Diplazon 195
Thaumatogelis 21, 105, 266
Therion 17
Therion circumflexum 127
Therion morio 127, 128
Theronia 21, 269, 270, 271, 275
Theronia atalantae *273*, 275, 280
Theronia maculosa 275
Theroniini 7, 21, 268, 270, *273*, 275
Theroscopus 21, 266, 267
Thrybius 18
Thrybius brevispina 169
Thrybius praedator 169
Thrybius togashii 169
Thymaris 22, 52, 309, 312, 316
Thymaris tener *106, 108, 318, 319*
Thyrateles 19
Thyrateles camelinus 210
Thyreodon 245
Thyreodonini 243
tibialis, Exochus 140, 230, 231
Tobiasitini 201
togashii, Thrybius 169
tomentosum, Ctenopelma *109*
tomostethae, Glyptorhaestus 180
torvus, Eurylabus *204*, 206, *214*
tournieri, Enicospilus 245
Townesia 21, 272
Townesion 136
Townesioninae 136
Townesitinae 331
Trachyarus 20, 210
Tranosema 17, 156
Tranosema rostrale 153, 154
Tranosemella 17, 155, 156
Tranosemella citrofrontalis 156
Trathala 162
Trathala flavoorbitalis 163
Trematopygodes 18, 183
Trematopygus 18, 183
Triancyra galloisi 291
triangulifera, Lycorina 75, *106*, *108*, *217*, 218, *219*
tricarinatus, Trieces *108*, 230, 231, *233-235*
Tricholabus 19
Tricholabus strigatorius 210
Tricholinum 21
Trichomma 17, 129
Trichomma enecator 128
Trichomma circumflexum 129

Trichomma fulvidens 129
trichophthalma, Collyria *108, 109, 158-160*
tricinctorius, Syrphophilus *109*, 195, *196*
Triclistus 20, 230, 232
Triclistus anthophilae 230
Triclistus areolatus 230
Triclistus epermeniae 230
Triclistus globulipes *108, 234, 235*
Triclistus lativentris *108*
Triclistus podagricus 231
Triclistus pygmaeus 231
Triclistus yponomeutae 231
tricolor, Hyposoter *148*
Trieces 20, 229, 232
Trieces tricarinatus *108*, 230, 231, *233-235*
tripartitus, Tersilochus 301
Triptognathus 19
tristis, Eurylabus 206
Trogus 19, 212
Trogus lapidator 208, *211*
Tromatobia 21, 273, 274, 278
Tromatobia lineatoria *277*
Tropistes 21, 266
Trychosis 18, 169, 171
Tryphon 22, 305, 306, 311, 316
Tryphon latrator *107- 110*, *317, 318*, 320
Tryphon relator *110*
Tryphoninae 5, 9, 12, 22, 55, 62, 64, 66, 83, 86, 88, 96, 98, 103, *106- 110*, 117, 125, 135, 136, 152, 175, 176, 178, 195, 197-199, 218, 221, 242, 257,271, 294, 299, **305**, 306-308, 310, 312, 314, 316, *317-320*, 329
Tryphonini 22, 64, 88, 103, *106- 110*, 176, 305-307, 310, *311, 312*, 314, 316
tunetana, Itoplectis 212
turionellae, Pimpla *109*, 277, 278, *281*
Tycherus 20, 210, 213
Tycherus elongatus 213
Tylopius 229
Tymmophorus 19, 192

Uchidella 21, 266
undulatus, Enicospilus 245
unicinctus, Orthomiscus 307, 311
unispina, Euceros 199
Urotryphon 308

vanus, Aperileptus 251
varians, Campoletis 157
variator, Perilissus 179, *184*
variegata, Monganella 328
variegatorius, Hepiopelmus *206*
varitarsus, Polyblastus *109, 110, 319*
ventricosus, Ophion 245
Venturia 17, 147, 155, 156
Venturia canescens 150-152, 154
vernalis, Absyrtus 179
vesicarius, Scambus 276
vesparum, Sphecophaga 39, 41, 91, *108- 110*, 166, *168-170, 174*
vetula, Stilbops *108-110, 293*, 294, *296, 297*
vexata, Olesicampe 151
vexator, Phygadeuon 262
vicinator, Absyrtus 179

vicinus, Blapsidotes 263
vicinus, Gelis 263, 265
vinulae, Netelia 312, *318*, *319*
Virgichneumon 19
Virgichneumon albilineatus *106*, *107*
Virgichneumon callicerus 210
Virgichneumon tergenus 210
volutatorius, Banchus *107*, *108*, *109*, *134*, *141*, *142*
Vulgichneumon 19
Vulgichneumon suavis 210
vulnerator, Pristomerus *107*, *161*, 162, 163, *164*

Wahlamia 138
wahlbergi, Polyblastus *107*, *320*
Westwoodiini 175, 177, 181
Woldstedtius 19

Xenolytus 21, 266
Xenoschesis 18
Xenoschesis fulvipes *30*, *31*,
Xenoschesis resplendens *177*
Xenothyrini *327*
Xenothyris ferruginea *327*
Xestopelta 19
Xiphosomella 162

Xiphulcus 21, 266
Xorides 22, 321, 322, 323,
Xorides brachylabis 323
Xorides filiformis 324
Xorides fuligator *110*, *324*, *325*
Xorides irrigator 323
Xorides praecatorius 323
Xorides rufipes 324
Xoridinae 2, 7, 8, 22, 94, *110*, 285, **321**, 324, *325*, 327
Xoridini 321
Xylophrurus 18, 171

Zaglyptus 21, 273. 274, 278
Zaglyptus multicolor *15*
Zapedias 308
Zaplethocornia 18, 183
Zatypota 21, 74, 280
Zatypota albicoxa *108*, 279, *283*
Zele albiditarsus *106*
Zemiophora 18
Zemiophora scutulata 178
zeugophorae, Lathrolestes 178
Zimmeriini 20, 204, 209
Zoophthorus 21, 263, 266

Index to hosts of Ichneumonidae, food plants and associated organisms

Abies 138, 139
Abraxas grossulariata 148, 209
Acer pseudoplatanus 302
Acleris schalleriana 218
Acrobasis 273
Acrolepia autumnitella 230
Acrolepiopsis assectella 210
Acronicta 129
aculeate 1, 7, 39, 168-170, 272, 286, 327, 333
Adela reaumerella 294
Adelgidae 193, 302
Adelidae 137, 294, 295
aestivum, Triticum 158
Agrilus viridis 323
Agriopis 245
Agroeca 262
Agrotis exclamationis 244
Agrotis ripae 245
albicorne, Odontocerum 120
alder woodwasp 271
alea, Mycetophila 251
alfalfa weevil 152, 223
Aleiodes bicolor 224
alienus, Lasius 202
Allantinae 178, 311
Allodiopsis rustica 251
Allognosta fuscitarsis 237
Alnus 290
amerinae, Euura 273
Amphipyrinae 137
Amylostereum 289
Angelica 207
angiosperm 199
Anomaloninae 199
ant 13, 202, 211
Anthonomus pomorum 273
Anthophila fabriciana 139, 148, 230
Anthyllis 148
Apamea 139, 212
Apatura iris 208
Apethymus 178
aphid 5, 193
Aphididae 223
Aphidiinae 223
Aphidoidea 193
Apiaceae 123, 137, 210
Apidae 155
Apocheima 245
Apoda limacodes 310, 313, 314, 315
Apoderus quadripunctatus 330
apple 273
approximator, Rhyssella 271
Arachnida 5, 11
Araneae 5, 11, 12, 168, 169, 222, 261-264, 275, 277, 328
Araneidae 274, 277
Araneus diadematus 277
Arctiinae 127, 137, 230
Arge 181, 308

Arge ustulata 311
Argidae 181, 199, 307, 308, 310, 311
argiolus, Celastrina 208, 211
Argyresthia 139
Argyresthiidae 210
Arhopalus rusticus 322
Aricia artaxerxes 149
arietis, Clytus 286
Aromia moschata 286
artaxerxes, Aricia 149
Artematopodidae 146
arvense, Cirsium 113
arvensis, Knautia 294, 295
assectella, Acrolepiopsis 210
astrarches, Cotesia 224
atalanta, Vanessa 148
atra, Erigone 264
Attelabidae 330
Aulacidae 269
Aulacus 269
autumnitella, Acrolepia 230
auricularia, Forficula 262
avellana, Corylus 285, 324

balteatus, Episyrphus 193
Banchinae 199
Banchus 225
Banksia 327
banksiana, Pinus 198
bark 250, 263, 272, 274, 275, 302, 322-324
Bathyplectes 224
Bathyplectes curculionis 223
bean 277
bedeguar gall 255
bedeguaris, Pteromalus 255
beetle 272, 286, 301
Betula 290, 314
Bibionomorpha 193, 250
bicolor, Aleiodes 224
bicruris, Hadena 207
bifasciatum, Rhagium 322
birch 314
Blasticotoma filiceti 181
Blasticotomidae 181
Bledius spectabilis 302
Blennocampinae 178, 179, 180, 311
Boletus 193
Bombus 155
Bombycoidea 245
Braconidae 221-225, 262-264, 265, 327
Brachypeza radiata 250
Bradysia giraudii 250
Brassica nigra 299
broadleaved tree 322
brumata, Operophtera 127, 137, 210
bumblebee 155
buoliana, Rhyacionia 163
Buprestidae 286, 322, 323, 326, 327

Bupalus piniarius 129
butterfly 209, 263, 274, 275
Byrrhidae 299

cabbage 195
Caliroa 178
Calliphoridae 155
Callitaera pudibunda 207
Callophrys rubi 208
Calluna 162
camelus, Xiphydria 271, 290
Campopleginae 141, 199, 223, 224, 262, 263, 272, 276
cardui, Vanessa 210
Celastrina argiolus 208, 211
Cepaea nemoralis 263, 264
cephalotes, Megalodontes 178
Cephidae 12, 158
Cephus 159
Cephus cinctus 159
Cephus pygmeus 158, 159
Cerambycidae 112, 146, 286, 291, 322, 323, 327
Cerura vinula 315, 316
Cetoniinae 169
Ceutorhynchus pleurostigma 302
chaerophyllella, Epermenia 230
Chalcidoidea 10, 11, 169, 255
Cheilosia longula 193
chlorana, Earias 230
Chloromyia formosa 237
choragella, Morophaga 272
Choreutidae 137, 139, 210, 230
Choristoneura fumiferana 139
Choristoneura murinana 139
Chrysomelidae 146, 153, 163, 189, 224, 225, 299, 300
Chrysopidae 143
Ciidae 299
Cimbex 179
Cimbex femorata 151
Cimbicidae 151, 168, 178, 179, 307, 310, 311
cinctus, Cephus 159
cinxia, Melitaea 152, 224
Cirsium arvense 113
Cleonis pigra 113
Cleroidea 147
Clubionidae 274
Clytus arietis 286
codling moth 264
coecus, Otiorhynchus 113
Coeloglossum viride 123
Coenonympha 210
coeruleipennis, Tomapoderus 330
Colaphellus sophiae 225
Coleophora 309
Coleophoridae 309
Coleoptera 7, 11-14, 112, 113, 126, 146, 147, 153, 163, 168-170, 178, 179, 222, 224, 262, 263, 272, 273, 275, 286, 299, 300, 302, 322, 330
cone, conifer 308
cone, pine 272, 302, 309
conifer 127, 129, 193, 289, 307, 308, 322
conifer cone 308
Conotrachelus nenuphar 300
cornucopiae, Pleurotus 250

Corylus avellana 285, 324
Cossidae, 136, 137, 138
Cossus cossus 137, 138
cossus, Cossus 137, 138
Cotesia 223, 263, 265
Cotesia astrarches 224
Cotesia glomerata 265
Cotesia marginiventris 224
Cotesia melitearum agg. 224
Cotesia saltatoria 224
Cotesia tenebrosa 224
Crambidae 137, 163, 210, 218, 230, 263
Crabronidae 12, 240, 271, 286
Cryptinae 199
Cryptocephalus moraei 163
Ctenopelmatinae 199, 225
Cucullia 210, 245
Cuculliinae 137
cunea, Hyphantria 127
Cupido minimus 148, 208
Curculio 273
Curculionidae 113, 146, 263, 273, 299, 300, 302, 327
curculionis, Bathyplectes 223
Cydia 273
Cydia pomonella 162, 163, 164, 264
Cynipidae 12, 255
Cynipoidea 289

dealbana, Gypsonoma 162
decidua, Larix 232
deciduous tree 323
Deilephila elpenor 208
Deilephila porcellus 208
Dendrolasius 202
Depressaria pastinacella 210
Depressaria radiella 210
Depressariidae 210, 230, 309
Dermaptera 222, 262
Dermestidae 146
Derodontidae 302
diadematus, Araneus 277
Diamphidia nigroornata 163
Diaparsis jucunda 224
Diaphora 210
Diarsia 210
Diastrophus 255
Dicallomeria fascelina 149
differens, Spilomena 240
dimidiata, Rondaniella 251
diniana, Zeiraphera 231, 232
Diplolepis 255
Diplolepis nervosa 255
Diplolepis rosae 255
Diprionidae 169, 178, 179, 182, 198, 307, 310, 311, 312
Diprioninae 171, 183
Diptera 5, 11, 12, 14, 155, 170, 187, 192, 193, 194, 221, 222, 236, 237, 250, 257, 261-264, 273, 275, 280, 327
dispar, Lymantria 276
Dolerini 117, 178, 311
Dolerus 179
Dolichovespula 169
Drepanidae 208, 245, 310
Dryobotodes eremita 245

Dusona 199, 225

Earias chlorana 230
earwig 262
eastern larch 178, 179
Elachistidae 210
Elateridae 146
elpenor, Deilephila 208
empetrella, Scythris 162
Enicospilus 225
enslini, Spilomena 240
Entypotrachelus meyeri 263
Epermenia chaerophyllella 230
Epermeniidae 230
Ephestia 150
Epinota tedella 139, 141, 230
Episyrphus 193
Episyrphus balteatus 193
Epuraea melanocephala 302
Erebidae 127, 137, 199, 207, 209, 210, 230, 276
eremita, Dryobotodes 245
Eremobia ochroleuca 207
Erica 162
erichsonii, Laricobius 302
erichsonii, Pristiphora 152, 178, 179, 225
Erigone atra 264
Eriocrania 299, 310
Eriocraniidae 177, 299, 310
Eriogaster lanestris 208
Esperia sulphurella 137
Ethmiidae 218, 230
Euclidia mi 210
Eumeninae 262
Eumeta minuscula 132
Euphorbia 112
Euphorinae 222
Eupithecia venosata 148
Eupsilia transversa 245
Eupteropidae 245
Eurytomidae 11, 169
Euthrix potatoria 262
Euura 116, 117, 178
Euura amerinae 273
excerptalis, Scirpophaga 263
exclamationis, Agrotis 244

fabriciana, Anthophila 139, 148, 230
fagi, Stauropus 245
fascelina, Dicallomeria 149
femorata, Cimbex 151
Fenusa pusilla 179, 314
Fenusini 179, 180, 310
ferchaultella, Luffia 140
Ferdinandea 193
Fergusonina 327
Fergusoninidae 327
fern 181
festaliella, Schreckensteinia 230
filiceti, Blasticotoma 181
Filipendula ulmaria 250
filipendulae, Zygaena 168, 225
fimbriata, Noctua 210
fir 140

flammea, Panolis 139
flavago, Gortyna 112
Forficula auricularia 262
Formica 202
Formicidae 13, 202, 211
formosa, Chloromyia 237
fruit 154, 162, 272
frugiperda, Spodoptera 126, 153, 243
fuliginosus, Lasius 202
fumiferana, Choristoneura 139
fungus 168, 250, 289
fungus gnat 250
fusca, Phyllophaga 243
fuscatella, Lampronia 294, 295
fuscitarsis, Allognosta 237

gabrieli, Tetropium 323
gall 116, 117, 137, 138, 193, 255, 273, 276, 294, 295, 327
gall, bedeguar 255
gall wasp 12
Galleria mellonella 155
Gelechiidae 230, 310
geniculata, Pristiphora 152
Geometridae 127, 129, 137, 148, 199, 208-210, 221, 225, 230, 245, 310
germanica, Vespula 170
gigas, Urocerus 290
giraudii, Bradysia 250
glomerata, Cotesia 265
Glyphipterigidae 210, 230
Goera pilosa 120
Goeridae 120
Gonepteryx rhamni 149
Gonocephalum rusticum 126
Gortyna flavago 112
gothica, Orthosia 137
Gracillariidae 222, 273
grandis, Lasius 202
grass 117, 137, 1398, 212
grossulariata, Abraxas 148, 209
gymnosperm 199
Gypsonoma dealbana 162
Gyrinidae 263

Hadena bicruris 207
Hadeninae 137, 245
hazel 285, 324
Hemiptera 5
Hepialidae 11, 123, 209
Hepialus lupulinus 123
Hepialus sylvinus 123
Heracleum 159, 207
Hesperiidae 129, 163, 209
Heterarthrus 146, 178
heterarthrines 180
Heteroptera 222
Homoptera 222
Hoplocampa testudinaria 180
horsetail 311
horticola, Hyposoter 224
hoverfly 195
Hyblaeidae 291
Hymenoptera 5, 7, 11-13, 112, 116, 117, 146, 150, 151,

158, 168-170, 177, 181, 198, 222, 225, 240, 255, 263, 271, 273, 276-278, 289, 290, 299, 305, 307, 308, 312, 322, 326
Hypera postica 152, 223
Hypera 153, 224
Hyphantria cunea 127
Hyposoter 263, 272
Hyposoter horticola 224
Hyposoter notatus 224

Ibalia 269, 289
Ibaliidae 269, 289
Ichneumonidae 199, 221, 223, 224, 225, 262, 263, 272, 276
Ichneumonoidea 5, 11, 262
Incurvariidae 137
Incurvarioidea 294
inquisitor, Rhagium 286
iris, Apatura 208
irrorata, Tipula 187

jucunda, Diaparsis 224

Keroplatidae 251
kerri, Mesostoa 327
Knautia arvensis 294, 295

Lampronia fuscatella 294, 295
lanestris, Eriogaster 208
larch 192, 231, 232, 316, 323
larch, eastern 178, 179
laricina, Larix 178, 179
laricis, Lethades 225
Laricobius erichsonii 302
Larix 192, 323
Larix decidua 232
Larix laricina 178, 179
Lasiocampa quercus 168, 232, 244
Lasiocampidae 137, 168, 208, 209, 230, 232, 244, 245, 262, 272, 274
Lasius 202
Lasius alienus 202
Lasius fuliginosus 202
Lasius grandis 202
Lasius niger 202
Lasius nipponensis 202
Leiophron 222
Lemophagus 224
Lepidoptera 5, 11-14, 112, 123, 126, 127, 129, 132, 136, 137, 145-50, 152-154, 161-163, 167-170, 176- 179, 187, 198, 199, 206, 207, 210, 211, 212, 218, 221-225, 228-230, 243, 244, 262, 263, 272-276, 280, 294, 295, 299, 305, 307-310, 313, 315, 316, 328
Lethades laricis 225
leucotrochus, Nematus 116
lignicola, Xylosciara 250
lilii, Lioceris 153
Lioceris 224
Lioceris lilii 153
limacodes, Apoda 310, 313, 314, 315
Limacodidae 163, 310, 313
lineola, Thymelicus 209
Linyphiidae 264, 274

longula, Cheilosia 193
luniger, Metasyrphus 193
lupulinus, Hepialus 123
luridiventris, Platycampus 151
Luffia ferchaultella 140
Lycaenidae 129, 150, 207, 210, 211, 224
Lymantria dispar 276
Lymantriinae 149, 199, 207, 274, 276
Lypha ruficauda 221

machaon, Papilio 211
macrolepidoptera 136, 148, 150, 206, 309
Macrophya 179
Maculinea 211
Malus 273
marginiventris, Cotesia 224
mealworm 275
Megalodontes cephalotes 178
Megalodontesidae 178
Megalopodidae 178
Melaleuca 327
Melandryidae 112, 299
melanocephala, Epuraea 302
melanogaster, Olesicampe 225
Melitaea cinxia 152, 224
melitearum agg., Cotesia 224
mellonella, Galleria 155
Mesostoa kerri 327
metallica, Nemophora 294, 295
Metasyrphus 193
Metasyrphus luniger 193
meyeri, Entypotrachelus 263
mi, Euclidia 210
Microdiprion pallipes 182
Microgastrinae 221-224, 262-265
microlepidoptera 129, 148, 150, 163, 169, 206, 210, 218, 230, 274, 309
Microplitis ocellatae 264
minimus, Cupido 148, 208
minor, Molorchus 286
minuscula, Eumeta 132
Miridae 222
mirifica, Ovalisia 286
Molorchus minor 286
moraei, Cryptocephalus 163
Morophaga choragella 272
moschata, Aromia 286
moss 137
moth, codling 264
Muscidae 264
murinana, Choristoneura 139
Mycetophila alea 251
Mycetophilidae 187, 250, 251
Myrmicinae 202
Mythimna 245

nebris, Papaipema 138
Nematinae 147, 178, 223, 224, 308, 310, 311, 315, 316
Nematopogon schwarziellus 294
Nematus 225
Nematus leucotrochus 116, 117
Nematus pavidus 313
nemoralis, Cepaea 263, 264

Nemophora metallica 294, 295
nenuphar, Conotrachelus 300
Neodiprion sertifer 178, 182, 312
Neodiprion swainei 198
Neozephyrus quercus 129
Nepticulidae 150, 223
nervosa, Diplolepis 255
Neuroptera 11, 12, 143, 262
niger, Lasius 202
nigra, Brassica 299
nigricornis, Silo 120
nigriventris, Sarcophaga 263
nigroornata, Diamphidia 163
nipponensis, Lasius 202
Nitidulidae 299, 302
Noctua 210
Noctua fimbriata 210
Noctuidae 127, 129, 137, 139, 148, 151, 153, 163, 207, 209, 210, 212, 225, 230, 243-245, 274, 310
Noctuoidea 310
Noctuinae 137
Nolidae 230
notatus, Hyposoter 224
Notodontidae 148, 207, 230, 245, 310, 315
nubilalis, Ostrinia 163
Nymphalidae 129, 152, 208, 210, 224

oak 273
Oberea 112
ocellatae, Microplitis 264
ocellatus, Smerinthus 264
ochroleuca, Eremobia 207
Odontoceridae 120
Odontocerum albicorne 120
Oecophoridae 137, 230
oilseed rape 299
Olesicampe 151, 224, 225
Olesicampe melanogaster 225
Operophtera brumata 127, 137, 210
Ophion 199
Ophioninae 199
Orchidaceae 123
Orgyia 149
Orthosia 137, 244
Orthosia gothica 137
Orthotylus 222
Ostrinia nubilalis 163
Otiorhynchus coecus 113
Ovalisia mirifica 286

Pachythelia villosella 162
pallipes, Microdiprion 182
pallipes, Pristiphora 116, 117
pallipes, Silo 120, 121
Pamphiliidae 35, 117, 169, 178
Pamphilioidea 146, 150, 169, 177, 178, 222, 307
Pamphilius 35
Panolis flammea 139
Papaipema nebris 138
paper wasp 280
Papilio 208
Papilio machaon 211
Papilionidae 208, 211

Passaloecus 286
pastinacella, Depressaria 210
pavidus, Nematus 313
pavonia, Saturnia 168
Pergidae 181, 307, 326
Phasmida 263
Pholetesor 222
Phragmites 212
Phycitinae 154
Phyllocolpa 117, 178
Phyllophaga fusca 243
Phymatodes testaceus 286
Phytodietini 199
Phytodietus 221
Phytodietus polyzonias 221
Phytoecia 112
Pieridae 225
Pieris 265
pigra, Cleonis 113
pilosa, Goera 120
pine 198, 250
pine cone 272, 302, 309
piniarius, Bupalus 129
Pinus 302
Pinus banksiana 198
Pinus virginiana 309
Pipizella 193
Pipizini 193
Platycampus luridiventris 151
Platycheirus 193
pleurostigma, Ceutorhynchus 302
Pleurotus cornucopiae 250
plum 300
Plutella xylostella 210, 225
Plutellidae 210, 225
Poaceae 272
Polistinae 280
Polyommatini 208, 224
polyzonias, Phytodietus 221
pomonella, Cydia 162, 163, 164, 264
pomorum, Anthonomus 273
Pontania 116, 117, 178
Populus 127
porcellus, Deilephila 208
postica, Hypera 152, 223
potatoria, Euthrix 262
Pristiphora 225
Pristiphora erichsonii 152, 178, 179, 225
Pristiphora geniculata 152
Pristiphora pallipes 116
Prodoxidae 294
prolongata, Xiphydria 290
Pseudocephaleia 35
pseudoplatanus, Acer 302
pseudoscorpion 11, 222, 263
Psocoptera 5, 222
Psychidae 11, 132, 137, 140, 162, 210, 279
Pteromalidae 255
Pteromalus bedeguaris 255
Pterophoridae 210
pudibunda, Callitaera 207
pusilla, Fenusa 179, 314
pygmeus, Cephus 158, 159

Pyralidae 137, 150, 154, 163, 169, 230, 273, 309, 310, 328
pyrastri, Scaeva 193

quadripunctatus, Apoderus 330
Quercus 273, 274, 315
quercus, Lasiocampa 168, 232, 244
quercus, Neozephyrus 129

radiata, Brachypeza 250
radiella, Depressaria 210
rape, oilseed 299
Raphidioptera 11, 13, 147, 263
reaumerella, Adela 294
reed 278
Rhagium 322
Rhagium bifasciatum 322
Rhagium inquisitor 286
rhamni, Gonepteryx 149
Rhyacionia buoliana 163
Rhyssella 21, 271, 286, 290
Rhyssella approximator 271
ribesii, Syrphus 193
ripae, Agrotis 245
robustus, Scirtetes 262, 276
Rondaniella dimidiata 251
Rosa 255
rosae, Diplolepis 255
rubi, Callophrys 208
Rubus 230, 255
ruficauda, Lypha 221
rustica, Allodiopsis 251
rusticum, Gonocephalum 126
rusticus, Arhopalus 322

Salix 117, 290
saltatoria, Cotesia 224
Salticidae 274
Sarcophaga 264
Sarcophaga nigriventris 263
Sarcophagidae 263, 264
Saturnia pavonia 168
Saturniidae 137, 168, 245
Satyrinae 129, 211
sawfly 5, 11-13, 168, 169, 176, 177, 198, 199, 222, 224, 225, 261, 262, 273, 276, 299, 305, 307, 308, 326
Scaeva pyrastri 193
Scarabaeidae 169, 243
schalleriana, Acleris 218
Schreckensteinia festaliella 230
Schreckensteiniidae 230
schwarziellus, Nematopogon 294
Sciaridae 250
Sciaroidea 251
Sciomyzidae 264
Sciophila 250
Scirpophaga excerptalis 263
Scirtetes robustus 262, 276
Scopariinae 137
Scythrididae 162
Scythris empetrella 162
sedge 117
Segestriidae 274
sertifer, Neodiprion 178, 182, 312

Sesamia vuteria 243
Sesiidae 112, 136, 137, 138, 163, 272
Silene uniflora 148
Silene vulgaris 148
Silo pallipes 120, 121
Silo nigricornis 120
Sirex 289, 290
Siricidae 5, 289, 290, 291
Siricoidea 322
Smerinthus ocellatus 264
snail 263, 264
sophiae, Colaphellus 225
spectabilis, Bledius 302
Sphaerophoria 193
Sphecidae 169
Sphingidae 151, 208, 245
spider 5, 11, 12, 168, 169, 222, 261-264, 273-279, 328
Spilomena 240
Spilomena differens 240
Spilomena enslini 240
Spilomena troglodytes 240
Spilosoma 210
spinipennis, Triarthria 262
Spodoptera frugiperda 126, 153, 243
spruce 250
Staphylinidae 299, 302
Stauropus fagi 245
Stratiomyidae 11, 237
sulphurella, Esperia 137
sylvinus, Hepialus 123
Syrphidae 170, 192, 193 195, 262
Syrphus ribesii 193
swainei, Neodiprion 198
sycamore 302

Tachinidae 221, 222, 262
tedella, Epinota 139, 141, 230
Tenebrio 276
Tenebrionidae 126
tenebrosa, Cotesia 224
Tenthredinidae 116, 117, 151, 169, 178, 180, 223, 225, 273, 307, 308, 310, 311
Tenthredininae 179, 311
Tenthredinoidea 146, 147, 150, 169, 177, 222, 305, 307
Tenthredo 178, 179, 313
Tenthredopsis 313
Tersilochinae 224
testaceus, Phymatodes 286
testudinaria, Hoplocampa 180
Tetragnatha 275
Tetragnathidae 274, 275
Tetramesa 169
Tetropium gabrieli 323
Theridiidae 274
Thyatirinae 245, 310
Thymelicus lineola 209
tibiale, Trichiosoma 307
Tineidae 230
Tipula 187
Tipula irrorata 187
Tipulidae 187
Tischeriidae 150
Tomapoderus coeruleipennis 330

Tomostethus 180
Tortricidae 129, 134, 136, 137, 138, 139, 147, 162, 163, 169, 170, 210, 218, 230, 231, 264, 273, 274, 309, 310, 316
Tortricoidea 218
transversa, Eupsilia 245
tree, broadleaved 322
tree, deciduous 323
Triarthria spinipennis 262
Trichiosoma tibiale 307
Trichoptera 11, 13, 119, 147, 150, 263
Triticum aestivum 158
troglodytes, Spilomena 240
Tryphoninae 199, 221
Typha 212

ulmaria, Filipendula 250
Ulmus 250
umbellifer 123, 207, 305
uniflora, Silene 148
Urocerus gigas 290
ustulata, Arge 311

Vanessa atalanta 148
Vanessa cardui 210
varia, Sciophila 187
venosata, Eupithecia 148
Vespidae 168, 169, 170, 280
Vespinae 169, 170
Vespula 169
Vespula germanica 170
Vespula vulgaris 168, 170
villosella, Pachythelia 162
vinula, Cerura 315, 316
virginiana, Pinus 309
viride, Coeloglossum 123
viridis, Agrilus 323

vulgaris, Silene 148
vulgaris, Vespula 168, 170
vuteria, Sesamia 243

wasp 262, 286
wasp, gall 12
wasp, paper 280
weevil 11, 149, 263, 273, 330
weevil, alfalfa 152, 223
wheat 158, 159
wood 92, 113, 137, 138, 146, 150, 163, 169, 170, 187, 272, 276, 285, 286, 288-291, 321-324, 326, 327, 342
woodwasp 1, 23, 271, 289-291, 322, 333
woodwasp, alder 271

Xestia 210, 245
Xiphydria 271, 290
Xiphydria camelus 271, 290
Xiphydria prolongata 290
Xiphydriidae 5, 271, 290
Xyela 299, 308, 309
Xyelidae 299, 307, 308
Xyeloidea 305, 307
Xyleninae 137
Xylosciara lignicola 250
xylostella, Plutella 210, 225

Yponomeuta 210, 231
Yponomeutidae 137, 139, 169, 210, 230, 231
Ypsolophidae 230, 309

Zeiraphera diniana 231, 232
Zeugophora 178
Zygaena 150, 221
Zygaena filipendulae 168, 225
Zygaenidae 150, 168, 169, 221, 225

www.ingramcontent.com/pod-product-compliance
Lightning Source LLC
Chambersburg PA
CBHW040538220526
45473CB00016B/2970